SQUID
Superconducting Quantum Interference Devices
and their Applications

SQUID
Superconducting Quantum Interference Devices and their Applications

Proceedings of the International Conference
on Superconducting Quantum Devices
Berlin (West), October 5-8, 1976

Editors
H. D. Hahlbohm · H. Lübbig

W
DE
G

Walter de Gruyter · Berlin · New York 1977

Editors

H. D. Hahlbohm
H. Lübbig
Institut Berlin
der Physikalisch-Technischen Bundesanstalt,
1000 Berlin (West), Germany

CIP-Kurztitelaufnahme der Deutschen Bibliothek

SQUID: superconducting quantum interference devices and their applications; proceedings of the Internat. Conference on Superconducting Quantum Devices Berlin (West), October 5–8, 1976/ed. H.-D. Hahlbohm; H. Lübbig. – 1. Aufl. – Berlin,
New York: de Gruyter, 1977.
ISBN 3-11-006878-8

NE: Hahlbohm, Hans-Dieter [Hrsg.]; International Conference on Superconducting Quantum Devices <01, 1976, Berlin, West>

Library of Congress Cataloging in Publication Data

International Conference on Superconducting Quantum Devices, 1st, Berlin, 1976
SQUID, superconducting quantum interference devices and their applications.

Includes bibliographical references and indexes.
1. Superconducting quantum interference devices – Congresses.
I. Hahlbohm, H. D., 1930 –
II. Lübbig, H., 1932 – III. Title.
TK7872.S8157 1976 621,39 77-5409
ISBN 3-11-006878-8

Printing: Karl Gerike, Berlin. – Binding: Lüderitz & Bauer, Buchgewerbe GmbH, Berlin.
Printed in Germany.

PREFACE

In form of the proceedings of a first international conference devoted to Superconducting Quantum Interference Devices (SQUID's), in this volume extended papers are published concerning specialized topics of fundamentals and applications of weak superconductivity.

Being the basis, the Josephson effect of phase-coherence in weakly connected superconductors brought insight into superconductivity in the presence of fields, and simultaneously, became starting point for revolutionary progress in a great number of technical applications, metrology and computer techniques for instance. The output accumulated in more than one decade suggested to attempt an international interchange between scientists working in this area. The aim was to summarize the development, to give a survey over the present state, with respect to results as well as with respect to problems, and to estimate future aspects.

This aim could be achieved only partially, since it soon became clear that it would be impossible to cover within four days the whole area of what may be called SQUID fundamentals and applications. Thus important applications as medical and geophysical ones could not be included.

Fourtyseven papers were presented at this meeting. The delegates represented thirteen countries including Austria, CSSR, Denmark, Finland, France, Great Britain, Italy, Japan, The Netherlands, Sweden, Switzerland, United States of America and Germany. Together with the contributed papers nine invited lectures demonstrated the high level of weak-superconductivity research. However, it becomes clear as well that theory and application in several branches are just at the beginning.

Hence it might be expected that the next years will be characterized by further intensive research activities.

The conference has been inspired by the European Physical Society.

The edition of the proceedings is a good occasion to acknowledge once more the generous support of the conference by the President of the Deutscher Bundestag, the Berlin Senate, the President of the Physikalisch-Technische Bundesanstalt, the Deutsche Physikalische Gesellschaft e.V., Fachausschuß Tiefe Temperaturen and the Siemens AG.

Furthermore, we are indebted to the De Gruyter Publishing Company for its valuable cooperation in preparing these proceedings.

Berlin, February 1977

H.-D. Hahlbohm

H. Lübbig

CONTENTS

THEORY OF SUPERCONDUCTING PHASE COUPLING

JUNCTION FABRICATION TECHNIQUES

REAL JUNCTION PROPERTIES

LOW FREQUENCY APPLICATIONS

HIGH FREQUENCY APPLICATIONS

NOISE

COMMUNICATION SYSTEMS

SIGNAL PROCESSING

COMPUTER ELEMENTS

SUMMARY AND CONCLUSIONS

LIST OF PARTICIPANTS

Adde, R.	Institut d'Electronique Fondamentale, Université Paris-Sud, Bat. 220, 91405 Orsay, France
Albrecht, H.	2. Physikalisches Institut, Technische Hochschule Aachen, Templergraben 56, 5100 Aachen 1, Germany
Andreone, D.	Istituto Elettrotecnico Nazionale Galileo Ferraris, Corso Massimo D'Azeglio 42, 10125 Torino, Italy
Andriesse, C.	Kapteyn Astronomical Institute, University of Groningen, Space Research, P.O. Box 800, Groningen, The Netherlands
Arri, E.	Istituto Elettrotecnico Nazionale Galileo Ferraris, Corso Massimo D'Azeglio 42, 10125 Torino, Italy
Arons, R.R.	Institut für Festkörperforschung, Kernforschungsanlage Jülich, Postfach 365, 5170 Jülich, Germany
Balsamo, P.E.	Comitato Nazionale per l'Energia Nucleare, Laboratori Nazionali di Frascati, C.P.N. 70, 00044 Frascati, Roma, Italy
Bannink, G.	Twente University of Technology, Lintveldebrink 262, Enschede, The Netherlands
Baratoff, A.	IBM Research Laboratory Zürich, Säumerstraße 4, 8803 Rüschlikon, Switzerland

Barone, A.	Laboratorio di Cibernetica del CNR, Via Toiano 2, 80072 Arco Felice (Napoli), Italy
Berchier, J.-L.	University of Geneve, Institut de Physique, DPMC, 32 Bd. d'Yvoy, CH 1211 Geneve 4, Switzerland
Berlincourt, T.G.	Office of Naval Research, Physical Sciences Division, 800 N Quincy St. Arlington, Virginia 22217, USA
Beyer, J.	Department of Electrical and Computer Engineering, University of Wisconsin-Madison, Madison 53706, USA
Brandt, R.	Office of Naval Research, 1030 East Green Street, Pasadena, California 91106, USA
Buchholz, F.I.	Physikalisch-Technische Bundesanstalt, Bundesallee 100, 3300 Braunschweig, Germany
Buhrman, R.A.	School of Applied and Engineering Physics, Cornell University, Clark Hall, Ithaca, New York 14853, USA
Carelli, P.	Laboratorio di Elettronica dello Stato Solido del CNR, Via Cineto Romano 42, 00156 Roma, Italy
Clarke, J.	Department of Physics, University of California, Berkeley, California 94720, USA
Colegrove, F.D.	Texas Instruments Inc., Mail Station 266, P.O. Box 6015, Dallas, Texas 75222, USA

Crum, D.B. — S.H.E. c/o Klaus Schäfer GmbH, Postfach 488, 6078 Neu-Isenburg, Germany
S.H.E. Corporation, 4174 Sorrento Valley Blvd., San Diego, California 92121, USA

Daalmans, G.M. — Laboratorium voor Technische Natuurkunde, Technische Hogeschool Delft, Lorentzweg 1, Delft, The Netherlands

Davis, R.J. — U.S. Naval Research Lab., Code 5460, Washington D.C. 20375, USA

Daunt, J.G. — Cryogenics Center, Stevens Institute of Technology, Hoboken, N.J. 07030, USA

Donaldson, G.B. — Department of Applied Physics, John Anderson Building, University of Strathclyde, Glasgow G4 ONG, Great Britain

Dousselin, G. — I.N.S.A., Department de Physique, 20, Av. des Buttes de Coesmes, 35031 Rennes-Cedex, France

Dunajski, Z. — Technische Hogeschool Twente, Campuslaan 36, Enschede, The Netherlands

Durcansky, G. — Institut für Festkörperforschung, Kernforschungsanlage Jülich, Postfach 365, 5170 Jülich, Germany

Duret, D. — Institut d'Electronique Fondamentale, Université Paris-Sud, Bat. 220, 91405 Orsay, France

Ehnholm, G.J. — Helsinki University of Technology, Department of Technical Physics, SF-02150 Otaniemi, Finland

Ernè, S.N.	Physikalisch-Technische Bundesanstalt, Institut Berlin, Abbestraße 2-12, 1000 Berlin 10, Germany
Finnegan, T.F.	Institut for Basic Standards, National Bureau of Standards, Room A-247, Bldg. 220, Washington D.C. 20234, USA
Fujita, T.	Department of Physics, Tohoku University, Katahira Sendai, 980, Japan
Gallop, J.	National Physical Laboratory, Queens Road, Teddington, Middlesex TW11 OLW, Great Britain
Geisler, R.	Institut für Geophysik u. Meteorologie Technische Universität Braunschweig, Mendelssohnstr. 1a, 3300 Braunschweig, Germany
Giovannini, B.	University of Geneve, Department of Physics, 32, Boulevard d'Yvoy, CH-1211 Geneve 4, Switzerland
Good, J.A.	Cryogenic Consultants Ltd., Metrostore Building, 231 The Vale, London W3 7QS, Great Britain
Greidanus, F.J.A.M.	Kamerlingh Onnes Laboratorium, Nieuwsteeg 18, Leiden, The Netherlands
Grohmann, K.	Physikalisch-Technische Bundesanstalt, Institut Berlin, Abbestraße 2-12, 1000 Berlin 10, Germany
Gross, F.	Physikalisches Institut, Universität Graz, Universitätsplatz 5, 8010 Graz, Austria
Guèret, P.	IBM Research Laboratory Zürich, Säumerstraße 4, 8803 Rüschlikon, Switzerland

Gundlach, K.	Max-Planck-Institut für Physik und Astrophysik, Föhringer Ring 6, 8000 München 40, Germany
Gutmann, P.	Physikalisch-Technische Bundesanstalt, Bundesallee 100, 3300 Braunschweig, Germany
Hahlbohm, H.-D.	Physikalisch-Technische Bundesanstalt, Institut Berlin, Abbestraße 2-12, 1000 Berlin 10, Germany
Hahn, D.	Physikalisch-Technische Bundesanstalt, Institut Berlin, Abbestraße 2-12, 1000 Berlin 10, Germany
Halse, M.R.	Physics Department, University of Kent, Giles Lane, Canterbury, Kent CTL 7NR, Great Britain
Hansen, J.	2. Physikalisches Institut, Universität Köln, Universitätsstraße 14, 5000 Köln 41, Germany
Hasuo, S.	Advanced Technology Lab., Fujitsu Lab. Ltd., Nakahara-ku, 1015 Kamikodanaka, Kawasaki 211, Japan
Hechtfischer, D.	Physikalisch-Technische Bundesanstalt, Institut Berlin, Abbestraße 2-12, 1000 Berlin 10, Germany
Heiden, C.	Institut für Angewandte Physik, Universität Giessen, Heinrich-Buff-Ring 16, 6300 Giessen, Germany
Hillenbrand, B.	Research Laboratories of Siemens AG, FL MET 11, G. Scharowsky Str. 2, 8520 Erlangen, Germany
Hoenig, H.E.	Physikalisches Institut, Universität Frankfurt, Robert-Mayer-Str. 2-4, 6000 Frankfurt/M., Germany

Hoffmann, A.	Physikalisch-Technische Bundesanstalt, Institut Berlin, Abbestraße 2-12, 1000 Berlin 10, Germany
Jablonski, H.	Fa. Klaus Schäfer GmbH., Postfach 488, 6078 Neu-Isenburg, Germany
Jakschik, J.	Physikalisch-Technische Bundesanstalt, Institut Berlin, Abbestraße 2-12, 1000 Berlin 10, Germany
Jutzi, W.	Institut für elektronische Grundlagen der Informatik, Universität Karlsruhe, 7500 Karlsruhe, Germany
Kadlec, J.	Max-Planck-Institut für Physik und Astrophysik, Föhringer Ring 6, 8000 München 40, Germany
Kaiser-Dieckhoff, U.	Physikalisches Institut, Universität Würzburg, Röntgenring 8, 8700 Würzburg, Germany
Kessel, W.	Physikalisch-Technische Bundesanstalt, Bundesallee 100, 3300 Braunschweig, Germany
Kirschman, R.K.	Jet Propulsion Lab., California Institute of Technology, 4800 Oak Grove Drive, Pasadena, California 91103, USA
Klein, K.D.	Physikalisch-Technische Bundesanstalt, Institut Berlin, Abbestraße 2-12, 1000 Berlin 10, Germany
Klipping, G.	Fritz-Haber-Institut der Max-Planck-Gesellschaft, Faradayweg 4-6, 1000 Berlin 33, Germany
Kose, V.	Physikalisch-Technische Bundesanstalt, Bundesallee 100, 3300 Braunschweig, Germany

Kurkijärvi, J.	Helsinki University of Technology, Low Temperature Laboratory, SF-02150 Otaniemi, Finland
Kurti, N.	Clarendon Laboratory, University of Oxford, Parks Road,Oxford OX1 3PJ, Great Britain
Kusayanagi, E.	Department of Electronics, Kyushu University, Fukuoka, Japan
Leppin, H.P.	2. Physikalisches Institut, Universität Köln, Zülpicher Straße 77, 5000 Köln 41, Germany
Levinsen, M.T.	H.C. Ørsted Institute, Physics Laboratory I, University of Copenhagen, Universitetsparken 5, 2100 Copenhagen Ø, Denmark
Lindelof, P.E.	H.C. Ørsted Institute, Physics Laboratory I, University of Copenhagen, Universitetsparken 5, 2100 Copenhagen Ø, Denmark
Lindsay, N.M.	Cryogenic Consultants Ltd., Metrostore Building, 231 The Vale, London W3 7QS, Great Britain
Locatelli, M.	Centre d'Etudes Nucléaires de Grenoble Service des Basses Temperatures 85X, Av. des Martyrs, 38041 Grenoble-Cedex, France
Lübbig, H.	Physikalisch-Technische Bundesanstalt, Institut Berlin, Abbestraße 2-12, 1000 Berlin 10, Germany
Lüders, K.	Institut für Atom- und Festkörperphysik, Freie Universität Berlin, Königin-Luise-Str. 28, 1000 Berlin 33, Germany

Ludwig, W.	Max-Planck-Institut für Metallforschung, Institut für Physik, Büsenauer Str. 171, 7000 Stuttgart 80, Germany
Lukens, J.E.	Department of Physics, State University of New York, Stony Brook, New York 11794, USA
Lundgren, L.	Solid State Division, Institute of Technology, Box 534, 75121 Uppsala, Sweden
Luther, H.	Physikalisch-Technische Bundesanstalt, Institut Berlin, Abbestraße 2-12, 1000 Berlin 10, Germany
Malassis, R.	C.N.A.M., 292 Rue Saint Martin, Paris III, France
Matisoo, J.	IBM Thomas J. Watson Research Center, P.O. Box 218, Yorktown Heights/N.Y. 10598, USA
Mercereau, J.E.	California Institute of Technology, Low Temperature Physics, mail code 63-37, Pasadena, California 91125, USA
Meredith, D.F.	Physics Department, University of Lancaster, Lancaster LA1 4YW, Great Britain
Meyer, J.-D.	I. Physikalisches Institut, Universität Göttingen, Bunsenstraße 9, 3400 Göttingen, Germany
Modena, I.	Laboratorio di Elettronica dello Stato Solido del CNR, Via Cineto Romano 42, 00156 Roma, Italy

Mooij, J.E.	Laboratorium voor Technische Natuurkunde, Technische Hogeschool Delft, Lorentzweg 1, Delft, The Netherlands
Niemeyer, J.	Physikalisch-Technische Bundesanstalt, Bundesallee 100, 3300 Braunschweig, Germany
Nisenoff, M.	U.S. Naval Research Laboratory, Cryogenic and Superconductivity Branch, Code 6435, Washington D.C. 20375, USA
Ohta, H.	The Institute of Physical and Chemical Research, Hirosawa 2-1, Wako-shi, Saitama, 351, Japan
Orr, P.D.A.	University of Cambridge, 21 St. Regis Chesterton Road, Cambridge CB4 IB4, Great Britain
Pace, S.	Istituto di Fisica, University of Salerno, Via Vernieri 42, 84100 Salerno, Italy
Pals, J.A.	Philips Research Laboratories, Eindhoven, The Netherlands
Park, J.G.	The Blackett Laboratory, Imperial College of Science and Technology, Prince Consort Road, London SW7 2BZ, Great Britain
Pascal, D.	Institut d'Electronique Fondamentale, Université Paris-Sud, Bat. 220, 91405 Orsay, France
Paterno, G.	Comitato Nazionale per L'Energia Nucleare, Laboratori Nazionali di Frascati, 00044 Frascati, Roma, Italy

Pegrum, C.M.	Department of Applied Physics, John Anderson Building, University of Strathclyde, Glasgow G4 ONG, Great Britain
Pierre, J.	Centre d'Etudes Nucleaires de Grenoble, Service des Basses Temperatures 85X, Av. des Martyrs, 38041 Grenoble-Cedex, France
Ramin, H.	Institut für Fernmeldetechnik, Technische Universität Berlin, Einsteinufer 29, 1000 Berlin 10, Germany
Ricci, F.	Istituto di Fisica "G. Marconi", Universitá Degli Studi di Roma, Piazzale Delle Scienze No. 5, Roma 00100, Italy
Richards, P.L.	Department of Physics, University of California, Berkeley, California 94720, USA
Rogalla, H.	Institut für Angewandte Physik, Universität Münster, Roxeler Str.70-72, 4400 Münster/Wstf., Germany
Romani, G.	Istituto di Fisica "G. Marconi", Universitá Degli Studi di Roma, Piazzale Delle Scienze No. 5, Roma 00100, Italy
Russo, M.	Laboratorio di Cibernetica del CNR, Via Toiano 2, 80072 Arco Felice, (Napoli), Italy
Sandell, R.D.	Department of Physics, State University of New York, Stony Brook, New York 11794, USA

Sanitas,	Bureau National de Metrologique, 21 Rue Casimir Perier, 75007 Paris, France
Sauerbrey, G.	Physikalisch-Technische Bundesanstalt, Institut Berlin, Abbestraße 2-12, 1000 Berlin 10, Germany
Saur, E.	Institut für Angewandte Physik, Universität Giessen, Leihgesterner Weg 106, 6300 Giessen, Germany
Sauzade, M.	Institut d'Electronique Fondamentale, Université Paris-Sud, Bat. 220, 91405 Orsay, France
Soerensen, O.H.	Technical University of Denmark, Physical Laboratory I, 2800 Lyngby, Denmark
Suhr, H.	Physikalisch-Technische Bundesanstalt, Institut Berlin, Abbestraße 2-12, 1000 Berlin 10, Germany
Sullivan, D.B.	Cryogenics Division, Institute for Basic Standards, National Bureau of Standards, Boulder, Colorado 80302, USA
Swithenby, S.J.	University of Oxford, Department of Physics, Clarendon Laboratory, Oxford OX1 3PU, Great Britain
Schäfer, G.	Fa. Leybold-Heraeus GmbH. u. Co. KG, Bonner Str. 504, 5000 Köln 51, Germany
Schlup, W.	IBM Research Laboratory Zürich, Säumerstraße 4, 8803 Rüschlikon, Switzerland

Schultz, G.	Max-Planck-Institut für Radioastronomie, Auf dem Hügel 69, 5300 Bonn 1, Germany
Schultz-DuBois, E.O.	Institut für Angewandte Physik, Universität Kiel, 2300 Kiel, Germany
Schuster, G.	Physikalisch-Technische Bundesanstalt, Institut Berlin, Abbestraße 2-12, 1000 Berlin 10, Germany
Schwarz, W.	Institut für Experimentelle Kernphysik, Universität Karlsruhe, Postfach 3640, 7500 Karlsruhe, Germany
Takács, S.	Slovak Academy of Science, Institute of Electrical Engineering, Dubravska Cesta, Bratislava 80932, CSSR
Tidecks, R.	I. Physikalisches Institut, Universität Göttingen, Bunsenstraße 9, 3400 Göttingen, Germany
Ulrich, B.T.	Physics Department, University of Nijmegen, Toernooiveld, Nijmegen, The Netherlands
Vaglio, R.	Istituto di Fisica, Università di Salerno, Via Vernieri 42, 84100 Salerno, Italy
Van Duzer, T.	Department of Electrical Engineering and Computer Sciences, University of California, Berkeley, California 94720, USA
Van Haeringen, W.	Philips Research Laboratories, Eindhoven, The Netherlands
Van Maaren, M.H.	Philips Research Laboratories, Eindhoven, The Netherlands

Walker, E.A.	Physics Department, University of Lancaster, Lancaster LA1 4YW, Great Britain
Walter-Peters, M.	Afdeling der Technische Natuurkunde, Technische Hogeschool Twente, Postbus 217, Enschede, The Netherlands
Welker, N.	Laboratories for Physical Sciences, 4928 College Av., College Park, Maryland 20740, USA
Wolf, P.	IBM Research Laboratory Zürich, Säumerstraße 4, 8803 Rüschlikon, Switzerland
Zenatti, D.	Centre d'Etudes Nucleaires de Grenoble, B.P. No. 85, 38041 Grenoble-Cedex, France
Zimmerman, J.E.	Cryogenics Division, National Bureau of Standards, Boulder, Colorado 80302, USA

THEORY OF SUPERCONDUCTING PHASE COUPLING

THEORY OF PHASE COHERENCE IN SQUID'S

H. Lübbig
Physikalisch-Technische Bundesanstalt, Institut Berlin,
Abbestr. 2-12, D-1000 Berlin 10, Germany

1 INTRODUCTION

Superconducting microstructures, capable for carrying phase gradients in the presence of currents and fields, as well as electromagnetic circuits, based essentially upon the non-linearity of their current phase characteristic have now been studied over a range of fifteen years.

A large number of geometrical different weak link configurations, spreading the range from tunnel contacts with discontinuous gap structure to thin film microbridges connecting two superconductors in a continuous way, have been developed, evolving a multifarious picture of phase coherence effects. For tunnel junctions, the original analysis of B.D. JOSEPHSON and its version due to N.R. WERTHAMER provide an unified thorough description of the constitutive current-phase relationship, including the complete dynamical response. In microbridges the situation is more complicated. Up to now the appearance of phase gradients in various thin film structures is well understood only under static conditions. A dynamic description of phase coherence effects in microbridges has come into focus only in the last years, but no unified theory is available.

On the side of the electromagnetic circuits, the SQUID configuration, in particular, is characterized by the direct relation between the phase difference, φ , localized in the weak coupling structure and the macroscopic magnetic flux, $2\pi \emptyset/\emptyset_0 = -\varphi$, which is enclosed in the twofold connected topology containing the weak link. As a consequence, the electromagnetic behaviour of the SQUID, i.e. the exchange of flux between the SQUID and the surrounding circuit, is determined essentially by the current-phase relationship of the weak link. Thus, the dynamics of flux transition, being reversible or irreversible, involves both the response of the weak link and the dispersion due to the macroscopic circuit elements. A SQUID response theory, restricted to frequencies small in comparison with the characteristic energy-gap frequency, has been developed in recent years, a detailed thorough description of the SQUID dynamics, however, is not available at present.

As an introduction, this paper is concerned to the theory of phase coherence in tunnel junctions with isolating barriers and in thin film microbridges. By reason of simplicity, the links are treated in the one-dimensional limit by assuming the transvers dimensions to be smaller than twice the coherence length and the penetration depth. Through the whole article, the superconductors are assumed to be in thermal equilibrium. In a final section an approach is reported to treat the low frequency behaviour of an rf-SQUID as an example for a circuit analysis.

2 TUNNELING STRUCTURE

Several microscopic concepts of the theory of electron-tunneling through thin isolating barriers separating two bulk superconductors have been developed using different mathematical methods; e.g. pertubation theory [1,2], GREEN'S function method [3], GINZBURG-LANDAU theory, [4]. Concerning the pertubation method, the system is described by an effective Hamiltonian, [5,6],

$$H = H_1 + H_2 + H_T + \mu \quad . \tag{1}$$

The Hamiltonians H_1 and H_2 are defined to describe the characteristics of the unperturbed system consisting of the two seperated superconductors 1 and 2. Bringing the two superconductors into interaction, their wave functions become overlapping within the isolating barrier constituting a connection between electron states on both sides of the barrier. An effective Hamiltonian H_T is introduced to describe this interaction. The properties of the contact structure are included in the matrix elements of H_T. The linked superconductors are assumed to be in thermal equilibrium. One may think of H_1, H_2 as BCS Hamiltonians for the case of weak-coupling superconductors. The case of strong-coupling superconductors has been investigated in ref [7,8]. The influence of the external circuit in this model is taken into account by means of the voltage V applied across the junction leading to the chemical potential μ, equ.(1). Under physical conditions the junction is subject to an impressed current. Consequently, the voltage arising across the contact controlls the balance of the different additive terms involved in the expression of the tunneling current. Therefore, it seems intuitively consequent, to visualize the physical meaning of the distinguishable contributions of the total current by means of describing their interdependence due to their voltage dependence.

2.1 TIME-INDEPENDENT VOLTAGE

In the case that a time-independent voltage V_0 is assumed the resulting tunneling current I shows periodic behaviour with respect to the time

$$I(t) = I_{J1}(V_0,T)\sin\varphi(t) + I_{J2}(V_0,T)\cos\varphi(t) + I_{qp2}(V_0,T) \quad (2)$$

$$\varphi(t) = -(2e/\hbar)V_0 t + \varphi(0) . \quad (3)$$

Here φ denotes the gauge-invariant phase difference across the barrier. The characteristics of the tunneling transfer enter the expression (2) only by means of the voltage depending amplitudes I_{J1}, I_{J2}, I_{qp2} which are functions of the temperature T and the energy-gaps (Δ_1, Δ_2) of the linked superconductors. Several authors have studied the amplitudes numerically under various conditions [2,9,10-12].

In fig. 1 the amplitudes I_{J1}, I_{J2}, I_{qp2} are shown constituting a tunneling structure of two different superconductors ($\Delta_2 = 3\Delta_1$) at vanishing temperature. Because of the fact that I_{J1} is finite at zero voltage, the first term of equ. (2) can be interpreted as the tunneling current of COOPER pairs. At vanishing voltage the $\sin\varphi$-term exhibits the dc-JOSEPHSON effect. At finite voltages an oscillating current arises, usually denoted as the ac-JOSEPHSON effect. If the voltage equals the sum of the gap-voltages, the amplitude I_{J1} shows a logarithmic singularity, the so called RIEDEL peak, [9]. This singularity directly results from the singular behaviour of the BCS energy density at the gap edge. The last term in equ. (2) is independent of phase difference and only depends on the voltage. Consequently, this term can be related to the static characteristic of quasiparticle tunneling. The second term in equ. (2), usually phrased as the $\cos\varphi$-term, depends on both the phase-difference and the voltage. The amplitude I_{J2}, vanishing for zero voltage, shows a sharp rise at the gap voltage. This term can be identified as a quasiparticle-pair-interference term.

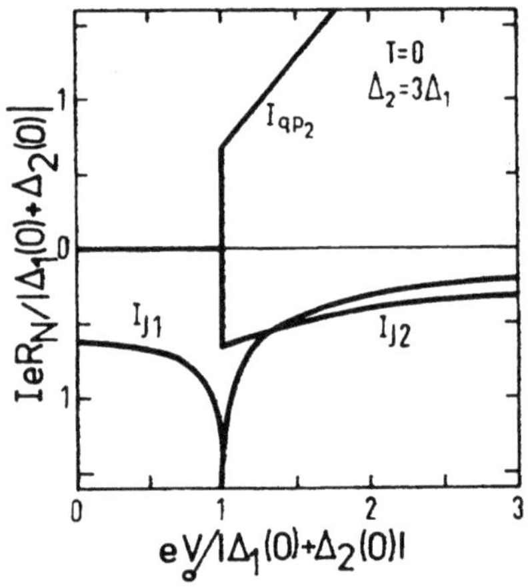

Fig. 1 The amplitudes of the current-phase-relation (2) for different superconductors and vanishing temperature, rf[12]; R_N denotes the normal resistance.

In general, the quasiparticles enter the expression (2) by means of a static characteristic I_{qp2}, while the terms which are related to the transfer of pairs show harmonic behaviour with respect to the phase difference $\varphi(t)$. In order to look for their dispersion relationship it is convenient to introduce the complex notation

$$I_J(t) = \mathrm{Im}\, \tilde{I}_J(\omega_0)\, e^{i\varphi(t)} \quad , \quad \hbar\omega_0 = eV_0 \ . \tag{4}$$

The amplitudes of the $\sin\varphi$ - and the $\cos\varphi$ -term appear as real- and imaginary part of the complex quantity

$$\tilde{I}_J(\omega_0) = I_{J1}(\omega_0) + i\, I_{J2}(\omega_0) \tag{5}$$

satisfying the following symmetry relations

$$I_{J1}(\omega_0) = I_{J1}(-\omega_0) \quad , \quad I_{J2}(\omega_0) = -\, I_{J2}(-\omega_0) \ . \tag{6}$$

It has been pointed out first by WERTHAMER , [13], and by WILKINS, [14], that the amplitudes $I_{J1}(\omega_0)$ and $I_{J2}(\omega_0)$ are correlated symmetrically by means of the KRAMERS-KRONIG-relationship, [15,16]:

$$I_{J1}(\omega) = \frac{1}{\pi}\, \mathcal{P} \int_{-\infty}^{\infty} I_{J2}(\omega') \frac{d\omega'}{\omega' - \omega} \tag{7}$$

$$I_{J2}(\omega) = \frac{1}{\pi}\, \mathcal{P} \int_{-\infty}^{\infty} I_{J1}(\omega') \frac{d\omega'}{\omega' - \omega} \ . \tag{8}$$

The symbol $\mathcal{P}$ stands for the CAUCHY principal value. In this way the frequency dependence of the one amplitude is determined as integral transformation of the other. Especially the RIEDEL peak, arising in the amplitude I_{J1} of the $\sin\varphi$ -term at the gap frequency,is correlated in a direct way to the step of the $\cos\varphi$ -amplitude I_{J2} at this frequency.

It is necessary to mention a mathematical restriction in this context: Deducing the KRAMERS-KRONIG relations one has to assume - besides other conditions - that the quantities to be correlated have to be bounded. Thus the logarithmic singularity in general violates the application of the theorem. Nevertheless, in real devices the singularities are broadened by quasiparticle damping, gap anisotropy and other influences as had been studied in detail by HASSELBERG, [17]. Therefore the mathematical difficulty is not present in real situations.

2.2 TIME-DEPENDENT VOLTAGE

This scheme of interpreting the interdependence of the pair current terms $I_{J1}(\omega)$ and $I_{J2}(\omega)$ still can be applied to the more complex case if an arbitrary time depending voltage V(t) is biasing the tunnel junction. HARRIS, [18], recently has generalized a calculation of SCHRIEFFER concerning this case. The tunneling current is found to be of the form:

$$I(t) = \mathrm{Im}\left\{ e^{-i\varphi(t)/2} \int_{-\infty}^{\infty} d\omega\, \mathcal{U}(\omega)\, \tilde{I}_J(\omega)\, e^{-i\omega t} + \right. \tag{9}$$

$$\left. + e^{i\varphi(t)/2} \int_{-\infty}^{\infty} d\omega\, \mathcal{U}(\omega)\, \tilde{I}_{qp}(\omega)\, e^{-i\omega t} \right\}.$$

The voltage influences this expression only by means of

$$\varphi(t) = -\frac{e}{\hbar}\int_{-\infty}^{t} V(t')\,dt' \tag{10}$$

and the corresponding spectral function

$$\mathcal{U}(\omega) = \int_{-\infty}^{\infty} (dt/2\pi)\, e^{i\omega t}\, e^{-i\varphi(t)} \quad . \tag{11}$$

The properties of the contact structure again enter the expression by means of the two complex valued amplitudes

$$\tilde{I}_J(\omega) = I_{J_1}(\omega) + i\, I_{J_2}(\omega) \tag{12}$$

$$\tilde{I}_{qp}(\omega) = I_{qp_1}(\omega) + i\, I_{qp_2}(\omega) \tag{13}$$

which are linear combinations of four real amplitudes. The generalization of this expression with respect to the case of time-independent voltage lies in the appearance of an additional fourth amplitude $I_{qp1}(\omega)$, whereas the other amplitudes remain unchanged. Reducing the general expression, equ. (9), to the time-independent case by setting $U(\omega) \propto \delta(\omega - \omega_0)$, $\hbar\omega_0 = eV_0$, the influence of the amplitude

$I_{qp1}(\omega)$ on I(t) is inhibited.

The dependence of I_{qp1} on the frequency as well as on the energy-gaps and the temperature has been studied numerically by HARRIS, [18]. Again the two components I_{qp1} and I_{qp2} which

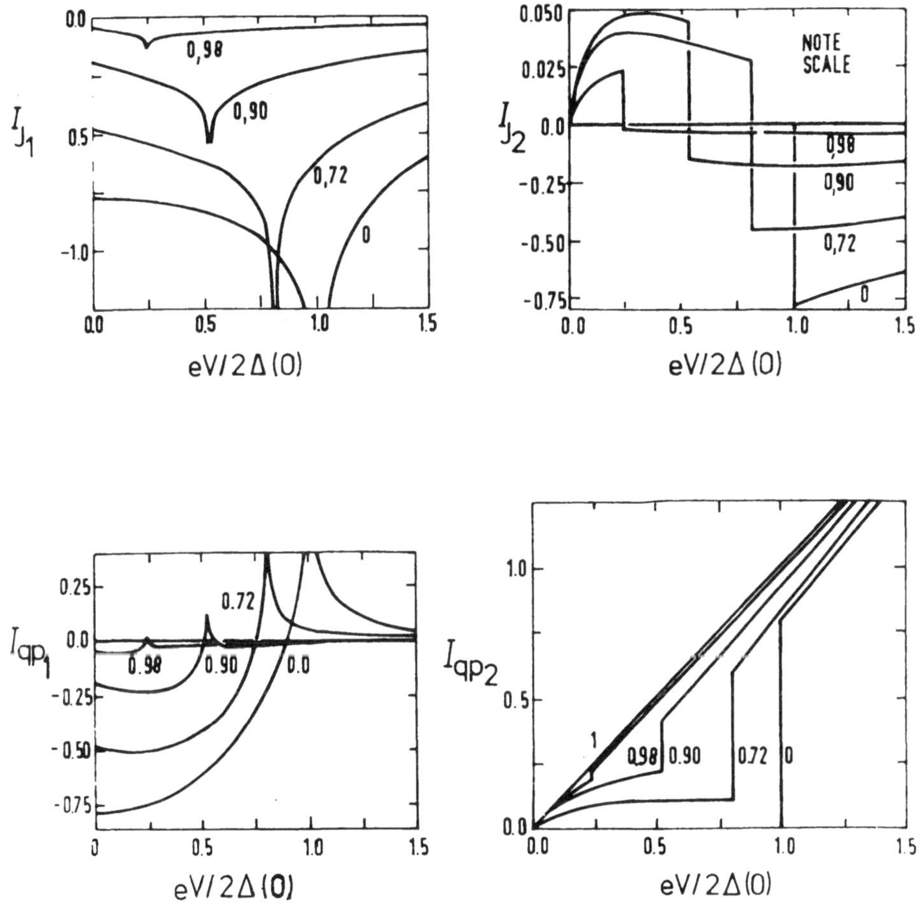

Fig. 2 The four response functions I_{J1}, I_{J2} and I_{qp1}, I_{qp2} (in units of $2\Delta(o)/eR_N$) in dependence of the (normalized) frequency $eV/2\Delta(o)$ for various values of the reduced temperature for identical superconductors, [18,19]. The pairs I_{J1}, I_{J2} and I_{qp1}, I_{qp2} satisfy the KRAMERS-KRONIG relation, resp..

are related to the transfer of quasiparticles are found to satisfy the KRAMERS-KRONIG-transform with respect to their frequency dependence. This had been mentioned earlier by WERTHAMER , [13]. Consequently, the symmetry relations

$$I_{qp_1}(\omega) = I_{qp_1}(-\omega) \quad , \quad I_{qp_2}(\omega) = -I_{qp_2}(-\omega) \tag{14}$$

hold.

In fig. 2 the four response amplitudes I_{J1}, I_{J2} and I_{qp1}, I_{qp2} are shown for a junction of identic superconductors and for various values of the temperature. The picture indicates the high symmetry of the relationship between the four amplitudes from the lowest harmonic order of the tunneling transfer problem.

In the upper part the amplitudes related to the flow of COOPER pairs, in the lower part the amplitudes identified as quasiparticle currents are shown. Each pair of amplitudes is correlated by the KRAMERS-KRONIG relationship. So, as an example, the logarithmic singularities appear as a natural consequence of the sharp rises of the other amplitudes and vice versa.

2.3 TIME DEPENDENT FORMULATION

To demonstrate the reciprocal correspondence between $I_{J2}(\omega)$ and $I_{qp2}(\omega)$ on the one hand and $I_{J1}(\omega)$ and $I_{qp1}(\omega)$ on the other hand in a general way, one has to transform the frequency representation into the time domain. This yields

$$I(t) = \int_0^\infty dt' \left\{ K_J(t') \sin \frac{\varphi(t-t') + \varphi(t)}{2} + K_{qp}(t') \sin \frac{\varphi(t-t') - \varphi(t)}{2} \right\} . \tag{15}$$

As a consequence of the KRAMERS-KRONIG relation, here the <u>four</u> characteristic tunneling response functions are involved completely in the <u>two</u> realvalued kernels (which have the unit of a current per unit time)

$$K_J(t') = \frac{1}{\pi} \int_0^\infty d\omega \left[I_{J_1}(\omega) \cos \omega t' + I_{J_2}(\omega) \sin \omega t' \right] \quad (16)$$

and

$$K_{qp}(t') = \frac{1}{\pi} \int_0^\infty d\omega \left[I_{qp_1}(\omega) \cos \omega t' + I_{qp_2}(\omega) \sin \omega t' \right] . \quad (17)$$

Thus, the frequency dependent response functions I_{J1}, I_{qp1} and I_{J2}, I_{qp2} appear as the (unilateral) FOURIER cos- and FOURIER sin - transform of the two corresponding retarding kernels K_J and K_{qp} resp..

The integral kernels K_J and K_{qp} describe the time delay caused by the frequency dependence of the COOPER pair- and the quasi-particle-response resp.. By this way HARRIS, [19], recently showed the direct relationship between the gap singularities arising in the frequency domain and characteristic oscillations of I_J and I_{qp} with the intrinsic gap response time $\tau_J = \hbar/2\Delta$ in the time domain.

2.4 TRANSITION TO BULK-LIKE BEHAVIOUR

Since in bulk superconducting configurations no time-independent phase difference can persist, it is intuitively near by hand to describe the removal of the coupling between the two linked superconductors simply by setting $\varphi(0) = 0$, i.e. $\varphi(t)=0$ for $-\infty \leq t \leq 0$, equ. (10). By this way the junction is incapable for storing the amount of time-independent energy necessary for producing the dc-JOSEPHSON effect. Under this condition one would expect the current-phase relation (15, 17) to describe the bulk-like behaviour. Indeed, if a small monochromatic voltage $V = v \cos \omega t$, $2ev \ll \hbar\omega$, is assumed, i.e.

$$\sin \frac{1}{2} \left[\varphi(t-t') \mp \varphi(t) \right] \cong \frac{ev}{\hbar\omega} \left[\sin \omega(t-t') \pm \sin \omega t \right] \quad (18)$$

the current I(t), equ. (15), follows to be nearly monochromatic

$$I(t) \cong v \left[\sigma_1(\omega) \cos \omega t - \sigma_2(\omega) \sin \omega t \right] . \quad (19)$$

There the conductivities

$$\sigma_1 = \frac{e}{\hbar\omega} \left[I_{J_2}(\omega) + I_{qp_2}(\omega) \right] \quad (20)$$

$$\sigma_2 = \frac{e}{\hbar\omega} \left[I_{J_1}(\omega) + I_{J_1}(0) + I_{qp_1}(\omega) - I_{qp_1}(0) \right] , \quad (21)$$

describing the active and reactive parts of the current resp., are determined by the two pairs of response functions ($I_{J2}(\omega)$; $I_{qp2}(\omega)$) and ($I_{J1}(\omega)$; $I_{qp1}(\omega)$), resp.. The frequency dependence of σ_1, σ_2 can be proofed to be equal to that of the components of the complex conductivity $\tilde{\sigma} = \sigma_1 - i\,\sigma_2$ introduced by the MATTIS and BARDEEN theory of the local electromagnetic response of a single superconductor, [18]. This result, mentioned first by M. TINKHAM, shows that one may consider the four response-functions I_{J1}, I_{J2}, I_{qp1}, I_{qp2} of the tunneling transfer problem as a special case of the more general theory on the response of bulk superconductors, originally discussed in the framework of the BCS theory.

To conclude, due to the second order perturbation theory the tunnel current through isolating barriers is found to be determined completely by two pairs of response functions, ($I_{J1}(\omega)$; $I_{J2}(\omega)$) and ($I_{qp1}(\omega)$; $I_{qp2}(\omega)$), which are associated to the flow of COOPER pairs and of quasiparticles resp.. The components of each function-pair are related symmetrically in the frequency domain by means of the KRAMERS-KRONIG relations. Applying the equivalent formulation in the time domain, the components of the function-pairs are found to determine the FOURIER-components of two time dependent kernels K_J, K_{qp} which describe the time delay associated with the dispersion of the current phase relation.

The tunneling theory is based essentially upon the assumption of equilibrium for the linked BCS superconductors as well as for the tunneling procedure itself. By this way the JOSEPHSON tunneling theory is defined and competing effects which may result from non-equilibrium deviations of the system are not included.

3 MICROBRIDGES

Concerning microbridge film structures the appearance of dc- and of ac-JOSEPHSON effects is not so well understood as in the case of tunnel barriers. Because of the fact that the gap parameter varies continuously across such configurations various versions of the GINZBURG-LANDAU-Equation, or a certain time-dependent extension, have been applied to thin film bridges of fixed thickness (DAYEM bridges) or of variable thickness.

In these models the influence of the external circuit is taken into account by means of an impressed current. Consequently, the resulting phase-difference depends on the current, a parameter characterizing the "weakening" of the bridge material, and geometric parameters reflecting the boundary conditions at the ends of the region of non-linear phase coupling.

Only the case of an impressed dc-current has been studied in detail up to now. Therefore, the resulting current-phase relationship is time dependent or not whether the weak link model involves a finite voltage implicitely or not.

3.1 TIME-INDEPENDENT CURRENT PHASE RELATIONS

The first evidence that JOSEPHSON effects should appear in narrow superconducting constrictions was provided by ASLAMAZOV and LARKIN , [20], who applied a simplified version of the time-independent GINZBURG-LANDAU-Equation. The phase change across the junction was assumed to be discontinuous. A satisfactory understanding of the static current phase characteristic based upon a continuous phase gradient within the weak coupling region has been developed recently by BARATOFF,

BLACKBURN and SCHWARTZ, [21,22], and GREGERS-HANSEN, LEVINSEN and PEDERSEN, [23]. The case of variable thickness bridges was studied by LIKHAREV and YAKOBSON, [24]. Their investigations are based upon the onedimensional GINZBURG-LANDAU-Equation.

$$[1 - |\Psi|^2 + \xi^2 \frac{d^2}{dx^2}]\Psi = 0 \qquad (22)$$

$$j = (\phi_0/2\pi\mu_0\lambda^2)|\Psi|^2 d\Theta/dx \quad . \qquad (23)$$

Here $\Psi(x) = |\Psi| \exp i\,\Theta(x)$ denotes the complex order parameter, normalized to its value $\Psi_\infty = (mc^2\varepsilon_0/2e^2\lambda^2)^{1/2}$ in the absence of boundary effects where λ is the temperature dependent magnetic penetration depth. The GINZBURG-LANDAU coherence length $\xi(T)$ characterizes the natural unit of length of variation of the order parameter. The current density j of the corresponding stationary superfluid component is independent of the spatial coordinate x because of the condition of continuity. Consequently, equs.(22/23) can be replaced by two real valued equations:

$$|\Psi| + |\Psi|^3 - \frac{\xi^2}{\xi_r^2}\frac{d^2}{dX^2}|\Psi| + \frac{\xi_r^2}{\xi^2}\frac{J}{|\Psi|^3} = 0 \qquad (24)$$

and

$$\varphi = \int d\Theta = \frac{\xi_r^2}{\xi^2} J \int dX/|\Psi|^2 , \qquad (25)$$

where a dimensionless spatial variable $X = x/\xi_r$ and the dimensionless current $J = j\,\phi_0/2\pi\mu_0 2H_c^2$ were used. Equ. (24) describes the spatial variation of the modulus $|\Psi|$ along the filament, while equ. (25) determines the corresponding phase-difference φ induced by the impressed current J.

Following the approach of BARATOFF, BLACKBURN and SCHWARTZ, [22], the perturbation due to the weak coupling region connecting two bulk superconducting segments, can be described by introducing a coherence length ξ_1 within the coupling region, which is much smaller than its value ξ_2 $(= \xi_r)$ in the bulk segments. Therefore, the ratio $\gamma = \xi_1^2/\xi_2^2 < 1$ has the meaning of the weakness parameter of the structure. Under such conditions the equations related to the different regions can be solved analytically in terms of the elliptic functions and the solutions can be matched by means of appropriate boundary conditions, [22].

As an expression for the phase change related to the whole filament, the excess phase difference

$$\varphi = J \int_{-\infty}^{\infty} \left(\frac{1}{\gamma |\Psi|^2} - \frac{1}{\Psi_\infty^2} \right) dX \qquad (26)$$

is computed as function of parameters J, and the reduced length $d = a/2\xi_2$ of the weak link.

In this way a comprehensive analysis of the continuous phase change due to various situations concerning the weakness and the length a of the coupling region is given in rf. [22]. There are three limiting cases for the dc current phase relationship, see fig. 3: (1) If the link is very short, $a/2\xi_2 \ll \gamma$, most of the phase change occurs outside the weakened region. The maximum of the phase characteristic is skewed from $\pi/2$ towards smaller angles, and the corresponding spectrum of $J(\varphi)$ exhibits a content of higher harmonic components with respect to $n\varphi$, n integer. (2) A nearly sinusoidal phase characteristic results under the condition $\gamma\xi_2 < a/2 < \xi_1$. The corresponding maximum current density $j_m = (\pi / 4\sqrt{4\pi\varepsilon_0}\, a\varrho) \times \Delta_0^2(T)/k_B T_c$, $T \approx T_c$, is found to be equal to that value characterizing the

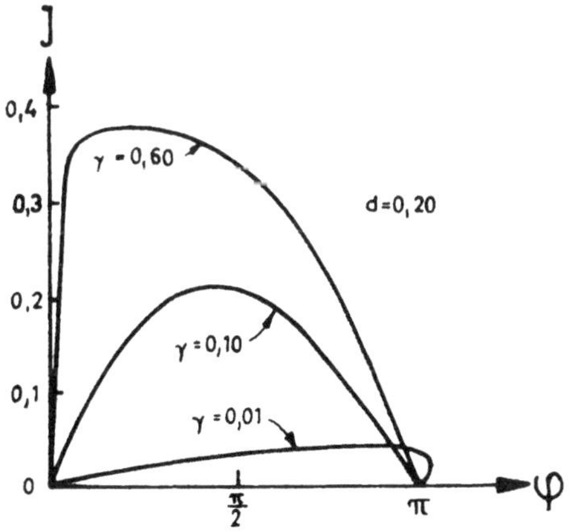

Fig. 3 Static current phase relations of a microbridge with the reduced length d = 0.2 for various values of the weakness parameter γ, [22].

tunnel junction. Thus, a JOSEPHSON-like behaviour is found for the microbridge in this case. (3) If the weakness is decreased furthermore, the phase characteristic can become reentrant. In particular, for a length $a > 5\xi_1$ the re-entrant characteristic tends to behave more like a filament of infinite length.

The model is able to explain a varity of different steady state properties of the current phase relationship of thin film bridges of fixed or variable thickness, [24]. Nevertheless, it must be emphazised, that the underlying GINZBURG-LANDAU equ. is restricted essentially to the case of vanishing voltage. In view of phenomenological similarities with phase-coherence effects arising in tunnel junctions, only the features of the dc-JOSEPHSON effect, i.e. the shape of the dc-phase characteristic in dependence of material and boundary parameters, can be used for a comparison.

3.2 TIME-DEPENDENT CURRENT-PHASE RELATION

For a rigorous description one needs a model which describes the variation of the order parameter in space and time in the presence of electromagnetic fields. Because of the great success of the steady state GINZBURG-LANDAU equation, several authors attempted to develope time dependent extensions, [25,26]; for a review please see rf. [27]. The main problem in this field arises from the fact that the singularity of the density of states at the gap edge results in slowly decaying oscillations in the time response. As a consequence, it is very complicated to deduce a local equation of sufficient generality, [27]. Up to now, only for the case of a gapless system, where magnetic impurities or other pair breakers are assumed to round off the singularity, a simple version of the time dependent generalization of the GINZBURG-LANDAU theory

$$\tau\left(\frac{\partial}{\partial t} + i\frac{2\mu}{\hbar}\right)\Psi + \left(|\Psi|^2 - 1 - \xi^2\frac{\partial^2}{\partial x^2}\right)\Psi = 0 \qquad (27)$$

$$j = \frac{\Phi_0}{2\pi\mu_0\lambda^2} |\Psi|^2 \frac{\partial\Theta}{\partial x} - \sigma \frac{\partial}{\partial x}\frac{\mu}{e} \quad , \tag{28}$$

has been used for the weak link analysis. Here 2μ denotes the electrochemical potential, σ the normal state conductivity, and τ the pair relaxation rate.

RIEGER, MERCEREAU, and SCALAPINO, [28], were the first who applied equ. (27/28) to an analysis of phase-slip processes in microbridges.

For an analytic soluble case, HØJGAARD JENSEN, and LINDELOF, [29], recently studied the influence of the pair relaxation rate τ on the phase characteristic of a short microbridge. Assuming a time independent voltage V_o, i.e. $\mu = e\,V_o\,x/a$, $o \leq x \leq a \ll \xi$, the (normalized) order parameter

$$\Psi(x,t) = g\left(\frac{x}{a}\right) + g^*\left(1-\frac{x}{a}\right) e^{i\varphi(t)} \tag{29}$$

is governed by an AIRY-equation

$$\left[\frac{d^2}{dx^2} + i\tau\omega_0\left(\frac{a}{\xi}\right)^2\frac{x}{a}\right] g\left(\frac{x}{a}\right) = 0 \,, \quad \begin{matrix} \hbar\omega_0 = 2eV_0 \\ g(0) = 1,\ g(1) = 0 . \end{matrix} \tag{30}$$

The spatial mean value of the superfluid component of the current can be calculated directly resulting in

$$I_s(\varphi,V_o) = I_o\,(V_o)\,\sin\varphi + I_1\,(V_o)\,\cos\varphi + I_2\,(V_o)\ . \tag{31}$$

In Fig. 4 the amplitudes I_o, I_1, I_2 are shown as function of the product $\tau\,\omega_o(a/\xi)^2$ combining the ratio of pair relaxation and phase oscillation, $\tau\omega_0$, and the parameter of the weakening a/ξ . Although this model is based essentially on the assumption of a time-independent voltage it clearly exhibits a $\cos\varphi$-term.

In principle, the dc-current controlled one-dimensional GINZBURG-LANDAU model, equs. (27/28), involves necessarily a time depending voltage and a spatial non-linear varying electric

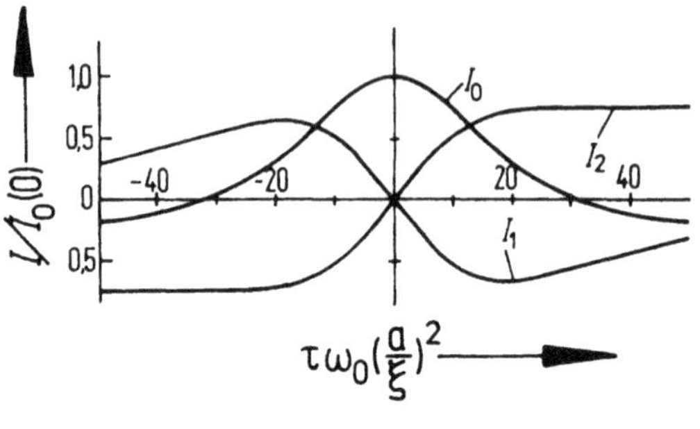

Fig. 4 The amplitudes of the current phase relation (31) as functions of $\tau\omega_0(a/\xi)^2$.

potential. A detailed numerical analysis based upon the complete equations (27/28) has been carried out by LIKHAREV and YAKOBSON, [30], for a microbridge of finite length. In order to adjust the normal conductivity σ, equ. (28), the voltage is scaled in units of $V_J = \pi\Delta_B^2/4ekT$, Δ_B being the energy-gap of the bulk segments. For this choice the "resistively shunted junction model", introduced by ASLAMAZOV and LARKIN, [20], is recovered in the limit of a dc phase-characteristic ($\tau = 0$) for short microbridges. As a consequence two time rates appear, the orderparameter relaxation time τ and the intrinsic gap response time $\tau_J = \hbar/2eV_J$ more closely related to the voltage. The relaxation time τ of the orderparameter is greater than τ_J; $\tau = 12\,\tau_J$ for superconductors with high content of paramagnetic impurities and $\tau = 5.79\,\tau_J$ for the superconductor with ordinary impurities in the dirty limit.

By studying the influence of the two characteristic times, τ, τ_J, the model is able to explain a varity of dynamical features of geometrical different structures. In particular, the model describes the appearance of "excess currents" in the corresponding current-voltage characteristic without ad hoc assumptions concerning the phase-slip-process, [28].

In spite of such success, it must be emphazised that the underlying version of the GINZBURG-LANDAU equation is valid only in the strong pairbreaking limit, and, consequently, does not include any gap response in the time domain. More general time depending extensions are available but have not yet been applied to microcontacts because of their complexness.

In summary, the analysis of localized phase gradients in microbridge structures based upon the dc-current controlled, one-dimensional GINZBURG-LANDAU equation, wether time-dependent or not, indicates inherent similarities with phase-coherence effects in tunnel structures. So far, the influence of geometric and material parameters on the dc-phase characteristic seems to be well understood. For the case of time depending phase characteristics, however, the analysis is based upon a simplified model. Mainly for the application at low temperatures and for frequencies in the range of the energy-gap frequency, a more detailed theory of the phase characteristic requires to take account of a finite gap response in the time domain. Thus, an attempt to compare phase coherence effects in microbridges and tunnel junctions in a phenomenological way, needs some artificial assumptions at date.

4. CIRCUIT ANALYSIS

Because of the non-linearity of the current-phase relationship, the observability of phase-coherence-phenomena essentially depends on the way of incorporating the junction into the electromagnetic circuit.

4.1 CURRENT BIASED JUNCTION

Commonly the influence of distinct weak-coupling structures on observable quantities is studied by means of the current-voltage characteristics resulting from the current biased junction. In this way only the relationship between the time-averages of current and voltage are observable.

Numerous theoretical studies have been carried out based upon different versions of the underlying current-phase relationship of the weak link structure.

Recently McDONALD et al., [31], have studied numerically current-voltage characteristics of a dc-current biased tunnel junction including the complete retarding integral kernels K_J, K_{qp}, equ. (16/17). An analytic approach to this general case has been given by SCHLUP, [32].

For the purpose of more qualitative investigations the adiabatic representation of the tunneling current expression

$$I(t) = I_{J_1}(V) \sin\varphi(t) + I_{J_2}(V) \cos\varphi(t) + I_{qp_2}(V) \tag{32}$$

$$\varphi(t) = -\frac{2e}{\hbar} \int_{-\infty}^{t} V(t')\,dt' , \tag{33}$$

or its low-voltage approximation

$$I(t) = \left[|I_{J_1}(0)| + \frac{\hbar}{2e} \frac{\alpha}{R_q} \frac{d}{dt} \right] \sin \frac{2e}{\hbar} \int_{-\infty}^{t} V(t')dt' + \frac{1}{R_q} V(t) , \tag{34}$$

$$1/R_q = (dI_{qp_2}/dV)_0 , \quad \alpha = R_q (dI_{J_2}/dV)_0 , \tag{35}$$

have been used as reasonable assumptions valid in the case of slow variations in time, and in the small voltage range $|V| < 2\Delta/e$.

The corresponding current balance equation

$$C\frac{dV}{dt} + \frac{1}{R_q} V + \left[|I_{J_1}(0)| + \frac{\hbar}{2e} \frac{\alpha}{R_q} \frac{d}{dt} \right] \sin \frac{2e}{\hbar} \int_{-\infty}^{t} V(t')dt' = I_{ext} , \tag{36}$$

where C denotes the junction capacitance and I_{ext} the impressed current, is known as the "resistively shunted junction model".

Analytic solutions are available only for the case $\alpha = 0$, $dI_{ext}/dt = 0$, and $C = 0$, [20, 33], or for finite capacitance by modification of the quasiparticle conductance, [34].

The influence of the capacitance on the hysteretic behaviour of the current-voltage characteristic has been investigated in the early work of STEWART, [35], and McCUMBER, [36], by numerical integration for the case $\alpha = 0$. SCHLUP, [32], has generalized this work by taking account for the $\cos\varphi$-term, $\alpha \neq 0$.

For a rigorous description of the resistively shunted junction model, please see ref. [37].

4.2 INDUCTIVELY COUPLED JUNCTION

A circuit analysis concerning the inductively coupled weak link by means of the rf-SQUID has been carried out by HANSMA, [38], NISENOFF and WOLF, [39], RIFKIN, VINCENT, DEAVER, and HANSMA, [40], and SOERENSEN, [41]. ERNÉ, HAHLBOHM, and LÜBBIG, [42], recently have reformulated HANSMA's theory of the low-frequency SQUID response.

In Fig. 5 the equivalent circuit of the rf-SQUID is shown. An ac-bias $I_{rf} = \hat{I}_{rf} \sin(2\pi\nu t)$ is impressed to the SQUID by means of the inductively coupled tank circuit. Assuming the frequency ν of the impressed current to be near the resonant

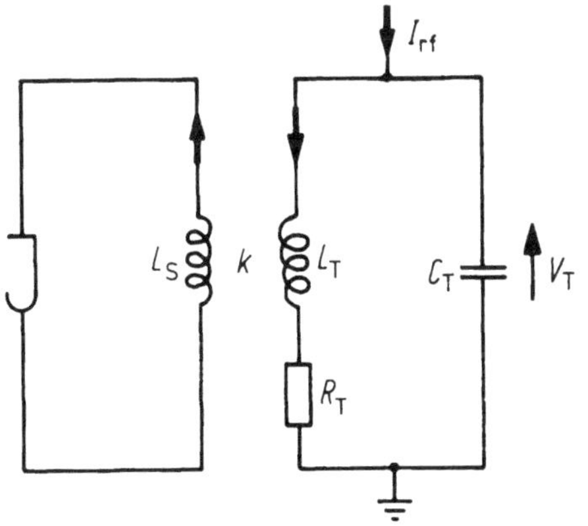

Fig. 5

Model circuit for the rf-SQUID. The symbol J stands for the weak link. The coupling coefficient K of the inductances L_T, L_S is given by $K = M/\sqrt{L_T L_S}$, where M is the mutual inductance coefficient.

frequency ν_T of the high Q_T-tank circuit, the excitation $\phi_a(t)$ seen by the SQUID circuit is nearly monochromatic

$$\phi_a(t) = \phi_{dc} + \hat{\phi}_{rf} \sin(2\pi\nu t + \delta) \quad . \tag{37}$$

Here ϕ_{dc} denotes a superimposed dc-flux component and δ the phase shift with respect to the bias signal. The response of the SQUID on this excitation is studied by means of the voltage V_T at the tank circuit as function of the bias parameters $\hat{I}_{rf}$, ν/ν_T. Amplitude and phase angle of this voltage are influenced sensitively by the parameters of the weak link enclosed in the SQUID ring.

4.3 RESPONSE OF THE SQUID CIRCUIT ON LOW-FREQUENCY EXCITATIONS

Assuming the frequency of the excitation ϕ_a to be much smaller than the characteristic energy-gap frequency $1/\tau_J = 2\Delta/\hbar$, i.e. $\nu\tau_J \lesssim 10^{-4}$, the current through the junction can be represented in the low-frequency limit; equ. (34).

The corresponding balance of the flux in the SQUID loop follows:

$$\left(\frac{1}{\omega_s^2}\frac{d^2}{dt^2} + \tau_q\frac{d}{dt} + 1\right)\frac{\phi}{\phi_o} + \left(1 + \tau_\alpha\frac{d}{dt}\right)\beta \sin 2\pi\frac{\phi}{\phi_o} = \frac{\phi_a}{\phi_o} . \tag{38}$$

The circuit parameters C_s, L_s, R_q, (i.e. the junction capacitance, the SQUID-loop inductance, and the quasiparticle resistance), are involved in the characteristic times

$$\frac{1}{\omega_s} = \sqrt{L_s C_s} \quad , \quad \tau_q = \frac{L_s}{R_q} \quad , \tag{39}$$

while the coefficients α, I_{J1} of the phase depending parts of the current-phase relation, equ. (34), are included in the parameters

$$\beta = \frac{L_s I_{J1}}{\phi_o} \quad , \quad \tau_\alpha = \frac{\alpha}{2\pi\beta}\tau_q \quad . \tag{40}$$

Applying commonly used values of the SQUID circuit:

$$\omega_s > 10^{11}\, s^{-1} \qquad \tau_q \leq 10^{-10}\, s \tag{41}$$
$$|\alpha| < 1.0 \qquad 2\pi\beta \geq 0.1 \quad ,$$

the dispersion of the system can be estimated to be very small and, consequently, the adiabatic limit of the flux balance, equ. (38), i.e.

$$\frac{\phi}{\phi_0} + \beta \sin 2\pi \frac{\phi}{\phi_0} = \frac{\phi_a}{\phi_0} \quad , \tag{42}$$

can be applied under these conditions.

Unique solutions of the adiabatic SQUID magnetization relationship, equ. (42), exist only for $2\pi\beta < 1$ - a case known as nonhysteretic SQUID operation mode - and these can be derived analytically. Using the periodicity of $(\phi - \phi_a)/\phi_0$ with respect to ϕ_a/ϕ_0, i.e.

$$\phi(\phi_a) = \phi_a + \sum_{n=1}^{\infty} \phi_0 M_n(\beta) \sin 2\pi n \frac{\phi_a}{\phi_0} \quad , \tag{43}$$

the coefficients $M_n(\beta)$ follow

$$M_n = \frac{1}{2\pi^2 n} \int_0^{2\pi} \cos\left(2\pi n \frac{\phi_a}{\phi_0}\right) d\left(2\pi \frac{\phi}{\phi_0}\right) = \frac{(-1)^n}{n\pi} J_n(2\pi n\beta) . \tag{44}$$

Here J_n denotes the BESSEL function of the first kind.

The set of coefficients M_n, n = 1,2,..., determines the nonlinearity of the adiabatic magnetization as function of the SQUID parameter β , $0 < 2\pi\beta < 1$.

$2\pi\beta$	$\lvert M_2/M_1 \rvert$	$\lvert M_3/M_1 \rvert$	$\lvert M_4/M_1 \rvert$
0.1	$5 \cdot 10^{-2}$	$4 \cdot 10^{-3}$	$3 \cdot 10^{-4}$
0.6	0.28	0.11	0.06
0.9	0.38	0.21	0.14

Tabel 1 Ratios of the structure coefficients $\lvert M_n/M_1 \rvert$ of the adiabatic SQUID magnetization, equ. (43), as function of $2\pi\beta$.

The influence of the parameter $2\pi\beta$ on the non-linearity of $\phi(\phi_a)$ can be illustrated by means of a complete mechanical analog: the ellipticity of the satellite motion. Using the eccentric anomaly u(t), which is related to azimuth angle ϑ by $u = 2\,\mathrm{arc\,tg}\,(\sqrt{(1-e)/(1+e)}\,\mathrm{tg}\,\vartheta/2)$, and the mean anomaly $\omega_m t = \sqrt{Gm/a^3}\,t$, where m denotes the mass of the planet, G the gravitation constant, and $e = \sqrt{1 - b^2/a^2}$ the numeric eccentricity of the ellipse, the satellite equation of motion

$$u + e \cdot \sin u = \omega_m t \qquad (45)$$

follows [43,44]. Thus, the identifications $u \mathrel{\hat{=}} 2\pi\phi/\phi_o$, and $\omega_m t \mathrel{\hat{=}} 2\pi\,\phi_a/\phi_o$ lead to the analogy $2\pi\beta \mathrel{\hat{=}} e < 1$, which gives an impression of the influence of the magnitude of the screening currents, involved in the parameter β , on the shape of the magnetization. This formal problem has been studied by LAGRANGE in 1769.

Because of the small dispersion of the system under consideration, the adiabatic solution, equ. (43), can be used as a first approximation in order to treat the non-linearity in the original flux balance, equ. (38). This approach leads to a simple integration,

$$\phi(t) = \int_0^t dt'\, G(t-t')\left[\phi(\phi_a) + \tau_\alpha \frac{d}{dt'}(\phi(\phi_a) - \phi_a)\right], \qquad (46)$$

where the kernel

$$G = \frac{\omega_s^2}{\bar{\omega}}\, e^{-t/\tau_o} \sin\bar{\omega}t \quad , \bar{\omega} = \sqrt{\omega_s^2 - \frac{1}{\tau_o^2}} \quad , \tau_o = 2R_q C_s \,, \qquad (47)$$

is determined completely by the characteristic parameters ω_s and τ_o of the SQUID circuit. The characteristic non-linear correlation between the internal flux ϕ and the applied flux ϕ_a here is taken into account by means of the adiabatic SQUID magnetization $\phi(\phi_a)$, equ. (43).

4.4 SIGNAL ANALYSIS

Applying a monochromatic excitation, equ. (37), to the adiabatic magnetization, equ. (43), the ω-component of the (reduced) mutual magnetic flux ϕ_M/ϕ_a in the inductivity L_T of the tank circuit follows to be a generalized SCHLÖMILCH series, [44],

$$\frac{\phi_M}{\hat{\phi}_a} = K\sqrt{\frac{L_T}{L_s}} \sum_{n=1}^{\infty} a_n J_1(2\pi n \hat{\phi}_a/\phi_o)/\pi n \hat{\phi}_a/\phi_o \qquad (48)$$

with respect to the (normalized) amplitude $\hat{\phi}_a$ of the ac-signal. The SQUID parameter $2\pi\beta$ and the dc-flux component ϕ_{dc}/ϕ_o enter the expression (48) only by means of the coefficients

$$a_n = (-1)^n J_n(2\pi n\beta) \cos(2\pi n \phi_{dc}/\phi_o) . \qquad (49)$$

The relations (48/49) represent the result of distortion of ϕ_a by means of the non linear SQUID magnetization (43). Using equs. (48/49) the voltage at the tank circuit

$$V_T = \hat{V}_T \cos(\omega t + \eta) \qquad (50)$$

can be determined by means of a numerical analysis, [42]. Both, the amplitude $\hat{V}_T$ and the phase shift η are influenced significantly by the parameters $\omega\tau_q$, α, $2\pi\beta$. In Fig. 6 the first modulation signal maximum

$$V_M = \hat{V}_T(\phi_{dc} = 0) - \hat{V}_T(\phi_{dc} = \tfrac{1}{2}\phi_o) \qquad (51)$$

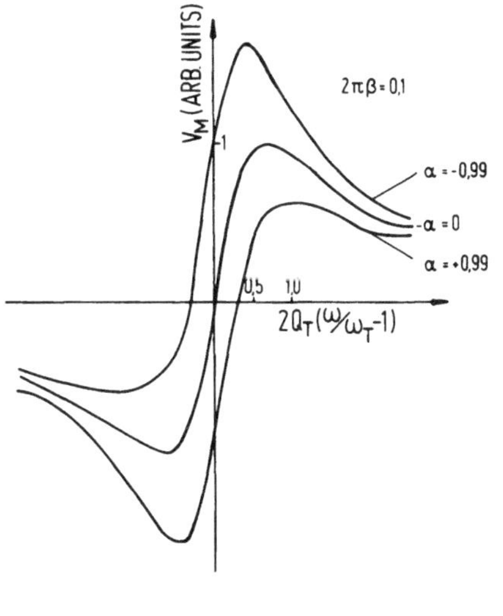

Fig. 6 Modulation depth V_M as function of the (normalized) frequency shift for various values of the $\cos\varphi$ -coefficient α. The values of the parameters are $\tau_q = 4 \cdot 10^{-10}$ s, $2\pi\beta = 0.1$, $\omega_s = 2 \cdot 10^{10}\,s^{-1}$, $\omega_T = 1.2 \cdot 10^8\,s^{-1}$ $Q_T = 100$, K = 0,1 .

is shown versus the normalized frequency shift $2\,Q_T(\omega/\omega_T - 1)$ for $2\pi\beta = 0.1$ and different values of the $\cos\varphi$-coefficient α. As a demonstration of the use of the rf-SQUID for the determination of the parameters τ_q, α, $2\pi\beta$, the result of a best fit calculation concerning experimental results of NISENOFF and WOLF,[39],is shown in Fig. 7. The three parameters have

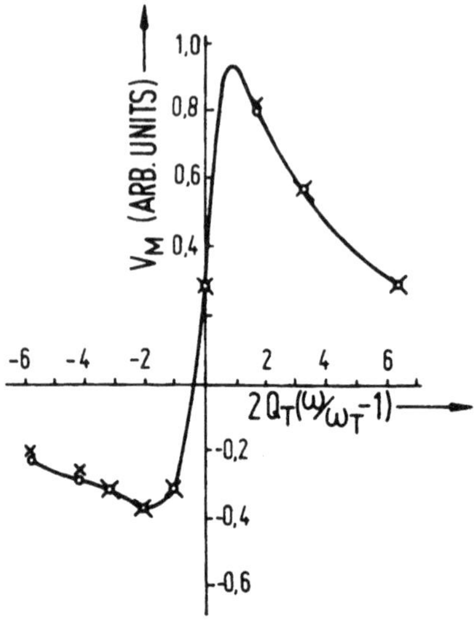

Fig. 7 Comparison of experimental results and a best fit calculation including the first ten M_n. The crosses denote experimental results, [39]; the circles denote calculated values, [42].

been varied independently in this program. The calculation results in $\tau_q = 1.9 \cdot 10^{-9}$ s , $\alpha = -0{,}235$ and $2\pi\beta = 0{,}6$. Furthermore, using the relation $I_{J1}(0)R_{qp} = \pi\Delta/2e$, which seems to provide for an universal intrinsic voltage scale for a great variety of weak coupling structures, the parameters of the low frequency approximation of the current phase relation of the weak link, equ. (34), can be deduced by this method.

In a similar way, the modulation of the phase shift η, equ.(50), is affected by the SQUID parameters as has been studied in detail by ERNÉ,[45].

As an example for a more general SQUID circuit theory, this simple analysis indicates a way of describing the characteristic nonlinear correlation of the SQUID loop, coupled inductively to a resonant circuit. It must be emphazised, however, that this simple example is restricted essentially to the use of the low-frequency approximation of the SQUID magnetization $\emptyset(\emptyset_a)$, $\nu \leq 10^8 s^{-1}$, as well as to the condition of the non-hysteretic operation mode, $2\pi\beta < 1$. Extensions in both directions have to take account for the finite dispersion of the system as caused by the frequency dependence of the current phase relationship, and no detailed description seems available at date.

5. CONCLUSION

By reason of their compactness, the theory of phase-coherence in tunnel junctions is used commonly for classifying similar phenomenons arising in weak coupling structures at all. Indeed, the striking analogies seem to indicate that a superior phenomenological description of phase coherence due to different weak link structures should exist. Under static conditions, the (non-reentrant) current-phase characteristics $I(\varphi) = \sum I_\kappa \sin\kappa\varphi$ may be ordered by means of the phase-coupling spectrum I_κ, κ integer, which depends on material and geometrical parameters of the different structures. Concerning dynamical properties, however, a general description of dispersion relations would be speculative at present, because of the restrictive assumptions concerning the time dependent GINZBURG-LANDAU equation, so far used for the analysis in microbridges. Hopefully, the enormeous difficulties in deducing treatable versions of the GINZBURG-LANDAU equation, which take account for the time response of the energy-gap, may be overcome in the next years.

In spite of the great success of the resistively shunted junction model, a rigorous circuit analysis needs the inclusion of time-depending current-phase relationships. Concerning tunnel-junctions, the investigations at present mainly suffer from mathematical problems of treating the complete integro differential equation describing the current controlled junction. An analysis of this type seems to be unavoidable for a detailed understanding of the high frequency response of weak link arrangements. This applies in particular for the dynamics of the irreversible flux passage in SQUID's operating in the hysteretic mode.

REFERENCES

[1] B.D. Josephson, Phys. Lett. 1, 251 (1962);Rev.Mod.Phys. 36, 216 (1964); Advan. Phys. 14, 419 (1965)

[2] V. Ambegaokar and A. Baratoff, Phys.Rev.Lett. 10, 486 (1963) , E 11, 104 (1963)

[3] B.D. Josephson, Advan. Phys. 14, 419 (1965)

[4] P.G. de Gennes Superconductivity of Metals and Alloys; New York, Amsterdam (1966)

[5] M.H. Cohen, L.M. Falicov and I.C. Phillips, Phys.Rev.Lett. 8, 316 (1962)

[6] R.E. Prange, Phys.Rev. 131, 1083 (1963)

[7] T.A. Fulton and D.E. McCumber, Phys. Rev. 175, 585(1968)

[8] S.B. Nam, Phys. Rev. 156, 470 (1967); 156, 487 (1967)

[9] E. Riedel, Z. f. Naturforschung 19a, 1634 (1964)

[10] W. Schlup, Solid State Comm. 12, 631 (1973) ; (Tables of functions entering the current-phase relation of a Josephson junction in the weak-coupling limit (IBM Thomas Watson Research Center, 1972).

[11] U.K. Poulsen, Phys.Lett. A 41, 195 (1972)

[12] R.E. Harris, Phys.Rev. B 10, 84 (1974)

[13] N.R. Werthamer, Phys.Rev. 147, 255 (1966)

[14] J.W. Wilkins in Tunneling Phenomena in Solids, ed. by E. Burstein and S. Lundqvist (Plenum, New York, 1969)

[15] F. Stern in Solid State Phys. 15, ed. by F. Seitz and D. Thurnbull (Academic Press, New York and London, 1963)

[16] J.S. Toll, Phys. Rev. 104,1760 (1956)

[17] L.E. Hasselberg, J. Phys. F: Metall Phys. 4, 1433 (1974)

[18] R.E. Harris, Phys. Rev. B 11, 3329 (1975)

[19] R.E. Harris, Phys. Rev. B 13, 3818 (1976)

[20] L.G. Aslamazov, A.I. Larkin, JETP Lett., 9, 87 (1969)

[21] A. Baratoff, J.A. Blackburn, B.B. Schwartz, Phys. Rev. Lett. 25, 1096 (1970); E 25, 1738 (1970)

[22] J.A. Blackburn, B.B. Schwartz, A. Baratoff, J. of Low Temp. Phys., 20, 523 (1975)

[23] P.E. Gregers-Hansen, M.T. Levinsen, G. Fog Pedersen, J. of Low Temp. Phys., 7, 99 (1972)

[24] K.K. Likharev, L.A. Yakobson, Sov. Phys. Techn. Phys., 20, 950 (1975)

[25] L.P. Gorkov, G.M. Eliashberg,JETP 27, 328 (1968)

[26] A. Schmid, G. Schön, J. Low Temp. Phys., 20, 207 (1975)

[27] M. Cyrot, Repts. Prog. Phys., 36, 103 (1973)

[28] T.J. Rieger, D.J. Scalapino, J.E. Mercereau, Phys. Rev. B 6, 1734 (1972)

[29] H. Højgaard Jensen, P.E. Lindelof, J. of Low Temp. Phys. 23, 469 (1976)

[30] K.K. Likharev, L.A. Yakobson, JETP 41, 570 (1975); IEEE Trans. Magn. MAG-11, 860 (1975)

[31] D.G. McDonald, E.G. Johnson, R.E. Harris, Phys. Rev. B 13, 1028 (1976)

[32] W.A. Schlup, J. Phys. C 7, 736 (1974)

[33] F. Auracher, T. van Duzer, Appl. Phys. Lett., 21, 515 (1972)

[34] W.C. Stewart, J. Appl. Phys., 45, 452 (1974)

[35] W.C. Stewart, Appl. Phys. Letters, 12, 277 (1968)

[36] D.E. McCumber, J. Appl. Phys., 39, 3113 (1968)

[37] V.N. Belykh, N.F. Pedersen, O.H. Soerensen, The Shunted Josephson Junction Model, Part I, II, Report No. 163, 164, Fysik. Lab. I, Denmarks Tekniske Hoegskole, Lyngby (1976)

[38] P.K. Hansma, Phys. Rev. B, 12, 1707 (1975)

[39] M. Nisenoff, S. Wolf, Phys. Rev. B, 12, 1712 (1975)

[40] R. Rifkin, D.A. Vincent, B.S. Deaver, P.K. Hansma, J. Appl. Phys., 47, 2645 (1976)

[41] O.H. Soerensen, J. Appl. Phys., 47, 5030 (1976)

[42] S.N. Erné, H.-D. Hahlbohm, H. Lübbig, J. Appl. Phys., 47 5440 (1976)

[43] E.T. Whittacker, Analytische Dynamik der Punkte und starren Körper, Berlin 1924

[44] G.N. Watson, A. Treatise on the Theory of Bessel Functions, Cambridge 1958

[45] S.N. Erné, this volume

INVESTIGATION OF THE VALIDITY OF GENERALIZATIONS OF THE ADIABATIC JOSEPHSON EQUATION FOR A LOCAL JUNCTION BY ASYMPTOTIC EXPANSIONS

W.A. Schlup
IBM Zurich Research Laboratory, 8803 Rüschlikon, Switzerland

ABSTRACT

Qualitatively, the adiabatic approximation is known to be good for parameter ranges, where the voltage change is slow or where its ac amplitude is small. This range coincides with the validity range of asymptotic expansions. The asymptotic solutions of different versions (McCumber equation, adiabatic Josephson equation and higher derivative generalizations, Werthamer equation) of the Josephson equation are investigated and compared. The adiabatic approximation is incorrect, even in the lowest order of the pair current.

1 INTRODUCTION

The adiabatic approximation of the Josephson equation[1] (AAJE) and its linear voltage approximation, the McCumber equation[2,3,4] (MCE), is a predominantly used approach to deal with differential equations and to avoid the complexity of the full integro-differential Josephson equation which has been explicitly given by Werthamer[5] and shall be named Werthamer equation (WE) for clarity. This WE has been discussed for a local junction and dc input current by McDonald, Johnson and Harris[6]. They found a I-V characteristic substantially different from that obtained by the AAJE by Hahlbohm, Lübbig and Luther[7]. The characteristics agreed rather well for large mean voltage V, whereas for finite voltage differences occur, which arise essentially from the Riedel singularity in the Fourier transforms of the kernel functions G(t) for quasiparticles, and F(t) for pairs in the WE.

It is the purpose of this note to prove that the AAJE and WE do not agree analytically anywhere in the lowest nontrivial order [beyond the quasiparticle charcteristic $i_0(V)$] of an asymptotical expansion with respect to large capacitance.

2 GENERALIZATION OF THE ADIABATIC APPROXIMATION

The usual adiabatic approximation (AA_1) of the WE is obtained by assuming $\ddot{\varphi} = 0$ in evaluating the retarded integral terms. This is justified qualitatively for solutions $\varphi(t)$, whose first derivative (or instantaneous voltage) depends slowly on time or is constant apart from a small ac amplitude. This occurs for large capacitance or large mean voltage in agreement with the numerically found range of coincidence[6,7]. It would be more consistent to include linear contributions of $\ddot{\varphi}$, which give rise to a phase-dependent correction of the capacitance in the AAJE and yield an improved adiabatic approximation (AA_2). If second and higher [up to $\varphi^{(N)}(t)$] derivatives are taken into account linearly, this would give an N-th order adiabatic approximation (AA_N) of the WE.

3 ASYMPTOTIC EXPANSIONS FOR LARGE CAPACITANCE

Asymptotic solutions of the WE can be found for large capacitance β, large current α or large voltages V. The $1/\beta$ expansions are more fundamental, since they hold for all $\alpha > j_1(0)$ and V not to small > 0; therefore large α or V expansions follow direct from the large β expansions. Their range of validity coincides with the range where WE and AA_1 apparently have the same characteristic.

The WE is supposed in a dimensionless form

$$(1)\qquad \beta\ddot{\varphi} + \int_0^\infty dt'\left[G(t')\sin\frac{\varphi-\varphi(t-t')}{2} + F(t')\sin\frac{\varphi+\varphi(t-t')}{2}\right] = \alpha,$$

where the kernel G(t) defines the quasiparticle-current α_Q, and F(t) the pair-current α_P. Since $\dot{\varphi}(t)$ is a periodic function, the average WE gives $\alpha = \alpha_Q + \alpha_P$ because of $\langle\ddot{\ }\rangle = 0$ ($\langle\ldots\rangle$ time average). According to the rules of asymptotic expansions of nonlinear oscillations[8] α_Q and α_P become

$$(2)\qquad \alpha_Q = i_0(V) - \frac{j_1^2(V)+i_1^2(V)}{256\,V^4\,\beta^2}\,(3i_0(V) - i_0(3V))$$

$$(3)\qquad \alpha_P = -\frac{i_1(V)j_1(V)}{8V^2\,\beta} + \frac{j_1(V)}{128V^4\,\beta^2}\cdot\left[j_1(V)(i_0(V)+i_0(3V)) - i_1(V)(j_0(V)-j_0(3V))\right],$$

where $j_0(\omega) + i\,i_0(\omega)$ is the unilateral Fourier transform of G(t), and $j_1(\omega) + i\,i_1(\omega)$ that of F(t).

These expansions are practical for large β, as long as V is not too small, or is near to the Riedel-singularities $V = 1$ or $V = 1/3$ (in $1/\beta^2$ order), where the coefficients become large. For $\beta = \infty$ the quasiparticle characteristic $i_0(V)$ results, whereas the lowest nontrivial order $1/\beta$ arises from the pair-term F(t). For large mean voltages V the lowest-order correction term gives, for all temperatures below T_c,

$$(4)\qquad \alpha = i_0(V) - \pi\,(\ln 4V)/16\,V^4\,\beta$$

with a logarithmic singularity for $V \to \infty$.

A similar asymptotic expansion will be given for the generalized adiabatic approximation AA_N. In this case the total current can be expressed by

$$(5) \quad \alpha = i_0(V) - \frac{1}{16 V^2 \beta} \sum_{n=0}^{N} \frac{(-2V)^n}{n!} \begin{vmatrix} i_1(V) & j_1(V) \\ i_1^{(n)}(V) & j_1^{(n)}(V) \end{vmatrix} + \dots ,$$

where the second term is due to the pair term alone. If $i_1(V)$, $j_1(V)$ were entire holomorphic functions the limit $N \to \infty$ would be, because of a convergent Taylor-series,

$$(6) \quad \alpha = i_0(V) - \frac{1}{16 V^2 \beta} \begin{vmatrix} i_1(V) & j_1(V) \\ i_1(-V) & j_1(-V) \end{vmatrix} = i_0(V) - \frac{i_1(V) j_1(V)}{8 V^2 \beta} ,$$

in agreement with Eq. (3) up to $1/\beta$ terms. This would prove the validity of the AA_∞ in the lowest order in $1/\beta$, if $i_1(V)$, $j_1(V)$ were holomorphic in the entire complex V plane. This cannot be fulfilled, since not even moments of $F(t)$, $G(t)$ exist, and as a consequence, no Taylor series of $i_1(V)$, $j_1(V)$ for small V. The logarithmic singularities in the Fourier transforms or the weak decrease of $F(t)$ and $G(t)$ for $t \to \infty$ are the reasons for a discrepancy of AA_∞ and WE. For large mean voltages the AA_N gives the correction term

$$(7) \quad \alpha = i_0(V) - (\pi/32 V^4 \beta) \sum_{n=1}^{N} 2^n \sum_{\ell=1}^{n} 1/\ell ,$$

which is without logarithmic term and therefore definitely different from (4) for all $N \geq 1$. Numerical results of AA_1 and WE coincide only in the trivial case $\beta = \infty$, where $\alpha = i_0(V)$, but not in lowest nontrivial order $1/\beta$.

These asymptotic solutions can also be extended to hold for $V \gtrapprox 0$ being power series with respect to $1/\beta^{1/2}$ containing logarithmic terms eventually. In the small voltage range the AA_1 also breaks down in the order $1/\beta^{1/2}$, because of logarithmic terms in the small voltage expansions. This is plausible, since the instantaneous voltage V(t) though very small for a long time, changes rapidly when the phase is away from its unstable equilibrium value.

4 CONCLUSION

It has been shown that even a generalization AA_N of the usual adiabatic approximation AA_1 does not agree with the characteristic derived from the Werthamer equation, if the lowest nontrivial order $(1/\beta)$ is considered. The apparent agreement of characteristics for large mean voltages originates from the quasiparticle characteristic to which both solutions converge $(\beta = \infty)$. Especially the AA_1 breaks down in the lowest order for extended asymptotic solutions, which determine the zero-voltage current (critical current).

The main reason for this disagreement is the nonexistence of the moments of the kernel-functions F(t) and G(t), which in BCS theory osciallate with the gap frequency and slowly $(\sim 1/t)$ vanish for $t \to \infty$.

REFERENCES

1. Josephson, B.D.: Supercurrents through barriers, Adv. Phys. 14, 419-451 (1965).

2. McCumber, D.E.: Effect of ac impedance on dc voltage-current characteristics of superconductor weak-link junctions, J. Appl. Phys. 39, 3113-3118 (1968).

3. Stewart, W.C.: Current-voltage characteristics of Josephson junctions, Appl. Phys. Lett. 12, 277-280 (1968).

4. Schlup, W.A.: The characteristic and stationary dynamics of a time retarded (non-adiabatic) Josephson equation with applied dc current, J. Phys. C 7, 736-748 (1974).

5. Werthamer, N.R.: Nonlinear self-coupling of Josephson radiation in superconducting tunnel junctions, Phys. Rev. 147, 255-263 (1966).

6. McDonald, D.G., Johnson, E.G., Harris, R.E.: Modeling Josephson junctions, Phys. Rev. B 13, 1028-1031 (1976).

7. Hahlbom, H.D., Lübbig, H., Luther, H.: Einfluss des cos φ Terms und des Riedelpeaks auf die Strom-Spannungs-Charakteristik eines Josephson-Tunnelkontaktes, Verhandl. DPG 5, 650 (1975).

8. Bogoliubov, N.N., Mitropolski, Y.A.: Asymptotic methods in the theory of nonlinear oscillations, Hindustan Publishing Company, Delhi, 1961.

A SELF-CONSISTENT MODEL OF THE JOSEPHSON JUNCTION

Hiroshi OHTA
The Institute of Physical and Chemical Research
Wako-Shi, Saitama, 351 Japan

1 Introduction

A self-consistent model of the Josephson junction has been presented emphasizing the role of an external circuit. We have taken account of the interaction between the Josephson junction and an external circuit in order to get the self-consistent solution. It has been proved by this model that the Josephson voltage frequency relation $\hbar\dot{\phi} = 2eV$ holds strict in the "weakly coupled two-level quantum system" model. This self-consistent model is an extension of Feynman's model and equivalent to Ferrell & Prange's model of the Josephson junction.

There are several models of the Josephson junction such as Feynman's model[1)2)3)], Ferrell & Prange's model[4)5)] and the model using the tunneling Hamiltonian[5)]. They all treat the effect of an external circuit on the Josephson junction only implicitly.

In the present model the total Hamiltonian has a term of the interaction between the Josephson junction and an external circuit. It is because of this term of the interaction that this model is self-consistent and gives the strict Josephson voltage frequency relation.

The number of superconducting electron pairs is a constant in a superconductor unless the temperature changes. This puts a constraint on the motion of the superconducting electron pairs. Our model is an extension of the "coupled two-level system" model presented by Feynman. The coupled two-level system making the constrained motion has turned out equivalent to Ferrell & Prange's model of the Josephson junction.

Let me refer to what has motivated me to develop this model.

Feynman has succeeded explaining the most essential part of the Josephson effect by his very simple model or the weakly coupled two-level system model. The thing which has bothered me long is that his model does not give us the strict Josephson voltage frequency relation $\hbar\dot{\phi} = 2eV$ unless the number density or the number in the superconductor on one side of the junction are equal to that on the other side. This restriction on the particle numbers in the superconductors cannot go along with both the law of conservation of the total particle number and the equations of motion.

In the following sections we use the coherent state representation[6)7)] to transfer our problem in quantum mechanics to the equivalent c-number problem without any loss of quantum mechanical features.

2 The Coupled Two-Level System Isolated from the External Circuit

We first confine ourselves to the superconducting electrons in the two-fluid model. We regard the superconducting electron pairs as ideal bosons of the charge -2e for simplicity.

We treat the superconductors on each side of the junction as a many-boson system and call the boson operators of each superconductor $\hat{a}_k$ and $\hat{b}_k$ respectively.

The commutation relation of the operator $\hat{a}_k$ is

(1) $$[\hat{a}_k , \hat{a}_{k'}{}^{\dagger}] = \delta_{k,k'}.$$

The number operator on one side of the junction is given by

$$\hat{N}_1 = \int \hat{\Psi}_1^{\dagger}\hat{\Psi}_1 dv = \sum_k \hat{a}_k^{\dagger}\hat{a}_k .$$

The wave function describing the superconductor S_1 is

$$\hat{\Psi}_1 = \sum_k \hat{a}_k \Phi_k .$$

where Φ_k is an eigen function of the differential equation

$$H_0^{(1)}\Phi_k = E_k^{(1)}\Phi_k .$$

The unperturbed Hamiltonian $H_0^{(1)}$ is the sum of the kinetic energy and the potential energy of the bosons in the superconductor S_1.

The wave function is normalized as

$$\int \Phi_k^* \Phi_{k'} \, dv = \delta_{k,k'} .$$

When Φ_k s form a complete set, we get the commutation rule of the wave functions

$$[\hat{\Psi}_1(\underline{x}), \hat{\Psi}_1^{\dagger}(\underline{x}')] = \sum_k \Phi_k(\underline{x})\Phi_k^*(\underline{x}') = \delta(\underline{x} - \underline{x}') .$$

Similarly for the superconductor S_2 we have

$$\hat{\Psi}_2 = \sum_q \hat{b}_q \Phi_q , \quad H_0^{(2)}\Phi_q = E_q^{(2)}\Phi_q .$$

When the momenta vanish, we get

(2) $$E_0^{(1)} - E_0^{(2)} = \mu_1 - \mu_2 ,$$

where μ_1, μ_2 are the chemical potentials of the boson system on each side of the junction.

The Hamiltonian of a coupled two-level system is

(3) $$\hat{H}_e = \int [\hat{\Psi}_1{}^{\dagger} H_0{}^{(1)} \hat{\Psi}_1 + \hat{\Psi}_2{}^{\dagger} H_0{}^{(2)} \hat{\Psi}_2]\, dv + \hat{H}_c$$

and

(4) $$\hat{H}_c = \int [\hat{\Psi}_1{}^{\dagger} \hat{K}\, \hat{\Psi}_2 + \Psi_1{}^{\dagger 2} \hat{L}\, \hat{\Psi}_2{}^2 + \ldots + c.c.]\, dv$$

The operators $\hat{K}$ and $\hat{L}$ are in normal order. Each term in $\hat{K}$ and $\hat{L}$ has the same number of $\hat{\Psi}_1$ as that of $\hat{\Psi}_1{}^{\dagger}$ and the same number of $\hat{\Psi}_2$ as that of $\hat{\Psi}_2{}^{\dagger}$.

The expectation value of the Hamiltonian is given by

(5) $$H_e = \langle \alpha\beta | \hat{U}^{\dagger}(t) \hat{H}_e \hat{U}(t) | \alpha\beta \rangle = E_1 N_1 + E_2 N_2 + E_c(N_1, N_2, \phi)$$
$$= \frac{1}{2}(E_1 + E_2)(N_1 + N_2) + \frac{1}{2}(\mu_1 - \mu_2)(N_1 - N_2) + E_c$$

where

(6) $$E_1 - E_2 = \mu_1 - \mu_2 = -2eV, \quad (e > 0)$$

and

(7) $$\phi(t) = \phi_1(t) - \phi_2(t), \quad E_0{}^{(1)} = E_1, \quad E_0{}^{(2)} = E_2 .$$

The coupling energy E_c is

(8) $$E_c = \langle \alpha\beta | \hat{U}^{\dagger} \hat{H}_c \hat{U} | \alpha\beta \rangle = \sum_{\ell=1}^{\infty} C_{\ell}(N_1, N_2) \cos \ell\phi$$

which meets two conditions, the periodicity condition

$$E_c(\phi + 2\pi) = E_c(\phi)$$

and

$$\frac{\partial E_c}{\partial \phi} = 0 \quad \text{for } \phi = 0$$

The second condition means that E_c has the local minimum at zero phase-difference.

We have also used the expectation values of the operators $\hat{a}_0$ and $\hat{b}_0$

(9) $$\alpha(t) = \langle \alpha\beta | \hat{U}^{\dagger} \hat{a}_0 \hat{U} | \alpha\beta \rangle = \sqrt{N_1(t)} \exp[i\phi_1(t)],$$
$$\beta(t) = \sqrt{N_2(t)} \exp[i\phi_2(t)] .$$

The total number of the bosons is a constant of the motion

(10) $$N_1(t) + N_2(t) = N .$$

When the numbers N_1 and N_2 are large enough in the coherent states, we can measure the number with a standard deviation of $\sqrt{N_1}$ and the phase ϕ_1 with a standard deviation of $(2\sqrt{N_1})^{-1}$.

We are using the coherent state representation to get the c-number equations which have the same mathematical form as the corresponding operator equations. In order to get them, we require the eigenstate conditions[8)].

(11) $$\begin{aligned} \hat{a}(t)|\alpha\beta\rangle &= \hat{a}(t)|\alpha\rangle|\beta\rangle = \alpha(t)|\alpha\beta\rangle \\ \hat{b}(t)|\alpha\beta\rangle &= |\alpha\rangle\hat{b}(t)|\beta\rangle = \beta(t)|\alpha\beta\rangle \end{aligned}$$

where we let

$$\hat{a}(t) = \hat{U}^{\dagger}\hat{a}_0\hat{U} , \quad \hat{b}(t) = \hat{U}^{\dagger}\hat{b}_0\hat{U} .$$

Equation (11) may be written as

(12) $$\hat{U}|\alpha\beta\rangle = |\alpha(t)\beta(t)\rangle$$

We have been confining ourselves to the superconducting component in the two-fluid model and therefore it has been assumed that the momenta of all the bosons have vanished. Namely we have set the very important condition that

(13) $$\begin{aligned} \hat{a}_k(t)|\alpha\beta\rangle &= 0 \quad (k \neq 0) , \\ \hat{b}_k(t)|\alpha\beta\rangle &= 0 \quad (k \neq 0) . \end{aligned}$$

When the volumes of the superconductors are microscopically large and macroscopically small, the superconductors can be homogeneous spatially.

We define the coherent state vector $|\alpha\beta\rangle$ as

(14) $$|\alpha\beta\rangle = \hat{A}(\alpha)\hat{B}(\beta)|0\rangle$$

where we let

(15) $$\hat{A}(\alpha) = \exp(\alpha\,\hat{a}_0^{\dagger} - \alpha^*\hat{a}_0) = \hat{A}^{\dagger}(-\alpha)$$
$$B(\beta) = \exp(\beta\,\hat{b}_0^{\dagger} - \beta^*\hat{b}_0) = B^{\dagger}(-\beta)$$

and

(16) $$\alpha = \alpha(0), \qquad \beta = \beta(0)$$

The operator $\hat{U}$ is the solution of the equation

(17) $$i\hbar\frac{d}{dt}\hat{U}(t) = \hat{H}_e\hat{U}(t) \;,$$

where $\hat{U}$ is unitary

$$\hat{U}^{\dagger} = \hat{U}^{-1} \text{ and } \hat{U}(0) = 1.$$

From Eq.(17) we get Heisenberg's equation of motion

(18) $$i\hbar\frac{d}{dt}\hat{a}(t) = [\hat{a}(t), \hat{H}_e(t)]$$

From Eq.(18), we see that

(19) $$\begin{aligned} i\hbar\frac{d}{dt}\alpha(t) &= \mathrm{Tr}\left(\hat{\rho}\;[\hat{a}(t), \hat{H}_e(t)]\right) \\ &= \mathrm{Tr}\left(\hat{\rho}\;[\hat{a}(t), \hat{H}_e(\hat{a}(t), \hat{b}(t)]\right) \\ &= \mathrm{Tr}\left(\hat{\rho}\;\frac{\partial}{\partial a^{\dagger}(t)}H_e(\hat{a}^{\dagger}(t), \hat{a}(t); \hat{b}^{\dagger}(t)b(t))\right) \\ &= \frac{\partial}{\partial\alpha^*(t)}H_e(\alpha^*(t), \alpha(t); \beta^*(t), \beta(t)) \end{aligned}$$

since $\hat{H}_e$ is in normal order.

We have used the density operator

$$\hat{\rho} = |\alpha\beta\rangle\langle\alpha\beta|$$

Similarly we get

(20) $$i\hbar\frac{d}{dt}\beta(t) = \frac{\partial}{\partial\beta^*(t)}H_e$$

Equations (19) and (20) determine the expectation values $\alpha(t)$ and $\beta(t)$.

From Eqs.(9)(19) and (20) we readily obtain the equations of motion.

$$\text{(21)} \qquad \dot{\phi}_j = -\frac{1}{\hbar}\frac{\partial H_e}{\partial N_j}\,, \qquad \dot{N}_j = \frac{1}{\hbar}\frac{\partial H_e}{\partial \phi_j}$$

which tell that $-\phi_j$ and $\hbar N_j$ are canonical conjugate.

From Eqs.(5) and (21) we have that

$$\text{(22)} \qquad \dot{N}_1 = -\dot{N}_2 = \frac{1}{\hbar}\frac{\partial E_c}{\partial \phi}$$

and

$$\text{(23)} \qquad \begin{aligned} \hbar\phi_1 &= -E_1 - \frac{\partial E_c}{\partial N_1} \\ \hbar\phi_2 &= -E_2 - \frac{\partial E_c}{\partial N_2} \end{aligned}$$

the simplest case of which was obtained by Feynman for the first time.[1)]

From Eqs.(6)(7)(8) and (23) we get the Josephson voltage frequency relation

$$\text{(24)} \qquad \hbar\,\frac{\partial \phi}{\partial t} = 2eV$$

only if all the time

$$\text{(25)} \qquad \frac{\partial C_\ell}{\partial N_1} = \frac{\partial C_\ell}{\partial N_2}$$

In order to see that the condition of Eq.(25) cannot be met self-consistently, we have only to take the simplest case of the coupling energy E_c of Eq.(8):

$$\frac{C_1(N_1,\,N_2)}{\sqrt{N_1\,N_2}} = K = \text{const.} \qquad \text{for } \ell = 1$$

and all the other C_ℓ s are zero.

In this case Eq.(25) reduces to

$$\text{(26)} \qquad N_1(t) = N_2(t)$$

The condition of Eq.(26), however, cannot go along with both the law of conservation of the total particle number in Eq.(10) and the equations of motion in Eq.(22). This discrepancy disappears when we take account of the work on the Josephson junction done by an external circuit.

3 The Josephson Junctions Connected to an External Circuit

In this section we are going to get the Hamiltonian which gives a self-consistent description of the system. To do this, it is necessary to take account of the effect of the external circuit on the coupled two-level system. When we take account of the effect of the external circuit, we do not need the condition of Eq.(25) anymore to get the strict voltage-frequency relation of Eq.(24).

We connect a certain external circuit to the Josephson junction when we try to observe the Josephson effect. In order to get the work done by any kind of external circuit, we have only to integrate the electric power with regard to time.

(27) $$W = \int VI \, dt = 2e \int V\dot{M} \, dt = 2e \int V \, dM$$

where the current is

(28) $$I = 2e\dot{M}$$

Either the work W or the number M is a function of ϕ_1 ϕ_2 N_1 and N_2, and the voltage V depends on time.

The energy of the positive ions on the background of the superconductors is given by

(29) $$H_i = -E_1\bar{N}_1 - E_2\bar{N}_2 ,$$

Either the number $\bar{N}_1$ or $\bar{N}_2$ of the positive ions which cancels the charge of the superconducting electron pairs in the two-fluid model is a constant at a given temperature.

At the same rate that superconducting particles become normal ones around the interface of the superconductor S_1 and the external circuit, normal particles become superconducting ones around the interface of the superconductor S_2 and the external circuit. Therefore these changes cancel out each other and have been ignored in the present argument.[9)]

Now Hamilton's principle for our system is

(30) $$\delta \int (-\hbar \sum_j N_j \dot{\phi}_j - H_e - H_i + W) dt = 0.$$

Hamilton's principle gives the canonical equations

(31) $$\dot{\phi}_j = -\frac{1}{\hbar} \frac{\partial H}{\partial N_j}, \quad \dot{N}_j = \frac{1}{\hbar} \frac{\partial H}{\partial \phi_j},$$

where we have introduced the Hamiltonian H by

(32) $$H = H(\phi_1 N_1, \phi_2 N_2) = H_e + H_i - W$$
$$= E_1(N_1 - \overline{N}_1) + E_2(N_2 - \overline{N}_2) + E_c - W$$

Equations (31) and (32) say that

(33) $$\hbar\dot{\phi}_1 = -E_1 - \frac{\partial E_c}{\partial N_1} + \frac{\partial W}{\partial N_1}$$

(34) $$\hbar\dot{\phi}_2 = -E_2 - \frac{\partial E_c}{\partial N_2} + \frac{\partial W}{\partial N_2}$$

(35) $$\hbar\dot{N}_1 = \frac{\partial E_c}{\partial \phi_1} - \frac{\partial W}{\partial \phi_1}$$

(36) $$\hbar\dot{N}_2 = \frac{\partial E_c}{\partial \phi_2} - \frac{\partial W}{\partial \phi_2}$$

The time derivative of the function W of ϕ_j and N_j is given by

(37) $$\dot{W} = \sum_j \left(\frac{\partial W}{\partial \phi_j} \dot{\phi}_j + \frac{\partial W}{\partial N_j} \dot{N}_j \right).$$

The total number of the electron pairs is still conserved in our system. Eq.(10) gives

(38) $$\dot{N}_1 + \dot{N}_2 = 0$$

From Eqs.(8)(35)(36) and (38) we see that W satisfies the equation

(39) $$\frac{\partial W}{\partial \phi_1} = - \frac{\partial w}{\partial \phi_2} = \frac{\partial W}{\partial \phi} .$$

Namely W is a function only of $\phi = \phi_1 - \phi_2$ as far as the phase variables are concerned.

We get the equation

(40) $$\dot{N}_1 = \dot{n} - \dot{M} .$$

when we volume-integrate the equation of continuity

$$\dot{\rho} = - \operatorname{div} \rho \underline{v},$$

with respect to the space co-ordinates within the volume enclosed by the dashed line in Figure 1.

$\dot{N}_1$ is the rate of increase in the particle number in the superconductor S_1.

$\dot{n}$ is the rate of increase in the number of the particles which come into the superconductor S_1 through the junction.

$\dot{M}$ is the rate of increase in the number of the particles which leave the superconductor S_1 for the external circuits.

We easily obtain the law of conservation of energy

(41) $$W = -e\int V\ (\dot{N}_1 - \dot{N}_2)\ dt + E_c$$

(42) $$= -e\int V\ dN_1 + e\int V\ dN_2 + E_c$$

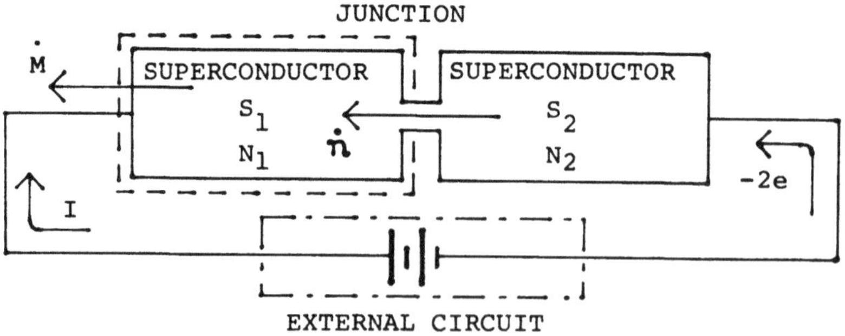

Fig.1. The Josephson junction connected to a voltage source. The voltage source may be exchanged by an arbitrary external circuit.

when we eliminate all the W s on the right hand side of Eq. (37) using Eqs. (33) (34) (35) (36) and (38).

Eqs. (27) and (41) say that the work done by the external circuit is equal to the energy of the system of the bosons.

We also see that

$$(43) \qquad \frac{dH}{dt} = \dot{E}_1 (N_1 - \bar{N}_2) + \dot{E}_2 (N_2 - \bar{N}_2),$$

from Eqs. (32) and (41). Equations (40) and (41) give us a constant of the motion

$$(44) \qquad -2e \int V\dot{n}\, dt + E_c = \text{const.}$$

which has the dimension of energy.

We have been confining ourselves to the superconducting component in the two-fluid model. When the temperature stays constant, both of the numbers N_1 and N_2 have to be constant

$$(45) \qquad \dot{N}_1 = -\dot{N}_2 = 0$$

which sets a constraint to the motion of our system.

Then Eq. (40) reduces to $\dot{n} = \dot{M}$ which means that the particle number entering the superconductor $\mathcal{E}_1$ is equal to the particle number leaving it.

And Eq. (41) reduces to

$$(46) \qquad E_c - W = \text{const.}$$

where N_1 and N_2 do not depend on time anymore and they are arbitrary parameters for the time being.

When we substitute Eq. (46) into (33) (34) (35) and (36), we get

$$(47) \qquad \dot{\phi}_j = -\frac{E_j}{\hbar} ,$$

and Eq. (45)

The strict Josephson voltage frequency relation comes from Eq. (47)

$$(48) \qquad \dot{\phi} = \dot{\phi}_1 - \dot{\phi}_2 = \frac{2eV}{\hbar}$$

Equation (41) is the law of conservation of energy. Without the law, the strict voltage frequency relation (48) would not come out. Equations (35)(36)(39) and (45) just require that the left hand side of Eq.(46) is an arbitrary function of N_1 and N_2.

Equations (32) and (46), or Eqs.(43) and (45), give

$$(49) \qquad H = E_1(N_1 - \bar{N}_1) + E_2(N_2 - \bar{N}_2) ,$$

which does not include ϕ_j. Namely the phase ϕ_j is a cyclic variable.[10)]

When we substitute Eq.(49) into Eq.(31) we obtain the same result as Eq.(47)
It follows from Eqs.(27)(37)(39)(40)(45)(46) and (48) that

$$(50) \qquad \dot{W} = 2eV\dot{M} = \frac{\partial E_c}{\partial \phi}\dot{\phi}$$

or

$$(51) \qquad \dot{n} = \dot{M} = \frac{1}{\hbar}\frac{\partial E_c}{\partial \phi}$$

From Eqs.(27)(45) and (48) we have that

$$(52) \qquad \frac{1}{\hbar}\frac{\partial W}{\partial \phi} = \frac{2eV}{\hbar}\frac{\partial M}{\partial \phi} = \dot{\phi}\frac{\partial M}{\partial \phi} = \dot{M} ,$$

which is consistent with Eqs.(35)(39) and (51).
The charge of the normal electrons in the two-fluid model is not always canceled by that of the positive ions in each superconductor because there is a stray capacity between the superconductors S_1 and S_2. The charge of the superconducting electron pairs in the two-fluid model are always canceled by that of the positive ions in either the superconductor S_1 or S_2.

$$(53) \qquad \begin{aligned} N_1 &= \bar{N}_1 , \\ N_2 &= \bar{N}_2 = N - \bar{N}_1 . \end{aligned}$$

The condition (53) is a little more severe than Eq.(45) but

it is consistent with Eq.(45)

Then the current is given by

$$I = \frac{2e}{\hbar} \frac{\partial}{\partial \phi} E_c(\bar{N}_1, \bar{N}_2, \phi) \tag{54}$$

$$= \sum_{\ell=1}^{\infty} I_\ell \sin \ell\phi$$

and

$$I_\ell = - \frac{2e\ell}{\hbar} C_\ell(\bar{N}_1, \bar{N}_2) \tag{55}$$

from Eqs.(8)(28) and (51)

Eq.(43) and (53) say that

$$\frac{dH}{dt} = 0 \tag{56}$$

The Hamiltonian is a constant of the motion.

It should be noted again that we have not used the assumption of Eq.(25) or Eq.(26) to derive Eqs.(48) and (54). Instead we have used a more realistic assumption of Eq.(45) or Eq.(53). And this model is a self-consistent one.

Incidentally it can be proved that these solutions are equivalent to Ferrell & Prange's model of the Josephson junction.

By Eqs.(27)(51) and (53) the Hamiltonian H in Eq.(32) reduces to

$$H = -2e\int V\dot{n}\, dt + E_c(\bar{N}_1, \bar{N}_2, \phi) \tag{57}$$

$$= -2e \int V\, dn + E_c(\phi)$$

which, by Eq.(48), gives

$$\dot{\phi} = - \frac{1}{\hbar} \frac{\partial H}{\partial n} \tag{58}$$

We readily see that

$$\dot{n} = \left(\frac{\partial n}{\partial \phi}\right)_H \dot{\phi} = - \frac{1}{\hbar}\left(\frac{\partial H}{\partial n}\right)_\phi \left(\frac{\partial n}{\partial \phi}\right)_H , \tag{59}$$

$$= - \frac{1}{\hbar} \frac{\partial (H\ \phi)}{\partial (n\ \phi)} \cdot \frac{\partial (n\ H)}{\partial (\phi\ H)} ,$$

$$= \frac{1}{\hbar} \frac{\partial (n\ H)}{\partial (n\ \phi)} = \frac{1}{\hbar} \frac{\partial H}{\partial \phi} .$$

This equation and Eq.(58) say that $-\phi$ and $\hbar n$ are canonical conjugate. From Eqs.(57) and (59) we have

$$\dot{n} = \frac{1}{\hbar} \frac{\partial H}{\partial \phi} = \frac{1}{2e} \sum_{\ell=1}^{\infty} I_\ell \sin \ell\phi \tag{60}$$

which is consistent with Eq.(54)

Eqs.(58) and (59) are equivalent to Ferrell & Prange's model of the Josephson junction.

Finally we take account of the normal as well as the superconducting components in the two-fluid model. When this junction is connected in parallel with an equivalent capacitance C and an equivalent resistance R and it is driven by a constant-current source I_{DC}, a microwave current i $\cos(\omega t + \theta_0)$ and a noise current i_n, the current flowing through the junction is given by

$$I = I_{DC} + i\cos(\omega t + \theta_0) + i_n - C\frac{dv}{dt} - \frac{V}{R} \tag{61}$$

When we substitute Eq.(61) into Eq.(54), we get differential equation of the so-called Resistively Shunted Junction (RSJ) model.[11)12)]

$$C \frac{dV}{dt} + \frac{V}{R} + \sum_{\ell=1}^{\infty} I_\ell \sin \ell\phi = I_{DC} + i\cos(\omega t + \theta_0) + i_n , \tag{62}$$

where V and ϕ are connected by Eq.(48).

4 Conclusion

A self-consistent model of the Josephson junction has been presented emphasiging the essential role of an external circuit. Either Feynman's model or Ferrell & Prange's model is a special case of this self-consistent model. In this model the Josephson voltage-frequency relation holds strict without the assumption that the particle number in the superconductor on one side of the junction stays equal to that on the other side.

References

1) P.R. Feynman, R.B. Leighton & M. Sands: The Feynman Lectures on Physics Vol.III, Addison-Wesley Publishing Co., Massachusetts, 1965.
2) R.P. Feynman: Statistical Mechanics, W.A. Benjamin, Inc., Massachusetts 1972.
3) D. Rogovin and M.O. Scully: Ann. Phys. (USA) 88 371 (1974).
4) R.A. Ferrell and R.E. Prange: Phys. Rev. Letters 10 479 (1963).
5) B.D. Josephson: Superconductivity, Vol.1 ed. by R.D. Parks (Dekker, 1969).
6) W.H. Louisell: Quantum Statistical Properties of Radiation, John Wiley & Sons, New York, 1973.
7) P. Carruthers: Rev. Mod. Phys. 40 411 (1968).
8) R.J. Glauber: Quantum Optics, Proc of the International School of Physics <Enrico Fermi> Course 42, 32 - 36 (1969), Academic Press, New York.
9) D. Rogovin: Phys. Rev. 12B 130 (1975).
10) H. Goldstein: Classical Mechanics, Addison-Wesley Publishing Co. 1950.
11) P.W. Anderson: Progress in Low Temperature Physics V ed. by C.J. Gorter, North-Holland Publishing Co., 1967.
12) F. Auracher and T. Van Duzer: Appl. Phys. Letters 21 515 (1972).

ORDER PARAMETER RELAXATION AND JOSEPHSON EFFECTS IN SUPERCONDUCTING WEAK LINKS

A. Baratoff and L. Kramer*

IBM Zürich Research Laboratory, 8803 Rüschlikon, Switzerland

INTRODUCTION

Until the last three years experiments on SQUIDs and other superconducting devices incorporating weak links (WL), such as "point" contacts and thin film microbridges, have been interpreted in terms of the resistively shunted junction (RSJ) model.[1,2] According to this adaptation of the two-fluid model the current through a WL is

(1) $I = I_m \sin\phi + V/R,$

R being its resistance in the normal state and I_m the maximum supercurrent that it can support. The voltage drop V is related to the phase difference ϕ through the Josephson relation

(2) $d\phi/dt = 2eV/\hbar \quad (e > 0)$

which holds quite generally provided that both quantities are measured between two points where the local chemical potential is constant.

Deviations from the predictions of this model have been observed in dc I-V characteristics (hysteresis at low enough temperatures,[3-5] excess supercurrent at high voltages,[5-8] subharmonic gap structure[9]), as well as in the microwave-induced step structure (subharmonic steps,[3,7,8] frequency dependence,[8] enhancement of maximum supercurrent[3]). In many cases some of these deviations can be attributed to external circuit elements,[1] heating,[6] and vortex motion.[10,11] By proper design such effects can be avoided, leaving deviations of a more fundamental nature, e.g. due to

*Visiting scientist; permanent address: Physik Department der Technischen Universität München, D-8046 Garching b. München, Germany.

breakdown of the two-fluid model, to the energy gap in the electrodes, to inelastic quasiparticle processes, and to the spatial dependence of the superconducting order parameter and its temporal evolution. Since our understanding of the relative importance and interplay of such effects in WLs is still in its infancy, it seems worthwhile to elucidate the influence of one or at most a few of them at a time.

Recently Likharev and Yakobson (LY)[12] investigated the influence of a finite relaxation time τ of the order parameter ψ within a one-dimensional WL model based on the simplest time-dependent Ginzburg-Landau (TDGL) equations. Subsequently Højgaard Jensen and Lindelof[7] produced a dynamic supercurrent-phase relation (RSJτ model)[8] based on an Ansatz proposed by LY for WLs shorter than a coherence length $\xi(T)$. Besides confirming most of LY's results, calculations performed by us within their model clear up some ambiguities and reveal interesting new features, some of which we report here. In the range where a good fit has been claimed between the RSJτ model and the measured characteristics and frequency dependence of microwave-induced steps in submicrometer microbridges[8] our computed I-V characteristics deviate, however, significantly from those calculated in Ref. 7. The agreement achieved with the RSJτ model should therefore be viewed as an empirical fact which remains to be justified.

DESCRIPTION OF THE MODEL, LIMITING RESULTS

Consider a superconducting WL such that the dominant contribution to the phase difference ϕ is localized inside the link. This occurs if the electronic mean free path in the link is much shorter than in the banks that it bridges,[13] or if the weak section is sufficiently thin.[12,14] Thick banks also ensure more efficient heat transfer.[15] A dc source supplies a current which is assumed uniformly distributed inside the WL both below and above I_m (no vortex penetration). This can be achieved if the thin film containing the WL is close to a superconducting ground plane. Otherwise the WL width should be smaller than the coherence length or the effective penetration depth, whichever is smaller, a requirement which is difficult to meet in practice except very close to the transition

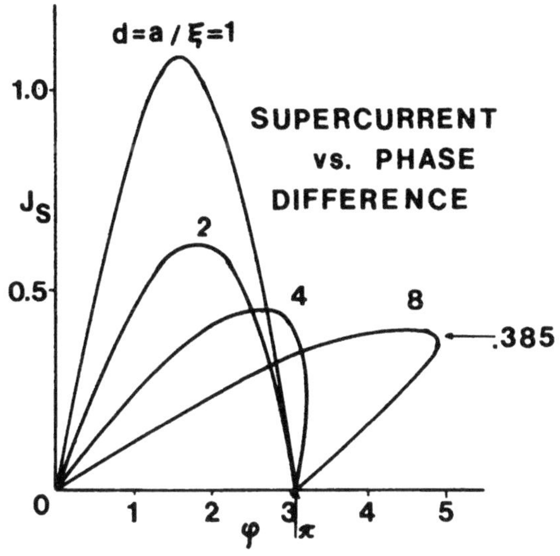

Fig. 1 Static supercurrent-phase relation of a one-dimensional superconducting weak link for different values of the ratio length a/coherence length ξ. The supercurrent is normalized such that $J_s \simeq \phi d$ for small ϕ.

temperature T_c.

The dependence of the static ($V = 0$) supercurrent-phase relation $I_s(\phi)$ of such a WL on its length a is well understood within Ginzburg-Landau (GL) theory.[13,14] For sufficiently small ϕ, $I_s/\phi \propto a$; hence it is convenient to define a normalized current J such that $J_s \simeq \phi d$ with $d = a/\xi$ in that range. For an applied current $J < J_m$ there are two static solutions; that with the larger value of ϕ corresponds to a saddle point of the GL free energy functional while the other one is stable.[16] $J_s(\phi)$ is sinusoidal if $d < 1$ whereas, as may be seen in Fig.1, J_m approaches the constant value 0.385 appropriate for an infinite homogeneous filament when $d > 8$. In the former limit $J_m \simeq d^{-1}$; moreover, if the WL consists of dirty material, the product RI_m has the same value as in a tunnel junction,[2,3] i.e. $V_J = 635(T_c - T)\mu V/{}^\circ K$ for weak coupling superconductors close to T_c. Ohm's law may then be written as $V_N = RI = V_J Jd$ (also if $d > 1$). Both V_J and the associated time $\tau_J = \hbar/(2eV_J)$ provide natural scales for electromagnetic phenomena in dirty superconductors. In particular, the solution of Eqs. (1-2)[2]

(3) $V = RI(1 - \alpha^2)/\{1 + \alpha\sin(\omega t - \phi_0)\}$, where

(4) $\alpha \equiv \sin\phi_0 = I_m/I > 1, \quad \hbar\omega/2e = V_{dc} = RI(1 - \alpha^2)^{1/2}$,

consists of Lorentzian spikes of height $\simeq 2V_J$ and halfwidth τ_J when $I \to I_m$ although the period of the Josephson voltage oscillations diverges.

The RSJ model, as well as its generalization to a nonsinusoidal $I_s(\phi)$, are expected to break down if the relaxation time τ of the order parameter is sufficiently long. Indeed the spatial dependence of ψ which determines $I_s(\phi)$ in the static case[13,14] cannot then be maintained, at least during the important part of the cycle in which V(t) varies rapidly, even if $\omega\tau < 1$!

Our calculations are based on a stable implicit difference approximation scheme applied to the one-dimensional TDGL equations

(5) $u(\dot{\psi} + i\mu\psi) = \psi'' + (1 - \psi^2)\psi$,

(6) $J = -\mathrm{Im}\psi^*\psi' + \mu'$,

together with the bondary conditions[12]

(7) $\psi = e^{\mp i\phi/2}, \quad \mu = \pm\dot{\phi}/2 \quad \text{at} \quad x = \pm d/2.$

The complex order parameter ψ is normalized to its absolute value for zero current which it is assumed to take at the ends of the link. This is a good approximation if the contribution to ϕ from the connecting banks is negligible.[13,14] The general case of slightly different T_c outside is easily treated,[12] but is not considered here for simplicity. Time t, distance x (referred to the WL center) and electrochemical potential μ are expressed in units of τ_J, ξ, and eV_J respectively; thus Eq. (1) becomes $\dot{\phi} = V/V_J$. Eq. (6), the local two-fluid model, is valid close to T_c in a gapless superconductor, while Eq. (5) can only be justified if additional restrictions are satisfied (dirty limit, strong depairing); one then has $u \equiv \tau/\tau_J = 12.$[17] For weak depairing $u = 5.79$, but microscopic theory predicts additional terms whose effect is not considered here. Despite these limitations Eqs.(5-6) lead to nontrivial features which probably have general validity. We simply treat u as a parameter whose effective value might be quite large if a gap is present.[8]

As emphasized by LY, it is important to work with Reψ and Imψ to avoid numerical instabilities whenever $f \equiv |\psi|$ vanishes, althogh the equivalent equations for f and $\chi \equiv \arg\psi$,

(8) $u\dot{f} = f'' - \chi'^2 f + (1 - f^2)f,$

(9) $u(\dot{\chi} + \mu)f^2 = (f^2\chi')' = \mu'',$

(10) $J = -f^2\chi' + \mu' \quad J_s + J_n,$

are more amenable to physical interpretation. In particular Eqs. (9-10) imply that (i) the local Josephson relation $\dot{\chi} = -\mu$ is satisfied only where $J_s' = 0$, and (ii) J_s is independent of x, $\mu = (J - J_s)x$ when u = 0; $\chi' = J_s/f^2$ with f determined by the static solution of Eq.(8) then leads to the generalized RSJ model.[12]

For sufficiently high currents the WL is in the normal state ($\psi << 1$) except near its ends. Since the two sides are then decoupled, μ is time-independent and[12]

(11) $\psi(x,t) \simeq g(x + d/2)e^{+i\phi(t)/2} + g^*(-x + d/2)e^{-i\phi(t)/2},$

where $g(\bar{x})$ is a stationary solution of Eqs.(5-6) satisfying $g(0) = 1$ and decaying to zero for large $\bar{x}$. When nonlinear terms can be neglected, $\mu \simeq Jx$ and g must be an Airy function. The proper choice is

(12) $g(\bar{x}) \propto \mathrm{Ai}\{e^{-i\pi/6}(uJ)^{1/3}(\bar{x} - x_s - i/(uJ))\}$, where

(13) $Jx_s = \mathrm{Im}\int_0^\infty g^* g' \, d\bar{x}$

is a constant independent of u if the argument of Ai is negligible for x = 0. This leads to the asymptotic dependence[12]

(14) $V/V_J \simeq V_N/V_J - 0.749$ provided that

(15) $uJ >> 1, \quad d/2 >> x_0 \equiv (uJ)^{-1/3} >> x_s \simeq 0.375/J.$

Eq.(11) may still hold and voltage oscillations remain suppressed even if all the inequalities above arenot satisfied; thus if

(16) $\quad |d/2 - x_s - i/(uJ)| >> x_0, \quad V/V_J \simeq J(d - 2x_s),$

where x_s can be determined via numerical integration by shooting from the asymptotic expression (12). The I-V characteristic approaches the asymptotic dependence (14) from above since Jx_s increases with decreasing J. In agreement with Likharev[18] we find that x_s diverges at a critical current J_c which decreases from 0.385 to zero as u varies between 0 and ∞. J_c = 0.335 and 0.291 for u = 5.79 and 12, respectively. As discussed elsewhere,[19] the normal state is globally unstable below J_c in an infinite homogeneous superconducting filament.

RESULTS - SHORT WEAK LINKS ($d = a/\xi < 1$)

Most of the remaining discussion centers on short WLs since they exhibit pronounced Josephson effects. When $d << 1$ and $u = 0$, ψ must vary so rapidly in space that the spatial derivative dominates in Eq.(5) or Eqs. (8-9). Thus both ψ and μ approximately obey Laplace's equation.[2] In the one-dimensional case considered here this implies

(17) $\quad \mu \simeq \dot{\phi}x/d, \quad \psi(x,t) \simeq (1 + 2x/d)e^{+i\phi/2} + (1 - 2x/d)e^{-i\phi/2}.$

Eq.(8) then leads to the RSJ model with $J_m \simeq d^{-1}$.

The similarity between Eqs.(11) and (17) prompted LY to claim that Eq.(11) is correct for all J with $g(\bar{x})$ a solution of

(18) $\quad u(g + i\dot{\phi}\bar{x}/d) = g'',$

going to zero at the right boundary and to 1 at the left one. This leads to a dynamic relation between ϕ and the supercurrent averaged over the length of the WL,

(19) $\quad \bar{J}_s \simeq J_0(\theta)\sin\phi + J_1(\theta)\cos\phi + J_2(\theta), \; \theta \equiv ud^2\dot{\phi}.$

The dependence of J_0, J_1, J_2 on θ is shown in Fig.2 of Ref.7. If $\theta < 1$ one obtains[12]

(20) $\quad J/J_m \simeq \sin\phi + \dot{\phi}\{1 + (1 - \cos\phi)ud^2/15\}.$

The same result can be obtained by perturbation theory, using Eq.(17) as a zero order approximation. The same approach can be used to show that the leading correction to the static $J_s \simeq \sin\phi/d$ is a factor[14] $1 - (1/2 - \cos\phi/3)d^2/5$. Since the time average of $\dot{\phi}\cos\phi$ vanishes, Eq.(20) yields a hyperbolic dc I-V characteristic just like the RSJ model, but with the slightly higher conductance $(1 + \Gamma/15)R^{-1}$. If

(21) $\Gamma \equiv ud^2 = (a/\xi)^2 \; \tau/\tau_J$

is sufficiently small, deviations from the RSJ model remain small up to large values of $V_{dc}/V_J \sim J/J_m$ for which the amplitude of voltage oscillations, i.e. $\sim V_J$ according to Eqs.(3-4), is relatively small and $\mu \simeq V_{dc}x/V_J d$. The derivation of Eq.(19) given in Ref. 7 is then justified. Høјgaard Jensen and Lindelof claimed that it should be acceptable up to $\Gamma \sim 100$, but we find this not to be the case.

Results of our computations for a WL of length $a = \xi$ and several values of Γ (=u in this special case) are shown on Figs. 2-6. Note that the static $J_s(\phi)$ is nearly sinusoidal with $J_m = 1.067$, in good agreement with the correction indicated previously. The reliability of our computations was checked by employing asymmetric boundary conditions,

(22) $\psi = e^{-i\phi}$, $\mu = \dot{\phi}$ at $x = -d/2$; $\psi = 1$, $\mu = 0$ at $x = d/2$,

and verifying that the same results are obtained with a finer time interval

Typical results are shown on Fig. 2. With $J > J_m$, a limit cycle is established in which the voltage executes periodic oscillations whose harmonic content decreases with increasing J. V(t) peaks up soon after the order parameter goes through zero in the center of the link. Although ψ looses a winding in each such <u>phase slip</u>, this has no observable effect on V(t). The same phenomenon already occurs if $u = 0$, i.e. whenever $\phi = \pi \bmod (2\pi)$ according to Eq.(17). If u is increased, keeping J fixed, the oscillations decrease in amplitude, the peaks in V(t) become broader and more skewed to the right, while the minimum of V(t) and the maximum of $|\psi(0,t)|$ are shifted towards each other. If, starting from $\phi = 0$, the applied current is switched from 0 to $J < J_m$, the corresponding stable static solution is approached after a characteristic time which increases with u and with proximity to J_m. The same final state is reached (at least if

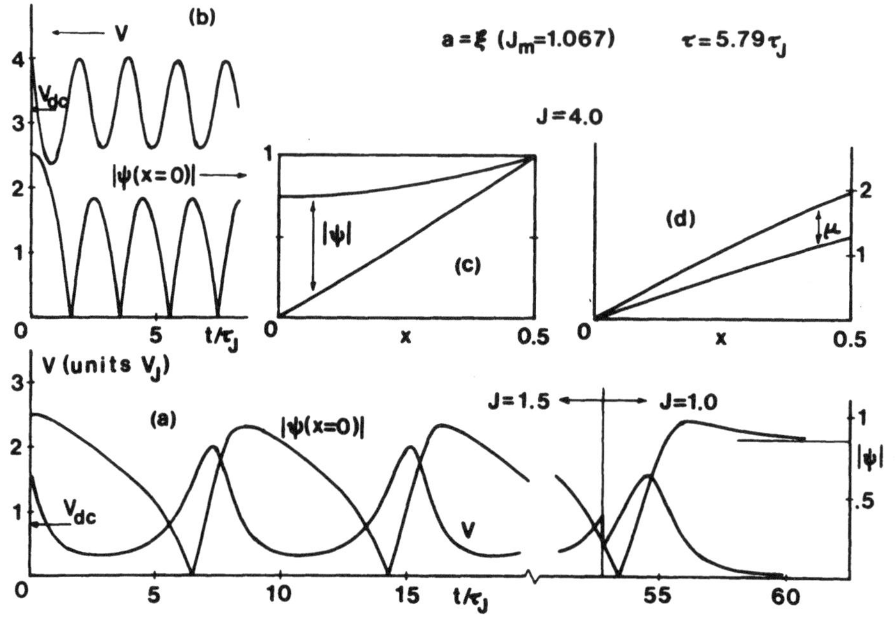

Fig. 2 (a) Time dependence of the voltage V in units of $V_J \simeq RI_m$ (left scale) across a WL of length $a = \xi$ and of the magnitude $|\psi(0,t)|$ of the order parameter at its center (right scale) for a relaxation time ratio of 5.79. With $\phi = 0$ initially, the applied current is switched from $J = 0$ to 1.5 at $t = 0$ and then back to 1.0 after 6.5 cycles. (b) Same as (a) but for $J = 4.0$. (c) Spatial dependence of $|\psi|$ between the center and the right end of the WL for $J = 4.0$; oscillations occur between the two extremals shown. (d) Same as (c) but for the electrochemical potential μ.

$\Gamma < 38$, see below) if J is switched from a value above to one below J_m, as shown in Fig. 2(a). Depending on the phase at the instant of switching an extra phase slip may occur afterwards.

In Fig. 3 J is plotted vs. the reduced dc voltage inferred from the period of V(t). Curve 1 ($\Gamma = 1$) agrees well with the hyperbolic characteristic predicted by Eq.(20). It eventually deviates from it, crosses the universal asymptote given by Eq.(14) and approaches the latter from above at currents above the range shown on the figure. Curve 2 ($\Gamma = 5.79$), which agrees with the corresponding one in Fig. 1c of LY,[12] already exhibits these trends and deviates from the appropriate hyperbola above $V \sim V_J$; it then stays close to the asymptote. As may be seen from the inset, Eq.(20)

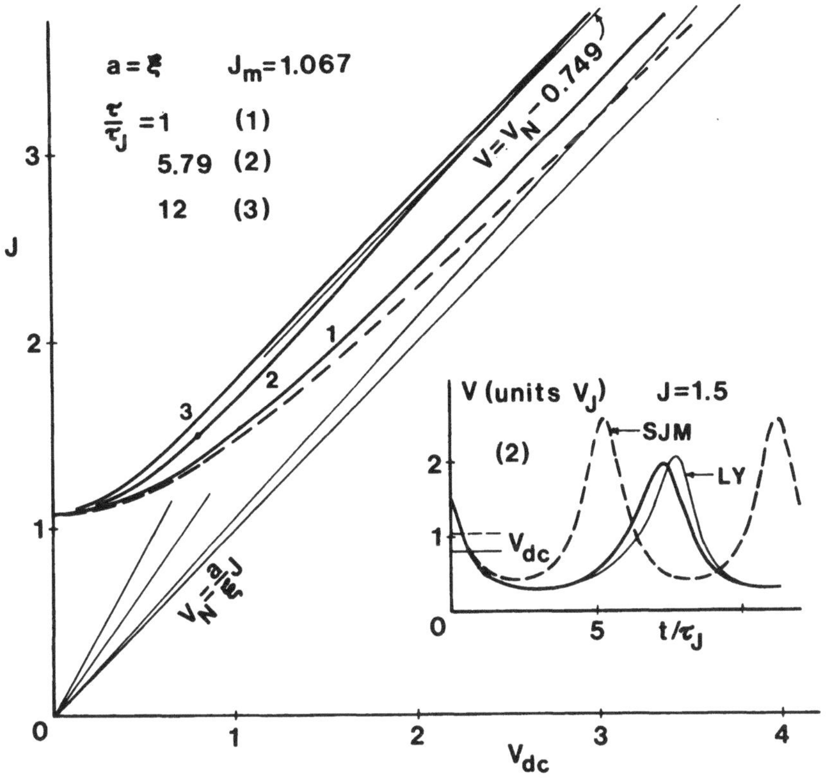

Fig. 3 Thick curves: dc I-V characteristivcs (same units as in Figs. 1-2) for a WL of length $a = \xi$ and different values of the relaxation time ratio $u = \tau/\tau_J$; the universal asymptote $V = V_N - 0.749$ is approached at sufficiently high values of J. The hyperbolic characteristic predicted by the RSJ model (u = 0, dashed curve) tends to the normal ohmic line V_N. Thin lines: asymptotes predicted from Eq.(20). Inset: time dependence of the voltage at J = 1.5 according to the RSJ model (dashed curve), to our computations for u = 5.79 (thick curve), and to Eq.(20) for $\Gamma = 5.79$ (thin curve).

still provides a reasonable approximation to the voltage waveform and its dc average at J = 1.5, certainly better than the RSJ model. Curve 3 ($\Gamma = 12$) also lies close to the asymptote for $V > V_T$, but remains above it for all J. The behavior of the maximum and minimum voltages obtained for the same values of Γ is shown in Fig. 4. According to Eq.(3-4) $V_{max} = R(I + I_m)$ and $V_{min} = R(I - I_m)$ in the RSJ model. If $u \neq 0$, however, the amplitude of voltage oscillations decreases with increasing J; both V_{max}

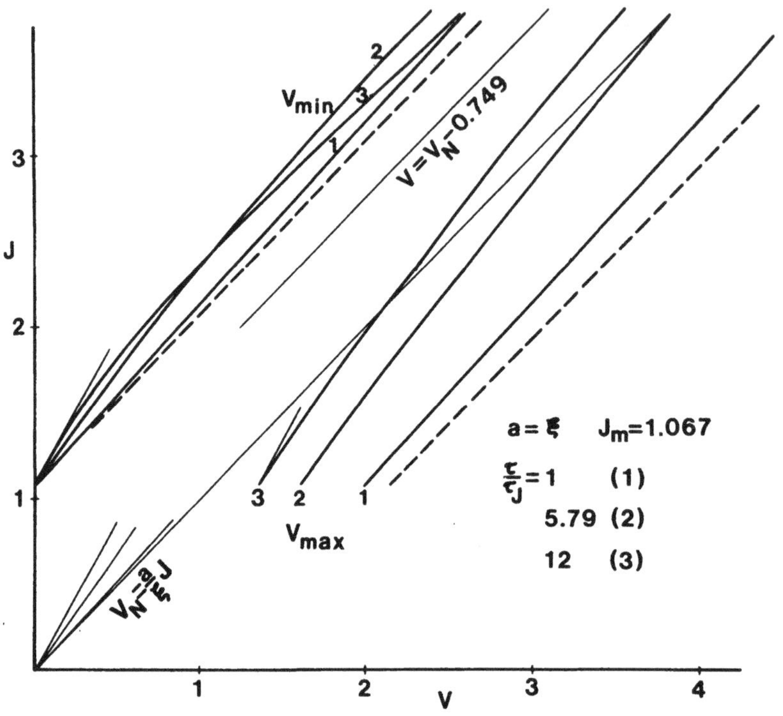

Fig. 4 Maximum and minimum voltages for the situations described on Fig. 3. Thin lines: asymptotes predicted by Eq.(20); according to this equation V_{max} and V_{min} should have the same initial slope as those lines at $J = J_m$.

and V_{min} approach the same universal asymptote at sufficiently high currents. As Γ is increased V_{max} decreases monotonically, while V_{min} is initially below the corresponding RSJ value, the maximum deviation shifting to lower J. Note that the initial slope predicted by Eq.(20) for $\Gamma = 12$ is significantly higher than the computed one for V_{max} while it is in good agreement for V_{min} $(\theta \to 0)$.

The behavior of V_{dc}, V_{max} and V_{min} for larger values of Γ, shown on Figs. 5-6, exhibits unexpected features. When $\Gamma = 30$ the dc characteristic is close to that for $\Gamma = 12$, but V_{max} and V_{min} are much closer to the common asymptote. Hysteresis appears if $\Gamma > 38$, i.e. the $V \neq 0$ state persists down to $J_{min} < J_m$. In contrast to V_{dc}, V_{min} has a finite slope at J_{min}, and can be conveniently used to accurately estimate J_{min} (1.014 and 0.883

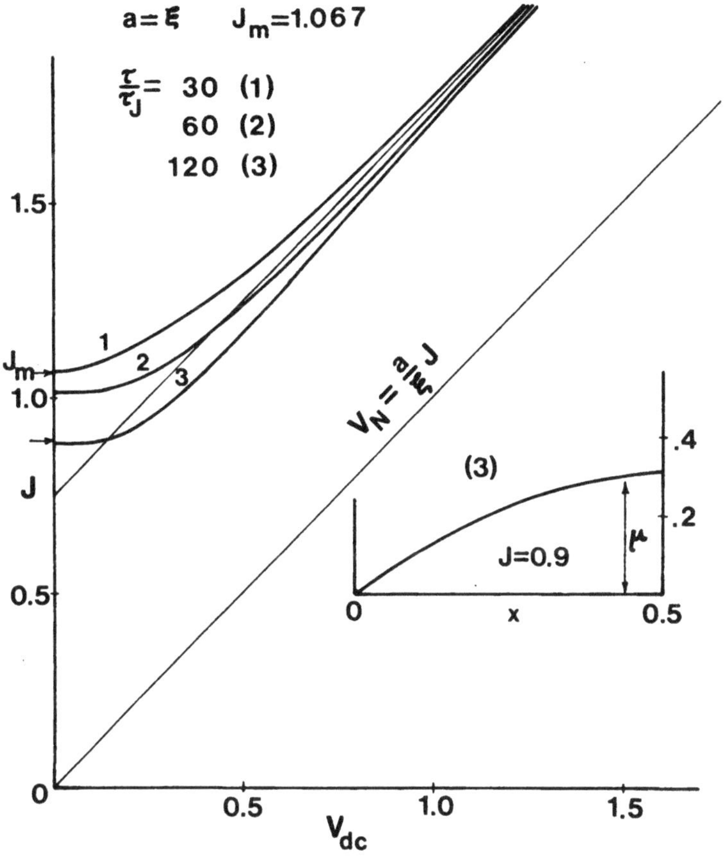

Fig. 5 Same as Fig. 3 (note larger scale, however), but for larger values of the relaxation time ratio $u = \tau/\tau_J$. Inset: spatial dependence of the electrochemical potential μ at $J = 0.9$, just above the lower limit $J_{min} = 0.883$ of the $V \neq 0$ state for $u = 120$.

for $\Gamma = 60$ and 120, respectively). The metastable $V \neq 0$ branch can also be reached by switching J from 0 to a value between J_{min} and J_m, starting with $\psi = 0$ in the center of the link. As $J \to J_{min}$, $|\psi|$ spends an increasing time near the upper extremal which tends to the unstable static solution[16] for $J = J_{min}$. As Γ is increased, V_{max} and V_{min} get closer to their common asymptote, but for $\Gamma > 38$ the dc characteristic drops below the low voltage range shown on Fig. 5, although it eventually crosses the former and approaches it from above at higher currents. This dependence differs considerably from that predicted by Eq.(19) which is shown for

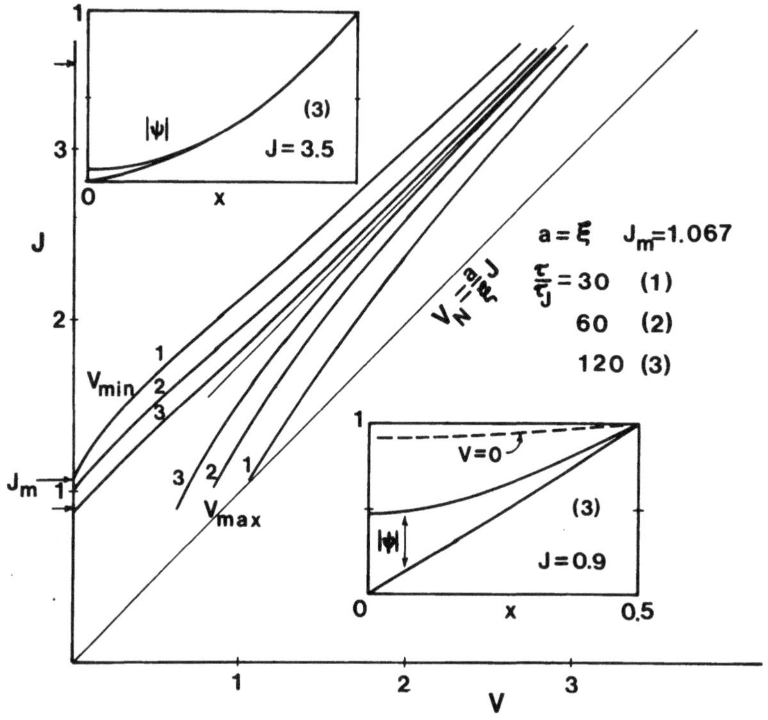

Fig. 6 Maximum and minimum voltages for the situations depicted on Fig.5. Insets: spatial dependence of the magnitude $|\psi|$ of the order parameter at J = 3.5 and 0.9 (values marked by arrows on the J-axis) for u = 120; when J = 0.9 the upper extremal curve shown is indistinguishable form the corresponding static saddle point solution which lies well below its stable counterpart (dashed curve).

Γ = 30, 60 and 120 on Fig. 3 of Ref. 7 (note that the voltage unit used there is V_J/Γ in our notation, while the current is only scaled by an arbitrary factor). Indeed the corresponding curves all lie above the asymptote and show a bump below $V = V_J$, while that for $\Gamma = 15$ is similar to our dc characteristic for $\Gamma = 12$; hysteresis occurs for $\Gamma = 120$, but is much smaller than in our computations. We note that Eq.(19) is justified if (i) the amplitude of voltage oscillations is small compared to V_{dc} or (ii) $\psi'' \simeq 0$ and (iii) μ varies linearly throughout the WL. The first condition is violated at small enough voltages; next $\theta_{max}(J_{min})$ equals 16 and 74 for Γ = 12 and 120, respectively, while (ii) holds if $\theta < 8$ at most. The last condition is not even satisfied at higher currents where the first one is, e.g. the situation depicted in the inset on Fig. 5 for $\Gamma = 120$ persists up to $J = 3.5$ where the two sides are still slightly coupled (see top inset on Fig. 6).

Without going into details we remark that the last two inequalities in Eq.(15) suggest a classification into <u>three regimes</u> separated by $\Gamma \simeq 8\eta$ and 19η. These values correspond to the conditions $d = 2x_0$ and $x_s = x_0$ being satisfied at $J = J_m$, with x_0 redefined as $(\eta/uJ)^{1/3}$. The reasonable choice $\eta \sim 2$ is consistent with the trends and peculiar features revealed by our computations. A full discussion including results obtained for long links and boundary effects will be given at a later date. We only note that the first condition in Eq.(15) is only relevant if $d > 2$ and $u < 2.33$.

CONCLUSIONS

Besides demonstrating that the dynamic current-phase relation Eq.(19) has a much smaller range of applicability than originally claimed,[7] our computations reveal a richer variety of behavior than was anticipated by Likharev and Yakobson.[12] In particular did not pay much attention to the possibility $x_s > x_0$ and failed to notice the intermediate and large Γ regimes.

We have deliberately limited our discussion of dynamics to short WLs which exhibit Josephson effects in a large current range and are more promising for applications. The reduction in amplitude of oscillations at a given current level and the appearance of hysteresis with increasing Γ is similar to the effect of a parallel capacitance within the RSJ model. In the latter case hysteresis appears when $\beta = RC/\tau_J > 1$.[1] This analogy was recognized by Song.[4]

In spite of the limited validity of Eqs.(5-6) on which our computations are based, we believe that the main qualitative features at low voltages and high currents will survive in more realistic models. Since measurements probing the local behavior of the order parameter or electrochemical potential appear extremely difficult, at least in short links, one must rely on models for interpretation. In this connection we note that significant changes in the shape of computed dc I-V characteristics are found only by varying the relaxation parameter Γ over a wide range. In practice this says that dc characteristics alone are difficult to interpret unambiguously. It is also worth keeping in mind that heating[5] as well as non-uniform vortex penetration and motion[10,20] can also give rise to hysteresis and excess supercurrents at high voltages. Careful experiments which avoid these and other complications and which probe the frequency spectrum and the frequency response of weak links over a judiciously chosen range where relaxation effects should show up are clearly desirable. From the theoretical side it will be interesting to check whether the fit to the frequency dependence of microwave-induced steps achieved by Clark and Lindelof[8] still makes sense within the full LY model.

REFERENCES

1. McCumber, D. E.: Effect of ac Impedance on dc Voltage-Current Characteristics of Superconductor Weak Link Junctions. J. Appl. Phys. 39, 3113-3118 (1968). Stewart, W. C.: Current-Voltage Characteristics of Josephson Junctions. Appl. Phys. Lett. 12, 277-280 (1968).

2. Aslamazov, L. G. and Larkin, A. T.: Josephson Effect in Superconducting Point Contacts. Zh. Eksp. Teor. Fiz. Pis'ma 9, 150-154 (1969) - Soviet Phys. - JETP Letters 9, 87-91 (1969).

3. Levinsen, M.T.: Rev. Phys. Appl. 9, 135-144 (1974).

4. Song, Y.: Origin of 'Capacitance' in Superconducting Microbridges. J. Appl. Phys. 47, 2651-2655 (1976).

5. Skocpol, W.J., Beasley, M. R., Tinkham, M.: Self-heating Hotspots in Superconducting Thin-Film Microbridges. J. Appl. Phys. 45, 4054-4066 (1974).

6. Notarys, H. A. and Mercereau, J. E.: Proximity Effect Bridge and Superconducting Microcircuitry. J. Appl. Phys. 44, 1821-1830 (1973).

7. Højgaard Jensen, H. and Lindelof, P. E.: Comparison between the Resistively-Shunted Josephson Junction Model Incorporating a Voltage-Dependent Current-Phase Relation and Experiments on Superconducting Thin-Film Microbridges. J. Low Temp. Phys. 23, 469 (1976).

8. Clark, T. D.: Frequency Dependence of the ac Josephson Effect in Small Dayem Bridges. Phys. Rev. Lett. 37, 368-371 (1976).

9. Gregers-Hansen, P. E. and Pickett, G. R.: Subharmonic Gap Structure and Gap-Enhancement by the Josephson Radiation in Small Dayem Bridges. Rev. Phys. Appl. 9, 145-152 (1974).

10. Likharev, K. K.: Vortex Motion and Josephson Effect in Superconducting Thin-Film Bridges. Zh. Eksp. Teor. Fiz. 61, 1700-1711(1971) - Soviet Phys. JETP 34, 906-914 (1972).

11. Aslamazov, L. G. and Larkin, A. T.: Josephson Effect in Wide-Area Superconducting Contacts. Zh. Eksp. Teor. Fiz. 68, (1975) - Soviet Phys. - JETP 41, 381 (1975).

12. Likharev, K. K. and Yakobson, L. A.: Dynamic Properties of Finite Superconducting Filaments. Zh. Eksp. Teor. Fiz. 68, 1150-1159 (1975) - Soviet Phys. - JETP 41, 570-575 (1975).

13. Baratoff, A., Blackburn, J. A., Schwartz, B. B.: Current-Phase Relationship in Short Superconducting Weak Links. Phys. Rev. Lett. 25, 1096-1099 (1970); Erratum ibid. 25, 1738 (1970).

14. Likharev, K. K.,and Yakobson, L. A.: Steady-State Properties of Superconducting Bridges. Zh. Tekhn. Fiz. 45, 1503-1509 (1975) - Soviet Phys. - Tech. Phys. 20, 950-954 (1976).

15. Klapwijk, T. M. and Mooij, J. E.: Extension of the Operating Range of Superconducting Microbridges, IEEE Trans. MAG 11, 858-859 (1975).

16. Christiansen, P. V., Hansen, A. B., Sjøstrøm, C. J.: Negative Self-Inductance in Superconducting Thin Wires and Weak Links. J. Low Temp. Phys. 4, 349-389 (1971). Langer, J. S. and Ambegaokar, A.: Intrinsic Resistive Transition in Narrow Superconducting Channels. Phys. Rev. 164, 498-510 (1967).

17. Gor'kov, L. P. and Eliashberg, G. M.: Dynamical Properties of Gapless Superconductors. J. Low Temp. Phys. 2, 161- 171 (1970).

18. Likharev, K. K.: Isothermal Domains in Quasi-One-Dimensional Superconductors. Zh. Eksp. Teor. Fiz. Pis'ma 20, 730-733 (1974) - Soviet Phys. JETP 20, 338-339 (1974).

19. Kramer, L. and Baratoff, A.: Lossless and Dissipative Current-Carrying States in Quasi-One-Dimensional Superconductors. To be published.

20. Clarke, J.: Finite Voltage Behavior of Lead-Copper-Lead Junctions Phys. Rev. B4, 2963-2977 (1971).

JUNCTION FABRICATION TECHNIQUES

JUNCTION FABRICATION TECHNIQUES

T. Van Duzer
Dept. of Electrical Engineering and Computer Sciences and the Electronics Research Laboratory, University of California, Berkeley, CA 94720

INTRODUCTION

The first demonstration of the Josephson effect employed an oxide-barrier thin-film structure in 1963. Since that time, a variety of other structures have been shown to exhibit the Josephson effect. These have included: (1) point contacts, (2) crossed, touching wires, (3) solder blobs on oxidized wires, (4) thin superconductive films with a superconductive region weakened by thinning, proximity effect, or ion implantation, (5) thin superconductive films with a narrow gap bridged by normal metal, semimetal, or semiconductor, and (6) sandwich structures with barries of normal metal or semiconductor. A great deal of useful work has been done with point contacts; indeed, they have been used in the only Josephson junction detection and mixing experiments that have given good results. Some of the other types suffer from fabrication or ruggedness problems or from such low resistances as to be useless in applications. It seems clear that only the types that can be fabricated with fixed and reproducible properties will find their way into practical applications. In this writer's opinion, only the oxide-barrier junction, the semiconductor-barrier sandwich-type junction, and the coplanar junctions with high-resistance bridges offer potential. These will be the subject of this paper.

PARAMETER CRITERIA FOR APPLICATIONS

We will discuss separately the desirable properties for junctions to be used in various types of systems, including detectors and mixers, parametric amplifiers, magnetometers, voltage standards, and switching circuits. It is assumed that simplicity of fabrication, reproducibility, and resistance to thermal and mechanical stresses are common desirable properties. For switching circuits and arrays for RF applications, where large numbers of junctions are required, the demands on reproducibility are higher than in applications requiring one, or a few, junctions.

Detectors and Mixers

The usual mode of operation of a Josephson junction for detection and mixing[1] is illustrated in Fig. 1. We assume the bias current is applied to

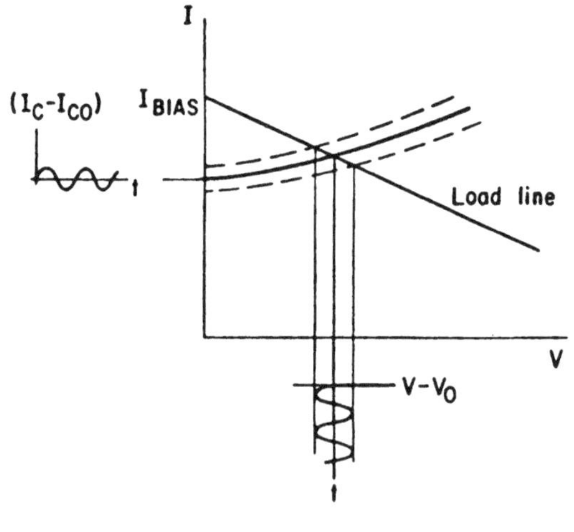

Fig. 1. Method of operating a nonhysteretic Josephson junction to convert modulation of critical current into voltage variations.

the junction in parallel with a load resistor and static operation is at V_0. Application of an RF signal depresses I_c so, if the RF signal amplitude is modulated, I_c will follow the modulation and so also will V_0; this voltage variation is the output signal.

In order to avoid hysteresis in the I-V characteristic and thereby permit operation in the generally accepted mode for mixing and detection, McCumber's parameter[2] β_c must be less than about unity:

(1) $$\beta_c = \frac{2e}{\hbar}(I_cR)(RC) \leq 1$$

where I_c is the junction critical current, R is the normal-state resistance of the junction at the operating temperature, and C is the junction capacitance. The equivalent circuit of a Josephson junction is shown in Fig. 2. The RC product is determined by the material and configuration of

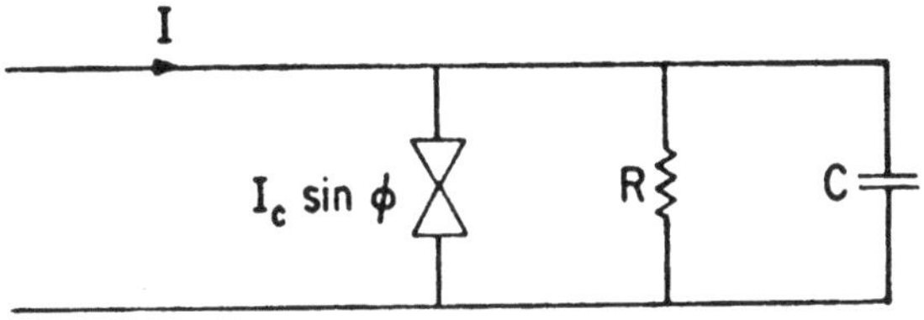

Fig. 2. Josephson junction equivalent circuit.

the region separating the electrodes of the junction. For tunnel junctions the I_cR product depends only on the electrode material and the operating temperature but, for junctions with resistive coupling regions I_cR can also depend on the coupling region.[3]

Mixing theory[4] makes it clear that it is advantageous, in achieving low receiver noise temperatures, to minimize the normalized operating frequency ω/ω_0, where $\omega_0 = (2e/\hbar)I_cR$. Thus, we must seek devices having as large a product I_cR as possible. The advantage resides not in the mixer noise temperature, which is largely independent of ω/ω_0, but rather in the conversion efficiency, which dictates the role played by the noise of the IF amplifier in the receiver noise temperature.

At the same time, it is necessary to keep the RC product small enough to satisfy Eq. (1). The RC product is also the inverse of the (angular) frequency at which the current shunted around the Josephson element by the capacitor in Fig. 2 becomes equal to that shunted by the resistor. It is desirable to make that frequency as high as possible as it constitutes a kind of cut-off frequency. For both reasons, we must seek devices with low RC products.

To facilitate matching and avoid excessive Johnson noise, the junction resistance should be about 20 ohms. Further, the I-V characteristic should have as small a slope as possible at the bias point. The inverse of the slope, the dynamic resistance, enters both the noise temperature of the mixer and the conversion efficiency. Finally, I_c should have a high sensitivity to applied RF power.

The usual mode of operation of Josephson detectors, like mixers, requires a lack of hysteresis in the I-V characteristic, so Eq. (1) must apply. The theory for the small-bias mode of detector operation[5] shows that the responsivity (ratio of detected voltage to signal power) increases with dynamic resistance and I_cR product. Again, the RC product should be low to avoid excessive capacitive shunting around the Josephson element and to satisfy Eq. (1) in the presence of the highest possible I_cR product. The noise equivalent power is reduced by increasing the I_cR product. Just as for mixers, the sensitivity of I_c to RF power should be high. The desired resistance level is probably about the same as in the case of the mixer (20 ohms) as there is a similar trade-off between impedance-matching problems and Johnson noise. Therefore, the optimization of junctions for mixer and detector applications involve the same factors.

Parametric Amplifiers

We consider here the situation in which the dc voltage on the junction is zero and it behaves like a nonlinear inductance.[6] Any capacitance inherent in the junction (Fig. 2) will bypass the signal and pump power around the active element ($I_c \sin \phi$) so it must be minimized. It is desirable to have high critical current to minimize the competitive effects of noise currents. Increasing the shunt resistor minimizes the shunting effect but increases the noise. The $I_c R$ product should be made as large as possible and the resistance adjusted for optimum noise conditions.

Magnetometers

Josephson-junction magnetometers employ either a dc SQUID, which is a superconductive loop containing two symmetrically placed junctions, or an RF SQUID, in which the loop is interrupted by only one junction.[7] For the dc SQUID, the operation is rather similar to the RF detection mode discussed above where the dc bias is applied to hold the nonhysteretic junctions in the finite-voltage regime as shown in Fig. 1. The sensitivity of the magnetometer is increased by increasing the voltage change for a given change of critical current. As seen from Fig. 1, the voltage change is enhanced by a higher dynamic resistance and a higher bias. To avoid hysteresis, Eq. (1) must be satisfied. Also, for a symmetric SQUID the modulation of the critical current increases smoothly from I_c for $\frac{LI_c}{\Phi_0} << 1$ to a limiting value of $\frac{\Phi_0}{L}$ for $\frac{LI_c}{\Phi_0} >> 1$, where L is the inductance of the connecting leads and Φ_0 is the flux quantum. For other reasons, reduction of the inductance leads to better resolution but its minimum value is limited by the need to couple to the SQUID loop; a typical value is 10^{-9} H. If I_c is too low, noise rounding of the I-V characteristic occurs and there is an attendant loss of dynamic resistance. A value of I_c such that $\frac{LI_c}{\Phi_0} \approx 1$ is sometimes used. This corresponds to a critical current of $\approx 2\ \mu A$. Larger values of I_c give slightly increased I_c modulation but, more importantly, are achieved at the expense of dynamic resistance since it is proportional to junction resistance R, and the maximum I_cR product is essentially a constant for a given superconductor (i.e., $I_cR \cong \frac{\pi\Delta}{2e}$). One should pick a superconductor with a large energy gap Δ and use a junction with as small a capacitance as possible so $\beta_c \leq 1$. Then I_c and R can be traded-off to avoid noise rounding and to maximize the dynamic resistance.

The electrical parameters of the RF SQUID seems less critical than for the dc SQUID, at least when used for low-frequency measurements. If the coupling problem can be solved for evaporated SQUID loops, some kind of thin-film junction would be desirable for convenience.

Voltage Standards

Junctions used in voltage-standard systems are current-biased to a high-order step induced in the I-V characteristics by an RF signal of precisely known frequency. The main criterion for this application is that the induced steps be sharp and large at as high a voltage as possible in order to facilitate instrumentation.

Switching Circuits

There are many different kinds of switching circuits so it is difficult to state any simple set of criteria. It is generally desirable to have the "control characteristic," the dependence of critical current on magnetic field (or equivalently, control current), of such a form as to permit gain. That is, an input control current switches a larger current into the output of the circuit. This is often achieved by using an in-line junction of length greater than the Josephson penetration length. However, it is possible to use small junctions with a bias line,[8] or to connect the current leads to the junction so as to shape its control characteristic,[9] or to use multiple-contact junctions to adjust the gain.[10]

The amount of hysteresis desired for different circuits varies. Logic circuits with perfectly matched loads will not reset to the zero-voltage state, as required in nonlatching logic, unless there is small enough hysteresis to allow the bias current to be set below the minimum current I_0 as shown in Fig. 3. Another possible mode of operation of a nonlatching logic circuit, which should give more consistent values for the output current, requires both some inductance in the output line and hysteresis in the junction.[8] The current transferred is optimized by choice of β_c. This, in turn, requires control of the I_cR and RC products in Eq. (1).

Switching times involve both RC and L/R, where L is the inductance of a memory loop or that of an unmatched output line in a logic circuit. The L/R ratio is usually dominant. There is some advantage, then, in keeping the junction resistance reasonably high; but at some point, the RC product can become limiting. It should also be pointed out that the current

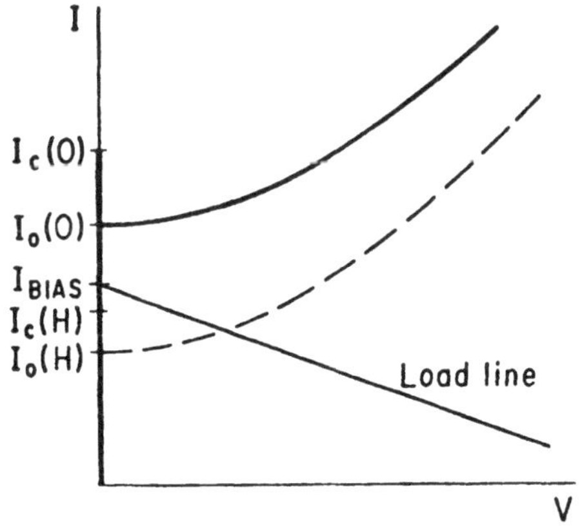

Fig. 3. I-V characteristics of a junction suitable for use in nonlatching operation of a Josephson junction logic circuit with a matched output line. $I_c(0)$ and $I_0(0)$ are the critical and minimum currents in the absence of an applied field and $I_c(H)$ and $I_0(H)$ refer to the same quantities with a control field applied. The bias current is lower than $I_0(0)$ so the circuit returns to the zero-voltage state upon removal of the control field.

transferred in a logic circuit, as a fraction of the gate current, is roughly equal to the ratio of the load resistance to the junction resistance so this may affect the choice of junction resistance.

FABRICATION TECHNOLOGY

These are two important aspects of the fabrication of thin-film Josephson devices, the materials and their patterning. It will be most advantageous to postpone discussion of the materials until later when they can be discussed in connection with the individual types of junctions. Here we will comment on the techniques that can be employed to pattern the films.

When the desired patterns are simple, like straight lines or circular holes, and the feature size is larger than about 0.2 mm, the simplest approach is often to have a metal evaporation mask made in the machine shop.

If the feature size is in the range of 2 μm to 0.2 mm and/or the shapes of the desired electrodes are complex, photolithographic methods are required. The procedure is to make the desired pattern on a double-layered plastic sheet called Rubylith with overall dimensions of about 80 x 80 cm^2. One layer is transparent and the other is semi-opaque; the latter is removed in the chosen pattern and the remaining pieces are supported by the transparent sheet. Typically, there are two stages of photographic reduction, giving final patterns that are reduced in size between 100 and 300 times from that on the Rubylith. In the usual use, one obtains 100 or more identical images on a 5 x 5 cm^2 glass slide. To get a 2 μm feature on the final mask, the corresponding feature on the Rubylith sheet has to be between 0.2 and 0.6 mm. Such small dimensions are difficult by hand but there are both hand- and computer-controlled cutting machines which have sufficient accuracy. In integrated circuit technology, 5 μm feature size is the industry standard for large-scale integration because of the need for a large and complex circuit free of defects. It is rather easy to do 5 μm photolithography in the laboratory, 1-2 μm work requires care, and features smaller than 1 μm can be achieved with a variety of special measures.

The patterns on the mask are transferred to the films on the sample substrate by exposing a thin film of photoresist with a mercury arc lamp through the mask which is placed in contact with the resist. The photoresist is a polymer which comes as a liquid and is dropped on the substrate when the latter is spinning at a high rate. Typically, the resist thickness is in the range of several thousand angstroms. There are two kinds of resists, positive and negative, depending on whether exposure softens or hardens the resist. The developer removes exposed positive resist and unexposed negative resist is removed by its developer solution. There are strategic reasons for choosing one or the other, depending on a variety of factors including feature size, available etchants for the materials being patterned, relative protective and adhesive properties of the resists, etc. Details of photolithographic processes may be found in Ref. 11.

When it is desired to produce circuits and devices with feature sizes in the range of 0.1 μm to 2 μm, one should use electron lithography.[12] Different resists are used from those in photolithography. The resists are exposed by a beam (typically, 200 Å diameter) in an electron microscope which is usually scanned in a discrete-step TV-like raster with the beam current turned on and off by a computer which is programmed to give the desired pattern of exposure. The patterns commonly consist of sets of rectangles but special programs can be written to produce any desired shape. There are situations where the use of electron lithography has another advantage over contact photolithography. This is where there are elevation variations of the sample surface which are greater than the desired feature sizes. We will see an example of this in our discussion of silicon etch pits.

In the use of lithographic techniques, there is often a need to etch metal or insulator films; this can be done either with chemicals or with energetic ions. In either case, a mask is needed for the regions not to be etched. The mask is usually a resist that has been patterned lithographically but somewhat more complex schemes involving deposition of another material for the mask are needed. Chemical etching is often the easiest. The chemical etch rate for different materials can be greatly different so only a thin layer of one is needed to protect a region of the one to be etched. However, the etchant has a tendency to penetrate underneath resist layers used for protection so it is difficult to make very small features by chemical etching. The use of energetic ions is more controllable and gives sharper edges on the etched film. There are two possible sources for the ions. The sample may be affixed to the target of a sputtering system and the ions (usually argon) of the discharge strike the sample and etch both the mask and the exposed regions. In either case, the mask must etch slowly enough that the desired etching takes place before the mask is eroded away. The alternative is to employ a beam of ions, usually neutralized, from an ion gun; this gives somewhat better resolution but the system is more expensive.

SANDWICH-TYPE JUNCTIONS

One of the common and important configurations of Josephson junctions is the sandwich form in which two superconductive films are separated by about 20-50 Å of oxide, hundreds of angstroms of semiconductor, or up to 1 μm of normal metal. We will concentrate on oxide- and semiconductor-barrier junctions; those with normal-metal barriers have extremely low resistance and their practical application is less likely than the others.

Oxide-Barrier Junctions

We have seen in above that most applications require minimization of the shunt capacitance so the high capacitance concommitant to having a thin oxide barrier is usually undesirable. It is possible to reduce the effective β_c of an oxide-barrier junction to a value less than unity to eliminate hysteresis by resistive shunting but this has the disadvantage of reducing the net I_cR product which, for most applications, should be as large as possible. The one application where oxide-barrier junctions are likely to find significant use is for switching circuits where hysteresis is usually desirable, as was pointed out in above. Thermal cyclability of oxide-barrier junctions has been difficult to achieve but some cyclable types do exist.

The most used superconducting elements with transition temperatures above 4.2 K are the transition metals Nb, V, and Ta and the soft metal Pb. The most used in the next-lower-temperature group between 4.2 K and about 1.8 K, the temperature achievable by pumping with a small mechanical pump on helium vapor, is Sn, which is a soft metal.

Lead and tin are commonly used in Josephson junction fabrication as they are easy to vacuum deposit with transition temperatures close to the bulk values. The difficulty with these materials is that they form hillocks and whiskers upon temperature cycling between room and cryogenic temperatures. (See Fig. 4 for an example of protrusions that grow on a lead film.) As the temperature of the sample is lowered, the metal contracts at a greater rate than the insulating substrate and the film is stretched inelastically. When the sample is returned to room temperature, the film is then compressed. Hillocks and whiskers form to relieve the resulting stress.

Fig. 4. Hillock grown on a Pb film after temperature cycling. Hillock dimensions are about 1.2 μm. (Courtesy Chi Yung Fu.)

Whiskers can be as long as several micrometers and hillocks are typically 0.2 - 0.5 μm high. Whiskers only grow under certain circumstances; hillocks are unavoidable in these soft elemental films. Even small hillocks are large enough to puncture the thin oxides used in tunnel junctions so those made with the soft metals are usually kept continuously at cryogenic temperatures. The junctions can be stored in liquid nitrogen between experiments in liquid helium so the stress-relief mechanism apparently requires raising the temperature above 77 K. (It has been reported[13] that some amelioration of the cycling problem is obtained by coating the completed junctions with a photoresist layer of a few tenths micrometer thickness and making the temperature transitions as rapid as possible. In this way, Pb-PbO-Pb junctions have been cycled between 4 K and 293 K eight times with no significant change in the critical current, but they cannot be stored more than one day at room temperature.) For practical applications where many temperature cycles must be anticipated, the junctions made of pure soft metals are not acceptable.

Alloys of soft metals have been employed to achieve more cyclability of Josephson junctions. These have been of both binary and ternary types. It has been shown[14] that hillock formation in Pb films can be reduced by adding small quantities of In. Cyclable junctions have also been produced using In and In-Sn (50:50) alloying.[15] The usual procedure is to deposit the In or In-Sn with Pb on top and to allow interdiffusion at room temperature or slightly elevated levels.

Indium apparently plays a two-fold role in addition to aiding in the prevention of hillock formation. It is usually deposited first and serves to improve bonding of the film to the substrate. More importantly, substantial decrease of the tunneling barrier is obtained by virtue of the presence of indium.[16] In the range 0-20 percent indium in a Pb-In alloy, the oxide is a mixture of PbO and In_2O_3. For a junction with a Pb-In base electrode and a 32 Å barrier thickness, the tunneling current density is a strong function of In content, as seen in Fig. 5, and can be higher than $10^4 A/cm^2$.

The use of other alloys for lowering the barrier height has also been

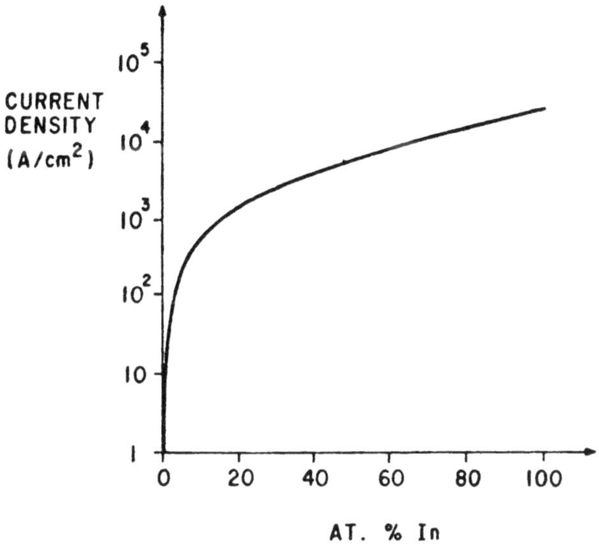

Fig. 5. Dependence of the critical current density on indium content in a Pb-In alloy oxide-barrier Josephson junction. The oxide thickness is 32 Å. (Ref. 16)

reported.[16] These include Te-Pb, Sn-Pb, and combinations of In, Te, Sn, and Bi with other superconductors, though no experimental results have been given.

There is reason to believe that the most effective element for the prevention of hillock and whisker formation in lead junctions is gold.[17] The stress relaxation phenomenon which leads to the growth of protrusions is apparently related to grain-boundary migration and diffusion. Measurements on bulk lead have shown that the best alloying elements for inhibiting grain-boundary motion are ones with atomic sizes much different from that of lead atoms and also with large valence difference. The valence difference seems to play the dominant role[18] so gold and silver (valence of 1 compared with Pb valence of 4) should be good choices. Experiments have shown that gold is more effective in inhibiting hillock growth than is silver.[17] This suggests that one might advantageously use gold to inhibit hillock formation and indium mainly for bonding to the substrate and lowering of the tunneling barrier.

The use of a ternary alloy of lead, gold, and indium has been reported.[19,20] The alloy used in the base electrode was formed by depositing 500 Å of In, 100 Å of Au, and 3500 Å of Pb in succession, and subsequently heating to aid interdiffusion. After formation of the oxide barrier, a counter electrode of Pb with a small amount of added Au to suppress the above mentioned stress-relaxation effect, was formed. It was made by depositing 3000 Å of Pb, 50 Å of Au, and 2000 Å of Pb in succession with subsequent annealing to homogenize the alloy. It should be pointed out that the layered alloys should be made without exposure to air between layers because a small layer of oxide can strongly inhibit inter-diffusion.

Another way to form reliable junctions with transition temperatures above 4.2 K is to use niobium (or, less commonly, tantalum or vanadium) for the base electrode. The formation of the niobium film can be done using electron-beam evaporation if special precautions are taken. The transition metals (Nb, V, Ta) are good getters and their transition temperatures are reduced by absorbed oxygen (e.g., T_c of Nb is reduced about 5 percent for 0.5 percent oxygen content[21]). The partial pressure of oxygen in the vacuum system must be kept at a very low level either by having a special very-high-vacuum system or by using the gettering property of niobium on large metal surfaces during the deposition of the film on the substrate. It is usually necessary to also raise the substrate temperature by hundreds of degrees. The more common procedure is to sputter deposit the niobium in either a dc[22] or RF[23] diode system or using getter sputtering.[24,25] Probably the method which most easily gives good results employs a magnetron-type unit[26,27] in which the rate of deposition is fast enough that inappreciable amounts of oxygen are absorbed even with a room-temperature substrate. Attempts to use niobium also for the counter electrode have generally failed due to superconducting shorts in the oxide but some success has been achieved.[27] Good stable junctions are obtained where the counter electrode is made of Pb or a Pb alloy.[26,28] The advantage of using a Nb counter electrode is that it is not subject to corrosion (a severe problem for Pb). Alloying the Pb in the top electrode with a small amount of In improves its corrosion resistance.[26] Junctions containing either Pb or alloys thereof should be coated with a thin layer of photoresist to prevent moisture from reaching the Pb.

A variety of methods of forming the oxide barrier have been described in the literature. The principal problem experienced with Pb is the formation of porous oxides that contain pinholes which lead to short circuits upon deposition of the top electrode. Oxidation in room environment with the sample either heated or not appears to be the least controllable for Pb, possibly because of climatic variations, but it is often used in experimental work because of its simplicity and has been used for Nb with rather reproducible results.[26,27] Oxides can also be formed in an oxygen flow with heated or unheated samples. A more reliable method is to form the oxide in the vacuum system immediately after deposition of the base electrode so the environment can be more completely controlled. This is done by introduction of oxygen and heating of the sample for a period of hours. A similar procedure employs a glow discharge in the vacuum system to form the oxide.[29] Oxidation strategies become more complex if a lithographic step is used after deposition of the first electrode.[30]

A recently developed method involves a competition between growth of an oxide and its sputter removal.[31] The sample is placed in a sputtering system, perhaps with the junction tunneling region defined by a resist mask, and the surface is first cleaned in a low-power discharge of an inert gas such as argon. Then with the argon replaced by oxygen or a combination of argon and oxygen, the oxide begins to form and sputtering of the oxide also proceeds. The rate of sputtering is independent of the oxide thickness. At a thickness which depends on the gas pressures and discharge power, an equilibrium is reached. Thus one can control the oxide thickness accurately by setting the sputtering parameters. This method has been employed for lead[31] and its alloys, niobium,[32] and nickel.[33]

Evaporated Semiconductor Barrier Junctions

As we saw above, it is often desirable for applications to maximize the

I_c/C ratio. This can be done by lowering the tunneling barrier. One approach to lowering the barrier of a sandwich-type Josephson junction is to replace the oxide with a deposited pad of a semiconductor. The semiconductor materials used have included CdS, CdSe, C, Ge, Te, PbTe, InSb, GaAs, and ZnS. Electrode materials have included Pb, Sn, Tl, Pb_xIn_{1-x}, Pb_xSn_{1-x}, Nb, Nb_xRe_{1-x}, and Al.[34] In no case has a current density comparable with that achievable with oxide-barrier junctions been obtained. In almost all cases, it has been necessary to patch the pinholes through the semiconductor by forming a heavy oxide on the base electrode in the holes. These junctions have some desirable properties such as the suppression of junction resonances by the semiconductor losses and the achievement of parameters equivalent to shunted oxide-barrier junctions for use in certain switching circuits. Generally, the properties of a thin deposited film of a semiconductor are difficult to control and this will probably limit their application.

Some of the best junctions with deposited semiconductor barriers have been formed with Pb electrodes and Te barriers.[35] These can be made with either hysteretic or nonhysteretic I-V characteristics, depending on the Te thickness and have been shown to have uniform, though low, current density. The highest I_cR product achieved was about 0.4 of the theoretical value for tunneling barriers and the maximum critical current density was about 10 A/cm^2. The junctions could only be temperature cycled a few times before failure.

Single-Crystal Silicon-Membrane Junctions

A recently developed fabrication technique forms a barrier for a Josephson junction by etching a silicon wafer in a localized region.[36] The etching is initiated through a mask on one side of a standard wafer and is stopped at a predetermined distance from the opposite face by a layer of boron doping. The result is a single-crystal membrane of uniform thickness which can subsequently be coated with a superconductor having the desired electrode pattern. This structure has an advantage over the deposited-

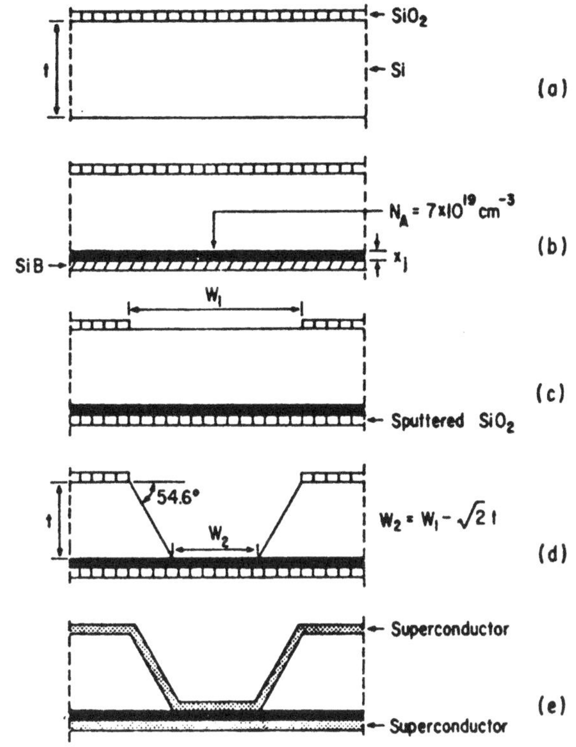

Fig. 6. Steps in the fabrication of a Josephson junction in which the barrier is an integral part of a silicon wafer.

semiconductor junction in that the membrane can be processed after fabrication and before electrode deposition to adjust its properties.

The fabrication procedure is shown in Fig. 6. In step (a) a SiO_2 layer is grown on one side of a (100) oriented wafer. In step (b) a boron diffusion is made in the opposite side. The later etching will stop at the plane of boron density $N_A = 7 \times 10^{19}$ cm^{-3}; the diffusion is designed so that density exists at the depth desired for the membrane thickness. The process is calibrated to account for layers of Si-B and boron glass that must be removed. In step (c), a window (preferably rectangular and aligned with the crystal axes for smoothness of the resulting etch pit) is opened on the top side and the diffused layer is protected with a sputtered SiO_2 layer after removal of a film of boron glass and the Si-B phase layer. Step (d) shows the result of subjecting the wafer to a heated etchant, ethylene-diamine-water-pyrocatechol. This so-called preferential etchant has the properties that it etches (100) planes many times faster than (111) planes, etches SiO_2 only very slowly, and will not etch silicon where the boron concentration exceeds 7×10^{19} cm^{-3}. Membranes as thin as 400 Å and as thick as about 4 μm have been made this way.

The process described above is limited with respect to the minimum achievable dimensions in the plane of the junction; a process has been developed to overcome this limitation.[37] As seen in Fig. 6d, the dimension w_2 of the membrane depends on the wafer thickness, a quantity which can only be known to an accuracy of several micrometers. Thus, if membranes with w_2 on the order of one micrometer are desired, as is the case for some applications, the process in Fig. 6 must be modified. This can be done by first using the process of Fig. 6 to make a membrane of about 2 μm thickness which can be measured to an accuracy of about 100-200 Å. The heavy boron doping is outdiffused and a second etching procedure like that in Fig. 7 is followed to make a second etch pit. It is estimated that its dimension (w_2' in Fig. 7c) can be controlled to better than 0.1 μm. This work is still at an early stage and the double-etching procedure has only been used for $w_2' \geq 7$ μm to date.

Josephson junctions are made by removing the remaining SiO_2 films on the Si membrane and depositing superconductive films. A 600 Å-thick membrane with an area of 50 μm^2 was uniformly doped to 1.2×10^{20} cm^{-3} and, when coated with Pb electrodes, had a critical supercurrent density of 1.1×10^4 A/cm^2, which is comparable with the best oxide barrier junctions. The resistance was 0.55 Ω so that the I_cR product exceeds that theoretically obtainable in oxide-barrier junctions. This may be explainable partially in terms of the one-dimensional theory developed for superconductor-normal metal-superconductor (SNS) coplanar structures[3] to be discussed below, though Schottky barriers form at the silicon-superconductor interfaces and this prevents exact applicability SNS theory.

One important reason for interest in the silicon-membrane junction is that

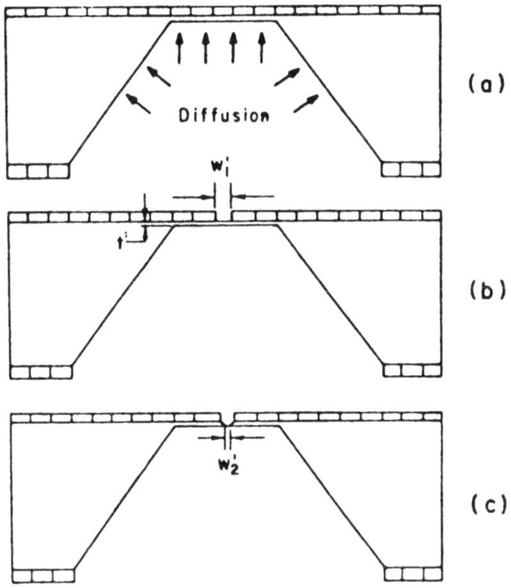

Fig. 7. A double-etching procedure for formation of barrier membranes of submicrometer dimensions. After outdiffusing the dopant from the 0.2-0.4 μm-thick membrane a new shallow diffusion is made and the etching procedure is as in Fig. 6.

the barrier is a single-crystal material supported in a frame which facilitates handling; it is therefore possible the use highly developed silicon processing techniques including diffusion, oxidation, and ion implantation to optimize the properties of the barrier. The Fermi level is normally pinned at a point one-third of the silicon energy gap above the valence band; this pinning leads to two different kinds of energy-band profiles. We are assuming here p-type doping but the same ideas can be applied to n-type silicon. For sufficiently thick silicon membranes, Schottky barriers form at the silicon-superconductor interfaces and are connected by a degenerate region of substantially constant Fermi degeneracy (E_V-E_F) as seen in Fig. 8a. For sufficiently thin membranes the Schottky barriers blend into one another and the entire silicon region is devoid of free carriers. The energy-band profile for this case is shown in Fig. 8b where it is clear that the coupling mechanism is tunneling. In addition to

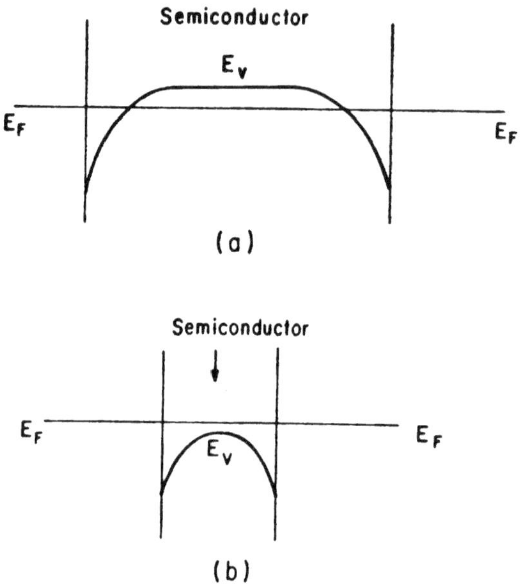

Fig. 8. Semiconductor barrier separating two superconductors. (a) Semiconductor thicker than Schottky-barrier thickness. (b) Semiconductor thickness comparable with Schottky-barrier thickness.

changing the thickness of the barrier, the doping profile can be used to control the energy-band profile. Optimization of the junction parameters by control of the doping and thickness has not yet been carried out.

Another potential advantage of the silicon membrane is that it should be possible to use high-temperature superconductors and therefore have higher I_cR products. The shape of the etch pits may also facilitate heat removal from the junction. However, the shape makes it hard to use for systems of many junctions such as memory arrays. Three possible electrode configurations are shown in Fig. 9. Figure 9a shows overlapping superconductive fingers on a singly etched membrane, Fig. 9b shows an electrode structure on a doubly-etched membrane which may have heat-removal advantages and Fig. 9c depicts a strip line connection to the junction.

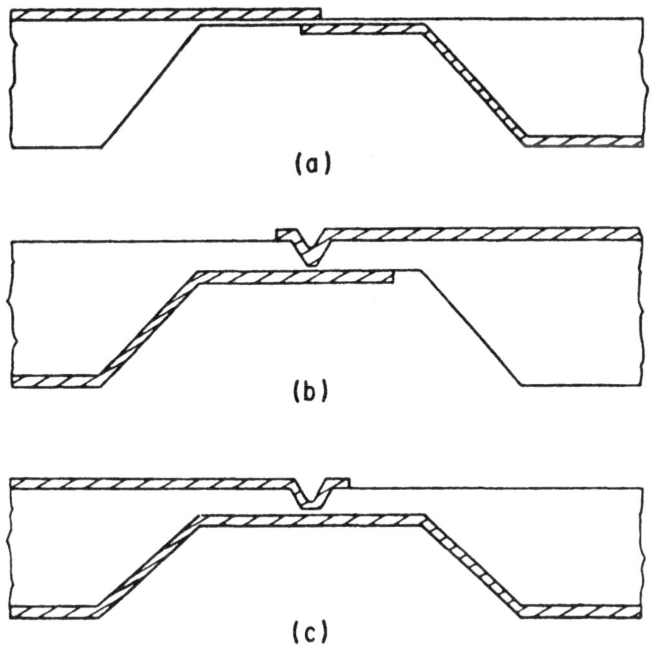

Fig. 9. Two possible ways of fabricating low-capacitance electrodes on silicon membrane Josephson Junctions. (a) Overlapping strip electrodes on a singly etched membrane. (b) Overlapping electrodes on a doubly etched membrane. (c) Stripline electrodes.

JUNCTIONS WITH COPLANAR ELECTRODES

Several different kinds of junctions having both electrodes in the same plane have been realized. These are shown in Fig. 10. The fabrication involves deposition of a strip of a superconductive material and a subsequent weakening of a short section L of the strip. This may involve changing the properties of the superconductor in the weakened gap, thinning of the material in the gap, narrowing but not thinning in the gap, or both narrowing and thinning in the gap. Also, the structure may be composite, with the connection between the main superconductive parts of the strip

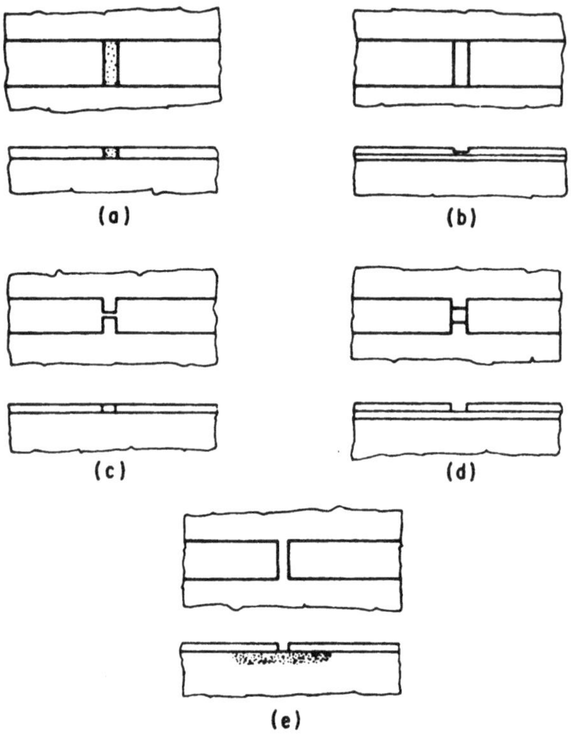

Fig. 10. Various possible configurations for junctions with coplanar electrodes, (a) Gap filled with material of same thickness and width as the superconductive strip. (b) Reduced thickness in gap region in (possibly composite) superconductive strip. (c) Gap region width constriction but uniform thickness. (d) Constriction of both width and thickness. (e) Coupling through a semiconductor substrate.

being a weakened superconductor, normal metal, semimetal, or semiconductor. The various possibilities are shown in Fig. 10. In the structure of Fig. 10c, the weakening of the coupling material is caused by an increase of the current density. In all the other structures, there are some geometrical and/or material factors that weaken the coupling (though there may also be a weakening due to an increase of current density).

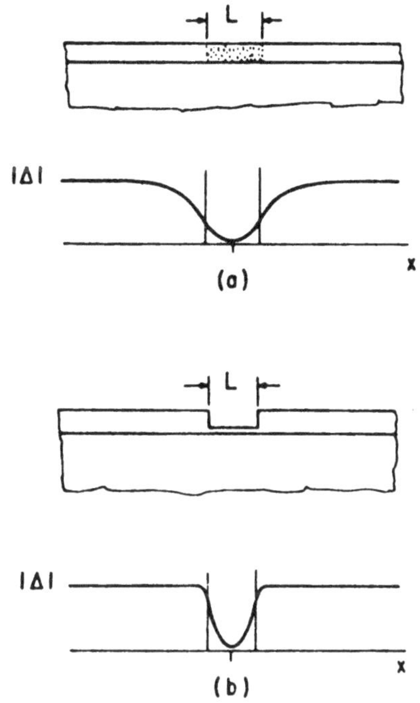

Fig. 11. (a) Coplanar structure with no thickness variation and the attendant depression of the order parameter in the region outside the gap region. (b) Coplanar structure with reduced thickness in the gap region showing less depression of the order parameter in the regions contiguous with the gap.

Recent theoretical work on both normal-metal connections[3] and superconductive connections[38] shows that I_cR products greater than those for tunnel junctions are obtained for short gaps. It is advantageous in obtaining high values of I_cR to avoid depression of the order parameter in the "banks" (the edges of the superconductive strip adjacent to the gap). This can best be achieved by having a minimum amount of lower-T_c or normal metal material in the gap. Thus, narrowing and thinning the material in the gap will lead to higher values of I_cR. Likewise, the use of semimetals or semiconductors for coupling will have less effect on the order parameter in the banks. Figure 11a shows the strong effect of the material in the gap on the order parameter in the banks for the type of junction shown in

Fig. 10a where a large quantity of normal metal or weak superconductor fills the gap. The situation in Fig. 11b is more typical of other types of connections where the material has been thinned and/or narrowed or replaced by a low-carrier-density material.

Junctions Coupled by Normal Metal or Weak Superconductor

The fabrication of the structure in Fig. 10a can be done in various ways. The weak connection can be provided by superposing a cross strip of a normal metal. For example, a gold cross strip about one micron wide is formed on the substrate using photolithographic methods. Then the superconductive strip is deposited across it. Where they cross, the proximity effect causes a reduction of the order parameter in the superconductive strip. Another possibility is to use ion implantation[39] to raise or lower the transition temperature. As an example, the transition temperature of Mo can be raised by implantation of sulphur or nitrogen. Thus, one could make a strip of Mo, produce crossed line of photoresist about one micron wide, and subject the strip to ion implantation which would raise the transition temperature outside the gap region protected by the photoresist. The result would be either a superconductor-weak superconductor-superconductor junction or an SNS junction, depending on the operating temperature.

The structure shown in Fig. 10b depends on a reduction of thickness to weaken the connection. It could be all of one type of superconductor with the gap region so thin as to get a low T_c there. Alternatively, the bottom layer of the two layer structure could be a low-T_c superconductor or a normal metal, and the top layer, a superconductor.[40] The transition temperature of superposed materials will depend on the materials and their relative thicknesses. The transition temperature of the gap region is reduced by removing part of the top material. To do this one coats the entire film with a photoresist and exposes and develops away a narrow strip perpendicular to the superconductive strip. The material is removed either by anodizing or by ion etching. A partial removal of the superconductor will reduce the T_c in the gap but it could still be operated with the gap material superconducting. One could also operate with the gap material in the normal state.

Any of the composite structures discussed above can also be made with a narrower connection as in Fig. 10d, with an attendant effective shortening of the connection. This should increase the I_cR product.[41]

Semimetal-Coupled Junctions

Related to the structure in Fig. 10a is the semimetal overlay.[42] The reported junctions of this type employed a film of Pb in which a narrow (0.3 μm) gap was cut. Subsequently, a 3000 Å-thick layer of Bi is deposited to make the connection. As in all uses of superposed films, it is necessary to assure the first metal presents a clean, bare surface when the second material is deposited. This can be assured by sequential deposition without opening the vacuum system or by cleaning the first with a glow discharge in an inert gas. Actually, the reported junctions used a narrow strip for the connection so it is more like Fig. 10c but the width of the strip is only selected to control the resistance. The I_cR product was found to be 0.6 of the theoretical value for a tunnel junction. The Pb-Bi-Pb structure suffers from interdiffusion of the Pb and Bi so its characteristics are not stable. A NbN-Bi-NbN junction of the same form was also reported. It offers the potential of greater stability but has apparently not been tested extensively.

Semiconductor-Coupled Junctions

A semiconductor-coupled coplanar junction has recently been reported.[43] In this structure (Fig. 10e), the surface of a silicon wafer was doped to saturation and the resulting boron-glass and Si-B phase layers were removed. An oxide etch procedure which leaves only 6-7 Å of oxide was used prior to deposition of a lead film. The film was patterned into a strip. The substrate was covered with an electron resist and a narrow line was exposed and developed across the lead strip. A narrow (0.1-0.3 μm) gap was ion etched in the lead in an RF sputtering system. Hysteretic and nonhysteretic I-V characteristics and I_cR products in excess of the theoretical value for tunnel junctions have been seen but the data are preliminary so far.

Almost all of the coplanar structures can be expected to have small capacitances. Those that can also be made with resistances in the range of tens of ohms and high I_cR products should be very attractive for applications. This could include the semiconductor and semimetal types as well as thin, narrow, short normal-metal bridges.

Microbridges

The structure in Fig. 10c depends for the weakness of its connection on its smallness which must be in the submicrometer range. These microbridges (or "Dayem bridges")[44] have been made in a number of ways. The first was by careful manipulation of metal evaporation masks. Later, electron lithography[12] and scratching techniques[45] were developed. It has become clear that microbridges can have thermally induced transitions in their I-V characteristics which makes them unsuitable for applications except parametric amplifiers where operation is at zero dc voltage. The problem can be solved by making the banks much thicker than the bridge to improve its heat-removal capability.[46]

ACKNOWLEDGEMENTS

The author would like to express his appreciation for helpful conversations with H. W. Chan, M. J. Feldman, M. B. Ketchen, K. K. Likharev, and C. B. Tesche and to D. R. Simpson and B. Kerekes for the preparation of the manuscript. Research sponsored by the U. S. Army Research Office Grant DAAG29-76-G-0191.

REFERENCES

1. Richards, P.L., Auracher, F., and Van Duzer, T.: Millimeter and submillimeter wave detection and mixing with superconducting weak links. Proc. IEEE 61, 36-45 (1973).
2. McCumber, D.E.: Effect of ac impedance on dc voltage current characteristics of superconducting weak-link junctions. J. Appl. Phys. 39, 3113-3118 (1968).
3. Likharev, K.K.: $J_s(\phi)$ relationship in SNS variable thickness bridges: Lett. J. Tech. Phys. (in Russian) 2, 29-34 (1976).
4. Taur, Y., Claassen, J.H., and Richards, P.L.: Josephson junctions as heterodyne detectors. IEEE Trans. Microwave Theory and Techniques MTT-22, 1005-1009 (1974).
5. Kanter, H. and Vernon, Jr., F.L.: High frequency response of Josephson point contacts. J Appl. Phys. 43, 3174-3183 (1972).
6. Feldman, M.J., Parrish, P.T., and Chiao, R.Y.: Parametric amplification by unbiased Josephson junctions. J. Appl Phys. 46, 4031-4042 (1975).
7. Clarke, J.: Low frequency applications of superconducting quantum interference devices. This volume.
8. Lum, W.Y., Chan, H.W., and Van Duzer, T.: Memory and logic circuits using semiconductor-barrier Josephson junctions. Proceedings of the 1976 Applied Syperconductivity Conference, Stanford, 17-20 August. To be published in IEEE Trans. Magn.
9. Kupriyanov, M.Yu., Likharev, K.K., and Semenov, V.K.: J. Low Temp. Physics (in Russian) 2, N10a (1976).
10. Zappe, H.H.: Josephson quantum interference computer devices. Proceedings of the 1976 Applied Superconductivity Conference, Stanford, 17-20 August. To be published in IEEE Trans. Magn.
11. Gregor, L.V.: Thin-film processes for microelectronic application. Proc. IEEE 59, 1390-1402 (1971).
12. Herzog, R.F., Greeneich, J.E., Everhart, T.E., and Van Duzer, T.: Computer-controlled resist exposure in the scanning electron microscope. IEEE Trans. Electron Devices ED-19, 635-641 (1972).
13. Finnegan, T.F.: Private communication.

14. Emmanuel, A., Donaldson, G.B., Band, W.T., and Dew-Hughes, E.: Barrier formation in lead-based tunnel junctions studied by surface techniques. IEEE Trans. Magn., MAG-11, 763-765 (1975).

15. Anacker, W., Grebe, K.R., Greiner, J.H., Lahiri, S.K., Park, K.C., and Zappe, H.H.: Lead alloy Josephson junctions. U.S. Patent No. 3,733, 526, May 15, 1973.

16. Eldridge, J.M. and Matisoo, J.: Fabrication of variable current density Josephson junctions. U.S. Patent No. 3,816,173, June 11, 1974.

17. Fu, C.Y. and Van Duzer, T.: Thermally cyclable lead-alloy Josephson junctions. Bulletin Amer. Phys. Soc. Series II, 21, 339 (1976).

18. Rutter, J.W. and Aust, K.T.: Kinetics of grain boundary migration in high-purity lead containing very small additions of silver and gold. Trans. Metalur. Soc. AIME 218, 682-688 (1960).

19. Ames, I.: Josephson tunneling circuits with superconducting contacts. U.S. Patent No. 3,852,795, Dec. 3, 1974.

20. Lahiri, S.K.: Metallurgical considerations with respect to electrodes and interconnection lines for Josephson tunneling circuits. J. Vac. Sci. Technol. 13, 148-151 (1976).

21. Neugebauer, C.A. and Ekvall, R.A.: Vapor-deposited superconductive films of Nb, Ta, and V. J. Appl. Phys. 35, 547-553 (1964).

22. Sosniak, J. and Hull, Jr., G.W." Superconductivity of niobium films deposited by dc diode sputtering. J. Appl. Phys. 38, 4390-4392 (1967).

23. Owen, S. and Nordman, J.E.: Application of integrated circuit technology to the fabrication of large numbers of niobium based Josephson junctions. IEEE Trans. Magn. MAG-11, 774-778 (1975).

24. Theuerer, H.C., and Hauser, J.J.: Getter sputtering for the preparation of thin films of superconducting elements and compounds. J. Appl. Phys. 35, 554-555 (1964).

25. Nordman, J.E.: Thin-film Josephson junctions using getter-sputtered niobium. J. Appl. Phys. 40, 2111-2115 (1969).

26. Hansma, P.K.: A new method for fabricating niobium oxide barrier Josephson junctions. J. Appl. Phys. 45, 1472-1473 (1974).

27. Hawkins, G. and Clarke, J.: Nb-Nb thin-film Josephson junctions. J. Appl. Phys. 47, 1616-1619 (1976).

28. Schwidtal, K. and Finnegan, R.D.: Durable Thin-film Josephson tunnel junctions. Proceedings of 1972 Applied Superconductivity Conference,

Annapolis, Md. May 1-3, 562-564, IEEE Pub. No. 72CHO682-5-TABSC.

29. Miles, J.L. and Smith, P.H.: The formation of metal oxide films using gaseous and solid electrolytes. J. Electrochem. Soc. 110, 1240-1245 (1963).

30. Schroen, W.: Physics of the preparation of Josephson barriers. J. Appl. Phys. 39, 2671-2678 (1968).

31. Greiner, J.H.: Oxidation of lead films by rf sputter etching in an oxygen plasma. J. Appl. Phys. 45, 32-37 (1974).

32. Hornbuckle, D.: Fabrication of metal-oxide-metal junctions for low-temperature detection and mixing. Research report for M.S., Electrical Engineering and Computer Sciences, University of California Berkeley, 1975.

33. Heiblum, M. and Wang, S-Y.: Private communication.

34. Cardinne, P., Nordman, J., and Renard, M.: On the use of low barrier potential materials to improve high frequency coupling to Josephson tunnel junctions. Rev. Phys. Appl 9, 167-171 (1974).

35. Seto, J. and Van Duzer, T.: Supercurrent tunneling junctions with tellurium barriers. Appl. Phys. Lett. 19, 488-491 (1971).

36. Huang, C.L. and Van Duzer, T.: Single-crystal silicon-barrier Josephson junctions. IEEE Trans. Magn. MAG-11, 766-769 (1975).

______________________________ : Josephson tunneling through locally thinned silicon wafers. Appl. Phys. Lett. 25, 753-756 (1974).

37. Huang, C.L. and Van Duzer, T.: Schottky diodes and other devices on thin silicon membranes. IEEE Trans. Electron Devices ED-23, 579-583 (1976).

38. Kulik, I.O. and Olemyanchouk, A.N.: Proceedings of the 14th Low Temperature Physics (LT-14), Helsinki, 1974, 116-119.

39. Harris, E.P.: Preparation of superconducting weak links in molybdenum films by ion implantation. IEEE Trans. on Magn. MAG-11, 785-788 (1975); Harris, E.P.: Properties of superconducting weak links produced by ion implantation. J. Vac. Sci. Technol. 12, 1383-1386 (1975).

40. Notarys, H.W. and Mercereau, J.E.: Proximity effect bridge and superconducting microcircuitry. J. Appl. Phys. 44, 1821-1830 (1973) and U.S. Patent 3,798,511, March 19, 1974.

41. Likharev, K.K.: Private communication.

42. Ohta, H., Feldman, M.J., Parrish, P.T., and Chiao, R.Y.: Semimetal barrier planar Josephson junction. Rev. Phys. Appl. 9, 187-190,(1974).

43. Schyfter, M., Maah-Sango, J, Raley, N., Ruby, R., Ulrich, B.T. and Van Duzer, T.: Silicon-barrier Josephson junctions in coplanar and sandwich configurations. Proceedings of 1976 Applied Superconductivity Conference, Stanford, 17-20 August. To be published in IEEE Trans. Magn.

44. Anderson, P.W. and Dayem, A.H.: Radio-frequency effects in superconducting thin-film bridges. Phys. Rev. Lett. 13, 195-197 (1964).

45. Gregers-Hansen, P.E., Hendricks, E., Levinsen, M.T., and Pickett, G.R.: Subharmonic energy-gap structure and a Josephson-radiation-enhanced gap in Dayem bridges. Phys. Rev. Lett. 31, 524-527 (1973).

46. Klapwijk, T.M. and Mooij, J.E.: Extension of the operating range of superconducting microbridges. IEEE Trans. on Magn. MAG-11, 858-859 (1975).

PREPARATION OF VARIABLE THICKNESS MICROBRIDGES USING ELECTRON BEAM LITHOGRAPHY AND ION ETCHING**

R.D. Sandell, G.J. Dolan* and J.E. Lukens
State University of New York at Stony Brook
Stony Brook, New York 11794, U.S.A.
*Bell Laboratories, Murray Hill, New Jersey, U.S.A.

I INTRODUCTION

We have developed techniques using electron beam lithography (EBL) and ion beam etching to fabricate indium Josephson microbridges. For the past several years we have been making the usual uniform thickness bridges using EBL.[1] These bridges are formed by constricting a single 100 nm thick film in one dimension. Typical dimensions of the bridges are 0.3μm square, however bridges as small as 0.2μm square have been made. More recently we have been making bridges of the variable thickness geometry in which the film is constricted in two dimensions, i.e. thickness as well as width. Variable thickness bridges (VTB) have been constructed by locally thinning a single thick indium film by ion beam etching and by a two step evaporation process in which a thin narrow strip which forms the bridge is overlayed with a thick film forming the banks of the bridge. The latter process is much more reliable in producing repeatable bridges with well defined geometry; however, it was found necessary to use an intermediate step in which the first film is cleaned with an ion beam prior to the deposition of the second film in order to insure good superconducting contact between the film.

II ELECTRON BEAM LITHOGRAPHY

We briefly review the techniques of electron beam lithography,[2] Fig. 1. Saphire substrates are spin-coated with a positive electron resist, poly methyl methacrylate (PMM). The resist is baked and coated with a

** Research supported by Office of Naval Research and National Science Foundation

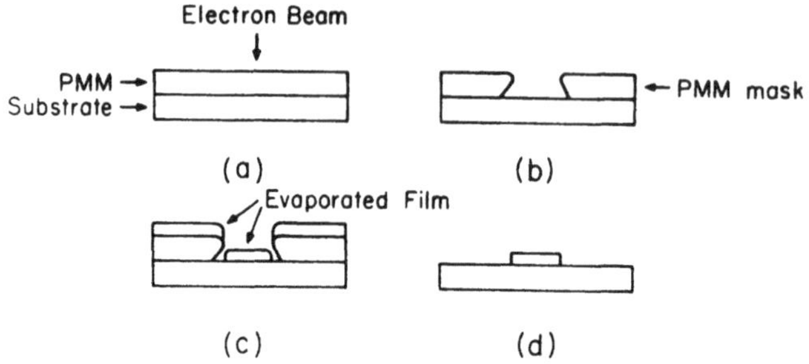

Fig. 1. EBL process. a) Substrate spin-coated with PMM and exposed to electron beam. b) Developed PMM. c) PMM mask with evaporated film. d) Final pattern after lift off.

thin layer (≈ 7.5nm) of Al to prevent charging during the exposure to the electron beam. The sample is then exposed by a 10nm diameter electron beam from a scanning electron microscope. The position, blanking and exposure time of the electron beam are computer controlled. The field is divided into a matrix of points and the pattern is written on the PMM by moving the beam to the desired points under computer control. Between points the beam is blanked and the time the beam is unblanked on each point is also controlled by the computer. Complex patterns are thus written by simple user written software. After exposure the Al is removed and the resist is developed in a dilute solvent. The effect of exposure to the electron beam is to break the molecular bonds of the PMM. Thus after development the PMM is removed in the areas which were exposed by the electron beam. Because of the scattering of the electrons from the substrate and the PMM during exposure, the mask also exhibits a significant amount of undercutting. This scattering also causes the exposure rate of an area to be dependent on the exposure of adjacent areas. It is often necessary to increase the exposure time of small regions such as the bridge relative to other regions. Next a thin film is evaporated through this mask. Because of the undercutting there is a break between the film on the substrate and on the mask so that when the unwanted film is now removed by dissolving the mask in acetone, the film lifts off cleanly without tearing.

While the undercutting is desirable it does limit the resolution of the technique when thick films and thus thick resist layers are used since the undercutting becomes more severe as the resist is made thicker. The computer controlled electron beam technique however allows one to fabricate repeatable complex submicron sized patterns such as arrays of microbridges. Fig. 2 shows a sample used to study the voltage locking and coupling interactions of two microbridges.[3]

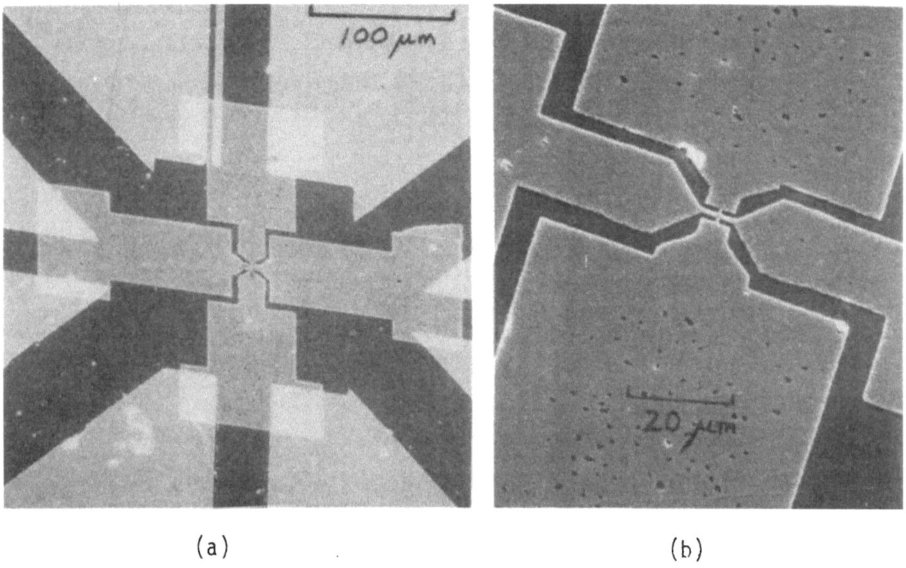

Fig. 2. Two microbridge series array made by EBL. a) Micrograph showing eight gold terminals to indium film. b) Micrograph of indium film containing two microbridges spaced ≈ 1μm apart.

III. ION BEAM ETCHING

In the ion beam etching process a indium thin film is first deposited on the substrate, then a mask usually made by EBL is placed on top of the film. The entire sample is then exposed to a 1 kV argon ion beam. The portion of the In film not protected by the mask is then selectively removed, Fig. 3. For some applications the PMM mask itself may be used

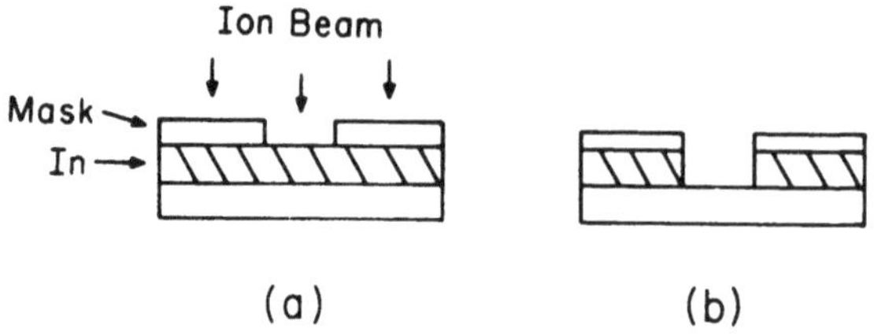

Fig. 3. Ion beam etching. a) Entire In film and mask exposed to ion beam. b) In film removed in region not protected by the mask.

as the mask for ion etching. PMM however etches rapidly and nonuniformly and after etching times of more than a few minutes polymerizes and is impossible to subsequently remove. For most applications we have found that Al film masks made by the EBL lift-off process to be very effective Then Al etches very uniformly and when etched in a oxygen environment (usually just the residual oxygen in the vacuum system) the etching rate approaches that of Al oxide which is very slow <10nm/min, about 7-8 times slower than indium. Because of the slow etch rate of Al relative to In, thin high resolution PMM masks may be used to construct the thin ≈ 30nm Al masks used in etching 200nm In films.

Another interesting masking method we have found is the use of the organic contamination produced by the electron beam of the scanning electron microscope. After an exposure of the sample to the electron beam of $\approx 10^{-7}$ coul/(μ m)^2, a contamination layer builds up which is nearly impervious to the ion beam. The resolution of this mask is only limited by the resolution of the electron microscope.

IV. PREPARATION OF MICROBRIDGES

One of our first attempts to make variable thickness bridges was to thin a single thick In film (≈ 200nm) in a localized region by partially etching with the ion beam through an Al mask. Unfortunately In does not etch uniformly as shown in Fig. 4. Upon further etching several crystalite bridges are formed between the banks as shown in Fig. 5. It is possible to produce a single microbridge by additional masking with either Al or contamination and save one bridge while removing the others with further etching. While this is not a reliable method of producing repeatable microbridges, bridges made by this method do exhibit the Josephson effects. From Fig. 5 however, it is clear that when the etching process is allowed to continue until all the In has been removed, the Al masking techniques is effective in fabricating structures with nearly vertical edges with an edge definition less than 0.1μm. We are pursuing techniques using Al and contamination masking to make uniform thickness bridges of dimensions smaller than is presently possible with PMM masks.

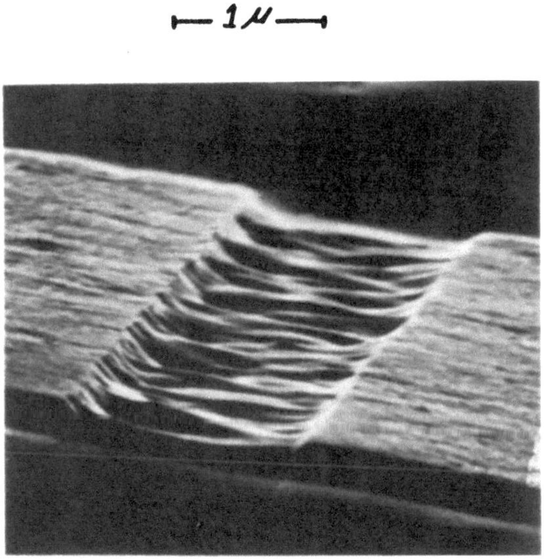

Fig. 4. Micrograph showing nonuniform etching of In film through Al mask.

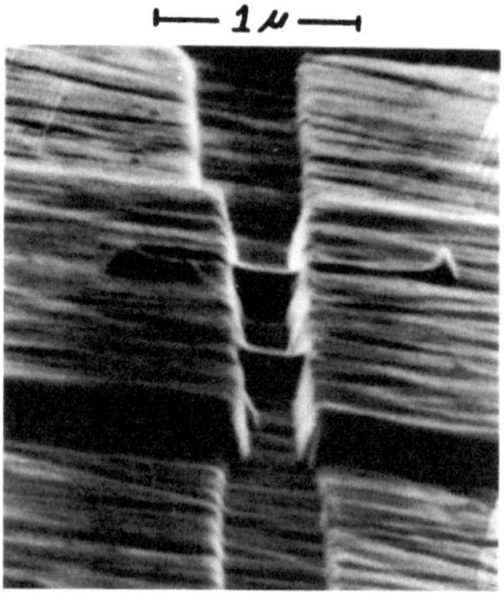

Fig. 5. Micrograph of ion etched In film and Al mask in which all In is removed in the gap except two crystallite bridges.

A more successful method of making VTB which yields bridges with well defined geometry and allows the systematic variation of the length, width and to some degree the thickness of the bridges is a two step process. First a thin 50nm, narrow (< 0.5μm) In strip which will form the bridge is deposited by EBL. Next a second PMM mask with a gap less than 0.5μm wide is constructed across this strip. A thick In film (200-300 nm) deposited through this mask then forms the banks and electrodes for the microbridge. Fig. 6. However before the second film is deposited it is essential that the first film be briefly etched by the ion beam to remove any oxide and contaminates and insure good superconducting contact between the films. In fact we found it necessary to perform the evaporation of the second film immediately after etching without exposing the sample to atmosphere. Earlier attempts in making bridges by this two step process in which the sample was briefly exposed to atmosphere in the process of transferring it from the etching to the evaporation chamber yielded bridges which exhibited the Josephson effect but had normal state resistances of several ohms and properties more characteristic of tunnel junctions. We therefore modified the equipment such that after deposition of the first film and construction of the second PMM mask the sample is placed in a vacuum chamber and pumped to less than 10^{-6} Torr. The film is then etched by a 1.3 kV,0.75 ma/cm^2 argon beam for 15 sec. at 10^{-4} Torr. The system is then pumped to 2 x 10^{-7} Torr and the sample rotated onto a liquid nitrogen cooled stage and the thick In film is evaporated through the mask. Bridges produced by this method have characteristics nearly identical to those produced by thinning a single film described above, indicated that good contact is made between the two films.

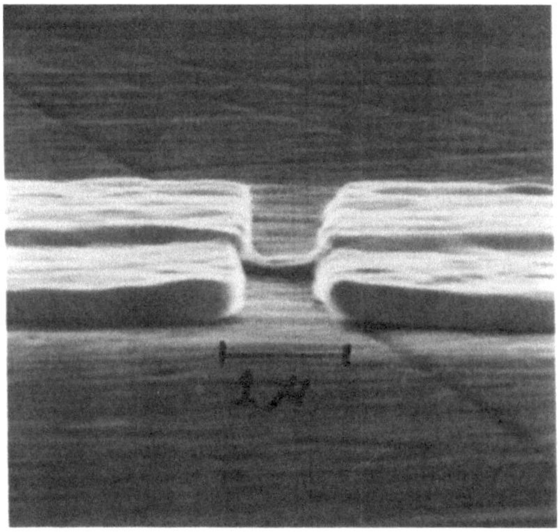

Fig. 6. Variable thickness bridge with film thickness of 250 and 50 nm made by the two step process.

The microbridges ≈ 0.5μm square and 50 nm thick have normal state resistance of 0.1 - 0.3 ohm. Typical current voltage characteristics are shown in Fig. 7a. At lower temperatures the characteristics deviate from the RSJ model as evidenced by the region of negative curvature. At lower temperatures the I-V curves become hysteric when the critical current reaches 1.5 - 2.0 ma, however the hysterisis occurs first not at zero voltages but in the region of negative curvature. The response of a bridge to 19.3 GHz radiation is shown in Fig. 7b. Microwave induced steps have been observed at 1mV and are observed in general up to and just beyond the gap voltage 2 Δ/e.

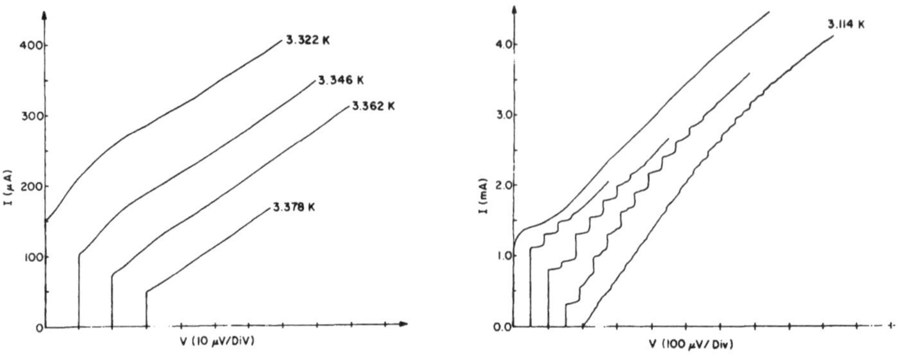

(a) (b)

Fig. 7. a) Current-voltage characteristics of a VTB.
b) Response of VTB to 19.3 GHz radiation.

V. CONCLUSIONS

The versatile electron beam lithography and ion beam etching techniques allow the reliable fabrication of uniform and variable thickness Josephson microbridges. An advantage of these techniques is the ability to systematically vary the dimensions of the bridge. We are currently investigating the characteristics of both types of microbridges such as intrinsic noise, current-phase relations, and coupling interactions of a two junction series array.

REFERENCES

1. D.W. Jillie, J. Lukens and Y.H. Kao, IEEE Trans. Mag. MAG 11 (1975) 671.

2. I. Haller, M. Hatzakis and R. Srinivasan, IBM J. Res. and Dev. 12 (1968) 251.

3. D.W. Jillie, J.E. Lukens, Y.H. Kao and G.J. Dolan, Phys. Letters 55A (1976) 381.

REAL JUNCTION PROPERTIES

PROPERTIES OF REAL JUNCTIONS

J. E. Mercereau
California Institute of Technology, Low Temperature Physics Mail Code 63-37
Pasadena, California 91125, USA.

INTRODUCTION

Josephson's prediction[1] of phase-dependent super and quasiparticle current between weakly coupled superconductors has now been followed by a decade and a half of intense analytic and experimental examination. All of the original predictions have been verified. In fact, the fundamentals have been sufficiently well established that the Josephson voltage-frequency relation is presently utilized in maintaining Voltage Standards in many countries and a new class of quantum interference electronics is being developed from the current-phase relation. The Josephson effect is pervasive: it always occurs between weakly connected superconductors. This generality is a consequence of the very basic nature of the phenomenon. However, since weakly coupling comes about in many forms, manifestation of the Josephson effect may occur differently in different circumstances. This paper will discuss the influence on the Josephson phenomena of some complicating and competing processes which occur in real junctions.

Controlled weak coupling has been achieved in many ways and by several techniques. The original, proposed by Josephson, was via an insulating tunneling barrier separating two superconductors to form a tunnel junction. Other barrier materials have been used including some semiconductors[2] and normal metals.[3] Other mechanisms for achieving weak coupling have also evolved, usually relying on a geometrically small contact to achieve the weakness required. This "small" type of junction generally takes the form of a point contact[4] or the Dayem bridge.[5] Another technique which can be applied to a geometrically homogeneous material is to locally weaken only the superconducting state. In this approach, weakening is usually via the proximity effect, anodization,[6] or ion implant.[7]

Basically all of these techniques produce the weak coupling required for the Josephson effect. However, the junctions which result from these various techniques each contain some unique modification of quantum interference, espeically at finite voltage, which is influenced by the particular mechanism or technique of coupling, the electronic and thermal characteristics of the material, and the junction geometry. It has been found that these modifications can be described in terms of two general classes of phenomena: one involves the influence of parallel currents on an ideal Josephson junction and the second involves some nonequilibrium aspect of superconductivity. It has been found that the real junction can be modeled as an ideal Josephson element in parallel (or series) with various resistive, capacitive, or inductive elements to form an equivalent circuit.[8] This sort of modeling has been quite effective in describing some junctions.[9] However, it seems probable that fundamental nonequilibrium processes will have to be included in a complete description of most "real" junctions at finite voltage. Nonequilibrium superconductivity is now relatively well understood,[10] but the nonequilibrium junction is not yet so mature and is an area of rapid development.[11]

Rather than attempt a review of the wide variety of effects which occur in the many different types of junctions this paper will focus on one type of junction and will attempt to demonstrate the influence of various parameters, both material and geometric, on its quantum behavior. Experimental data are presented on various aspects of one type of real junction but no detailed attempt is made here to correlate these results with a theoretical model of a real junction. However, some correlations are drawn between the quantum characteristics of the junction and the equilibrium and nonequilibrium characteristics of the superconducting materials.

MATERIALS CHARACTERISTICS

Sample Preparation

The type of junction to be discussed herein achieves weak coupling via a localized weakening of the superconducting state and has been termed a "proximity bridge." Measurements of the superconducting characteristics of the weakened state itself were made in order to correlate them with the quantum interference aspects of the junctions. Figure 1 illustrates this type of junction. A small region labeled S is created bisecting a strip

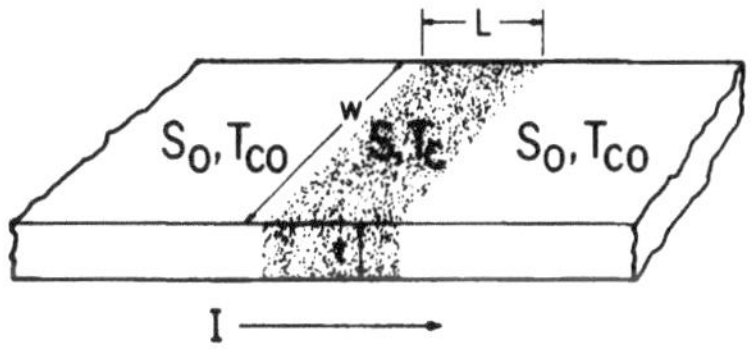

Fig. 1. Generalized structure and geometry of the proximity bridge. The main superconducting film, S_o, has transition temperature T_{co}. The shaded region has been modified to produce a material, S, with transition temperature T_c, and having width, w, thickness, t, and length, L, as shown. Note that L is the dimension along the current (I) direction.

labeled S_o by a processing technique to be described later. The material of region S has different superconducting characteristics than the material in region S_o. Basically, the transition temperature, T_c, of the material in region S is caused to be lower than the transition temperature, T_{co}, of the material in region S. It should be especially noted that the "transition temperature" of these junctions--the temperature at which a measurable supercurrent can first be detected--is neither of the above temperatures and depends in general on the geometry of the junction. The length, L, of the junction is defined along the current, I, direction while the width, w, and thickness, t, are as shown. For the junctions described here, the S material is made by modifying the S_o material in the narrow region by ion implantation, anodization of a layered film, or by sequential layering of films.[6]

Films for all the hard superconductors were deposited by electron beam evaporation onto sapphire substrates in ultra high vacuum, while films for the soft superconductors were deposited from boats onto sapphire or glass substrates in high vacuum. The ion implants were formed in Ta films with thickness, t = 46 nm, implanted with Cu or Fe ions at 60 KeV and dosages ranging from 10^{15} - 3×10^{16} ion/cm^2. The irradiated films were not annealed above room temperature, however, the junction characteristics were found to be quite stable with time. Doubled-layered S_o films for anodized junctions reported here were of 26 nm Ta on 10 nm Ti, 25 nm of Ta on 10 nm Hf, 28 nm Ta on 10 nm of W and 10 nm of Nb on 20 nm of Ta. Sequentially layered structures were usually made by overlaying a 25 nm

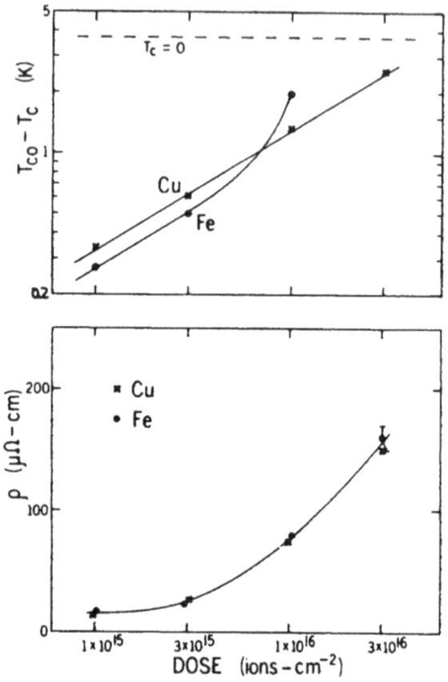

Fig. 2. Transition temperature depression $\Delta T_c = T_{co} - T_c$ and resistivity as a function of ion implant dose for implant materials.

thick Au film with 100 nm Sn. Regions of the films to be weakened by ion implant or anodization were defined by photolithography.[12] These weakened regions ranged in length, L, from 0.5 to 3 μm and had widths, w, typically 10 to 20 μm. Sequentially layered film bridge regions of the same dimensions were defined mechanically.

Superconducting Film Properties

A determination of the pertinent electronic and superconducting properties of the films were made to provide a general picture of the material properties and how they are effected by the weakening process. For example, Fig. 2 presents the dependence of the change in superconducting transition temperature $\Delta T_c = T_{co}-T_c$ and resistivity ρ, on ion dosage for Ta, weakened by ion implant. For ion implanted films the depression in transition temperature, ΔT_c, is seen to have approximately a square-root dependence on dose; and to be approximately the same at lower concentration for both Cu and Fe ions. The concentration-dependence and magnitude of ΔT_c are roughly the same as found by previous workers[13] and are consistent with the depression of transition temperature being primarily a result of

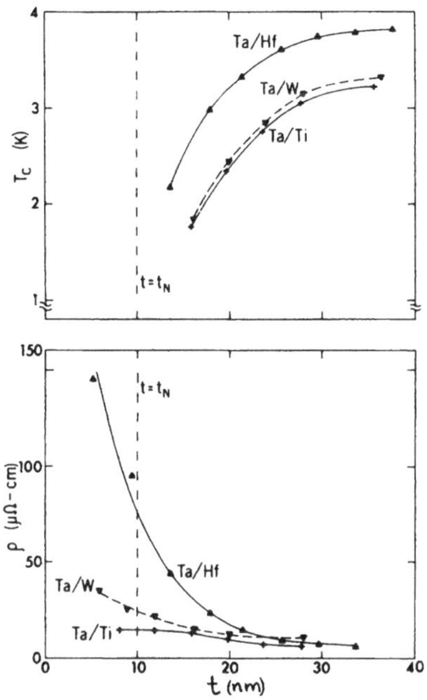

Fig. 3. Transition temperature and resistivity as a function of thickness of anodized film materials. t_N is the thickness of the bottom layer (Ti, W, Hf).

induced damage. For larger concentrations, Fe ions are seen to be more effective in depressing the transition temperature, even though there is no correspondingly anomaly in the resistivity, possibly indicating a magnetic depairing effect. There is also a large increase in resistivity with ion concentration, independent of implant ion element.

Figure 3 indicates the results of similar measurements on layered Ta films. For these films T_c decreases as the thickness of the top films (Ta) is decreased by anodization. Various bottom materials (Hf, W, Ti) of the same thickness (10 nm) were used with the same top material (Ta). As is shown in the figure, the effectiveness of the bottom material in depressing the transition temperature of the entire film strongly depends on its resistivity, as well as the ratio of the top to bottom film thicknesses. The lower the resistivity of the bottom film, the more effective it is in reducing the transition temperature of the entire film. A comparison of the resistivities of the ion implant and anodized films shows that the resis-

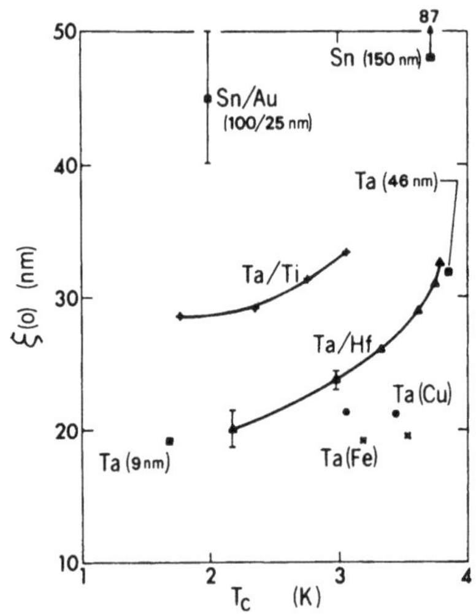

Fig. 4. Ginzburg-Landau zero-temperature coherence length, ξ(o), as a function of transition temperature, T_c.

tivity of the anodized films is appreciably lower, probably reflecting a shortened mean free path due to radiation damage. Thus, weakening by irradiation techniques is accompanied by a relatively larger increase in resistivity.

The coherence length and penetration depth for several of these film structures are shown in Figs. 4 and 5. The zero-temperature Ginzburg-Landau (G-L) coherence length, ξ (o), and zero-temperature G-L penetration depth, λ(o), are shown plotted versus T_c of the weakened material. Also shown are a few data points for simple, and as deposited films.

The G-L coherence length was determined by applying a magnetic field perpendicular to a strip of the film material and determining the field $H_\perp$ at which the resistance of the strip was one-half of its normal resistance.[14] Using $H_\perp(T) = \frac{\varphi_o}{2\pi\xi^2(T)} = \frac{\varphi_o}{2\pi\xi^2(o)} \frac{T_c - T}{T_c}$, near T_c, and plotting $H_\perp$ versus T produced a straight line whose slope yielded the value of the G-L coherence length. (The value of the slope did not change when $H_\perp$ was taken at 20% or 70% of the normal resistance.)

To determine penetration depth a flat coil was photo-etched from a film of

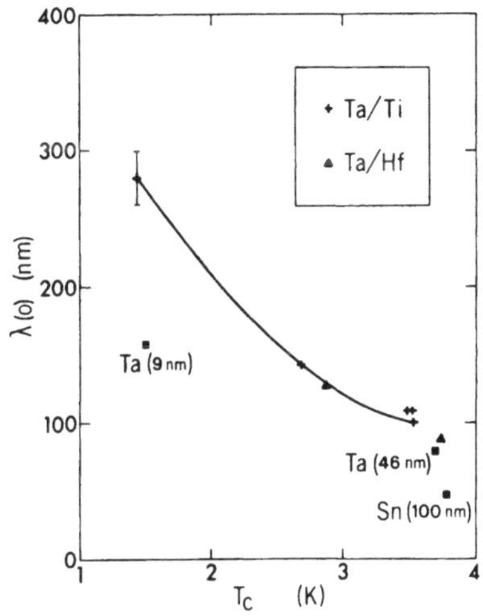

Fig. 5. Ginzburg-Landau zero-temperature penetration depth, λ(o), as a function of transition temperature, T_c.

the material. The inductance of the coil as a function of temperature, T, was then determined by measuring the resonant frequency of an L-C circuit formed by connecting a capacitor to the coil. The inductance thus measured was assumed to be the sum of a temperature-independent "classical" inductance and a kinetic inductance, $L_k(T) = \ell\mu_o \frac{\lambda^2(T)}{wt} = \ell \left(\frac{\mu_o}{wt}\right) \lambda^2(o) \left(\frac{T_c}{T_c - T}\right)$ near T_c, where μ_o is free space permeability and where w,t, and ℓ are the width, thickness, and length of the film strip.[15] The value used for t was the full thickness of the layered films. As in the case of the coherence length, plotting $1/L_K$ versus T produced a straight line whose slope gives λ(o).

Results for these superconducting parameters of the elemental film structures are consistent with previous measurements on these materials. And the variation of coherence length with T_c for the weakened materials is consistent with the variation of resistivity (and hence, mean free path, mfp) in these materials (Figs. 2 and 3). That is, the coherence length for the radiation damaged material is lowest while that for Ta/Ti lies higher than Ta/Hf. Both the coherence length and penetration depth reflect the change in resistivity (mfp) with T_c. These data will be used later to correlate with the quantum aspects of junctions.

Critical current density, j_c, measurements were also made on the weakened film material. In general, the temperature dependence was found to be as predicted from a mean-field theory, $j_c(T) = j_o\left(\frac{T_c - T}{T_c}\right)^{3/2}$. From the measured values of coherence length and penetration depth, such as Figs. 4 and 5, it is possible to evaluate j_o through the G-L relationship,[16] $j_c(T) = (c\varphi_o/12\sqrt{3}\pi^2\xi(o)\lambda^2(o))[(T_c-T)/T_c]^{3/2}$, where c is the velocity of light and φ_o is the flux quantum. It was found in general that this calculated value of j_o agreed with the measured value to within a factor of 2, with the main experimental uncertainty in this absolute correlation being the appropriate film thickness. Thus the indications are that a mean field theory seems to apply quite well even to these layered structures of different materials and that all measured quantities are reasonably self-consistent.

WEAK COUPLING

The junctions to be described herein were basically fabricated as illustrated in Fig. 1. Two superconducting regions, S_o, are separated by a region S whose intrinsic transition temperature, T_c, is less than that of region S_o. This lowering of the transition temperature has been accomplished by one of the weakening techniques described in the Materials Characteristics section. By these techniques two strong region S_o are coupled to one another via the superconductor S. The coupling region S is thus not a "barrier" to electrons in the sense of an insulator but is more of a "bridge" connecting the two strong regions. The material and geometric parameters of the bridge region will be shown to have profound influence on the Josephson effects of the junction. The terminology which will be used to describe the bridges is as follows: ion implant (for example, Cu ions in Ta) Ta(Cu), or double layered films (for example, Ta on Ti) Ta/Ti.

Since the bridge material is itself superconducting, the degree of coupling is a strong function of temperature. Junctions are usually made so that T_c is significantly less than T_{co}. In this case most of the temperature dependence of the Josephson effect comes through the variability of coupling with temperature rather than the change in the superconducting characteristics of S_o, as in the case of a tunneling junction. Conse-

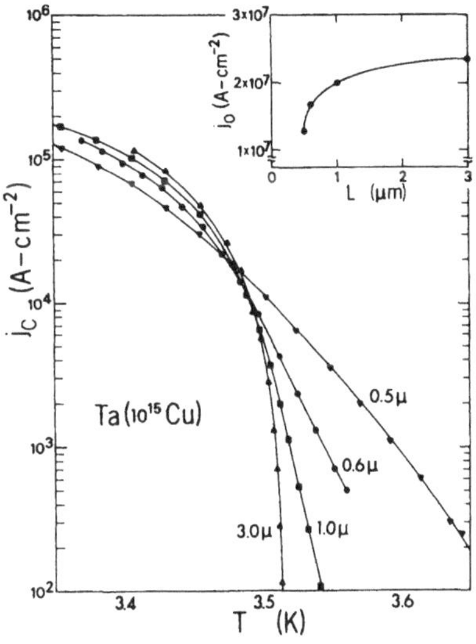

Fig. 6. Typical critical current density, j_c, versus temperature, T, behavior for proximity bridges of different lengths. Data shown are for four bridges of Ta implanted with 10^{15} ions/cm^2 Cu, having a bridge material T_c = 3.53K. Insert shows the $[(T_c-T)/T_c]^{3/2}$ coefficient dependence on bridge length.

quently, the degree of coupling achieved by the bridge region is a strong function of both its length and temperature.

For "long" bridges (L ≳ 3µm for most hard superconducting materials) the critical current is well described by a mean field theory as $j_c = j_o \left(\frac{T_c - T}{T_c}\right)^{3/2}$ at all temperatures as discussed in the Materials Characteristics section. For example, in Fig. 6, the temperature dependence of the critical current for the longest bridge (L = 3.0µm) is accurately fit by a mean field theory and the magnitude is consistent with separately measured values of $\xi(o)$ and $\lambda(o)$ to within experimental accuracy as discussed previously. As the bridge length is made shorter two effects begin to show up, as Fig. 6 illustrates. For temperatures much below T_c the temperature dependence of j_c remains $j_c = j_o\left(\frac{T_c - T}{T_c}\right)^{3/2}$ as expected from a mean field theory, for all lengths. However, the coefficient j_o becomes a function of L for short bridges. The insert in Fig. 6 indicates the typical length dependence; the coefficient j_o decreases rapidly with decreasing length below about 1 µm in this material.

As L is decreased, it is also found that supercurrent is developed at temperatures above T_c. For shorter bridges, in general, this extra supercurrent increases exponentially as L decreases. The extra current is phase sensitive and can roughly be described as due to interference between proximity induced states in S, arising from the two regions S_o.[17]

CRITICAL CURRENT

Figure 7 illustrates schematically the typical voltage-current (V-I) characteristic and differential resistance-current (dV/dI-I) characteristic for one of these junctions driven from a current source. There is a well-defined, temperature-dependent, critical current above which voltage appears. For currents much larger than critical the differential resistance is equivalent to that for the normal state of S at the same temperature. Also there is an "excess current"; i.e., the potential does not approach the "normal value" at high currents even though the differential resistance does. This feature can easily be masked by heating effects[18] which can occur (especially at high currents) if the film is not thermally

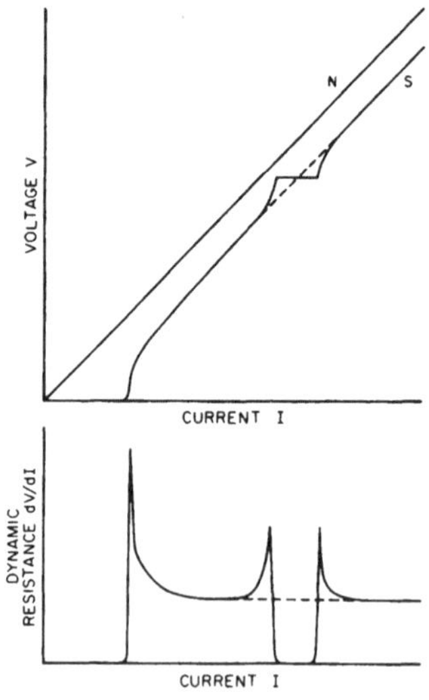

Fig. 7. Typical voltage-current (V-I) characteristic (upper graph) and differential resistance-current (dV/dI-I) characteristic (lower graph) for current driven junctions irradiated by rf. N and S refer to the normal and superconducting state of the junction.

anchored to a substrate material of high thermal conductivity. However, for the data presented here, we believe such heating effects are negligible and the bridge temperature is near the bath temperature.

Temperature Dependence

The general features of the bridge critical current versus temperature are given in Fig. 6. Critical current density versus temperature are shown for a typical series of four bridges whose lengths are 0.5 μm, 0.6μm, 1.0μm and 3.0μm of ion implanted Cu 1 x 10^{15} in Ta. For temperatures well below T_c all the bridges reach a critical current density versus temperature dependence $j_c(T) = j_o \left(\frac{T_c - T}{T_c}\right)^{3/2}$ typical of the material in the form of a long strip of transition temperature T_c. However, for the bridges j_o is a function of length which decreases as the bridge length decreases as shown in the insert of Fig. 6.

Except for long bridges (in this case 3.0μm) j_c is not zero above T_c for this current density measurement range. For intermediate lengths (in this case 1.0μm and 0.6μm) it obeys $j_c(T)$ $e^{-K(T-T_c)/T_c}$; for the shortest bridges (in this case 0.5μm), j_c varies more rapidly with temperature. The temperature dependence for the shortest bridges deviates in the direction of higher powers of the exponential temperature argument. Data for all S_o-S bridge combinations were consistent with the pattern of length dependence described above, except that the range of lengths would be different depending on the bridge materials and transition temperatures T_c and T_{co}. Generally the smaller T_c/T_{co} is, the shorter must be the bridges corresponding to each type of $j_c(T)$ behavior.

E. P. Harris[19] has reported that the $j_c(T)$ of his ion implanted bridges above T_c are well described by the analysis of Likharev and Yakobson (LY)[20]. We find a similar agreement. The LY theory produces all the functional dependences for a given material mentioned above for the bridges of this paper. Furthermore, the magnitude of the measured critical current density agrees with the theory within experimental error using the measured material and geometric parameters of the bridges and films in the cases where all the parameters are determined. In those cases there are no adjustable parameters.

Comparison of $j_c(T)$ versus bridge length for various materials permitted an investigation of the influence of material characteristics. The functional behavior of $j_c(T)$ is the same for all materials, but the parameter K, is a function of bridge length and resistivity. Preliminary results indicate that $j_c(T)$ scales as $e^{-L\rho}$ for all materials. Approximate correspondence between materials can be achieved by making $L\rho$ (the length of the bridge, L, times ρ the bridge material resistivity) the same for a particular T_c/T_{co} ratio.

MAGNETIC FIELD DEPENDENCE

The critical current of these junctions, in the high temperature limit ($T > T_c$), is diffraction modulated by magnetic field, similar to the magnetic modulation of a conventional Josephson tunnel junction. In Figs. 8 and 9, the dependence of critical current on magnetic field is shown for bridges of two different lengths at several different temperatures. The diffraction-like dependence of critical current with field is evident in all cases. This behavior is evidence that the current-phase relationship

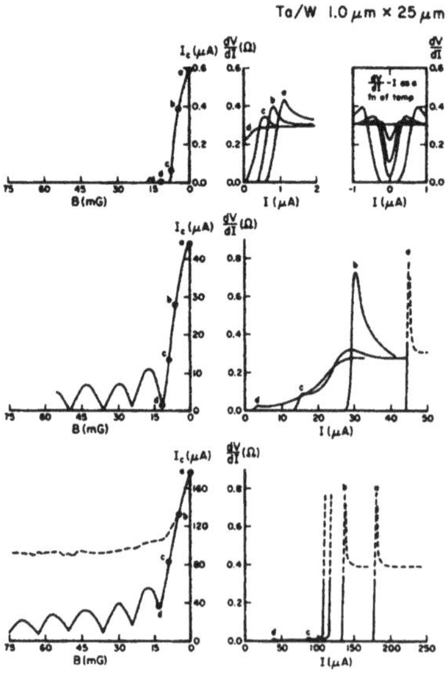

Fig. 8. Dependence of critical current on magnetic field (I_c-B) and differential resistance on current (dV/dI-I) for a 1μm long by 25μm wide Ta/W anodization bridge at various temperatures. Insert gives dV/dI-I as the bridge develops a critical current. For detailed discussion see text.

On the next page you will find an enlargement.

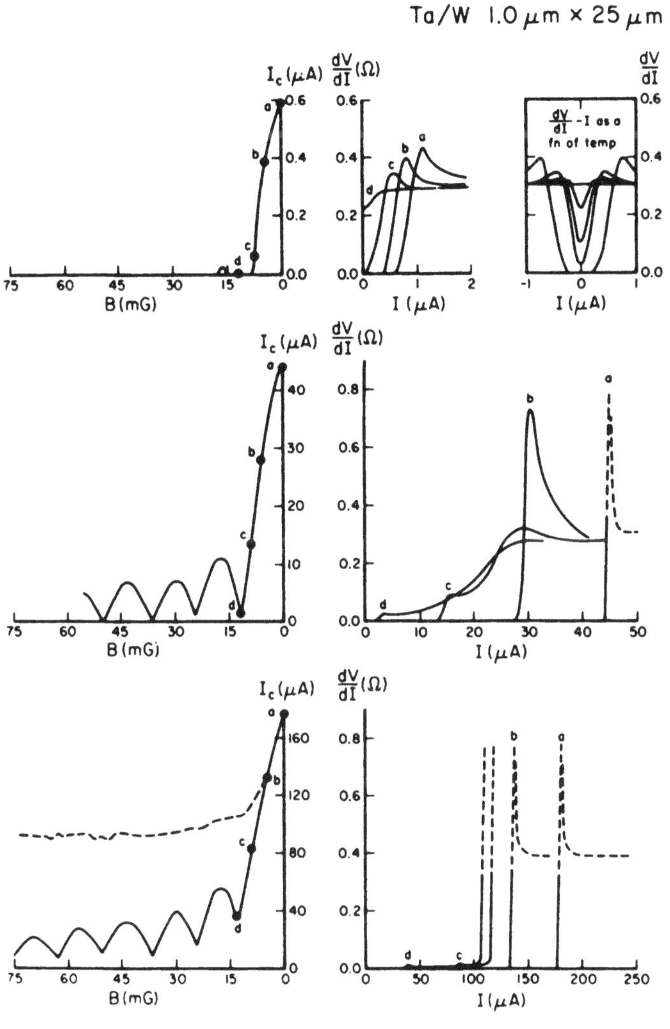

Enlargement of Fig. 8

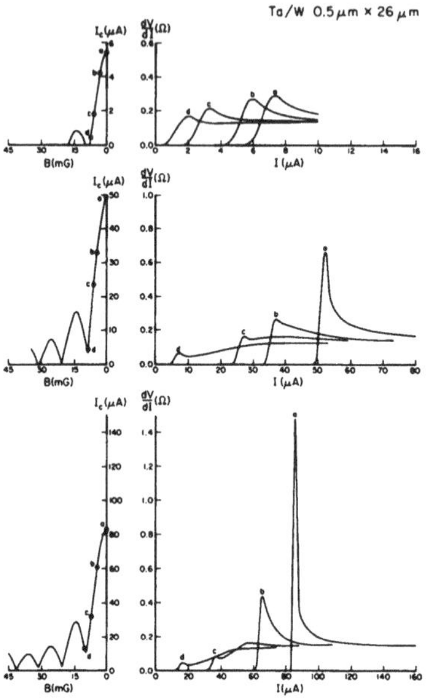

Fig. 9. Dependence of critical current on magnetic field (I_c-B) and differential resistance on current (dV/dI-I) for a 0.5μm long by 26μm wide Ta/W anodization bridge at various temperatures. For detailed discussion see text.
On the next page you will find an enlargement.

(at least at zero voltage) in these junctions is similar to the basic Josephson relation[1] $\cdot j = j_J \sin\varphi$. As the critical current is increased above the values indicated in Figs. 8 and 9 ($T < T_c$), the diffraction pattern becomes less well-defined in a manner similar to that for large tunnel junctions.[21]

For these magnetic field modulation effects the field is applied perpendicular to the plane of the film strip (see Fig. 1). Consequently, the film distorts the applied field and causes the period of the modulation to differ from what might be expected from the geometric area of the bridge. For the geometry shown in Fig. 1, these shielding effects for wide bridges (w/L>1) cause the effective aperture of the bridge to be approximately equal to the square of the width of the strip. Thus the field period, is $\sim\Delta B \sim \varphi_o/w^2$, where φ_o = h/2e is the flux quantum, and is nearly independent of bridge length.

As the field is increased to higher values than shown in Figs. 8 and 9, the

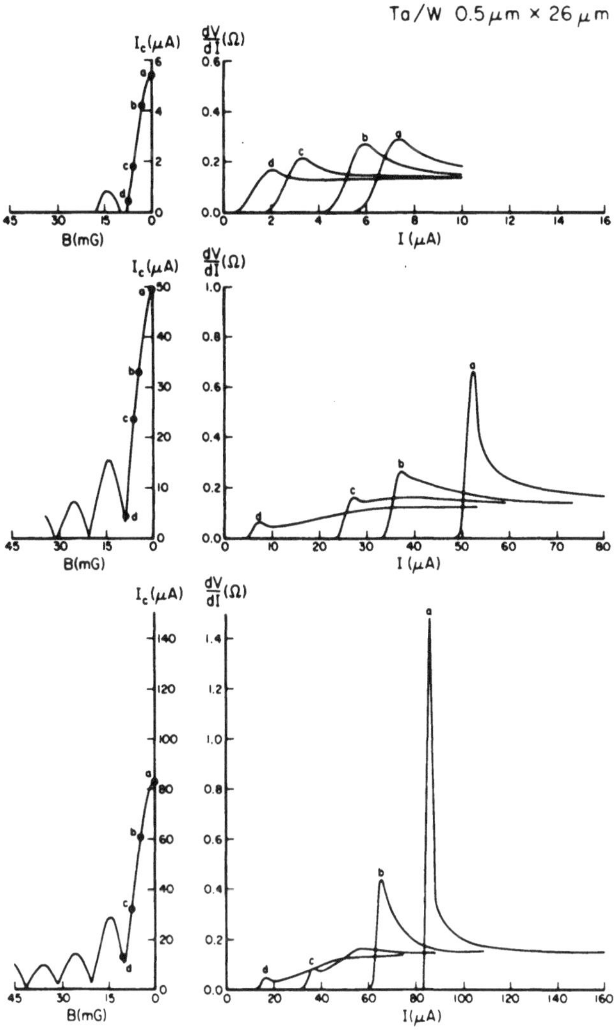

Enlargement of Fig. 9

modulation continues until the critical field H_{c1} of the film S_o is reached. If the magnetic field is increased beyond this value modulation of j_c continues but with a longer period as flux enters the film and lowers the shielding effects. In this way, values of H_{c1} can be determined which correspond quite well with the calculated values using the experimental data on $\lambda(o)$ (see Figs. 4 and 5) for S_o.

QUASIPARTICLE CURRENT

When the junctions are operated at finite voltage there will be quasiparticle currents in addition to the oscillating supercurrent. This section describes experiments to determine the quasiparticle resistance, r_{qp}, developed in these junctions. Essentially, what is done is to suppress the supercurrent in the bridge by a magnetic field and measure the dynamic resistance as a function of current. The insert at the upper righthand side of Fig. 8 indicates the differential resistance dV/dI of a Ta/W bridge as a function of current, at various temperatures through the transition temperature of the supercurrent. As the temperature is lowered, a supercurrent develops and the differential resistance changes from a constant value as a function of current, equal to the normal resistance R_n, (horizontal line in insert at 0.3 Ω) to a strongly current dependent resistance. The temperature interval indicated by this insert is ~ 0.05°K. At the lowest temperature indicated in this insert a critical current of about 0.3μA has been developed.

Figure 8 also shows the variation of differential resistance as a function of magnetic field. This is shown on the righthand side of Fig. 8 as dV/dI - I, with B as a parameter, at various temperatures. At the top of Fig. 8 dV/dI - I is shown at a temperature slightly below the lowest temperature indicated in the insert. Here the critical current is 0.6μA. Curves a,b,c, and d are taken at the field values similarly labeled on the I_c-B curve at the top left of Fig. 8. In this case, it is seen that when the critical current is completely suppressed the dynamic resistance returns to nearly its "normal" value.

However, as the temperature is lowered further (middle and bottom of Fig. 8) it is found that the dynamic resistance below I_c does not return to the normal value when the supercurrent is suppressed and in fact, decreases rapidly as the temperature is reduced. Contrariwise, above I_c the dynamic resistance always retains its "normal" value. It should be noted that all aspects of the supercurrent are modulated by magnetic field as is the critical current. That is, any rf induced "step structure" on the V-I or dV/dI-I curves also modulate with field similarly to the critical current. Thus both the ac and dc supercurrents can be suppressed by magnetic field, leaving only the phase-independent quasiparticle current I_{qp}.

Figure 9 shows similar results on a shorter bridge also indicating a decreasing quasiparticle resistance with temperature below I_c. At similar critical currents, the longer bridge has a higher normal resistance (above I_c) but a somewhat lower quasiparticle resistance (below I_c). By going to higher order minima in the I_c-B curve the critical current can be completely suppressed. Differential resistance in this limit is shown in Fig. 10 for the bridge and temperatures indicated in Fig. 9.

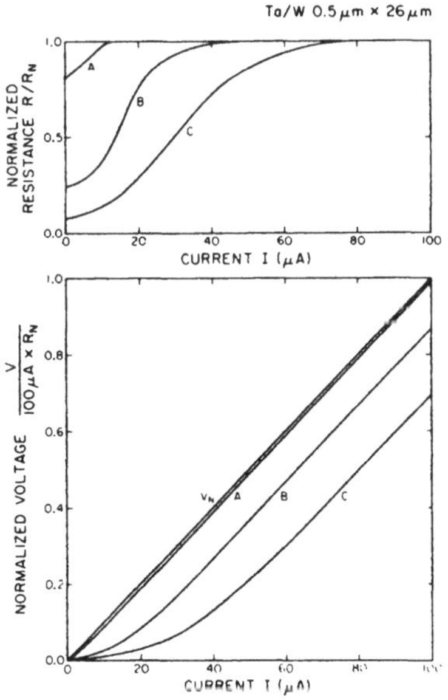

Fig. 10. Quasiparticle properties of 0.5μm long by 26μm wide Ta/W anodization bridge. Upper graph — Normalized quasiparticle resistance as a function of current. Lower graph — V-I curves resulting from the effective quasiparticle potential. V_N refers to the normal state. A,B,C correspond to the temperatures of the upper, middle, and lower data of Fig. 9.

We interpret this limiting differential resistance to be the current-dependent, quasiparticle resistance of these junctions; i.e., $(dV/dI)_{I_{su}=0} = r_{qp}$. With this assumption, it is possible to define a quasiparticle potential as $V = \int r_{qp}\, dI$. In the bottom half of Fig. 10, V is shown as a function of I, as derived from the r_{qp} data above. The nonlinearity of r_{qp} with current shows up here as a displacement (at high currents) of the V-I curve from its "normal" value $V = R_n I$. When the magnetic field is removed (and the supercurrent is thus restored) the simplest assumption is that the super and quasiparticle currents add linearly. If we also assume that I_{qp} is still related to V, as in Fig. 10, then the apparent excess current discussed previously is actually a manifestation of a nonlinear quasiparticle resistance. Our interpretation of these results in the high temperature ($T > T_c$) limit is that for $V = 0$ the supercurrent is of the Josephson form $j = j_J \sin\varphi$. At finite voltage both super and quasiparticle currents exist: the supercurrent retaining the Josephson form while the quasiparticle current is determined by a nonlinear resistance such as illustrated in Fig. 10.

As the temperature is further reduced, so that $T \ll T_c$, the magnetic modulation ceases except for a small effect around $B = 0$. The high current data of Fig. 8 indicates the onset of this behavior with the dashed curve. The dashed curve indicating $I_c(B)$ was measured at a resistance of 0.05Ω while the solid $I_c(B)$ was taken at zero resistance. As the temperature is reduced the dynamic resistance below I_c gradually approaches zero and the magnetic modulation ceases. This feature is probably due to the onset of "strong" coupling below T_c, as is also inferred from the differing temperature dependence of j_c in this region. In this strongly-coupled region, the current-phase relation is no longer sinusoidal and phase-dependent processes occur via phase-slip.[22]

JOSEPHSON OSCILLATION

In addition to the current-phase relationship, $j = j_J \sin\varphi$, Josephson also predicted an oscillating supercurrent of angular frequency $\omega = 2e/\hbar V$ at finite voltage V. This self-oscillation was first observed by Yanson et al.[23] and shortly thereafter by Langenberg et al.[24] It was also soon discovered[25] that the self-oscillation could be synchronized to an external oscillator. And in particular, this synchronized oscillation produces a "step structure" in the dc, voltage-current curve (see Fig. 7) at voltages $V_n = n\,\hbar\omega/2e$. This step structure is now used as the basis for maintaining the Standard Volt in many countries.[26]

In the oscillating mode, the Josephson effect competes with other electrodynamic aspects of the junction. One of the first of these electrodynamic effects was discovered by Fiske[27] who showed that the tunnel junction itself could act as a resonant cavity to stabilize the Josephson oscillation at the cavity resonance frequencies. Although the Josephson oscillation was predicted[28] to extend to frequencies far above the gap frequency, the structural capacity of most tunnel junctions limits their response to external radiation to a relatively low frequency. Consequently, attention has turned to devices such as point contacts and thin film bridges. The frequency response of point contacts currently extends to ~ 17 THz,[29] with thin film devices lagging behind[30] at about 1.3 THz. For the thin film devices it appears that heating, relaxation effects and other nonequilibrium phenomena must still be circumvented if very high frequencies are to be achieved. This section will outline some of the high frequency behavior of the proximity bridges.

The Josephson oscillation of these junctions has been examined over a wide range of frequencies and by a number of techniques. Self-oscillation of single, current biased junctions was examined by Kirschman[31] to determine the line width and frequency spectrum at low frequency (10-30 MHz). Emission from current biased arrays of junctions was studied by Palmer[32] to determine conditions for the mode locking of large arrays. The current-phase relation has been examined as a function of critical current and frequency by strongly coupling junctions to resonant strip lines.[33] Responsivity to external radiation was studied as a function of frequency to nearly 200 GHz by Hauser.[34]

This paper will report some additional aspects of the oscillating behavior of these junctions to correlate with the previous sections and to relate with the materials and geometric aspects of the bridges. By synchronizing the Josephson self-oscillation to external radiation a determination of the magnitude of the oscillating supercurrent has been made. Also measurements of the microwave impedance and the Dayem-Wyatt[35] effect will be discussed.

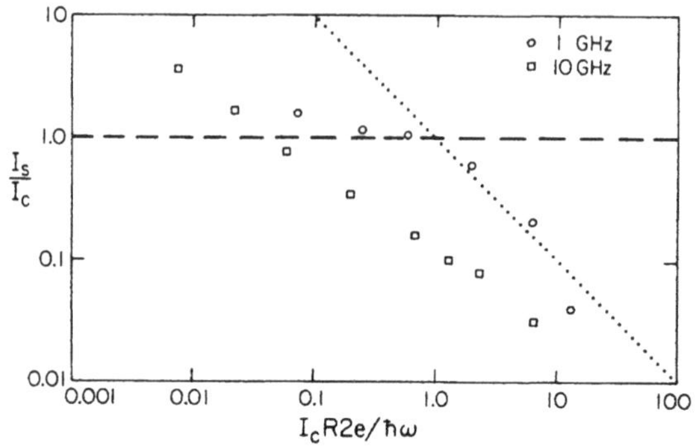

Fig. 11. Normalized first step maximum I_s/I_c as a function of reduced voltage $\frac{I_cR/2e}{\hbar\omega}$ for 1 GHz and 10 GHz irradiated 0.5μm long by 26μm wide Ta/Ti bridge. Dashed and dotted lines are asymptotic limits.

RF Response

The rf response of these bridges has been observed over a frequency range extending from about 1 MHz to 300 GHz. Up to about 1 GHz the response has been found[36] to be very well described by a resistively shorted junction model such as proposed by Russer[37] except that the resistance is the current-dependent quasiparticle resistance discussed in this Quasiparticle Current section, rather than R_n. Figure 11 shows some typical results of the relative magnitude of the maximum width of the first induced step J_s (normalized to the critical current I_c) plotted versus $RI_c/\frac{\hbar\omega}{2e}$. This behavior is typical of bridges of all materials examined. Obviously, the length cannot be too long or quantum interference cannot occur. In the hard superconducting materials, this length limit is 2-3μm while in the soft materials, it is about 6-8μm.

At 10 GHz and above, the rf response becomes more complex. For example, the maximum amplitude of the first step is a function of bridge material, temperature, and length. Figure 12 shows some typical results. The current density associated with the maximum amplitude of the first step is defined as j_s and the dc critical current as j_c. In Fig. 12a these data are shown plotted for bridges of two different lengths (2μm and 0.5μm) of the

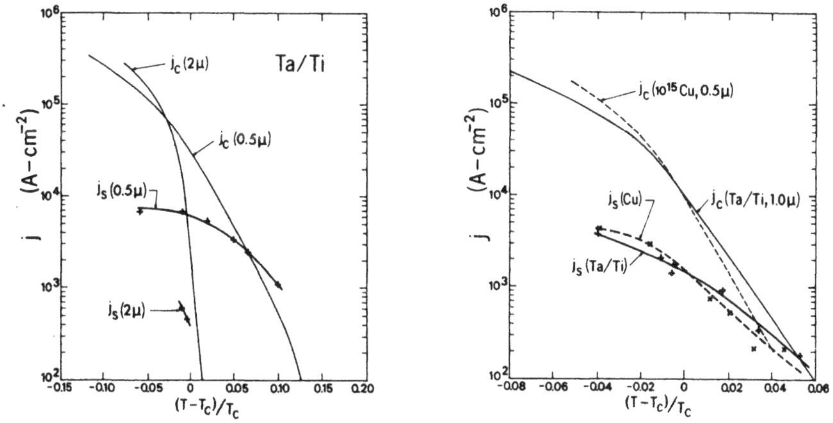

(a) (b)

Fig. 12. Comparisons of response of bridges to applied radiation at 10 GHz; j_s is current density corresponding to full width of maximum first step; j_c is critical current density without radiation. (a) data for two Ta/Ti anodization bridges, lengths 0.5μm and 2.0μm. (b) data for Ta/Ti anodization bridge, length 1.0μm, and Ta(10^{15} Cu) implant bridge, length 0.5μm.

same material. As indicated the amplitude of the rf supercurrent j_s is about an order of magnitude less than j_c at $T-T_c = 0$. At higher temperatures, j_s approaches j_c. In addition, for a given bridge length and material the rf response, j_s, decreases as $T_c/T_{co} \rightarrow 0$.

Figure 12b indicates the rf response of two different materials: one, a Ta/Ti composite anodized bridge and the second a Ta(Cu) ion implant. The resistivities of these materials differ by about a factor of 2, ρ (Ta/Ti) = 7μΩ-cm, ρ Ta(Cu) = 15μΩ-cm. As indicated in 12b, the quantum interference behavior of a 1 μm Ta/Ti bridge is equivalent to that of a 0.5μm Ta(Cu). Similar results were obtained in a comparison between 2μm Ta/Ti and 1μm Ta(Cu) bridges. Thus the rf response (as well as the critical current) depends parametrically on the product of bridge length and resistivity.

These results and data on the high frequency responsitivity[34] imply that some relaxation process is occurring in proximity coupled bridges which limits the high frequency response. From the experimental data the characteristic relaxation time τ_r ranges from 10^{-10}-10^{-11}sec and quantum interference effects are inhibited at frequencies $\omega > \tau_r^{-1}$. Functionally, the relaxation time depends on ρL for the bridge (as does the critical current) decreasing as ρL increases. This form indicates that the relaxation bottleneck may lie in quasiparticle relaxation processes rather than the pair state.[38] Also no single, simple current-phase relation will be adequate to describe "real" junctions of this type. The amplitude of both the supercurrent and quasiparticle currents are functions of voltage (frequency) and temperature. Over a limited range of these parameters a simple $\sin\varphi$ or $1 + \cos\varphi$ etc. is adequate, but the general behavior is more complex because it involves dynamic nonequilibrium aspects of superconductivity which are just now beginning to be understood.

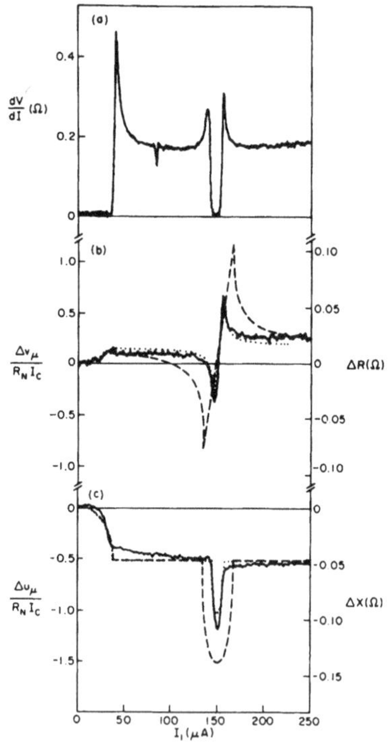

Fig. 13. Effects of the dc bias current I_1 on the impedance of a proximity effect bridge. (a) dc differential resistance, dV/dI, as a function of I_1. (b) Change in the real part of the microwave impedance ΔR, due to a change in bias current from zero to I_1. (c) Change in the imaginary part of the microwave impedance, ΔX, due to a change in bias current from zero to I_1.

Impedance

The microwave impedance of these bridges has been measured[39] at ~ 10 GHz. Typical results are shown in Fig. 13. These results are as expected from a quasiparticle resistively-shorted junction model but with the amplitude of the oscillating supercurrent j_s about 2/3 j_c. This ratio is consistent with the data given in the previous section on j_s versus j_c as a function of material, frequency, length, etc. Figure 13a shows the differential resistance of a typical junction--here a 0.5μm bridge anodized into Ta/W with a normal state resistance $R_n \sim 0.2\Omega$. Figure 13b indicates the real (resistive) part of the microwave impedance as a function of current. Note that the resistance is negative on the low frequency side of the step. Figure 13c shows the imaginary (reactive) part of the impedance. Below I_c, this impedance is inductive (the Josephson inductance). From I_c to the step, the reactive components is zero (but shown on the curve as -0.05) and at the step the impedance is capacitive. For details please see the original publication.

NONEQUILIBRIUM EFFECTS

Experimental data on some nonequilibrium aspects of the bridges are included here: The Dayem-Wyatt (D-W) effect[35] (the enhancement of critical current by means of an external rf field) and a phenomena discovered by Yu[40], (the development of a quasiparticle potential by external rf).

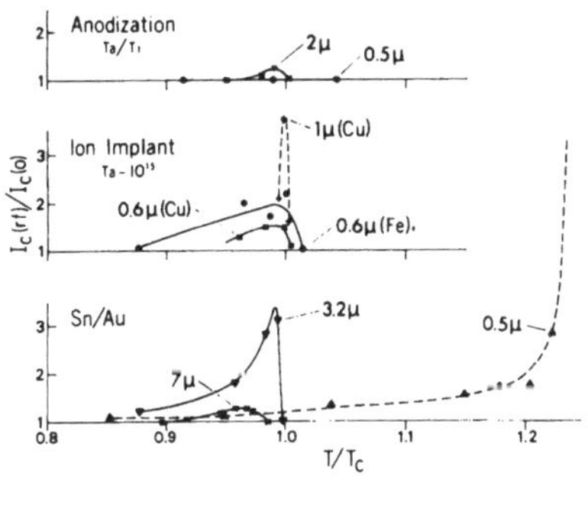

Fig. 14. Comparison of the maximum critical current for 10 GHz applied radiation, $I_c(rf)$, to the unirradiated critical current, $I_c(o)$, (Dayem-Wyatt effect) for bridges of various materials and lengths as a function of reduced temperature.

Figure 14 compares the ratio of maximum critical current with 10 GHz radiation to that without radiation for several bridges. Data are shown plotted versus T/T_c, where T_c is the intrinsic transition temperature for the bridge material as defined previously. As is seen from the figure, the Sn/Au bridges show the maximum D-W effect while the implant bridges showed less and the anodized Ta/Ti bridges showed practically no effect. In addition, the Sn/Au D-W effect was largest for short bridges whereas both the implant and anodized bridges had the largest effect for the longest bridges. For the hard superconducting bridges, enhancement up to ~ 4X is evident concentrated around $T \sim T_c$. But for the short Sn/Au bridge, supercurrent could be developed with radiation at temperatures where none existed without radiation,[1] as indicated by the dashed curve tending to infinity in Fig. 14. However, in no case did the enhancement exist above T_{co}.

The D-W results in "hard" bridges contrast with the Josephson effect in the same bridges at the same frequency. That is, the D-W effect increases with the bridge length while the Josephson effects decrease. It has also been noted[41] that the D-W effect is periodic with rf power in these bridges. Some theories indicate that the D-W effect is due to the rf excitation of quasiparticles to higher momentum states,[42] from which the pairing probability is increased thus enhancing the gap and critical current. In this case, the periodic rf power dependence may reflect the shielding effects of the phase (voltage)—dependent Josephson inductance on the quasiparticles. In any event, these results with the hard films indicate that the rf radiation influences the superconducting properties of the bridge region--enhancing the critical current while decreasing the Josephson effects. However, in the short, soft Sn/Au bridges the D-W effect is en-

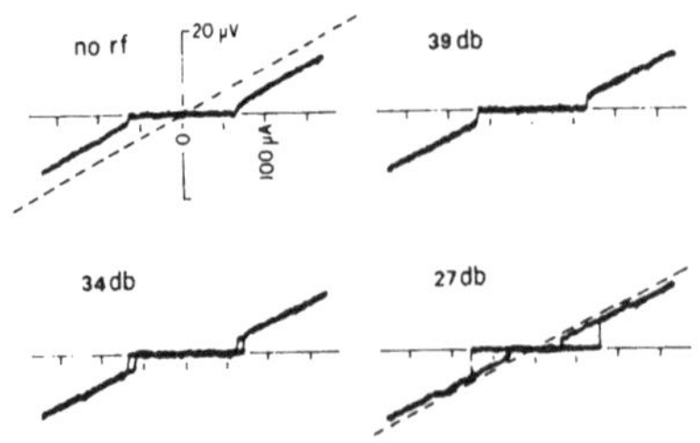

Fig. 15. I-V characteristics for a 0.5μm long bridge of Ta implanted with 10^{15} ions/cm^2 Cu for increasing 10 GHz radiation power, showing the hysteresis associated with the Wyatt effect.

hanced by decreasing the length, as is the Josephson effect. For these soft materials, the implication is that the D-W phenomena encompasses the entire film.[43]

As shown in Fig. 15, the appearance of the D-W effect is always associated with an increase in hysteresis in the V-I curve. In this figure, the V-I curves for an ion implant Ta(Cu), (1×10^{15}), 0.5μm bridge are shown as a function of the intensity of 10 GHz rf radiation. In this the critical current is enhanced from about 62μA with no radiation to a maximum about 80μA with radiation. Another feature evident in Fig. 15 is a decrease in the "excess current" (or an increase in r_{qp}) evident in the V-I curve with increasing rf power. Thus the effect of 10 GHz radiation on these bridges is a relatively small Josephson effect accompanied by an enhancement of the critical current and a tendency for $r_{qp} \to R_n$. For lower frequencies (1 GHz and below) r_{qp} was found to be independent of rf radiation. That is, suppressing I_c to zero by magnetic field or rf radiation results in the same quasiparticle resistance r_{qp}, (Fig. 10).

Figure 16 shows some results by Yu[40] on measurements of the quasiparticle potential <u>inside</u> the bridge, relative to the pair potential on either side

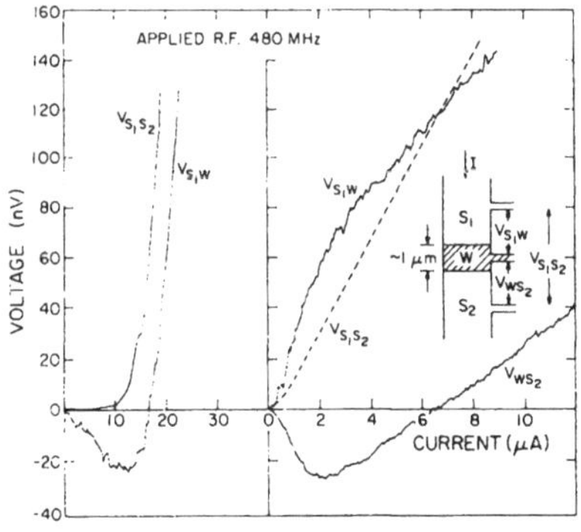

Fig. 16. RF induced potential anomalies within the bridge region. Left — The dc voltage, V_{S_1W}, across one of the S/W boundaries is negative while the whole bridge $(V_{S_1S_2})$ is still in the "zero" voltage state. Right - The potential, V_{S_1W}, increases above $V_{S_1S_2}$ to compensate V_{WS_2} giving back the "normal" V-I curve $V_{S_1S_2}$ for the entire bridge. The critical current in this case is suppressed nearly to zero by the applied rf.

of the bridge. These measurements were made by placing very small ($< 0.5\mu m$) voltage probes in and near the bridge as indicated in the insert to Fig. 16. Only the feature of these results will be discussed here: the quasiparticle potential within the bridge could be stimulated by rf radiation to be higher (or lower) than the potential on either side of the bridge depending periodically on the intensity of the radiation. When biased with a dc current below critical so that no voltage was developed across the bridge ($V_{S_1S_2} \equiv 0$ in Fig. 16) a relative potential V_{WS_1} (or V_{WS_2}) periodic in rf power could be developed as indicated in Fig. 16.

These results indicate that the external radiation stimulates a localized potential within the bridge whose amplitude is a function of both the dc bias current and the intensity of the radiation. Note that this local potential can exist even when the potential drop across the entire bridge $V_{S_1S_2}$ is zero, (however, $V_{WS_1} + V_{WS_2} \equiv V_{S_1S_2}$). For dc bias current much below the critical value I_c, the amplitude V_{WS_1} (or V_{WS_2}) was found to vary periodically with the intensity of the radiation and with magnetic field. This dependence of the potential on magnetic field and radiation was very similar to the dependence of the amplitude of the first step on the same quantities.

DISCUSSION

It should be emphasized that some of the data reported herein is preliminary and hence, any conclusions are speculative at best. Nevertheless, these results indicate a few clues to physical processes occurring in these junctions. We include here some of our speculations, hoping to stimulate further analysis and examination of this topic. In summary, in the weak coupling limit of these bridges ($T > T_c$): (a) The supercurrent obeys the Josephson relation ($j \propto \sin\varphi$); (b) At finite voltage there is a quasiparticle current for which the resistance r_{qp} does not equal the normal state resistance R_n; (c) There is a relaxation time, τ_r, which inhibits the Josephson phenomena at frequencies above about 1-10 GHz for the bridges fabricated thus far; and (d) At higher frequency a D-W effect appears accompanied by $r_{qp} \rightarrow R_n$.

Result (a) indicates that a short section of superconductor, above its intrinsic transition temperature, acts as a bridge across which Josephson weak coupling can occur. However, this type of bridge introduces several complicating features into the Josephson phenomena which are not present in tunnel junctions. In the present discussion, these show up as items b, c, and d. The complexity of these results relative to the tunnel junction is probably related to nonequilibrium electronic states in and near the bridge. It appears that quantum interference at finite voltage in these bridges is fundamentally accompanied by nonequilibrium electron processes. These processes determine the potential and its spatial distribution via the various electronic relaxation rates in the metal. This situation thus contrasts with a tunnel junction in which potential is developed via a capacitive charge distribution. The processes which determine the potential also act to perturb the electron distribution thus effecting the quantum interference.

We suspect that many of the complications of the Josephson effect in these bridges can be described in terms of quasiparticle branch imbalance in and near the bridge. A branch imbalance results from conversion, in the bridge, from pair to quasiparticle current and the subsequent relaxation of the finite momentum from the pair current via branch crossing. Because of the finite branch crossing time, τ_Q, a nonequilibrium distribution will be set up and result in a quasiparticle potential, V_{qp}, similar to that near a single super-normal (S/N) boundary as discussed by Yu.[44] At the S/N boundary supercurrent is converted to normal current. However, in a bridge conversion of a different sort may be possible i.e., supercurrent from one superconductor converts to quasiparticles which then recombine into supercurrent in the other superconductor.

For example, the nonlinear quasiparticle resistance, r_{qp}, suggests that two separate mechanisms are responsible for resistance in these bridges. One mechanism dominates at low current, below the critical current, and gives a temperature dependent resistance considerably lower than the normal state value. By contrast, at currents much above critical the differential resistance returns to its normal value R_n. We interpret the low current limit to the resistivity as a manifestation of branch imbalance in the bridge. That is, in addition to Josephson coupling, pair current, between the two superconductors adjoining the bridge, is accomplished via an intermediate quasiparticle state at potential V_{qp} where the potential V_{qp} is proportional to a relaxation process involving branch crossing. The low current resistivity discussed in the Quasiparticle Current section may reflect pair transfer by this process in which the voltage across the bridge is proportional to V_{qp} and thus, to the temperature dependent branch crossing time, τ_Q. Applying a similar concept to Yu's results implies that V_{qp} is also phase-dependent.

This branch imbalance accompanying pair transfer is maintained by the bias current and is determined by two characteristic times: one, the time for the quasiparticles to thermalize in each branch, τ_{th}, and the second is the branch crossing time, τ_Q. Our interpretation is that the relaxation time, τ_r, in the Josephson effect is related to τ_{th}. Estimates[44] of τ_{th} in Ta are about 2×10^{-10} sec. As long as $\omega\tau_r < 1$, the quasiparticle distribution in the bridge will be primarily determined by the bias current and thus can be considered as a constant of the system. Hence, radiation at frequency ω can modulate the quantum interference in the usual way. However, at higher frequency $\omega\tau_r > 1$ the radiation fields themselves must cause perturbing effects in the quasiparticle distribution, such as suggested by Eliashberg. The effect of the high frequency perturbation of the quasiparticle states in the bridge show up as the D-W effect and a tendency for $r_{qp} \rightarrow R_n$. Under these conditions, the phase of the pair state becomes less well defined (since the pair density is a function of an external parameter, the rf field) and quantum interference gradually disappears with increasing frequency. From this perspective, high frequency Josephson effects will be possible only in very short bridges in which the pair coupling is enhanced relative to quasiparticle processes.

ACKNOWLEDGEMENTS

The evidence on which this paper is based has been developed in the Low Temperature Physics Laboratory at Caltech over a number of years. Associated with this work have been E. P. Boud, S. K. Decker, J. D. Franson, T. Ganz, R. K. Kirschman, W. Y. Lai, R. P. McNamara, H. A. Notarys, D. W. Palmer, R. H. Wang, and M. L. Yu.

Work supported by the Office of Naval Research, Contract N00014-75-C-0913.

REFERENCES

1. Josephson, B. D.: Possible New Effects in Superconducting Tunnelling. Phys. Lett. 1, 251-253 (1962).

2. Cardinne, Ph., Marti, M., Renard, M.: Jonctions Josephson A Barriere Semiconductrice. Revue de Phys. Appl. 6, 547-550 (1971). Seto, John, Duzer, Theodore Van: Supercurrent Tunneling Junctions with Tellurium Barriers. Appl. Phys. Lett. 19, 488-491 (1971).

3. Clarke, J.: Supercurrents in Lead-Copper-Lead Sandwiches. Proc. Roy. Soc. A308, 447-471 (1969).

4. Zimmerman, J. E., Silver, A. H.: Macroscopic Quantum Interference Effects Through Superconducting Point Contacts. Phys. Rev. 141, 367-375 (1966).

5. Anderson, P. W., Dayem, A. H.: Radio-Frequency Effects in Superconducting Thin Film Bridges. Phys. Rev. Lett. 13, 195-197 (1964). Dayem, A. H., Wiegand, J. J.: Behavior of Thin-Film Superconducting Bridges in A Microwave Field. Phys. Rev. 155, 419-428 (1967).

6. Notarys, H. A., Mercereau, J. E.: Proximity Effect Bridge and Superconducting Microcircuitry. J. Appl. Phys. 44, 1821-1830 (1973).

7. Harris, E. P.: Properties of Superconducting Weak Links Produced by Ion Implantation. J. Vac. Sci. Technol. 12, 1383-1386 (1975). Arrington, C. H. III, Deaver, B. S. Jr.: Superconducting Weak Links Formed by Ion Implantation. Appl. Phys. Lett. 26, 204-206 (1975).

8. McCumber, D. E.: Effect of AC Impedance on DC Voltage-Current Characteristics of Superconductor Weak-Link Junctions. J. Appl. Phys. 39, 3113-3118 (1968). Stewart, W. C.: Current-Voltage Characteristics of Josephson Junctions. Appl. Phys. Lett. 12, 277-280 (1968). Blackburn, J. A., Warman, J.: Capacitive Weak Links and Finite Load Lines. Appl. Phys. Lett. 20, 459-461 (1972).

9. Hansma, P. K., Rochlin, G. I., Sweet, J. N.: Externally Shunted Josephson Junctions-Generalized Weak Links. Phys. Rev. B 4, 3003-3014 (1971).

10. Langenberg, D. N.: Characteristic Times in Superconductors. Advances in Solid State Physics, Freudenstadt, Germany, April 1-5, 1974. Oxford, England, Pergamon 1974. p. 67-86.

11. Kaplan, S. B., Chi, C. C., Langenberg, D. N., Chang, J. J., Jafarey, S., Scalapino, D. J.: Quasiparticle and Phonon Lifetimes in Superconductors. Phys. Rev. B. to be published.

12. Palmer, D. W., Decker, S. K.: Microscopic Circuit Fabrication on Refractory Superconducting Films. Rev. Sci. Instrum. 44, 1621-1624 (1973).

13. Crozat, P., Adde, R., Chaumont, J., Bernas, J.: Ion Implantation Effects in Superconducting Niobium Thin Films. Applications of Ion Beams to Metals. ed. by Picraux, S. T., Eer Nisse, E. P., Vook, F. L. Plenum Press, p. 27-33.

14. Harper, F. E., Tinkham, M.: The Mixed State in Superconducting Thin Films. Phys. Rev. 172, 441-450 (1968). Aleksejewski, N. E., Sakosa-

renko, W. M., Blüthner, K., Kohler, H. J.: Superconducting Properties of Vanadium Films. Phys. Stat. Sol. A34, 541-544 (1976).

15. Little, W. A.: Device Application of Super-Inductors. Proceedings of the Symposium on the Physics of Superconducting Devices, University of Virginia, Charlottesville (1967). p. S-1–S-17.

16. Tinkham, Michael: Introduction to Superconductivity. McGraw-Hill (1975). p.112, 117.

17. Likharev, K. K., Yakobson, L. A.: Steady-State Properties of Superconducting Bridges. Sov. Phys. Tech. Phys. 20, 950-954 (1976). Notarys, H. A., Mercereau, J. E.: Proximity Effect Bridge and Superconducting Microcircuitry. J. Appl. Phys. 44, 1821-1830 (1973).

18. Skocpol, W. J., Beasley, M. R., Tinkham, M.: Self-Heating Hotspots in Superconducting Thin-film Microbridges. J. Appl. Phys. 45, 4054-4066 (1974).

19. Harris, E. P., Laibowitz, R. B.: Properties of Superconducting Weak Links Prepared by Ion Implantation and by Electron Beam Lithography. Applied Superconductivity Conference, Stanford, Cal. (1976) U-2 to be published.

20. Likharev, K. K., Yakobson, L. A.: Steady-State Properties of Superconducting Bridges. Sov. Phys. Tech. Phys. 20, 950-954 (1976).

21. Owen, C. S., Scalapino, D. J.: Vortex Structure and Critical Currents in Josephson Junctions. Phys. Rev. 164, 538-544 (1967).

22. Rieger, J. J., Scalapino, D. J., Mercereau, J. E.: Time-Dependent Superconductivity and Quantum Dissipation. Phys. Rev. B 6, 1734-1743 (1972).

23. Yanson, I.K., Svistunov, V. M., Dmitrenko, I. M.: Experimental Observation of the Tunnel Effect for Cooper Pairs with Photon Emission. Sov. Phys. JETP 21, 650-652 (1965).

24. Langenberg, D. N., Scalapino, D. J., Taylor, B. N., Eck. R. E.: Investigation of Microwave Radiation Emitted by Josephson Junctions. Phys. Rev. Lett. 15, 294-297, 842(E) (1965).

25. Shapiro, S.: Josephson Currents in Superconducting Tunneling - The Effect of Microwaves and Other Observations. Phys. Rev. Lett. 11, 80-82 (1963).

26. Taylor, B. N.: Cryogenic Metrology. Low Temperature Physics-LT13. Plenum Press, 1972. Vol. 4, p. 465-482.

27. Fiske, Milan D.: Temperature and Magnetic Field Dependences of the Josephson Tunneling Current. Rev. Mod. Phys. 36, 221-222 (1964). Coon, D. D., Fiske, M. D.: Josephson AC and Step Structure in the Supercurrent Tunneling Characteristic. Phys. Rev. 138, A744-A746 (1965).

28. Riedel, von Eberhard: Zum Tunneleffekt bei Supraleitern im Mikrowellenfeld. Z. Naturforsch. 19a, 1634-1635 (1964).

29. Akimenko, A. I., Solov'ev, V. S., Yanson, I. K.: High-Frequency Limit in the Nonstationary Josephson Effect. Sov. Phys. JETP Lett. 21, 278-279 (1975).

30. Octavio, M., Skocpol, W. J., Tinkham, M.: Improved Performance of Tin Variable-Thickness Microbridges. Applied Superconductivity Conference, Stanford, Cal. (1976). U-5. to be published.

31. Kirschman, R. K., Notarys, H. A., Mercereau, J. E.: AC and DC Potentials in Superconducting Phase-Slip Structures. Phys. Lett. 34A, 209-211 (1971).

32. Palmer, David William, Mercereau, J. E.: Coherent Effects in Series Arrays of Proximity Effect Superconducting Bridges. IEEE Trans. MAG-11, 667-670 (1975).

33. Ganz, Tomas, Mercereau, J. E.: Stripline Coupling to Josephson Oscillators. J. Appl. Phys. 46, 4986-4988 (1975).

34. Hauser, M. G., Palmer, D. W.: DC Responsivity of Proximity Effect Bridges to High Frequency Radiation. International Conference on Detection and Emission of Electromagnetic Radiation with Josephson Junctions, Sept. 3-5, 1973. Perros Guirec, France.

35. Wyatt, A. F. G., Dmitriev, V. M., Moore, W. S., Sheard, F. W.: Microwave-Enhanced Critical Supercurrents in Constricted Tin Films. Phys. Rev. Lett. 16, 1166-1169 (1966). Dayem, A. H., Wiegand, J. J.: Behavior of Thin-Film Superconducting Bridges in a Microwave Field. Phys. Rev. 155, 419-428 (1967).

36. Kirschman, R. K.: Weakly Superconducting Thin-Film Structures as Radiation Detectors. Applied Superconductivity Conference, Annapolis, Maryland. (1972) p. 707-708. Kirschman, R. K., Hutchby, J. A., Burgess, J. W., McNamara, R. P., Notarys, H. A.: Comparative Studies of Ion-Implant Josephson-Effect Structures. Applied Superconductivity Conference, Stanford, Cal. (1976). U-3, to be published.

37. Russer, P.: Influence of Microwave Radiation on Current-Voltage Characteristic of Superconducting Weak Links. J. Appl. Phys. 43, 2008-2010 (1972).

38. Clark, T. D., Lindelof, P. E.: Frequency Dependence of the AC Josephson Effect in Small Dayem Bridges. Phys. Rev. Lett. 37, 368-371 (1976).

39. Franson, J. D., Mercereau, J. E.: Microwave Impedance of Superconducting Weak Links. J. Appl. Phys. 47, 3261-3265 (1976).

40. Yu, M. L.: Thesis - An Experimental Study of Nonequilibrium Superconductivity. (1974).

41. Notarys, H. A., Yu, M. L., Mercereau, J. E.: Josephson Effects at High Current Density. Phys. Rev. Lett. 30, 743-746 (1973).

42. Ivlev, B. I., Lisitsyn, S. G., Eliashberg, G. M.: Nonequilibrium Excitations in Superconductors in High-Frequency Fields. J. Low Temp. Phys. 10, 449-468 (1973).

43. Klapwijk, T. M., Mooij, J. E.: Microwave-Enhanced Superconductivity in Aluminum Films. Physica 81B, 132-136 (1976).

44. Yu, M. L., Mercereau, J. E.: Nonequilibrium Quasiparticle Current at Superconducting Boundaries. Phys. Rev. 12B, 4909-4916 (1975).

SOME PROPERTIES OF LONG SUPERCONDUCTING THREADS FORMED IN MONOCRYSTALLINE SILICON

Š. Beňačka, Ľ. Bezáková, Š. Gaži and J. Palaj *)
Electrotechnical Institute, Slovak Academy of Sciences, 809 32 Bratislava, Czechoslovakia
*) Tesla, n. p. Rožnov, závod Piešťany, 921 01 Piešťany, Czechoslovakia

1 INTRODUCTION

It is well known that superconducting microbridges which are small with respect to the coherence length ξ (T), apparently exhibit classical Josephson behaviour (1): a sinusoidal current-phase relationship, a linear temperature dependence of the critical current (2) and an ac Josephson effect with a quasi-Bessel function dependence on microwave power (3). Further, the occurence of quasiparticle steps, subharmonic gap structure (at voltages approximately equal to submultiples of the superconducting energy gap 2Δ) and microwave stimulated satellites around this structure (4) were frequently observed in the U-I characteristics.

In the last years a considerable effort has been put into the attemps to understand the weak superconducting links with variable thickness bridges, i. e. connection of two bulk superconducting materials with a superconductor (or a metal) of large dimensions (5). Although these weak connections are considered to be more perspective in applications than the so-called Dayem bridges, the fundamental physical processes which give rise to Josephson effects in the long bridges are not completely understood yet. Experimental results are often interpreted with the aid of coherent vortex motion (6).

In order to obtain further experimental insight into the phenomena in these structures, we have studied the properties of very long ($L \approx 20\ \mu m$) superconducting threads formed by pulse electrical breakdown of a semiconductor contacted by a bulk superconductor. Our preliminary experimental results show that in these structures one may observe not only surprisingly well defined all the phenomena mentioned above, but also some unusual feature of subharmonic energy gap structures (reported here) and properties in static magnetic field (7). However, we must emphasize that only a small part of samples (about 5%) have good pronounced Josephson effects. Our experiments with longer threads ($L \approx 200\ \mu m$), prepared by a method earlier described by Bruck (8), show that the presence of ac Josephson effects is exceptional (and they occur up to 50 μV only).

2 SAMPLE PREPARATION AND EXPERIMENTAL SET UP

The configuration of the samples which we have prepared and tested in our experiments is shown in Fig. 1. A monocrystalline n-type silicon plate 250 μm thick (with room temperature resistance $\varrho \approx 1\Omega cm$) was first cleaned and through a photoresist mask chemically etched (4g $AgNO_3$, 400 ml H_2O, 500 ml HNO_3, 250 ml HF) on both the opposite sides of the plate in order to obtain a Si membrane of the thickness about 20 μm. Then, a thin film of titanium and copper was evaporated (approximately 1 000 Å thick) in vacuum of 10^{-6} torr.

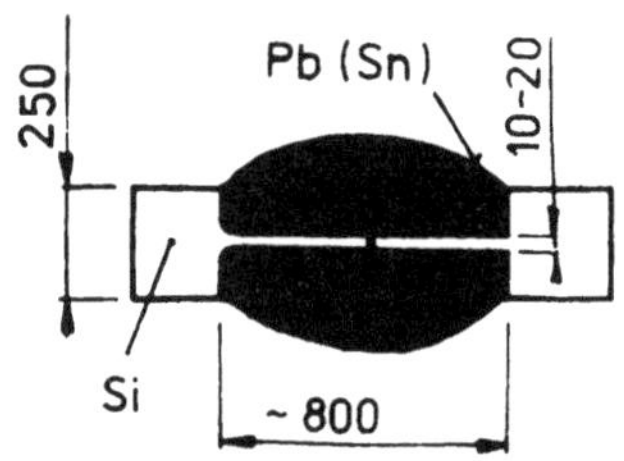

Fig. 1 Schematic of variable thickness weak superconducting link (dimensions in μm).

After evaporation the plate was immersed into a liquid lead ot tin, and then small chips (~ 1 mm^2) were cut. Further, the superconducting wires (NbTi, diameter 0.1 mm) were attached by soldering on surface of bulk superconductors. Finally, a thread was formed by pulse electrical breakdown of the semiconductor membrane (the current amplitudes were up to 100 A, with adjustable pulse duration from 1 µs to 500 ms) in liquid nitrogen or liquid helium.

Although the exact size of the thread could not be determined, the values of the critical currents and resistance indicate that the diameter of the thread is in most cases smaller than 3 µm. The ratio resistance at room temperature and low temperature ($T > T_c$) had a value about 10. At room temperatures, silicon represents a good protection against the chemical, mechanical and electrical destroying. Therefore, the samples have long life-time and stable properties at temperature cycling, but the questions of long-time ageing are not clear at present time.

The microwave system consists of a 3 cm klystron frequency stabilized on cavity resonator, a ferrite isolator, a directional coupler with crystal detector for monitoring the power output, and calibrated precision attenuator. The low temperature part of microwave system is formed by a matched coaxial line with the sample put in central conductor.

The double glass dewar system was surrounded by a two-layered µ- metal can in order to screen out the earth's magnetic field. A pair of superconducting Helmholtz coils providing magnetic fields up to about 0.14 T was placed into the liquid helium dewar. The temperature down to 1.3 K was stabilized with the aid of membrane manostat and determined from the He-vapour pressure and semiconductor diode thermometer.

The electrical characteristics of samples were measured by

drawing the U-I characteristics on an X-Y recorder. However, the standard ac-modulation techniques for recording the differential resistance (dU/dI) or differential conductance (dI/dU) were also used. In every case, the ac-modulation amplitude was chosen small enough so that the observed lineshape was independent of the modulation.

3 EXPERIMENTAL RESULTS

3.1 Typical dc Properties of Samples

The critical temperatures of lead and tin bridges with variable thickness were close to their bulk critical temperatures. The temperature dependence of resistance demonstrates a metallic connection between the bulk superconducting materials. Near the critical temperature T_c, a resistance plateau is observed, with higher T_{cb} for bulk material and lower T_{cf} for filament. This is illustrated in Fig. 2 for three different cases of tin bridges. One of the samples (Sn-Sn 15) has no transition to the superconducting state down to 3 K, the second one (Sn-Sn 17) has relatively large residual resistance and therefore rounded U-I characteristics in low voltage region (up to 100 μV). No interference effects were observed on samples with these two R(T) dependences. The samples with the (Sn-Sn 5) type R(T) dependence manifest well defined ac Josephson effects, however, only for

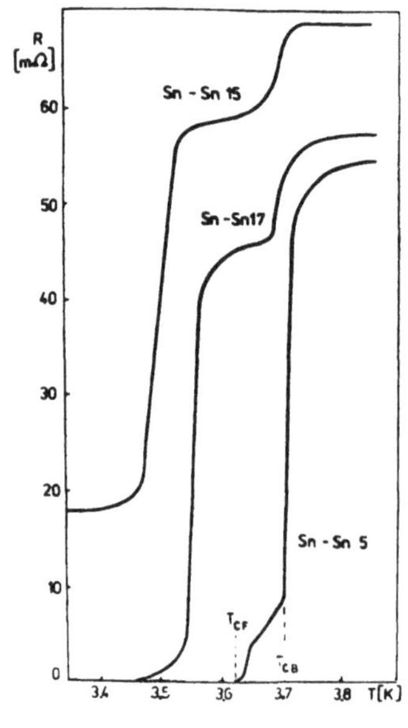

Fig. 2 The typical temperature dependence of the bridges' resistance near the critical temperature.

temperatures $T < T_{cf}$. Generally, we may conclude that for observation of coherent effects in this type of long bridges, the whole filament must be in superconducting state.

The value of resistance R_s determines not only the dynamical thread resistance but also the binding energy of the bulk superconductors. The ratio of binding energy to thermal energy is $\gamma = \hbar I_c / e k T$, where h is Planck constant, I_c the critical current, e - electron charge, k - Boltzmann constant, T - temperature. For all "good" samples $\gamma \gg 1$.

The temperature dependence of critical current should - according to the two-fluid model (2) - be like $I_c \sim (1-t)^{\alpha}$, where $t = T/T_c$ and, $\alpha = 3/2$ in bulk and $\alpha = 1$ for junction like connection, respectively. In our case, α has values from 1.2 to 1.3 within a wide temperature range $(1-t > 0.1)$.

3.2 ac Properties of Samples

The influence of external microwave field on the U-I characteristics were studied mainly in samples with non-hysteretic characteristics even at low temperatures (down to 1.4 K). In such samples, the existence of differential resistance maximum near $U = 0$ indicates small noise amplitude, and the ac Josephson effect is well defined by current steps up to ~ 0.5 mV ($\sim 3.10^{11}$ Hz).

With increasing fluctuations (gradual rise of R_D) the microwave power may either stimulate superconductivity (increase of I_c) or not, however, we did not find any relation of this effect to the amplitude of fluctuations. It seems that the fluctuations' size is proportional to that of resistance plateau just before thread transition into the superconducting state (Fig. 2).

The dependence of current steps amplitude on microwave power

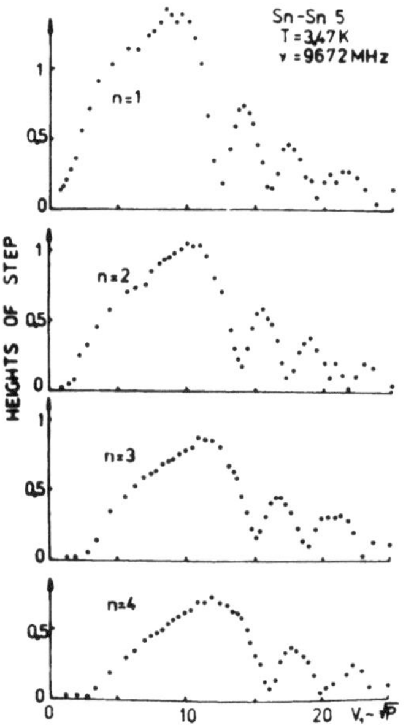

Fig. 3 The dependence of current steps amplitude on microwave power.

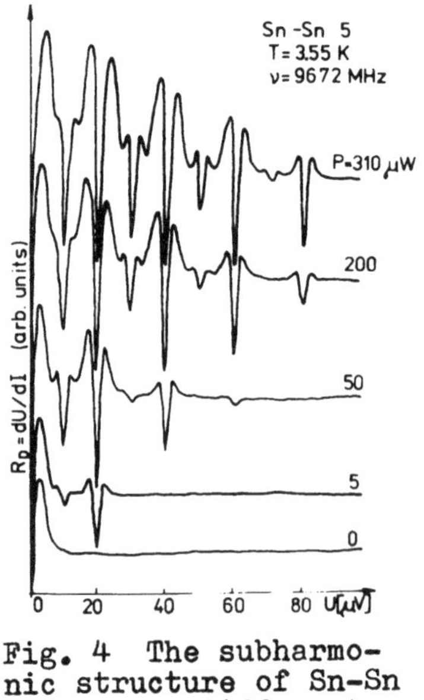

Fig. 4 The subharmonic structure of Sn-Sn bridge for different microwave power.

is in qualitative agreement with analysis of Russer (3). At temperatures near T_c (where the parameter $\eta = h\omega/2eI_cR_n > 0.1$) this dependence becomes quai-Bessel (Fig. 3), however, with decreasing temperature it tends to be non-monotonic.

Besides the basic current steps occuring at voltages $U = n\hbar\omega/2e$ there are, as a rule, observed some subharmonic current steps (9) in our samples. In the resistive model of weak link, the origin of these steps is usually prescribed to strong feedback through shunt resistance of the bridge and, possibly, as shown by Hamilton (10), by the reactive part of resistance.

The latter case can be true in our samples because of relatively large length of the thread. The subharmonic structure occurs at voltages $U = n\hbar\omega/2em$ (m, n are integers) and is well documented in Fig. 4. This figure also documents the high sensitivity of the element on microwave power (thus, we suppose that the interference effects are not stimulated by microwave field), since the power values are measured at the microwave generator output and the absorbed power can be several orders lower (10^3 - 10^4) due to mismatch of weak link to the microwave circuit.

3.3 Subharmonic Gap Structures

In the investigation of superconducting weak link much attention is paid to increasing the frequency limit, high spectrum purity and the power of Josephson radiation generated by the weak link. From this standpoint nonlinearities of differential resistance R_D in U-I characteristics due to quasiparticles are parasitic. It is known (4) that in U-I characteristics of weak links a subharmonic gap structure occurs at voltages $U = 2\Delta/me$ as well as its satellites created by microwave field excitation of quasiparticles at voltages

$$U = \frac{2\Delta \pm n\hbar\omega}{m e} .$$

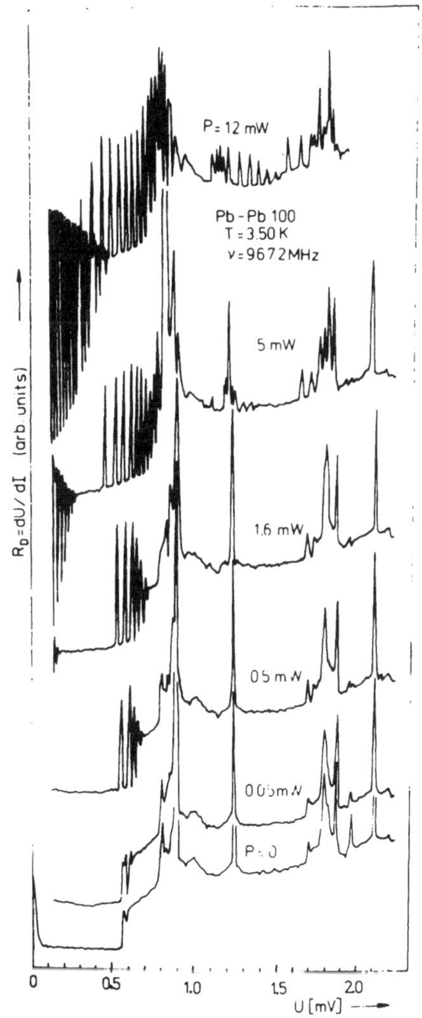

Fig. 5 The microwave induced satellites and the Josephson steps (the first five steps are not plotted for clarity).

In our samples (Fig. 7 in (7)) a rich subharmonic gap structure is often observed at voltages 0.6 mV < U < 7 mV. In this region the interference processes were not observed at any sample. Also the most expressive extrems of R_D are in region 0.5 mV < U < $2\Delta/e$, where 2Δ corresponds to the energy gap of bulk superconductor. The extrems of R_D exist also at higher voltages and they are caused by different properties of superconducting thread.

The microwave field with energy $\hbar\omega$ forms one-side satellites (Fig. 5), increasing with power of microwave field. The probability of excitations is higher at lower voltages (larger m). The distance between satellites is not equidistant - the density continuously increases. This is probably caused by energy contribution of dc current (magnetic field) that flows through the thread. In the range of higher powers the satellites penetrate into the region of Josephson current steps (for clarity, the first Josephson steps are not included in Fig. 5).

The differential resistance R_D increases in a step-like manner reaching the value which is 2 - 3 times greater than the original one, before the transition of the whole thread to the normal state ($\sim$ 7 mV).

By a closer inspection of Fig. 5 we find that in addition to the gap submultiples of the bulk material (up to $m = 4$) also their multiples (at voltages $U = k(2\Delta/m\,e)$, $k/m < 1$) are observed as sharp peaks of differential resistance. The structure is increasingly sensitive to the microwave power as the m-th submultiple is increasing.

The theoretical analysis of subharmonic gap structures and their satellites (4) predicts for the excitation mechanism (self-coupling and multiparticle tunneling) the same location of the series. Within the self-coupling formulation the amplitude of submultiples decrease much more slowly with increasing m than it would follow from the tunneling. However, if there exists a very close coupling between the weak link superconductors, as in our case, then the pair-breaking processes alone may produce structures for all m's (odd and even).

In the low-voltage part of Fig. 5 the Josephson interference current steps are well defined. In the lowest curve (microwave power P = 0), at voltage $U \approx 0.6$ mV, a small resistance dip arises and we assume that its origin is in excitations of quasiparticles induced by self coupling processes from Josephson radiation. As the microwave power increases, then at first a current step-like dependence arises around the 4-th submultiple, developing on both sides. However, in the direction of higher voltages the density of peaks is larger than on the opposite one. Structures at smaller m-numbers behave in the same way. The differential resistance oscillates around the mean value of R_D between the lower and higher resistance plateaus of R_D.

Since the satellite structure is very sensitive to the absorption of small energy, we anticipate that the non-equidistant distribution of the satellites originates from various energy contributions: Josephson radiation, external microwave field, dc current (static magnetic field), temperature, vortex structure losses, etc.

The temperature and static magnetic field dependence of all the features mentioned above approximately follow the BCS gap parameter $\Delta(T)$ and $\Delta(B)$, respectively.

With increasing microwave power (or temperature) the satellites may overlap the Josephson current steps region (Fig. 6). In contradistinction to Fig. 5 it is evident that the structure is less sensitive to the small energy contributions in the lower temperature range. At 1.4 K the satellites are almost equidistant, although the energy contribution from the dc current (or magnetic field) flowing in the thread is much larger ($I \approx 20$ mA). However, the influence of dc current fluctuations manifests the Josephson current step region. At higher temperatures, and near to U = 0, there is a very expressive R_D peak and the interference of the Josephson radiation with the external microwave field up to

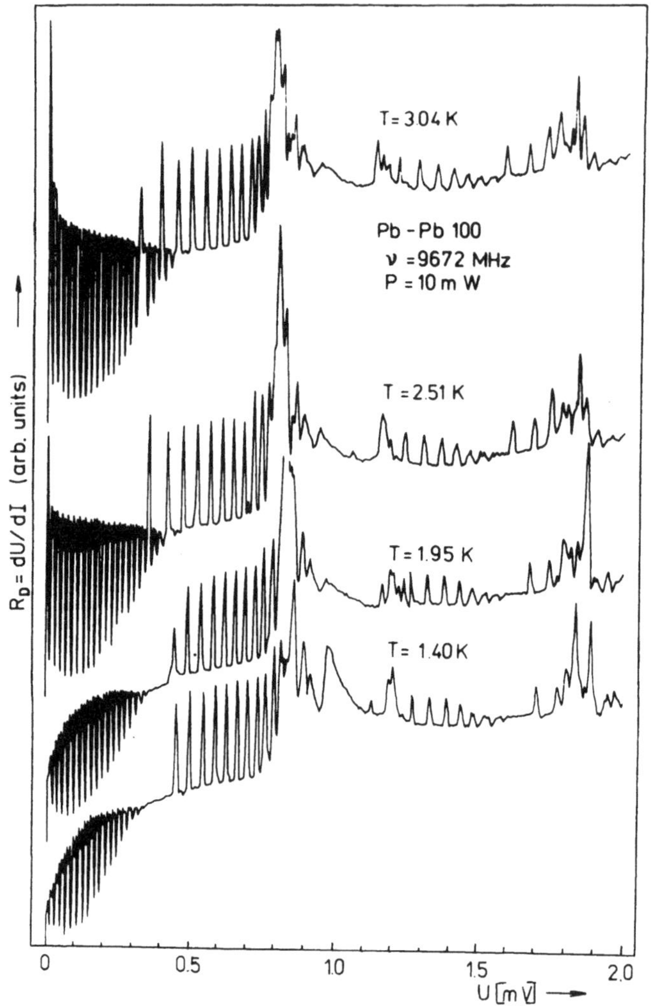

Fig. 6 The microwave induced one-side satellites of subharmonic gap structure in Pb-Pb bridge at different temperatures. The role of fluctuations by increasing the critical current at smaller temperatures is evident on Josephson current steps.

0.5 mV is well pronounced. At lower temperatures this peak disappears and R_D gradually increases, resembling to "bad" samples. The boundary of the interference effects also shifts to the lower voltages. The dark part of the plots ma-

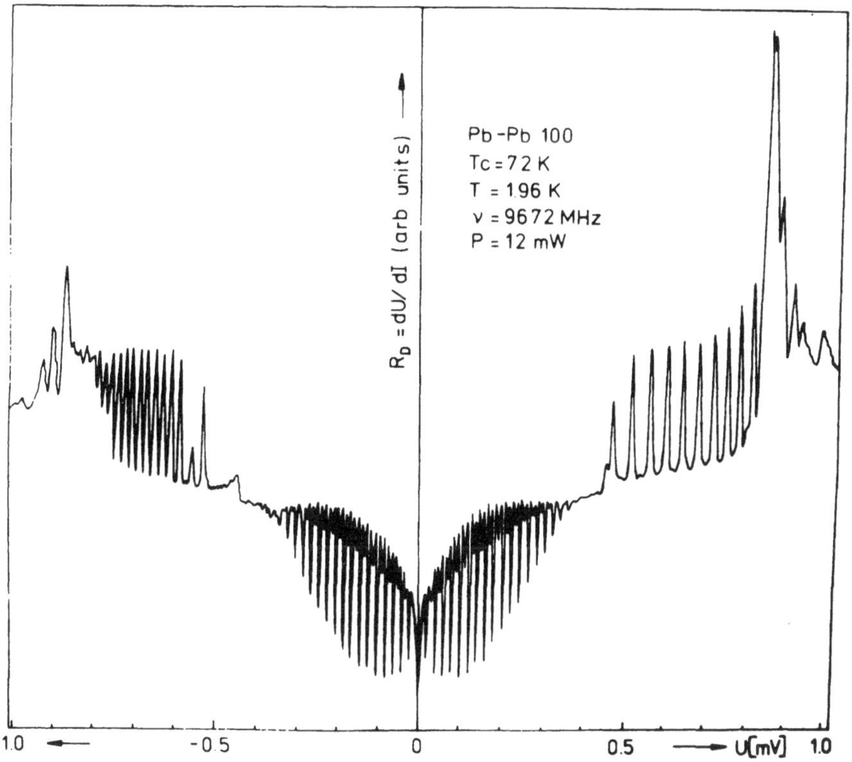

Fig. 7 The symmetry of Josephson current steps and the asymmetric displacement of the microwave induced satellites of superconducting gap structure.

nifests the existence of subharmonic structures of the Josephson current steps, which are well expressed, as a rule.

Nevertheless, for a detailed analysis more experiments on further samples are needed.

The interference effects of Josephson radiation with the external microwave field are exactly voltage located according to the Josephson relation $\omega = 2eU/\hbar$. However, the position of satellites of the subharmonic gap structures is strongly asymmetric with respect to the current direction (Fig. 7).

As we have mentioned above, the thread is formed by the

pulse electrical breakdown of a monocrystalline semiconductor, and connects two bulk polycrystalline superconducting materials (Pb). It is well known that properties of samples of small dimensions (thin films, whiskers, etc.) may be distinct from the bulk ones. Owing to mechanical strains, local injection of impurities, orientation of the threads to the "substrate" crystallographic axes, etc., they may show different properties. Of course, other influences - e. g. nonequilibrium quasiparticle distribution (11), the dependence of the gap parameter on nonequilibrium excess quasiparticle density (12), etc. - may also play a certain role. Also the pairing and depairing of electrons at a superconducting boundary in the current-carrying state (13), and/or current flows through superconductor-normal metal boundary (14) may create nonequilibrium quasiparticle states.

Recent theoretical considerations (15) confirm, in addition, that anisotropy of superconducting energy gap of Pb is large enough.

Our preliminary results about the asymmetry of satellites are as follows. The subharmonic gap structure is symmetric (within the experimental accuracy); the temperature and magnetic field, as well as the microwave power dependence of the asymmetry are the same as the plots on Fig. 5 and Fig. 6 are illustrating; the satellites may penetrate the "Josephson region" (at large microwave power densities), i. e. they are visible at small voltages (50 V/m); the effect is well pronounced even far from the critical temperature. From these facts we conclude that the asymmetry is inherent to the central part of the connection (thread) caused either by the anisotropy of the superconducting lead energy gap, or by the monocrystalline silicon anisotropy. The density ratio p of the satellites on both sides of plots ($p \approx 1.78$) is too high in comparison with the results of Tomlinson and Carbotte (15) for the anisotropy of Pb ($p \approx 1.3$). Therefore, we are concluding that the small temperature dependence of p suggests rather the relation with the silicon crystal lattice. However, the existence of the vortex structure may also play significant role, if one considers nonreciprocal conditions for the vortex motion across the thread. To determine this in more detail, further experimental and theoretical investigations are needed.

4 CONCLUSION

Our experimental results that we are presenting in this paper illustrate some interesting properties of long superconducting weak links - variable thickness bridges, with thread located in monocrystalline semiconductor. The bridges have good stability of properties for repeated experimental investigations. Some disadvantage of the samples is that the thechnological process of the sample preparation (electrical brakdown) is not well defined and the geometry of the thread and its physical properties are not well known.

A common property of "good" samples is that almost all the effects known with the weak superconducting links are well observable. Some unusual properties in static magnetic fields, which we report in the next paper (7), as well as the asymmetric gap structures, may contribute to understanding the basic properties and the interference effects in large-dimension superconducting weak links.

We cannot say exactly at the present time which of the processes give rise to coherent effects in our samples. These bridges behave as short bridges (quasi-Bessel dependence of the current steps amplitude, for example) near the critical temperature. On the other hand they behave as strongly coupled superconductors with well defined Josephson effects in high static magnetic fields and in wide temperature range.

In our opinion the fluctuations and other parasitic effects may be suppressed by shortening the thread to lengths of some micrometers. We believe that a further investigation of the properties and improvement of technology of the samples preparation may be promising for utilization in integrated cryoelectronic circuits.

1. Aslamazov, L. G., Larkin, A. I.: Zh. eksp. teor. fiz. Pis'ma 9,150(1969).
2. Jahn, M. T., Kao, Y. H.: J. Low Temp. Phys. 13,175(1973).
3. Russer, P.: J. Appl. Phys. 43,2008(1972).
4. Hasselberg, L. E., et al.: J. Low Temp. Phys. 21,567 (1975); Phys. Rev. B 9,3757(1974).
5. Likharev, K. K.: Zh. eksp. teor. fiz. 61,1700(1971); Likharev, K. K., Yakobson, L. A.: Zh. eksp. teor. fiz. 68,1150(1975).
6. Kulik, I. O., Omelyanchouk, A. N.: Zh. eksp. teor. fiz. 68,2140(1975).
7. Beňačka, Š., et al.: SQUID (next paper of this issue).
8. Bruck, A.: J. Phys. E 5,1047(1972).
9. Dayem, A. H., Wiegand, J. J.: Phys. Rev. 155,419(1967).
10. Hamilton, C. A.: Proc. LT 13, 2,585(1972).
11. Langenberg, D. N.: Proc. LT 14, 223(1975).
12. Owen, C. S., Scalapino, D. J.: Phys. Rev. Letters 28,1559(1972).
13. Clarke, J., Paterson, J. L.: J. Low Temp. Phys. 15,491 (1974).
14. Yu, M. L., Mercereau, J. E.: Phys. Rev. B 12,4909(1975).
15. Tomlinson, P. G., Carbotte, J. P.: Phys. Rev. B 13,4738 (1976).

AC JOSEPHSON EFFECT IN LONG SUPERCONDUCTING THREADS IN STATIC MAGNETIC FIELD

Š. Beňačka, S. Takács, Ľ. Bezáková and Š. Gaži
Electrotechnical Institute, Slovak Academy of Sciences,
809 32 Bratislava, Czechoslovakia

1 INTRODUCTION

One of the most important types of weak superconducting links are the S-C-S (superconductor-constriction-superconductor) bridges, which can be prepared by several methods. When narrowing the cross-section of the superconductor in a small length $L \lesssim \xi$ (microbridges), the existence of Josephson effects can be interpreted in terms of the "classical" thin film Josephson junctions (1), i. e. by assuming the overlapping of the wave functions of superconducting electrons from the "bulk parts" (banks).
The properties of these short bridges have been studied extensively, and one has no serious difficulty in understanding the phenomena connected with these types of bridges.

In some types of bridges (e. g., the Dayem bridges), the Josephson effects disappear rapidly by increasing the length over some multiples of the coherence length ξ.

Likharev (2) showed that bridges with variable thickness (i. e. the thickness of the constricted region is much smaller than the thickness of the banks) may exhibit the coherent properties also for $L > \xi$.

The coherent properties in these long bridges cannot be explained by the simple weak link theory. In such bridges, quantized vortices are created by applying the magnetic field and/or current, like in the type II superconductors (3). The existence of Josephson effects in these bridges can be then interpreted as a synchronous motion of the vortices (4,5).

The force, acting on the vortices in the presence of magnetic field equals the local Lorentz force (6), as in type II superconductors (7) :

$$f = \phi_0 J \quad ,$$

where $\phi_0 = h/2e$ is the flux quantum.

If the vortices are formed in a lattice (like in type II superconductors), the motion of a single vortex is very improbable, because of the strong interaction between the vortices - the vortex lattice moves coherently.

The existence of the coherent motion of the vortex structure was verified experimentally (8) by applying the high frequency electromagnetic fields on the bridges. The interference effects have been detected at fields

$$E_{n n'} = \frac{n' \nu}{n} \left(\frac{2}{\sqrt{3}} \phi_0 B \right)^{1/2}$$

(n, n′ are integers), ν is the frequency of the microwave radiation. The factor $2/\sqrt{3}$ indicates the existence of the triangular lattice.

Similar results have been obtained for the coherent flux flow (9) in superconducting films with periodically modulated thickness (this corresponds to periodically distributed pinning centres) where the so-called "matching" condition between the defects and the flux lines distances is fulfilled.

The coherent motion of the one-dimensional vortex structure is somewhat surprising. However, it is possible due to the macroscopically quantized character of the superconducting state (10).

Here, we report some unusual results on ac Josephson effect in long bridges and the influence of static magnetic field on this effect. Further, we try to explain some results on the base of the motion of quantized vortices in the bridges.

2 EXPERIMENTAL RESULTS

The bridges are prepared by pulse electrical breakdown of semiconductor plate (Si), conducted by Pb or Sn, as described in the preceding paper (11).

The U-I characteristics of the Pb-Pb bridges at different microwave powers are shown in Fig. 1.

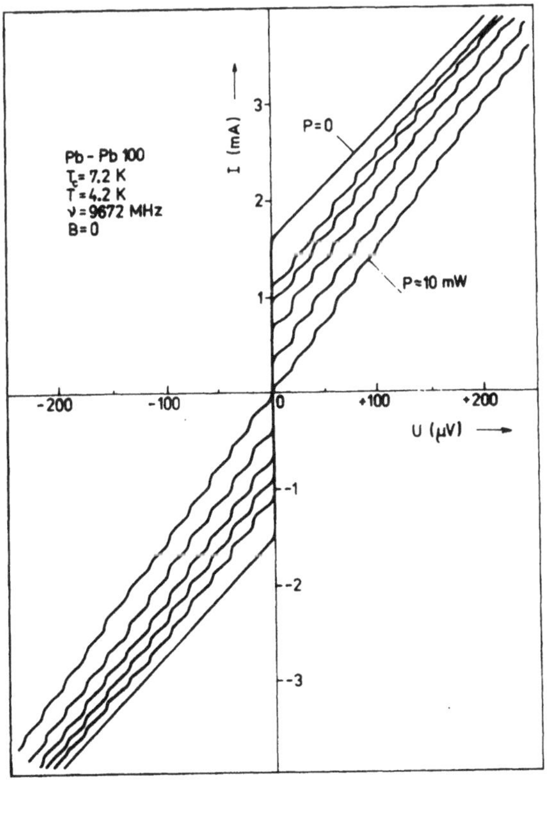

Fig. 1
The voltage-current characteristics of Pb-Pb bridge with increasing microwave power P.

The analysis of the dependence of the current steps amplitude on microwave power and temperature shows a good agreement with the results of Russer (12), as described in (11). We would like to point out here that the Josephson effects did not appear for bridges with gradual increase of differential resistance at voltages close to $U = 0$, although the U-I characteristics were almost like that for "good" junctions.

The aim of our investigations was to attest the properties of long superconducting bridges in which according to (2) the Josephson effects may exist also at extreme conditions (large magnetic fields, large temperature range, etc.). Therefore, the "parasitic" effects (from the standpoint of the Josephson effect), described in (11), are undesirable.

It is well known that magnetic fields of the order 10^{-3} T destroy almost completely the interference effects in all known superconducting weak links. From this point of view, our Pb-Pb as well as Sn-Sn bridges show very unusual and in some sense surprising properties.

The Pb-Pb bridge is relatively strongly coupled $(T \ll T_c)$. In Fig. 2, the first ten current steps of this bridge are shown for different magnetic fields oriented perpendicularly to the current in the bridge. For completness, we mention here that the critical magnetic field of pure bulk Pb at zero temperature $B_{cb}(0) \approx 0.0803$ T, and at the helium temperature $B_{cb}(4.2\,K) \approx 0.055$ T, respectively.

Up to very high magnetic fields (≈ 0.1 T), the field does not influence the shape and the location of the current steps at the U-axis, although the influence on the quasiparticle tunnel structure is remarkable (11). This is the reason why we can exclude the possibility that the complete screening of the magnetic field by the bulk parts can cause this effect.

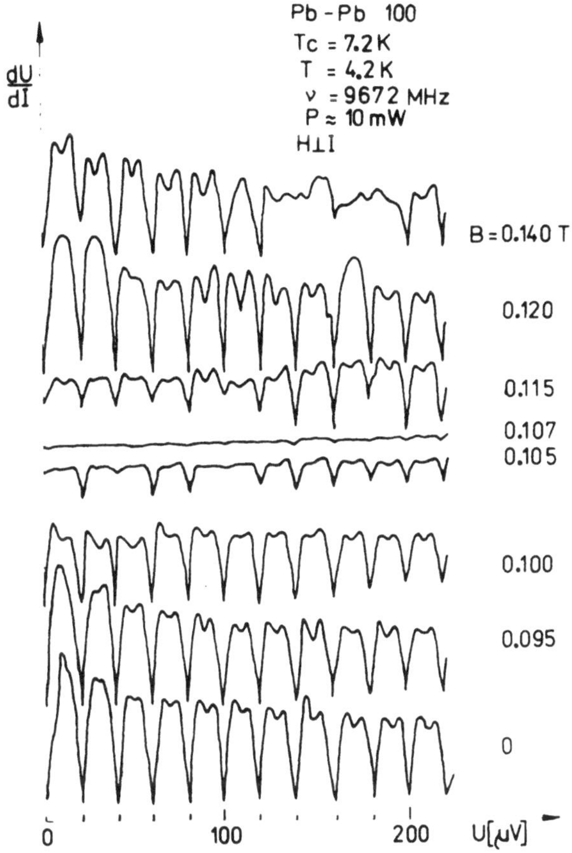

Fig. 2
The influence of static magnetic field on the Josephson effect of Pb-Pb bridge.

In a relatively small interval of the applied magnetic fields (≈ 0.106 ÷ 0.108 T), it seems that the interference of the Josephson radiation with the applied electromagnetic field vanishes completely. However, when increasing the magnetic field, the typical current steps of the ac Josephson effect appear surprisingly again, and are well defined. The steps have been detected up to B ≈ 0.14 T (limited by the maximum field of the employed Helmholtz coils).

One should note that the distance of the steps on the voltage axis remains constant (i. e. independent of the magneti field), the "Josephson frequency" $\omega_0 = 2eU/\hbar$ is therefore constant in the whole range of magnetic fields.

A probable explanation of these effects is given in the next Section.

It is interesting that the bridge is superconducting above the critical magnetic field of the bulk material. Namely, the "thin film effect" is very small at our dimensions of the thread ($d \lesssim 3\ \mu m$). Therefore, the thread cannot consist of pure Pb. However, the relatively small "incoherent" interval of magnetic fields shows that a very large part of the "weak link" has identical or very similar properties (e. g. the diameter, the Ginzburg-Landau parameter $\varkappa$, etc.).

The Sn-Sn bridges have also some unexpected properties in static magnetic fields. In contradistinction to the Pb-Pb junctions, they are only weakly connected (the corresponding temperature $T = 3.55$ K is only slightly below the critical temperature $T_c = 3.7$ K).

As the critical magnetic field of pure bulk Sn at zero temperature $B_{cb}(0) \approx 0.0306$ T and $B_{cb}(3.55\,K) \approx 0.0026$ T , the current steps are registered up to the values nearly 20-times larger (≈ 0.045 T) than the critical magnetic field of tin at the working temperature (we note that due to demagnetization effects, the magnetic field is even larger near the thread). The interference effects are vanishing at larger magnetic fields (Fig. 3) and they do not re-appear by further increasing the applied magnetic field (in contradistinction to the Pb-Pb structures).

Another interesting effect in these structures is the field dependence of the distances between the current steps. This distance is constant in the common Josephson structures, and is given as $\omega = 2eU/\hbar$ (very important for some applications of the Josephson effect, e. g. in metrology).

The distance between the current steps is constant also in the case of our Pb-Pb bridges.

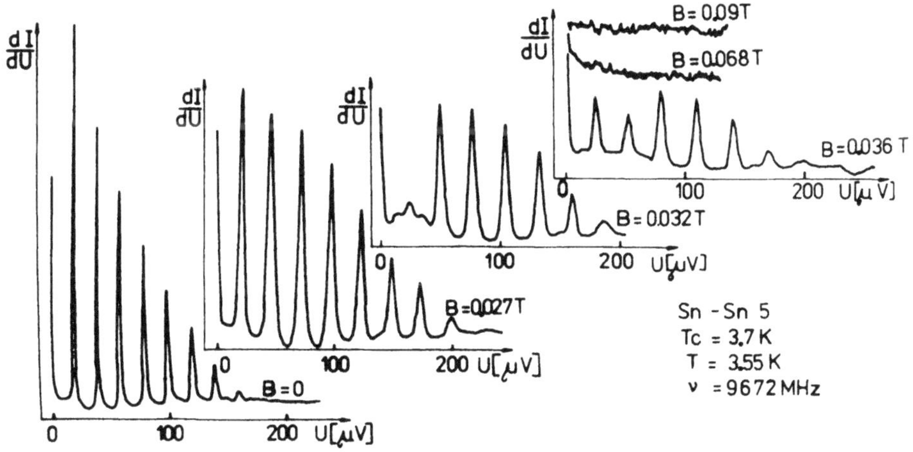

Fig. 3 The differential conductance - voltage plots for Sn-Sn bridges at different magnetic fields.

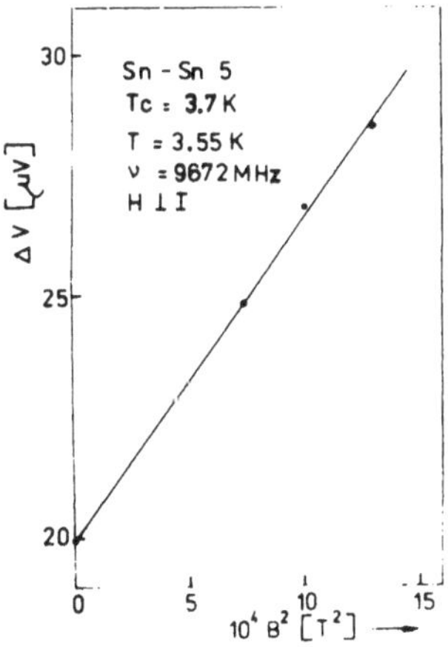

Fig. 4 The magnetic-field dependence of the current steps distance.

Due to the synchronous motion of the vortex lattice, the expected increase of the current steps distance would be proportional to $B^{1/2}$ (8). However, our bridges show a very different dependence, the current steps distance is proportional to B^2 (Fig. 4). The possible mechanisms which can lead to this effect, are given in the following Section.

3 THEORETICAL CONSIDERATIONS AND DISCUSSION

As already mentioned, the ac Josephson effect in our structures is connected with the coherent motion of vortices in the thread. The very high magnetic fields, up to which the threads show superconducting properties, indicate that the material has a large Ginzburg-Landau parameter $\varkappa$. Thus, we can assume that in the thread a vortex structure is created (due to the applied field and/or current) which resembles the mixed state of type II superconductors.

In the following considerations, we therefore profit from the calculations of the vortex structure in these superconductors.

The Josephson effects in long (i. e. the length $L \gg d$, the diameter of the thread) bridges have some special features (10). The current-phase (or the energy-phase) relation is many-valued, corresponding to different quantum states (i. e. different vortex numbers) of the bridge. In the Josephson regime the transitions between these quantum states occur. Depending whether the relaxation time of the states is larger or smaller than the Josephson period $T_D = h/2eU$ (10), this transition can be of different form.

The "intermediate" region between Abrikosov (vortex with normal kernel) and Josephson (vortex with kernel) type vortex structure has been calculated in (13) for wide junctions ($d \gg L$). According to this work, at $L > L_c$ (where $\pi < L_c/\xi < 5.31$ for different bank materials) the Abrikosov type solution with normal kernel of the dimension ξ (the characteristic dimension of the order parameter variations) and the surrounding supercurrents (with the characteristic length λ) appears. For $L < L_c$, the kernel type vortex structure cannot exist and the vortices are the Josephson type (the only characteristic length being the Josephson penetration depth λ_J).

The flux line structure in infinite type II superconductors is known already nearly twenty years (3). However, the vortex lattice is far from the "ideal" one in finite superconductors. So e. g., the superconductors' surface acts very strongly on the flux lines near the surface. This interaction causes that the flux lines cannot appear at magnetic fields $B \approx B_{c1}$ (the lower critical magnetic field), but only at higher fields near the thermodynamical magnetic field B_c (14), or they cannot penetrate into the volume of the superconductor, although their formation would be already energetically favourable (15). Even for magnetic fields $B > B_c$, the motion of flux lines is further "hindered" by their interaction with the surface (surface pinning).

Although the problem of vortex formation, their structure and the energy balance of isolated (i. e. non-interacting) vortices near surfaces have been calculated very extensively (15-18), the problem of interacting vortices near surfaces is a very complicated task. For instance, the existence of the relatively small energy barrier near the surface even at fields larger than B_c can lead to serious complications in calculations of the vortex structure. Therefore, our considerations are only approximate (or of the "model" nature).

As the thread diameter and mainly its length are much larger than the coherence length (but the diameter being already comparable with λ), we use the results for the free energy of isolated vortices from (15). The "surface" of the thread is by no means homogeneous, therefore we assume that by applying the magnetic field the vortices appear in the centre of the thread (minimum free energy). We denote this field B_{s1}. The contribution ΔF of the isolated vortex with the coordinate $x \in (-d/2, d/2)$ to the free energy per unit length of the thread is

$$(1) \quad \Delta F = \frac{\pi}{\varkappa^2 ad}\left[\sigma + \ln(1 + \cos\frac{2\pi x}{d}) - \frac{2\pi B}{\varkappa ad}(\frac{d}{2} + x)(\frac{d}{2} - x)\right],$$

where

$$\sigma = 2(\ln \varkappa d - 0.49) - \ln 2$$

and a is the vortex distance.

In our calculations, we use the known relative units (19,3) in which the length is measured in units of the penetration depth λ, the magnetic field in units of $\sqrt{2}\, B_c$, the energy in units of $B_c^2/2\mu_0$, etc.

For magnetic fields not too close to the upper critical magnetic field B_{c2}, the interaction energy U_{12} of vortices is given by the interaction of their magnetic fields.

The fact that the "transport current" acts in the same way on the Josephson type vortices and the Abrikosov type vortices (namely with the local Lorentz force) indicates (6) that the above considerations will be also true for the Josephson vortex structure where the normal kernel is unimportant. Then, we have in the relative units

$$(2) \quad U_{12} = \frac{8\pi}{\varkappa^2} K_0(D) \quad ,$$

where $D = |\bar{r}_1 - \bar{r}_2|$ is the distance between the vortices and K_0 is the 0th order Bessel function.

Because the free energy contribution of the vortex has minimum for $x = 0$, the vortex structure is one-dimensional for applied fields $B_0 \gtrsim B_{c1}$: the vortices are placed in the centre of the thread (Fig. 5a). The free energy per unit length is then

$$f_1 = \frac{1}{y_0}\left[\Delta F(x = 0) + U_{12}(y_0)\right] \quad ,$$

or

$$f_1 = \frac{1}{y_o} \frac{\pi}{\varkappa^2 d} \left[\sigma + \ln 2 - B_o \frac{\varkappa d^2}{2} + 8 y_o d \, K_o(y_o) \right] .$$

By increasing the magnetic field, the vortex distance y_o in the direction of the thread axis becomes comparable with the thread diameter d and the two-dimensional vortex structure can arise (Fig. 5b).

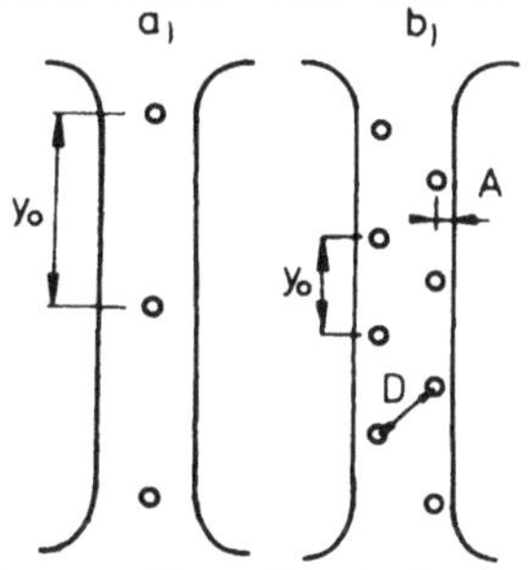

Fig. 5 The equilibrium positions of vortices in the thread for vortex distances much larger than the diameter of the thread (a) and after arising the two-dimensional lattice (b). The distance of the vortices from the surface is A.

Then, the free energy contribution of the two-dimensional lattice per unit length will be

$$f_2 = \frac{1}{y_o} \left[\Delta F(x = \frac{d}{2} - A) + U_{12}(2y_o) + U_{12}(D) \right] ,$$

or

$$f_2 = \frac{1}{y_o} \frac{\pi}{\varkappa^2 d} \left\{ \sigma + \ln \left[1 + \cos \pi (1 - \frac{2A}{d}) \right] + 8 y_o d \left[K_o(2y_o) + K_o(D) \right] - 2 B_o \varkappa A (d - A) \right\} ,$$

where

$$D = \sqrt{y_o^2 + (d - 2A)^2} .$$

The function $\Delta F(x)$ is symmetric about $x = 0$, as obtained in (15).

The equation $f_1 = f_2$ determines the field at which the two-dimensional lattice is energetically favourable (under the assumption that the magnetization curve of the thread is smooth).

However, the actual vortex positions are to be determined from the force balance equation

$$(3) \qquad 2\left|\frac{dU_{12}}{dD}\right|\frac{y_o}{D} = \left|\frac{d(\Delta F)}{dx}\right| \qquad ,$$

or

$$\frac{8y_o}{D}K_1(D) = \frac{\pi}{y_o d}\left[2B_o\varkappa x - \frac{\pi}{d}\,\mathrm{tg}\,(\pi\frac{x}{d})\right] .$$

The magnetic field B_{s2} , at which the two-dimensional vortex structure is possible, is given by the equation

$$(4) \qquad 8y_o d\left[K_o(2y_o) + K_o(D) - K_o(y_o)\right] = $$
$$\ln 2 - \ln\left[1 + \cos\pi(1 - \frac{2A}{d})\right] - B_{s2}\frac{\varkappa d^2}{2}(1 - \frac{4A}{d} + \frac{4A^2}{d^2}) .$$

There are many ways to avoid the difficulties connected with the divergence of ΔF near the surface (14,15) due to the fact that the vortex cannot exist closer to the surface than the coherence length. We do this by the simple transformation $x' \to x(1 + 1/\varkappa)$, i. e. $x' \to x(1 + \xi/\lambda)$ in conventional units. Then, we have the vortices appearing just on the surface instead of creating them effectively in the distance $\approx\xi$. At first, we do not investigate the role of the surface barrier, but we shall use this fact later (namely, that ΔF on the surface is smaller than in its vicinity, i. e. we take $A = 0$ in equation (4)) and make some quantitative calculations for the model case (15) $\varkappa d = 20$ (and, if necessary, $d = 0.2$).

We have in this case for $A = 0$

$$(5) \qquad 8y_o\left[K_o(2y_o) + K_o(\sqrt{y_o^2 + d^2}) - K_o(y_o)\right] = 7.6 - 5\alpha ,$$

where $\alpha = B_o/B_{s1}$.

The Bessel functions of the equations (4) and (5) are negligible for $y_o \gg 1$, therefore we obtain the field $B_{s2}(y_o \to \infty)$ by setting the left hand side of the equations (4) and (5) equal zero. We have then

$$\alpha(y_0 \to \infty) = 1.52 \quad , \quad B_{s2}(y_0 \to \infty) = 1.52\, B_{s1} \quad .$$

This field is by no means the smallest one for the appearence of the two-dimensional lattice, as the absolute value of the interaction terms on the left hand side of the equations (4) and (5) can be smaller than the decrease of the free energy due to lowering the vortex distance. To prove this, we have calculated numerically the field B_{s2} for some finite values of y_0. The following results were obtained:

y_0	2	1	0.5	0.4	0.3	0.2	0.1
α_{s2}	1.456	1.43	1.406	1.404	1.405	1.413	1.433

These results support our ideas about the processes in Pb-Pb bridges. Having in mind also the screening of the magnetic field by the bulk parts of the junction (which is effective to some extent at smaller magnetic fields), the one-dimensional vortex structure exists in a relatively wide range of applied magnetic fields. The motion of the structure is coherent, the bridge behaves like an ordinary Josephson junction.

We believe that besides the properties of the macroscopic coherent state, the free energy contribution of the vortices is also important. Namely, for the one-dimensional lattice the fluctuations are too small for causing any considerable displacement of the vortices from their equilibrium positions.

This is not the case for fields near B_{s2} where the free energy changes due to the vortex displacements near the surfaces are not very large, and very sensitive to surface irregularities and local changes in the thread (e. g. the diameter, GL parameter, etc.).

The "activation energy" for the displacement of single vor-

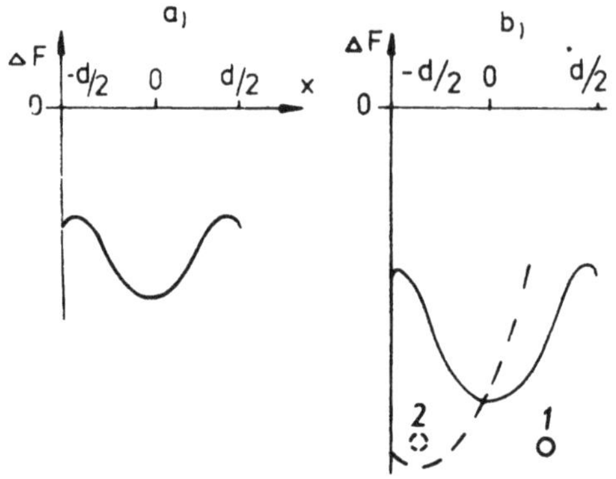

Fig. 6 The contribution ΔF of the vortices to the free energy in dependence on the vortex positions x. Full lines - one-dimensional lattice, dashed line - two-dimensional lattice in dependence on the vortex position (2) at the fixed vortex position (1). The case (b) corresponds to fields $B \gtrsim B_{s2}$.

tex with respect to other vortices is much smaller (Fig. 6) than in the previous case, and can be comparable with thermal fluctuations which act against the Josephson radiation. Therefore, the coherent motion of the vortices can be destroyed.

The interval of the incoherent motion is small, probably due to the fact that the vortex distance is relatively small just above B_{s2}. Therefore, the interaction terms in the energy and force balance equations are very important which means that the motion of a single vortex is again very improbable - the vortex motion will be again "coherent". This results then in the unexpected re-appearing of the ac Josephson effect.

The existence of the surface barrier, as well as some other pinning centres in the thread neglected in our considerations, can be another source of the incoherent motion.

There are two interesting effects in Sn-Sn bridges (the existence of current jumps up to very large magnetic fields and the B^2 dependence of the current steps distances). Although the existence of the Josephson effects in large magnetic fields is probably due to the presence of some "alloy" in the thread, this task requires more careful investigation.

The unusual field dependence of the current steps' distance

(which "changes" effectively also the Josephson frequency) can be caused by two effects:

a) some "weak" part of the thread is transformed into the normal state (the effective portion of the normal region is then increasing just proportional to the square of B); this normal region has then an effective resistance, resulting in larger voltage on the bridge at the given current density
b) the viscous flow of vortices leads to losses proportional to the magnetic induction B , as in type II superconductors (20) which can be combined with the $B^{1/2}$ dependence of Fiory (8).

Like the motion of the vortex lattice in (8), we have also a synchronous vortex motion. However, the effect seems to be in our case "autosynchronized", because the current jumps are detectable on the voltage-current characteristics also at very small microwave fields.

4 CONCLUSIONS

The Pb-Pb and Sn-Sn bridges, described in this paper, show some very peculiar and only partially explainable properties.

The common feature of both materials is the existence of Josephson effects up to much larger applied magnetic fields than the critical magnetic field of the given pure bulk materials.

In the Pb-Pb structure, the vanishing and re-appearing of current steps is explainable by means of coherent motion of the vortices in the thread. The lacking $B^{1/2}$ dependence of the current steps distances supports our ideas of the one-dimensional vortex structure up to very large magnetic

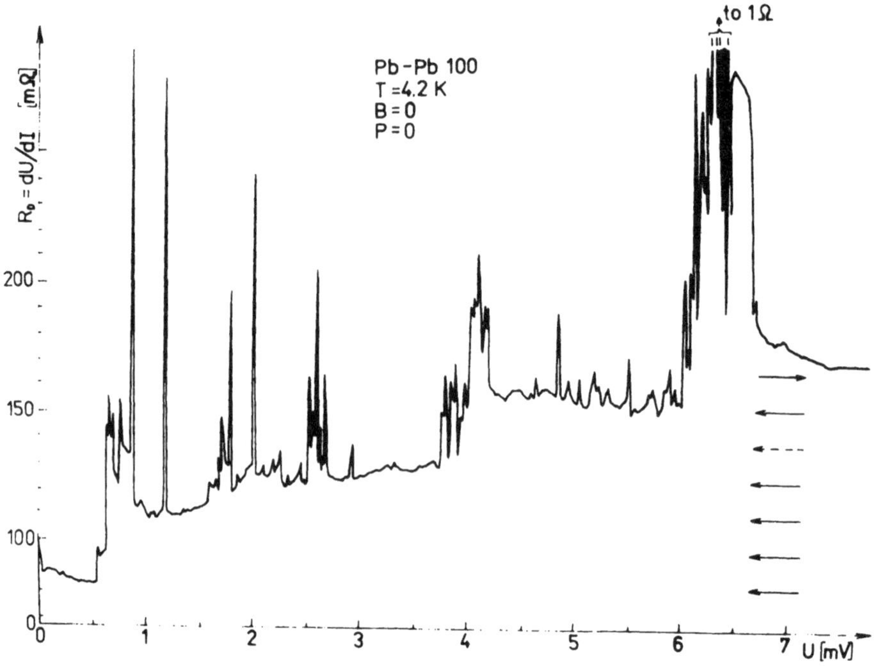

Fig. 7 The subharmonic gap structure of Pb-Pb bridge.

fields. At larger applied voltages, we have some evidence for creation of new "rows" of vortices, as the differential resistance increases gradually (Fig. 7). This gradual increase is strongly overlapped with very rich structure of subharmonics and other peaks. Such gradual increase of R_D by new "rows" in the vortex lattice is reported in (21) for wide Sn-Sn bridges, but without any detailed informations.

Concluding, we would like to mention that some other Sn-Sn bridges, which behaved like those reported here, show more complicated structure of the current steps. It seems that in some bridges two or even more threads are present, with very analogous field dependence of the current steps. Therefore, for explaining the magnetic field dependence of the current steps distances, more detailed investigations (experimental, structural, and theoretical) are required.

1. Aslamazov, L. G., Larkin, A. I.: Zh. eksp. teor. fiz. Pis'ma 9,150(1969).
2. Likharev, K. K.: Zh. eksp. teor. fiz. 61,1700(1971).
3. Abrikosov, A. A.: Zh. eksp. teor. fiz. 32,1442(1957).
4. Guthmann, C., et al.: Phys. Rev. B 11,1909(1975).
5. Gubankov, V. N., et al.: Proc. LT 14, 4,120(1975).
6. Kupriyanov, M. Yu., et al.: Fiz. niz. temp. 2,716(1976).
7. Friedel, J., et al.: Appl. Phys. Letters 2,119(1963).
8. Fiory, A. T.: Phys. Rev. Letters 27,501(1971).
9. Daldini, O., et al.: Solid State Commun. 16,509(1975).
10. Kulik, I. O., Omelyanchuk, A. N.: Zh. eksp. teor. fiz. 68,2140(1975).
11. Beňačka, Š., et al.: SQUID (preceding paper of this issue).
12. Russer, P.: J. Appl. Phys. 43,2008(1972).
13. Likharev, K. K.: Fiz. tver. tela 17,2787(1975).
14. Bean, C. P., Livingston, J. D.: Phys. Rev. Letters 12,14(1964);
de Gennes, P. G.: Solid State Commun. 3,127(1965).
15. Takács, S.: Z. Phys. 199,495(1967);
phys. stat. sol. 21,709(1967).
16. Galaiko, V. P.: Zh. eksp. teor. fiz. 50,717(1966).
17. Kramer, L.: Z. Phys. 259,333(1973).
18. Takács, S.: Czech. J. Phys. B 26,901(1976).
19. Ginzburg, V. L., Landau L. D.: Zh. eksp. teor. fiz. 20,1064(1950).
20. Kim, Y. B., et al.: Phys. Rev. 139,A1163(1965).
21. Gubankov, V. N., et al.: Zh. eksp. teor. fiz. Pis'ma 18,292(1973).

Voltage Current Characteristics of Superconducting to Normal Metal Point Contacts in an RF-Field

U. Kaiser-Dieckhoff [+]
2. Physikalisches Institut der Universitaet, Zuelpicherstr.77, D-5000 Koeln 41, Germany

As shown previously {1-3}, superconducting to normal metal (SN) point contacts can exhibit characteristics similar to those of superconductor-superconductor (SS) point contacts. We report here, that a) the Josephson like characteristics are indeed due to the SN interface, b) the current steps in these SN point contacts conform better with the weak coupling limit than SS point contacts and c) SN point contacts can also exhibit an entirely different, Giaever like, behaviour which is explained qualitatively by an adiabatic relative shifting of the Fermi levels with frequency ν and amplitude eU_{rf}, when irradiated with an rf-field. Here U_{rf} is the potential difference due to the rf-field.

Application of an rf-field with frequency ν to an SS point contact causes the well known step structure in the U(I)-dc-characteristic with voltage spacing $\Delta U = h\nu/2e$. SN point contacts can exhibit characteristics similar to these, which we call type I. When an ohmic slope is subtracted, the voltage spacing is again related to the rf-frequency by the Josephson relation.

This behaviour is quite puzzling. In order to make sure, that the type I characteristic is not due to a crack in the superconducting needle which would result in a series circuit of an SS and an SN point contact, a systematic microscopic inspection was made of the two parts constituting the contact before and after each contacting procedure. During about 50 such runs we never detected any breaks, although the tip showed some deformation.

Pressure induced intermetallic diffusion between the superconducting material and the normal metal (Nb, Cu) can also be excluded, because contact is only made at He-temperatures and because H.P. Leppin {4} reported type I characteristics between a carefully melted and oxidized Nb-sphere of about 0.5 mm radius and a Cu- surface. In this arrangement the contact pressure was about four orders of magnitude lower than in ours. Thus type I characteristics seem indeed due to the interface between a superconductor and a normal metal.

According to Josephson {5} in the weak coupling limit, the maximum supercurrent of the n-th step as a function of the rf-amplitude should scale like the n-th order Besselfunction $|J_n(2eU/h\nu)|$, where n is the number and U the ac-voltage of n-th step, counting from $n=0$, ν is the rf-frequency. Grimes and Shapiro {5} found, that real SS point contacts follow this scaling law only approximately and explained this by deviations from the weak coupling limit.

Fig. 1 shows the type I characteristics and their derivatives dU/dI of a Nb- Nb oxide -Cu contact for various amplitudes of the rf-field. The maximum rf-voltage at the contact is 5.5 mV (U_o). Evaluation of the width of the n-th current step results in the experimental points of Fig. 2. Also shown are the Besselfunctions $|J_n(U_{rf}/U_{dc})|$. Note that the fit for amplitude and phase for six functions is made at one point of the $n=0$ and $n=1$ curves only. The resulting scaling behaviour is much better than for SS point contacts in the literature {5}. Note in particulare that there are zeroes observed in the step width. This is inconsistent with strong coupling. From this we draw the conclusion, that our SN contacts conform better with the weak coupling theory {6} than SS point contacts.

The reason for the Josephson like behaviour of SN contacts re-

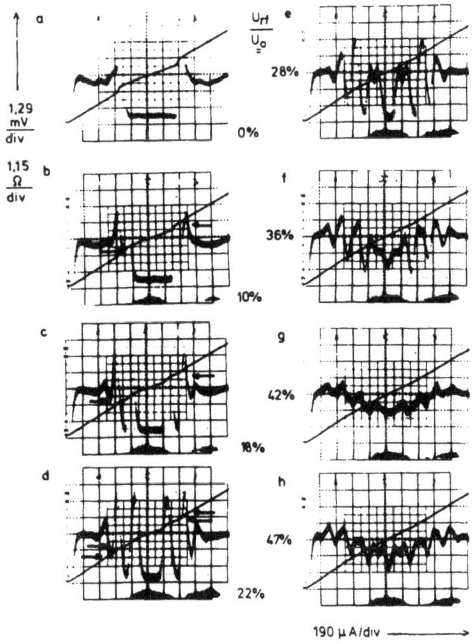

Fig. 1 Josephson like characteristic of Nb-Cu point contact for 8 different rf-amplitudes (steps marked by arrows)
Vert.: Voltage U and differential resistance dU/dI across junction
Hor. : Current
Maximum rf-amplitude U_O= 5.5 mV

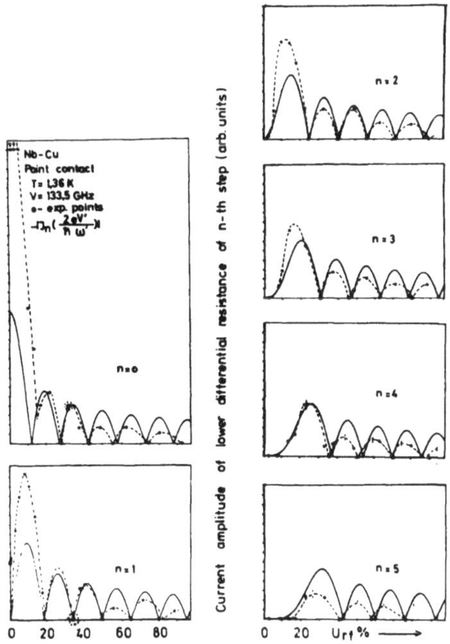

Fig. 2 Besselfunction dependence of width of n-th current step from rf-amplitude (note that the fit has been made for all figures at one point of the n = O and at one point of the n = 1 curve)

mains obscure. One might argue, that it is due to the proximity effect, if there are metallic links through the oxide layer. However, one should then expect strong coupling between the two "superconductors", which is inconsistent with the observed scaling behaviour. Mercereau et al. {7} reported, that proximity effect and ac-Josephson quantum interference effects do not exclude each other. They found microwave induced current steps in the characteristics of superconducting-normal metal boundaries, where superconductivity was induced in the normal metal by the proximity effect. In contrast to our result their characteristics showed a pronounced subharmonic structure as a consequence of strong coupling.

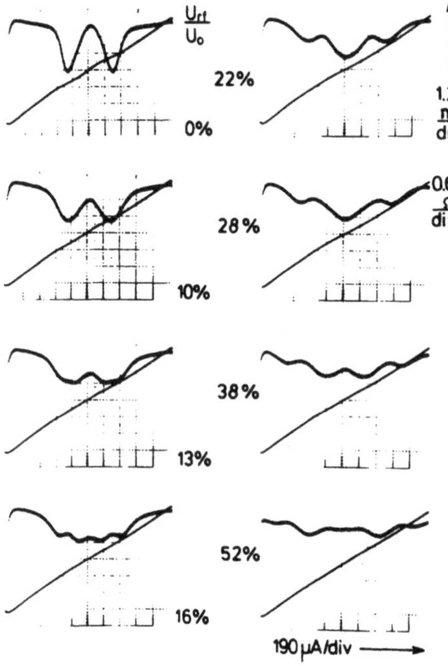

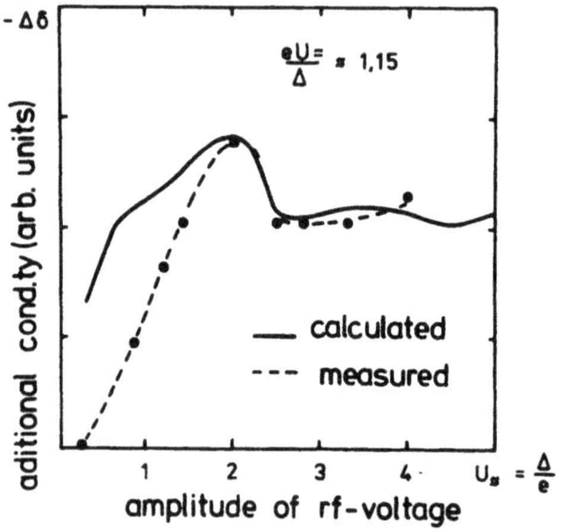

Fig. 4 Comparison between model and experiment for additional conductivity Δσ by rf-modulation-amplitude U_* (in units of gap Δ/e)

Fig. 3 Giaever like characteristic of Nb-Cu point contact for different rf-amplitudes (maximum amplitude $U_0 = 5.5$ mV); Vert.: Voltage U and differential resistance dU/dI across junction; Hor.: Current through junction

The Giaever (type II) behaviour of superconducting to normal metal point contacts is signaled by a higher differential resistance at zero bias than at high bias (by a factor of three in the best cases). RF-radiation (133.5 GHz) applied to a contact in this state (Fig. 3) modifies the dc-characteristic as an ac-voltage (150 kHz) superimposed on the dc-sweep voltage. This behaviour can be explained qualitatively by a simple model, according to which the Fermi levels of both metals are raised and lowered adiabatically with respect to each other with frequency ν by the amplitude of the rf-voltage between the metal surfaces (Fig. 4).

When the normal metal is doped with different concentrations of magnetic impurities (i.e. Co or Fe between 3 ppm and 50 ppm)+), the type I characteristics do not appear even with the smallest obtainable concentration. The Giaever like effect is not affected.

References:

1. Happ, H. and Kaiser Dieckhoff, U.: DC and MW-properties of SN and S(NS) point contacts, Phys. Let. 29A, 161 (1969)
2. Taguchi, I. and Yoshioka, H.: Josephson-effect-like behaviour of point contacts between normal and superconducting metals, J. Phys. Soc. Japan 27, 1074 (1969)
3. Taguchi, I. and Yoshioka, H.: Properties of superconducting point contacts, J. Phys. Soc. Japan 29, 371 (1970)
4. Leppin, H.P.: private communication, 2. Physik. Inst. Köln
5. Josephson, B.D.: Supercurrents through barriers, Adv. Phys. 14, 419 (1965)
6. Grimes, C.C. and Shapiro, S.: Millimeter wave mixing with Josephson junctions, Phys. Rev. 169, 397 (1968)
7. Mercereau, J.E. and Palmer, D.W.: Quantum interference of single superconducting - normal metal interfaces, Bull.Am. Phys. Soc. 20, 1493 (1975)

+ present address: Physikal. Institut, Roentgenring 8 D-8700 Wuerzburg, Germany

++ We thank Prof. D. Wollenberger, Institut für Metallkunde der RWTH Aachen for the samples with magnetic impurities and for reference samples.

MEASUREMENT OF THE COSϕ TERM IN JOSEPHSON TUNNELING

M. R. HALSE AND J. C. TAUNTON
Physics Laboratory, The University of Kent, Canterbury, England.

Falco et al (1,2) have examined the possibility of measuring the magnitude α of the cosϕ term from its effect on the thermal noise rounding of the I-V characteristic of a weak link near T_c. The theory showed that a phase dependent contribution to the Johnson noise would be expected to significantly modify a plot of the normalized slope at zero current $(dV/dI)_{I=0}$ versus $\gamma = \hbar I_c/ekT$, where I_c is the critical current of the junction and T is the temperature.

Unfortunately, more careful analysis of the theoretical results showed that the fitting of any one experimental thermally rounded I-V characteristic with the theoretical function $V = f(I, \gamma, \alpha)$ was ambiguous to within experimental error for a whole range of pairs of values of γ and α covering $-1 \leq \alpha \leq 1$. Thus no conclusion could be reached concerning the magnitude of α. Additional difficulties arose because of the extrapolation procedure used to determine $I_o(T)$ near T_c, and because of the inclusion of a low resistance shunt to reduce hysteresis effects which therefore dominated over any phase dependent conductivity.

In the present experiments, we believe that the above difficulties have been overcome by working much closer to T_c and by applying a new emphasis to the experimental results. Measurements were made of the maximum supercurrent I_s (fig. 1) as a function of temperature. Typical experimental results are shown in fig. 2. The junction current was in the range $\leq 1\mu A$. The junction voltage was measured with a DC photo-cell galvanometer preamplifier. In the region very close to T_c, the I-V characteristic is indistinguishable from a straight line, and a voltage difference technique employing a fixed resistor was used to measure I_s. The minimum detectable

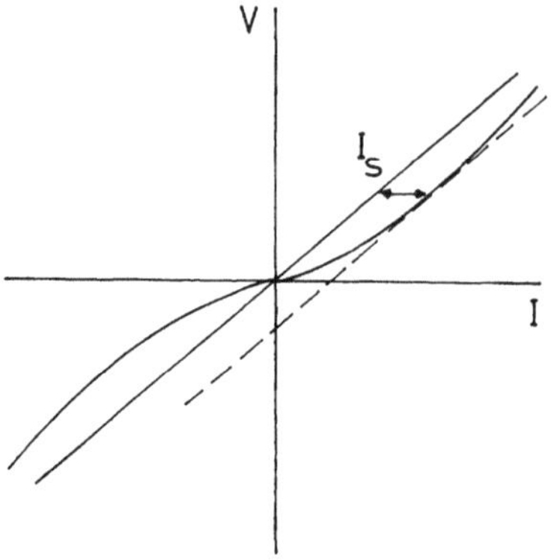

FIG. 1 Sketch of a typical I-V characteristic near T_c showing the definition of I_s.

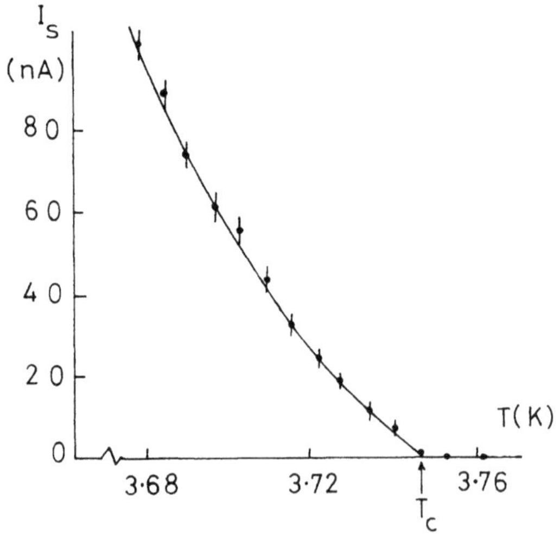

FIG. 2 Experimental plot of I_s versus T. T_c = 3.749 ± 0.001 K.

I_s was $\lesssim$1nA. By this means, the transition temperature T_c was determined to within 1 mK. The experimental I_s were converted to γ_s (= $\hbar I_s/ekT$) using the bath temperature T(≃3.7K) and then to γ. This latter step was accomplished by calculating the theoretical relationship between γ_s and γ for various values of α (fig. 3). Finally from γ, I_c was plotted as a function of temperature. This is shown in fig. 4 for three assumptions α = -0.75, 0, +0.75.

By a self-consistent argument, the correct choice of α is the one for which a linear extrapolation of the deduced values of I_c versus temperature reproduces the original T_c. As can be seen from fig. 4, the experimental results of this junction are consistent with $\alpha \simeq -0.75$. The average value for six weak links gives $\alpha = -0.75 \pm 0.25$. A linear extrapolation has been used in fig. 4 simply because the deduced critical currents lie on such good straight lines. As well as measuring α, we have in addition therefore experimental evidence that $I_c \propto T_c - T$ within 100 mK of T_c. This result is is agreement with previous work (3).

The self-consistent extrapolation procedure used above is equivalent to curve-fitting. We have emphasised the similarity between figs. 2 and 3 by reversing the x-axis in fig. 3. With $I_c \propto T_c - T$, the figures differ only by a horizontal scale factor x. We choose α and x so as to optimise the agreement between experiment and theory. It should in principle be possible to measure x experimentally from the asymptotic behaviour of I_s as I_s increases with decreasing temperature. When the critical current is large and the noise rounding is relatively insignificant, $I_s \to I_c$ (dotted line, fig. 3). Unfortunately the determination of x has not proved possible because of the type of weak link used in these experiments. These were Clarke blobs (4) of tin and niobium with a resistance above T_c of order 10Ω. Invariably the blobs showed multiple junction behaviour so that it was never possible to continue I_s measurements on a single junction down to low temperature.

In order to prevent any misinterpretation of I_s when two or more junctions were present, the data were rejected unless the transition temperatures

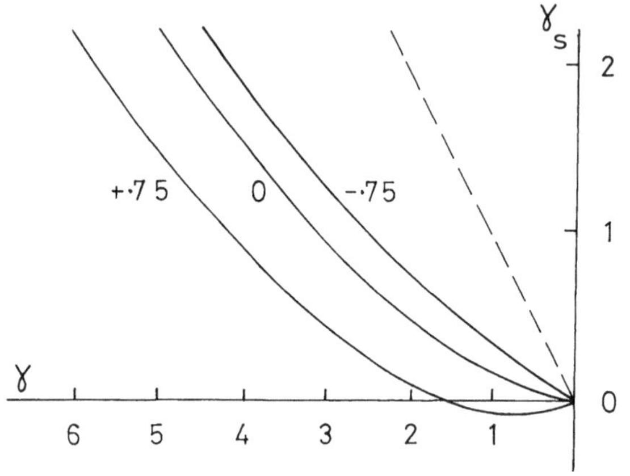

FIG. 3 Calculated relationships between γ_s and γ for three values of α.

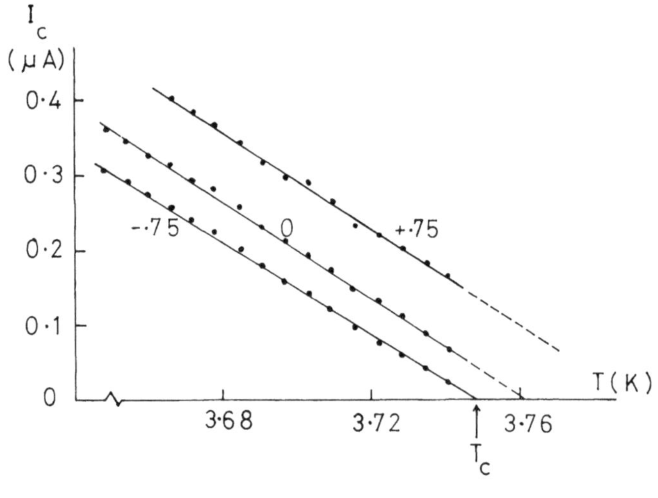

FIG. 4 Deduced critical currents versus T obtained from figs. 2 and 3 for three assumed values for α. A self-consistent T_c is obtained with α = -0.75.

of each junction were sufficiently well separated to allow unambiguous measurements to be made on the junction with the highest T_c (which we call $T_c(1)$). It is emphasised that I_s measurements in the temperature range $T_c(1)>T>T_c(2)>T_c(3)$ etc. are unaffected by the presence of the constant combined shunt resistance of the other junctions because they are not carrying supercurrent at this temperature. The presence of multiple junctions could be detected by the application of control current (4) along the niobium wire or inferred from the observation of discontinuous increases in gradient in the experimental I_s - T measurements at temperatures $T_c(2)$, $T_c(3)$ etc.

Returning to fig. 3, it can be seen that the ability to measure α depends critically on the detailed shape of the experimental I_s - T curve in the region $\gamma_s \ll 1$. With a minimum detectable I_s of $\leqslant$ 1 nA and a temperature $T \simeq 3.7$ K, this gives a minimum detectable γ_s of order 0.01. With this resolution, it is possible for example to distinguish between the curves for $\alpha = -.7$ and $-.5$ from shape alone.

The theoretical relationships between γ_s and γ shown in fig. 3 were computed from the theoretical expression for the noise-rounded I-V characteristic as given by Ambegaokar and Halperin (5) and as extended by Falco et al (1, eqn. 3). Our result is;

$$V = R\, I\, g(\gamma, \alpha)$$

$$1/g(\gamma, \alpha) = \sum_{K=-\infty}^{+\infty} (-1)^K \left[A(K) + \frac{\alpha}{2} B(K)\right]$$

$$A(K) = \frac{I_K^2\left(\frac{\gamma}{2}\right)}{\left[1 + \frac{4K^2}{\gamma^2 i^2}\right]}$$

$$B(K) = \frac{I_K\left(\frac{\gamma}{2}\right) I_{K+1}\left(\frac{\gamma}{2}\right)}{\left[1 + \frac{4(K+1)^2}{\gamma^2 i^2}\right]} + \frac{I_K\left(\frac{\gamma}{2}\right) I_{K-1}\left(\frac{\gamma}{2}\right)}{\left[1 + \frac{4(K-1)^2}{\gamma^2 i^2}\right]}$$

where I is the junction current, $i = I/I_c$ and I_K are the modified Bessel functions. The above expressions differ from that given by Falco et al (2, eqns. 12, 13).

The Ambegaokar-Halperin theory (5) is valid provided;

(i) $\beta \ll 1$ where $\beta = 2eI_cC/\hbar\sigma^2$ and σ is the total junction conductance,

(ii) $\nu_J \ll 2kT/h$ where ν_J is the maximum junction response frequency (plasma frequency, $(2eI_c/\hbar C)^{\frac{1}{2}}/2\pi$) and $2kT/h$ is roughly the highest frequency for which the simple Johnson noise approximation remains valid.

Condition (i) is satisfied because onset of hysteresis ($\beta \geqslant 1$) is typically observed in our junctions for $I_s \geqslant 10$ μA. Thus for a 10Ω junction $C \simeq 1$ pF and $\beta \leqslant 0.01$ for $I_s \leqslant 100$ nA. For condition (ii), $2kT/h = 1.6 \times 10^{11}$ Hz at T = 3.7 K while $\nu_J \leqslant 10^{10}$ Hz for C = 1 pF and $I_c \leqslant 1$μA.

In addition to the main experimental measurement of I_s we have also measured the differential slope dV/dI at I = 0. It is interesting to plot differential slope versus γ_s; typical experimental results are shown in fig. 5. Also shown in this figure are the theoretical calculations for $\alpha = -.75$, 0 and +.75; remarkably, the theoretical results for different α differ by such a negligible amount (at least for dV/dI (I = 0)<1) that only one continuous line is shown. We have made use of this fact to allow a test between experiment and theory independent of any knowledge of α. The apparent discrepancy shown in fig. 5 is, we believe, due simply to the combined shunt resistance of the other junctions as mentioned above. We can of course only measure the total current supplied to the blob. If due allowance is made for a shunt, the experimental results can be brought into excellent agreement.

The experimental measurement of I_s was performed as follows. A current ramp swept the blob current between ±1 μA and the junction voltage was detected with the D.C. photo-cell pre-amp. The voltage difference between the output of the pre-amp and a voltage generated by a decade resistance

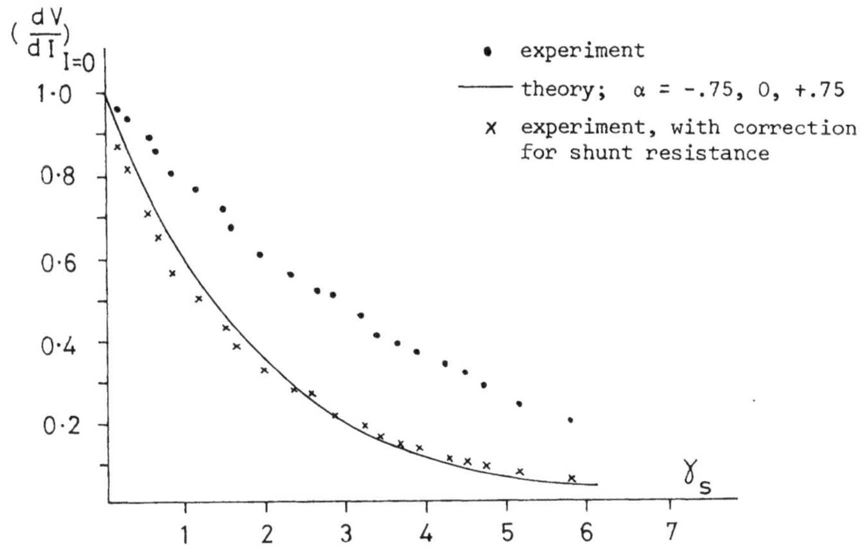

FIG. 5 Theoretical and experimental plots of differential slope at zero current versus γ_s. (The region of dV/dI (I = 0)>1 for α = +.75 has been omitted.)

box in series with another current ramp generator was detected by a differential milli-voltmeter. With the junction temperature held just above $T_c(1)$, the decade box was adjusted so that no voltage difference was detectable over the whole current sweep. The temperature was then reduced below $T_c(1)$ and the current sweep repeated. Any output is evidence of departures from linearity of the I-V characteristic of the junction. The maximum difference voltage is equal to GRI_s where G is the voltage gain of the system, R is the blob resistance (combined parallel resistance of junction and shunt) and I_s is the maximum supercurrent.
A phase-locked-loop analogue was built as an aid to understanding the effect of α on the mechanism of noise-rounding and incorporated two reference channels with 90° phase difference (sinϕ and cosϕ). The sinϕ term was used to lock the loop as usual, while the cosϕ output was arranged to modulate both the amplitude of a noise generator and the 'resistance' of the 'junction' so as to simulate values of α in the range $-.9 \lesssim \alpha \lesssim +.9$. It is hoped to publish details of this analogue elsewhere.

REFERENCES

(1) Falco, C. M., Parker, W. H., Trullinger, S. E.: Observation of a phase-modulated quasiparticle current in superconducting weak links. Phys. Rev. Lett 31, 933-936 (1973).

(2) Falco, C. M., Parker, W. H., Trullinger, S. E., Hansma, P. K.: Effect of thermal noise on current-voltage characteristics of Josephson junctions. Phys. Rev. B 10, 1865-1873 (1974).

(3) Gregers-Hansen, P. E., Levinsen, M. T., Fog Pedersen, G.: A sinusoidal current-phase relation for superconducting thin-film microbridges. J. Low Temp. Phys. 7, 99-117 (1972).

(4) Clarke, J.: A superconducting galvanometer employing Josephson tunneling. Phil. Mag. 13, 115-127 (1966).

(5) Ambegaokar, V., Halperin, B. I.: Voltage due to thermal noise in the d.c. Josephson effect. Phys. Rev. Lett. 22, 1364-1366 (1969).

Effects in High Critical Current Density Josephson Tunnel Junctions.

J. Niemeyer, V. Kose
Physikalisch-Technische Bundesanstalt
Braunschweig and Berlin

Abstract

The dependence of the switching hysteresis on the critical supercurrent density j_o is observed in small Josephson tunnel junctions of the type PbIn-$(PbIn)_xO_y$-Pb. The current density ranges from $2 \cdot 10^2$ A/cm^2 to $3 \cdot 10^5$ A/cm^2. For large j_o values the low voltage branch of the Riedel peak is directly observed in the dc current-voltage characteristic. The experimental results are compared with calculations by Stewart and McDonald et al..

Introduction

It is well known that in dc current biased Josephson tunnel junctions large hysteresis is usually observed in the current-voltage (I versus $\langle V\rangle$) characteristic, indicating an extremely small dc supercurrent for nonzero voltages (see Fig. 2a) for example. Stewart /1/ and McDonald et al. /2/ predicted large dc supercurrents to be present at $\langle V\rangle \neq 0$ for high critical current densities. Based on a lumped equivalent circuit model of a dc current biased tunnel junction, Stewart analyzed the I-$\langle V\rangle$ characteristics assuming various voltage-dependent normal conductances. This quasiparticle part I_{qu} of the total tunnelling current is expressed in the general form $I_{qu} = I_o(\langle V\rangle/\langle V\rangle_g)^n$, where $\langle V\rangle_g$ is the gap voltage, $\langle V\rangle$ the positive dc voltage, and I_o the critical supercurrent. Of particular interest in the following is the $n \rightarrow \infty$ case, which is an idealization of the zero temperature quasiparticle tunnelling characteristic. A useful measure of the dc supercurrent I_R - which is the lower switching current - for positive dc voltages $\langle V\rangle$ is the hysteresis parameter $a_m = I_R/I_o$ which is related to the dimensionless damping factor $k_{n\rightarrow\infty} = \phi_o I_o/2\pi C\langle V\rangle_g^2$ according to Ref. 1 by the implicit equation

$$(1)\quad k^{-1}_{n\rightarrow\infty} = 4(1-a_m^2)^{1/2} - 2a_m(\pi - 2\sin^{-1}a_m).$$

ϕ_o is the flux quantum and C the junction capacitance. By introducing $C = \varepsilon'\varepsilon_o A/t$ and the critical current density $j_o = I_o/A$, equation (1) yields

$$(2)\quad k^{-1}_{n\rightarrow\infty} = [\phi_o\, j_o\, t/2\pi\varepsilon'\varepsilon_o\langle V\rangle_g^2]^{-1} =$$

$$= 4(1-a_m^2)^{1/2} - 2a_m(\pi - 2\sin^{-1}a_m),$$

where A represents the junction area, $\mathcal{E}_o$ the permittivity of vacuum, $\mathcal{E}'$ the relative permittivity, and t the thickness of the insulating barrier between the superconducting electrodes. For large $k_{n\to\infty}$ a_m approaches 1, whereas for $k_{n\to\infty} \leq 0.25$ a_m is zero. Since $k_{n\to\infty}$ is proportional to j_o a large switching hysteresis exists for low critical supercurrent densities (light damping; see Fig. 2a) and for high j_o values (heavy damping, see Fig. 2c) there is small hysteresis.

Recently McDonald et al. /2/ calculated the I-$\langle V\rangle$ characteristic of a dc current biased tunnel junction at T = 0 by using Werthamer's tunnelling theory /3/. Drastic changes in the I-$\langle V\rangle$ characteristic are predicted, depending on the dimensionless parameter RC/τ_g, where R is the junction normal state resistance and $\tau_g = 2\,\phi_o/\langle V\rangle_g$. With decreasing RC/τ_g the hysteresis parameter also decreases. This means that the major current below the gap voltage comprises not quasiparticles but pairs. By introducing $I_oR = (1/4)\pi\langle V\rangle_g\, f$, where we assumed f to be 0.788 /4,5/ due to the strong coupling of the superconductors used in our experiment, $RC/\tau_g = \pi\,\mathcal{E}'\,\mathcal{E}_o\,\langle V\rangle^2_g \cdot f/8\,\phi_o t\, j_o$ is obtained. The dimensionless parameters introduced by Stewart and by McDonald are directly related to each other

$$(3) \qquad 1/k_{n\to\infty} = (16/f)\ RC/\tau_g$$

so that $(RC/\tau_g)^{-1}$ has the character of a damping factor. Indeed (τ_g/RC) qualitatively plays the same role with respect to the hysteresis parameter as used by Stewart, which can be verified by inspection of the figures in Ref.2 at dc voltages below the gap voltage. Besides this dependence of I_R - here the total current I_R comprises the dc pair current as well as the quasiparticle current - on j_o, the authors found that the enhancement of the dc supercurrent at the gap voltage (socalled Riedel peak /6/) should occur only for small values of RC/τ_g or large critical supercurrent densities. For $RC/\tau_g \geqslant 0.25$ the Riedel peak disappears and therefore cannot be directly observed in an I-⟨V⟩ characteristic when low critical supercurrent density tunnel junctions are used. Moreover, Ref. 2 describes the form of the subharmonic energy gap structure.

Switching Hysteresis

The only experimental data at present available for comparison with the theoretical results are from externally shunted tunnel junctions (small R·C) /7, 8/ and from tunnel junctions having small critical supercurrent densities /9/. However, for tunnel junctions which are not shunted the predicted effects on the I-⟨V⟩ characteristic are found at rather high critical current densities of more than 10^4 A/cm^2. The calculations in Ref. 2 exclude spatial variations of the supercurrent across the junction area, so that small-sized junctions are required. In order to obtain a wide range of j_o values, small SIS tunnel junctions of the type PbIn-$(PbIn)_xO_y$-Pb were used. Although some of the experimental results were published recently /10/, it is useful to discuss the subject in more detail. The mass content of In was about 10 % /11/. The junctions were prepared by using the oblique evaporation method /12/ and the insulating layer was formed by thermal oxidation. The oxygen pressure was 390 millibar, the oxidation time about half an hour at substrate temperatures within the region from 25 °C to 45 °C, depending on the desired j_o values. The in-line junctions are lenticular in shape (Fig. 1), with areas ranging from 100 μm^2 at $j_o = 10^2$ A/cm^2 to 3 μm^2 at $j_o = 10^5$ A/cm^2. The largest junction length was 6 μm for low critical supercurrent densities and the smallest length 1 μm for high j_o values. The dimensions of the junctions are therefore small compared with all relevant electromagnetic wavelengths such as that of the Josephson plasma frequency $\omega_p/2\pi = (I_o/2\pi\phi_o C)^{1/2}$ /13/, or the gap frequency $\nu_g = \langle V\rangle_g/\phi_o$. Even at a current density of 10^5 A/cm^2 the Josephson penetration depth $\lambda_J = (\phi_o/2\pi\mu_o j_o\, d)^{1/2}$ is about twice the junction length, where d constitutes the penetration depth of the superconductors and the thickness of the barrier /13/.

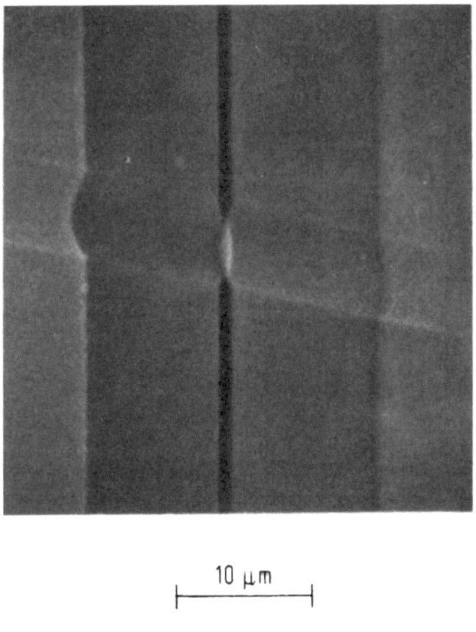

Fig. 1 Electron-microscopic photograph of a small tunnel junction; junction area $\approx 3\,\mu m^2$

Because the junction length is smaller than $4\lambda_J$, we exclude spatial variations of the supercurrent density across the junction area for zero magnetic field /14/. To achieve efficient heat transfer from the barrier, the adjacent superconducting electrodes have very large areas compared with the junction area. For the same reason the thickness of the electrodes is 600 nm for the high current density junctions. The experiments were carried out at a helium bath temperature of 4.2 K.

Fig. 2 shows the expected behavior: at increasing j_o the switching hysteresis decreases. The experimentally obtained relationship between j_o over three orders of magnitude, and the hysteresis parameter I_R/I_o, are given in Fig. 3. The solid line represents the experimental results (dots). A comparison with the theory of Ref. 2 (dashed curve a) in Fig. 3) and of Ref. 1 (dashed curve b) in Fig. 3) shows that the experimental data agree well qualitatively with the theoretical ones. The dashed lines were obtained by assuming $\mathcal{E}' = 8$ and a constant barrier thickness t = 2 nm /15/, neglecting the fact that t varies with the critical current density j_o and that the t values may differ slightly for the Pb-In alloy. With regard to the temperature dependence of the gap voltage, we used a mean measured value $\langle V \rangle_g = 2.2$ mV. We think that the deviations of the theoretical and experimental results, particularly at low current densities, can possibly be explained by the temperature dependence of the quasiparticle current and of the pair current at nonzero voltages. One should expect the largest deviation from Stewart's theory ($n \rightarrow \infty$) for low j_o values, because I_R does not include any quasiparticle current for dc voltages below $\langle V \rangle_g$. For low j_o values small deviations of the experimental data from McDonald's results are also expected, because the theory assumes a temperature T = OK, whereas the experiments were carried out at 4.2 K. There is also an influence of the quasiparticle-pair interference term on the hysteresis for T > OK /16, 17/.

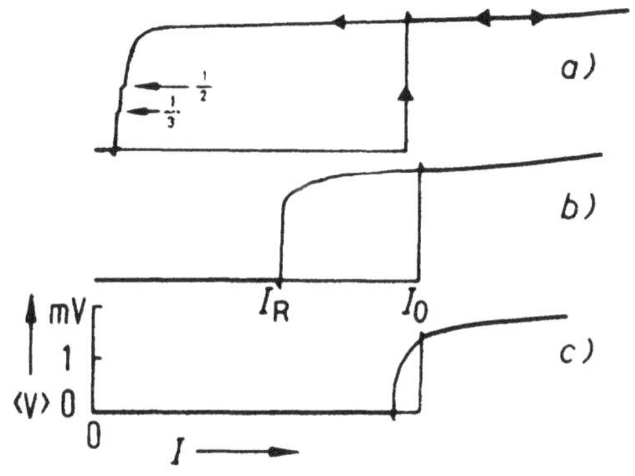

Fig. 2 Influence of the critical dc supercurrent density j_0 on the I-⟨V⟩ characteristic of tunnel junctions. The hysteresis is characterized by the ratio of the reverse switching current I_R and the critical dc supercurrent I_o. Temperature T = 4.2 K.
(a) $j_o = 3.1 \times 10^3$ A/cm^2; $I_o = 0.68$ mA.
(b) $j_o = 3.4 \times 10^4$ A/cm^2; $I_o = 3.4$ mA.
(c) $j_o = 2.1 \times 10^5$ A/cm^2; $I_o = 6.4$ mA.

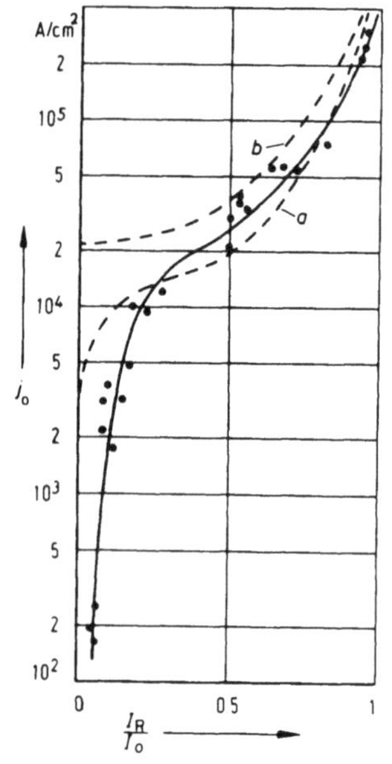

Fig. 3 Critical pair current density j_o vs. hysteresis parameter I_R/I_o of small PbIn-$(Pb\text{-}In)_xO_y$-Pb thin film tunnel junctions at T = 4.2 K. The solid line represents the experimental data (see dots).

(a) Theoretical results from McDonald et al. valid for T = 0K.

(b) Theoretical results from Stewart (nonlinear shunt resistance).

Magnetic Field Dependence of the Dc Supercurrent

In Fig. 4 the magnetic field dependence of curve b in Fig. 2 is displayed. The magnetic field at about $H = 5\ H_o$ corresponds to an absorption of one flux quantum in the junction. With increasing field the subharmonic energy gap structure appears at voltages $\langle V\rangle_m = \langle V\rangle_g/m$, with m = 2,3, the beginning of the even m and the odd m series /18, 19/. Without a magnetic field this structure is only seen in the characteristics of junctions with low critical current densities (see Fig. 2a). In this case a large dc-supercurrent does not exist at voltages $\langle V\rangle$ larger than zero with a strong negative slope $-\frac{dI}{d\langle V\rangle}$ /2/, in consequence of which the subharmonic energy gap structure remains hidden. Only when the pair current is supressed by means of a magnetic field, does the positive slope $dI/d\langle V\rangle$ of the quasiparticle current appear in the $I-\langle V\rangle$ characteristic, and the subharmonic energy gap structure becomes visible.

It is possible to derive an excess pair current I_p from the difference between the characteristic with an applied field of $9\ H_o$ or larger and, for example, the characteristic with an applied field $H = 0$. A comparison of the large field characteristic with the curves at $0 \leqslant H \leqslant 9\ H_o$ yields the magnetic field dependence of I_p (Fig. 5). For small applied magnetic fields, I_p is strongly enhanced at the gap voltage $\langle V\rangle_g$ and at the subharmonics of the energy gap for $\langle V\rangle_m = \langle V\rangle_g/m$ with m = 2, both predicted by the calculations of McDonald et al. /2/. The accuracy of the method for obtaining I_p is not high enough to observe peaks at values $\langle V\rangle_m$ with $m \geqslant 2$. The large peak in the vicinity of the gap voltage is the low voltage branch of the Riedel singularity /3, 6/. With increasing applied magnetic field I_p decreases. The magnetic field of

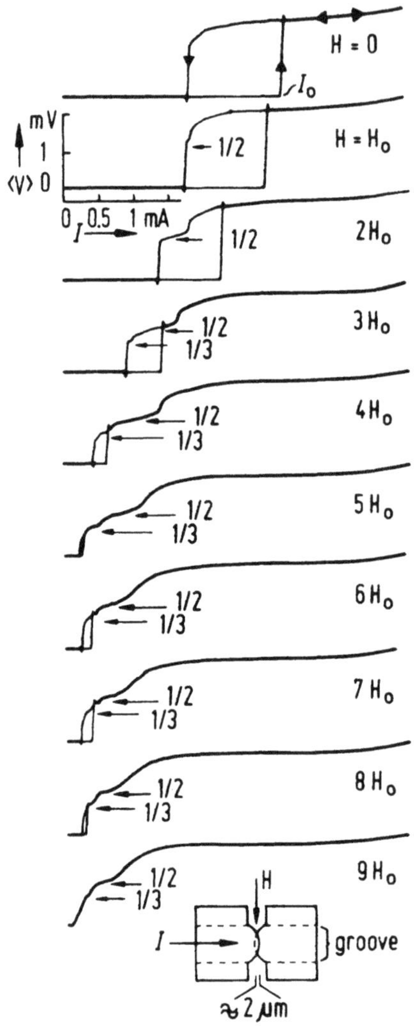

Fig. 4 Magnetic field dependence of the pair current of a tunnel junction.
$H_o = 1.2 \cdot 10^3$ A/m;
$j_o = 3.4 \cdot 10^4$ A/cm^2

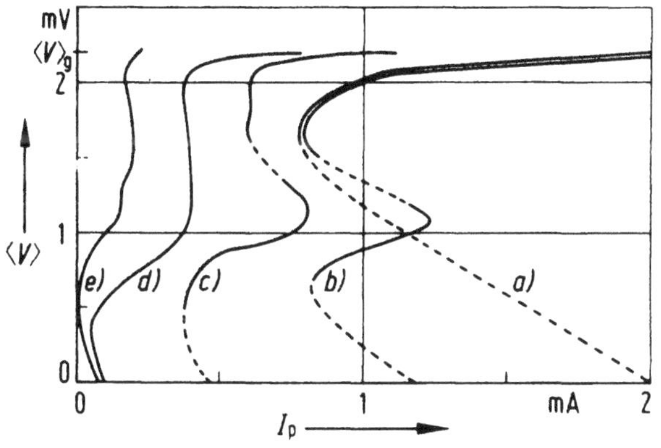

Fig. 5 Voltage dependence of the pair current I_p. The applied magnetic field is the parameter.

(a) H = 0

(b) H = 1.8 x 10^3 A/m

(c) H = 3 x 10^3 A/m

(d) H = 4.2 x 10^3 A/m

(e) H = 5.4 x 10^3 A/m

curve e) in Fig. 5 corresponds to the absorption of about one flux quantum in the junction. With decreasing critical current densities the pair current at the gap voltage diminishes and disappears for critical current densities $j_o \lesssim 10^4$ A/cm^2. At voltages higher than the gap voltage the pair current I_p decreases within a voltage region of about 100 μV. The decay is much steeper than predicted by the theory. We think that the effects associated with the rise of the quasiparticle current and the pair current at voltages larger than the gap voltage are responsable for the suppression of I_p in this region. These effects are the self heating caused by the quasiparticle current and self magnetic fields originating from both pair current and quasiparticle current. Furthermore, it is not clear whether the pair current at voltages $\langle V \rangle \neq 0$ is completely eliminated by applying a magnetic field, as is the case for the pair current at $\langle V \rangle = 0$. Hence the difference between the large magnetic field characteristic and the characteristic at low magnetic fields may possibly yield a qualitative picture of I_p.

The authors wish to acknowledge stimulating and useful discussions with D.G. McDonald and R.E. Harris.

/ 1/ Stewart, W.C.: J. appl. Phys. 45, 452 (1974)

/ 2/ McDonald, D.G., Johnson, E.G., Harris, R.E.: Phys. Rev. B 13, 1028 (1976)

/ 3/ Werthamer, N.R.: Phys. Rev. B 147, 255 (1966)

/ 4/ Ambegaokar, V., Baratoff, A.: Phys. Rev. Lett. 10, 486 (1963); 11, 104 (E) (1963)

/ 5/ Fulton, T.A., McCumber, D.E.: Phys. Rev. 175, 585 (1968)

/ 6/ Riedel, E.: Z. f. Naturforsch. 19 A, 1634 (1964)

/ 7/ Hansma, P.K., Rochlin, G.I.: J. Appl. Phys. 43, 4721 (1972)

/ 8/ Fulton, T.A.: Phys. Rev. B 7, 1189 (1973)

/ 9/ Scott, W.C.: Appl. Phys. Lett. 17, 166 (1970)

/10/ Niemeyer, J., Kose, V.: Appl. Phys. Lett. 29, 380 (1976)

/11/ Emmanuel, A., Donaldson, G.B., Band, W.T., Dew-Hughes, D.: IEEE Trans. Magn. MAG-11, 763 (1975)

/12/ Niemeyer, J.: PTB-Mitt. 84, 251 (1974)

/13/ Josephson, B.D.: Adv. Phys. 14, 419 (1965)

/14/ Johnson, W.J., Barone, A.: J. Appl. Phys. 41, 2958 (1970)

/15/ Basavaiah, S., Eldrige, J.M., Matisoo,J.: J. Appl. Phys. 45, 457 (1974)

/16/ Auracher, F., Richards, P.L., Rochlin, G.I.: Phys. Rev. B 8, 4182 (1973)

/17/ Harris, R.E.: Phys. Rev. B 10, 84 (1974)

/18/ Hasselberg, L.E., Levinsen, M.T., Samuelsen, M.R.: Phys Rev. B 9, 3757 (1974)

/19/ Taylor, B.N., Burstein, E.: Phys. Rev. Lett. 10, 14 (1963)

SUPERCURRENT INTERFERENCE PATTERNS IN A TUNNELING JUNCTION STRUCTURE.

E.P. Balsamo and G.Paternò
Laboratori Nazionali del CNEN, 00044 Frascati, Italy

A. Barone
Laboratorio di Cibernetica del CNR, 80072 Arco Felice, Naples, Italy
Istituto di Fisica, Università di Napoli, 80100 Napoli, Italy

M. Russo
Laboratorio di Cibernetica del CNR, 80072 Arco Felice, Italy

R. Vaglio
Istituto di Fisica, Università di Salerno, 84100 Salerno, Italy.

It is recognized how various features of a Josephson junction influence the dependence of the maximum zero-voltage current I_1 upon the applied magnetic field B. Barrier disuniformities can produce significant deviations of the I_1 vs. B dependence from the ordinary Fraunhofer-like pattern expected for "small" rectangular junctions (1,2). In particular in the usual oxide tunneling junctions of cross geometry edge effects can occur (3) in connection to a "thinning" of the oxide. Such effect results in a peaking of the tunneling current distribution at the edges of the junction causing the appearence of an interferential component in the I-B pattern. A detailed investigation of this topic is reported elsewhere (4).

In Fig. 1a is presented the experimental (dots) I_1 vs. B dependence of a $Pb\text{-}Pb_xO_y\text{-}In$ cross type junction ($0.30 \times 0.15\ mm^2$). The data are normalized to the maximum zero field supercurrent ($I_1(0) = 5.6$ m A). The interferential character of the pattern is evident. In Fig. 1b it is also reported the temperature dependence of the Josephson current for the same junction. Such curve recalls the I_1 vs. T dependence obtained for a double junction (5) (Clarke's solder blob).

The magnetic behaviour can be accounted by assuming a current density distribution J(x,y) as sketched in Fig. 2a. It is in fact (1) $I_1(B) = |\mathfrak{J}[B]|$ where $\mathfrak{J}[B]$ is the Fourier transform of $\mathcal{J}(x)$ which is the integral of J(x,y) along the applied field direction. Thus, for the assumed current density profile, we have

$$I_1 = J_1 W L \left| \frac{\xi l}{L} \frac{\sin \beta \frac{l}{L}}{\frac{\beta l}{L}} + \frac{2S}{L} \frac{\sin \frac{\beta S}{L}}{\beta \frac{S}{L}} \cos \beta \left(1 - \frac{S}{L}\right) \right|$$

with $\beta = \frac{e d L}{\hbar} B$

where d is the effective magnetic thickness, $l = L-2S$ and the other parameters are explained by Fig. 2a. The theoretical fitting (Fig.1, solid line) of the experimental data is obtained assuming in 1 the values $S/L=S'=0.02$ and $\xi = 0.02$.

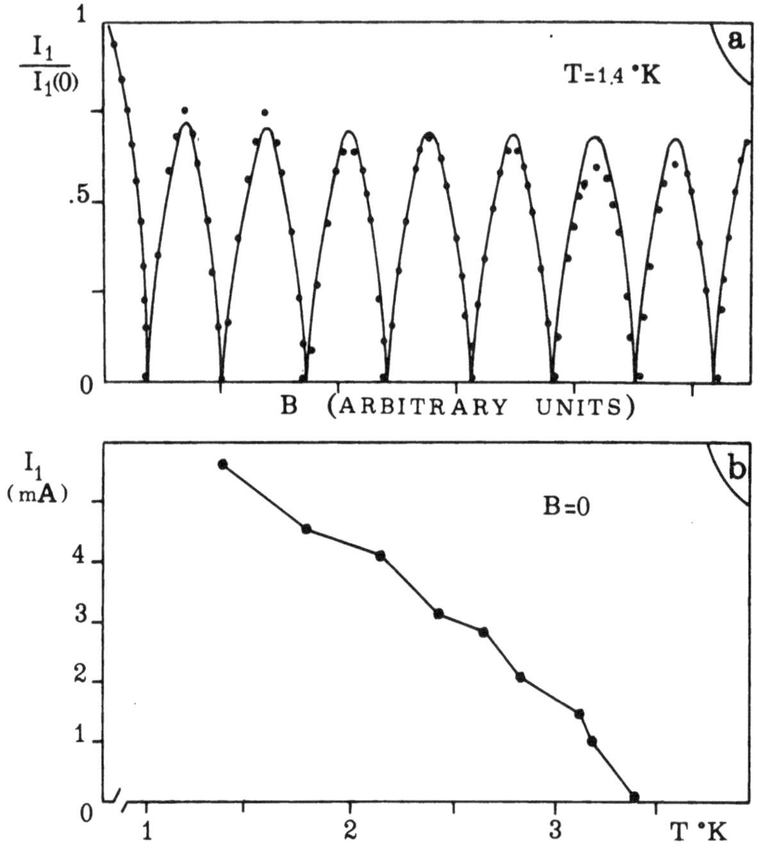

Fig.1 a) Normalized critical current vs. applied magnetic field for a $Pb-Pb_xO_y-In$ sample. Dots represent experimental data. Solid line indicates the theoretical curve corresponding to a step-like current density distribution with $S'=0.02$ and $\xi = 0.02$.
b) Qualitative temperature dependence of the same sample.

From the equation we see that in the limit of small S' ($S \ll L$) the envelope of the current secondary maxima I_m approaches zero at a value of the field corresponding to $1/S'$ flux quanta. In the case of Fig. 1a, for instance, $I_m \rightarrow 0$ after 50 flux quanta. Furthermore, we can see that the amplitude of I_m depends on both S' and ξ. From Fig. 3 it is clear that sufficiently high values of the ratio between I_m and the maximum zero field supercurrent $I_1(0)$ require values of ξ and S' of the same order of magnitude.

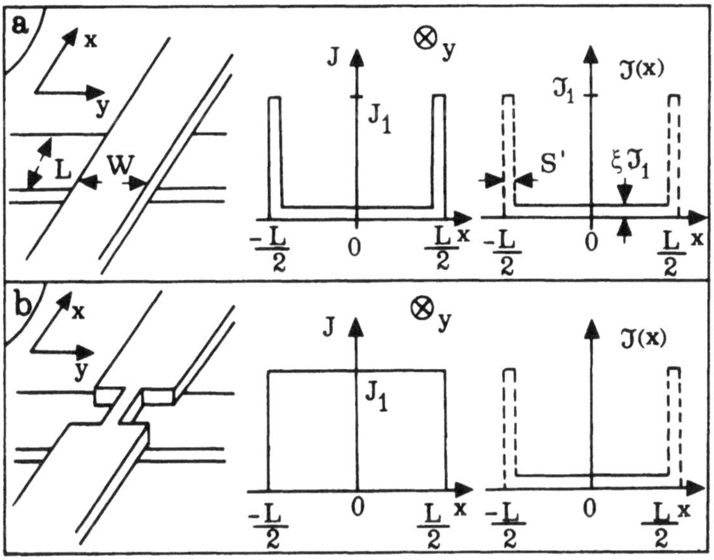

Fig.2 Junction configuration, current density distribution and integral of the current density along the y direction. a) cross geometry; b) "H" geometry.

Let us observe that what "plays" in determining the I_1 vs. B dependence is indeed $\mathfrak{J}(x)$. On the other hand a given $\mathfrak{J}(x)$ can be achieved either by a rectangular junction with a non uniform $J(x,y)$ or by assuming an uniform current density distribution with a suitable choice of the electrode geometry. In particular a step-like $\mathfrak{J}(x)$, as that of fig. 2a can be obtained for a junction with uniform $J(x,y)$ adopting an "H" shape electrode configuration (see Fig. 2b). Another reliable procedure which realizes this situation is achieved by using light-sensitive junctions (6). In this case the enhancement of the current at the junction edges is induced by light and controlled by varying the transparency of the superconducting electrodes. Again an $\mathfrak{J}(x)$ such as that of Fig. 2 can be obtained reflecting now the circumstance that the illumination is confined to the perimetrical region of the junction. Such devices may be of some interest whenever a double-junction like behaviour is required to work at "adjustable" maximum current level. Recent investigations (7,8) have shown an increasing reliability of semiconductor light sensitive junctions. In Fig. 4 it is reported a typical voltage current characteristics of a In-CdS-In junction at T =1.25 K. No zero-voltage current was present in "dark"

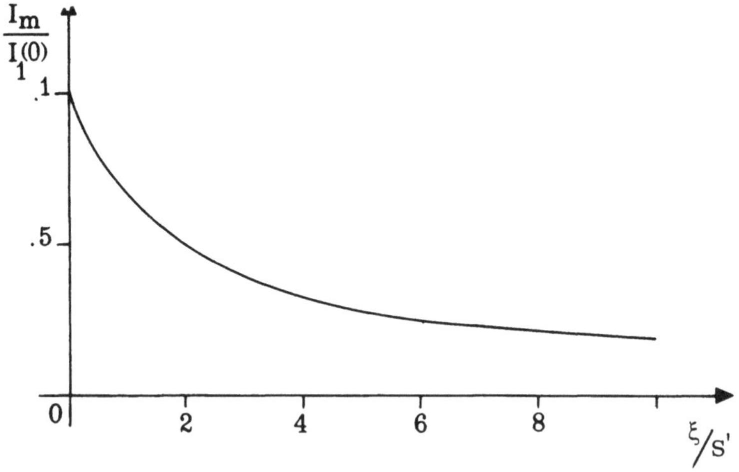

Fig. 3 Normalized "envelope current" I_m as a function of the ratio ξ/S' $(S \ll L)$

conditions. Furthermore a careful measurement of the light induced Josephson current vs. the applied magnetic field was performed giving the correct Fraunhofer-like pattern. In the complete V-I curve the gap structure was well defined.

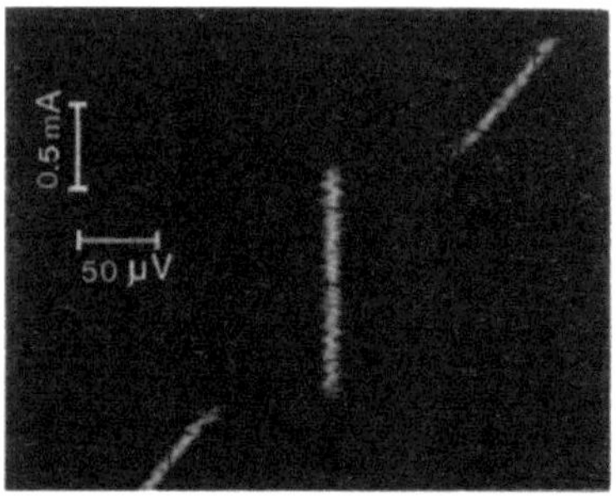

Fig. 4 Voltage current characteristics of a In-CdS-In junction after illumination. Horizontal: 50 μV/div; vertical: 500 μA/div.

It is worth to note in Fig. 4 the fully non hysteretic behaviour (single valued V-I dependence) of the junction which is of relevant importance in various kinds of applications.

Acknowledgements

The authors are grateful to G. Della Volpe, S. Piantedosi, C. Salinas and C. Salvia for their technical assistance.

References

1. Dynes, R.C., Fulton, T.A.: Supercurrent density distribution in Josephson junctions. Phys. Rev. B 3, 3015-3020 (1971)

2. Zappe, H.H.: Determination of the current density distribution in Josephson tunnel junctions. Phys. Rev. B11, 2535-2538 (1975)

3. Schwidtal, K., Finnegan, R.D.: Barrier-thickness dependence of the d.c. quantum interference effect in thin-film lead Josephson junctions. J.Appl. Phys. 40, 2123-2127 (1969).

4. Barone, A., Paternò, G., Russo, M. and Vaglio, R.: (to be published).

5. Halse, M.R., Salleh, K.M.: Temperature dependence of the critical current of a double Josephson junction. Proc. LT13 (Eds. K.D. Timmerhaus, W.J. O'Sullivan, E.F. Hammel Vol. 3, 293-296, Plenum Press, Boulder (1972).

6. Barone, A., Paternò, G., Russo, M. and Vaglio, R.: Experimental results on temperature and magnetic field dependence of light induced Josephson current. Proc. LT14 (Eds. M. Krusius, M. Vuorio) Vol. 4, 88-91, North Holland, Helsinki (1975).

7. Barone, A., and Russo, M.: Observation of light induced Josephson current through "thick" CdS barriers. Phys. Letters 49A, 45-46 (1974).

8. Barone, A., Paternò, G., Russo, M. and Vaglio, R.: Light induced transition from "small" to "large" Josephson junctions. Phys. Letters 53A, 393-394 (1975).

ANOMALOUS THERMOELECTRIC EFFECTS IN POINT CONTACT TUNNEL JUNCTIONS

Nancy K. Welker and Fernand D. Bedard
Laboratory for Physical Sciences
College Park, Maryland 20740, USA

INTRODUCTION

As SQUID detectors are pushed to higher and higher levels of sensitivity in magnetometry, metrology and electromagnetic wave detection, it becomes increasingly imperative that a thorough analysis of limiting noise sources be accomplished. To this end a great deal of work has been reported recently on the analysis of the noise limitations of total SQUID systems consisting of either oxide or weak link junctions closed by superconducting paths.[1,2] Noise in these systems arises from mechanisms intrinsic to the SQUID itself and from external sources such as room-temperature electronics, electromagnetic interference and thermal fluctuations in the bath. These bath temperature changes produce noise by the consequent changes in I_c and, in some systems, by the movement of trapped flux. Another mechanism which may produce noise is of thermoelectric origin, wherein a temperature gradient across the Josephson junction produces a current.

It has long been accepted and shown experimentally,[3,4] that the normal metal thermoelectric effects--the Thomson, Peltier and Seebeck effects--do not exist in superconductors. However, it has been suggested[5] that based on the London two-fluid model[6] of superconductivity and an analogy with superfluid helium, a thermoelectrostatic effect should exist in superconductors. This model leads to a thermally generated supercurrent flow in a superconducting sample or, of greater interest here, through the junction in an unequally heated point contact, a common component in SQUID detectors. This paper describes experimental work which addresses the question of the existence of this supercurrent.

The analogy between superconductors and superfluid helium is quite good. If two vessels of superfluid helium are connected and held at different temperatures, there will be a flow of superfluid from the colder to the hotter to balance the concentration of normal and superfluid components in each. If the two vessels are connected by a very narrow capillary, there is no corresponding flow of normal fluid from the hotter to the colder, and a hydrostatic pressure difference will be established. In a superconductor the clamping of the normal electron flow by lattice scattering corresponds to the restriction of the normal helium fluid flow by the thin capillary. This allows us to make the approximation that the normal electron velocity $v_n = 0$. For the stationary state where $\partial v_s/\partial t = 0$ (v_s = superelectron velocity), this leads to the conclusion that the gradient of the electrochemical potential is zero. In other words there exists an electrostatic field

(1) $\bar{E} = \nabla(\mu/e)$

where μ is the chemical potential. If the two ends of a superconducting wire are held at different temperatures, the resulting potential difference can only be measured by electrostatic means and has a magnitude $\leq 10^{-7}V/^{o}K$. Another means of detection arises from the description of the London two-fluid model of superconductivity in which the normal current density in the presence of a temperature gradient is given by

(2) $\bar{J}_n = \sigma(\bar{E} - \nabla\mu/e) - \sigma\alpha\nabla T$

where σ is the normal conductivity and α is the normal thermoelectric coefficient. Substituting Equation (1) into Equation (2) leads to a normal electron current density and, by the London model, a counterflowing superelectron current density of the form

(3) $\bar{J}_n = -\sigma\alpha\nabla T = -\bar{J}_s.$

The implications of this supercurrent flow must be considered when operating point contact SQUIDs in the presence of a possible temperature gradient. For example, for niobium the thermoelectric coefficient $\alpha \sim 5 \times 10^{-7} V/^{o}K$. With a typical junction resistance of 10 Ω, a current offset of 5×10^{-8} amps would result from a $1^{o}K$ temperature gradient. Correspondingly greater critical current shifts should be observable for lower resistance junctions.

A number of experiments have been reported recently whose positive results tend to confirm this picture of thermoelectrostatic effects in superconductors. In one[7] the experimenters looked for the electrostatic potential difference by making measurements on a pair of unequally heated capacitor plates made up of the superconducting samples. However, the results of this experiment were more than an order of magnitude larger than predicted by the theory. Another experiment by Clarke and Freake[8] was based on the assumption that less external current should be needed to switch a point contact Josephson junction out of its zero resistance state when it is in the direction of the thermally generated counterflow supercurrent than when it is opposed to it. This experiment yielded shifts in critical current that were always in the direction predicted by the theory, but there was some inconsistency in the magnitude of these shifts.

Further evidence that the effect exists comes from experiments looking for thermomagnetic effects. Here, by a description similar to that given above, it can be argued that, if one has two dissimilar superconductors joined together to form a loop with the junctions held at different temperatures, then magnetic flux should be generated in the loop. This experiment has been performed with positive results by several groups, but again the magnitude of the effect has not been consistent with the theory.[9] If, because of the general lack of agreement with the magnitude of the predicted effect, we look at all of these experiments to see what could be the cause, we find one common element in the experimental procedures. In all cases the samples are prepared in atmosphere so that there is always some sample contamination due to oxides and other impurities.

EXPERIMENT

In order to resolve the issue of contamination, we have performed a set of experiments using point contact Josephson junctions freshly formed at ultra-high vacuum and 4.2 K using the apparatus shown in Figure 1. The measurements

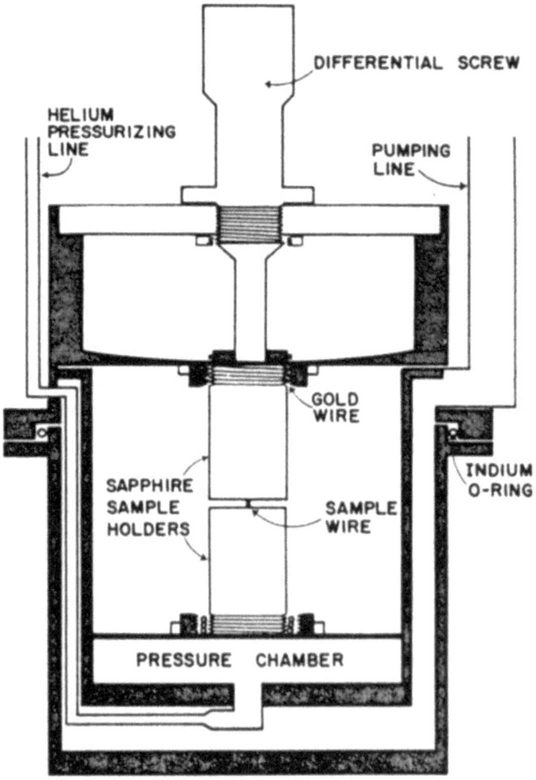

Figure 1 Cross Section of Experimental Chamber

made are modeled after those of Clarke and Freake described above.

Here the sample consists of a single piece of .25 mm diameter superconducting wire which is mounted in two cylindrical sapphire sample holders spaced $\leq$ .25 mm apart. This is done by slicing and grooving the sapphire cylinders, placing the wire in the grooves, pressing the halves of the sapphires together and securing them with clamps as shown in Figure 2. One half of each clamp is wound with a heater, and the other half contains a carbon resistance thermometer which is calibrated using the transition temperatures of Nb and Va wires and the boiling point of liquid helium. The two halves are thermally insulated at the hinge so that heat will be transmitted through the sapphire, not around the clamp.

Returning to Figure 1 we can observe the means for actually forming the point contact. The top plate of the chamber can be pulled back using a differential screw with a total

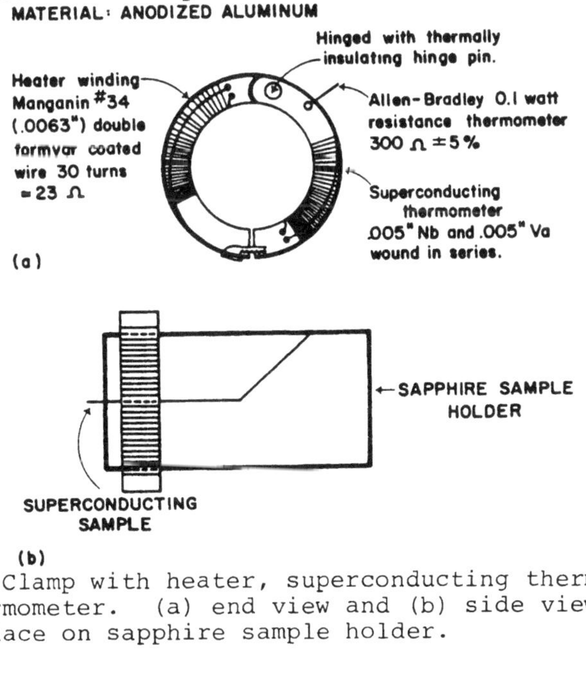

Figure 2 Clamp with heater, superconducting thermometer and carbon thermometer. (a) end view and (b) side view showing clamp in place on sapphire sample holder.

travel of 1.25 mm. This plate has been exponentially curved to allow ± .625 mm deflection with no plastic deformation. The broken ends of the wire can then be brought back to nearly touching, and a fine adjustment on their relative positions can be made by changing the helium gas pressure behind the ~5 mm thick aluminum diaphram which forms the bottom plate of the chamber. Changes in the position of this bottom plate can be made in steps as small as .1 nm. The aluminum walls of the chamber and the sapphire sample holders thermally tie the sample to the bath.

The experimental procedure consists of sealing the chamber, pumping to a high vacuum (~10^{-8} Torr) and immersing it in liquid helium. The additional cryopumping produces a vacuum of at least 10^{-10} Torr in the chamber. At this point the wire is broken to form the two sides of a point contact, and the contact is adjusted to have the desired critical current. We then heat neither, one or both sides of the junction and proceed with the electrical measurements to determine the shift in critical current under various heating conditions compared with the case when neither side is heated. In practice we adjust the junction to have a critical current under 25 μa and then use a dc current sweep slightly greater than the critical current. Superimposed on this is a 500 Hz ac current of the same peak-to-peak amplitude. The ac voltage across the junction is measured by means of a lock-in amplifier tuned to 1000 Hz to detect the second harmonic. Since the voltage is an odd function of the current through a Josephson junction, the second harmonic goes to zero when the ac sweep is exactly centered on the zero voltage part of the V vs I curve. Thus, finding the shift in the center of the zero voltage portion of the curve is exactly the same as finding the shift in the critical current. Figure 3 shows the block diagram of the data collection system used. The

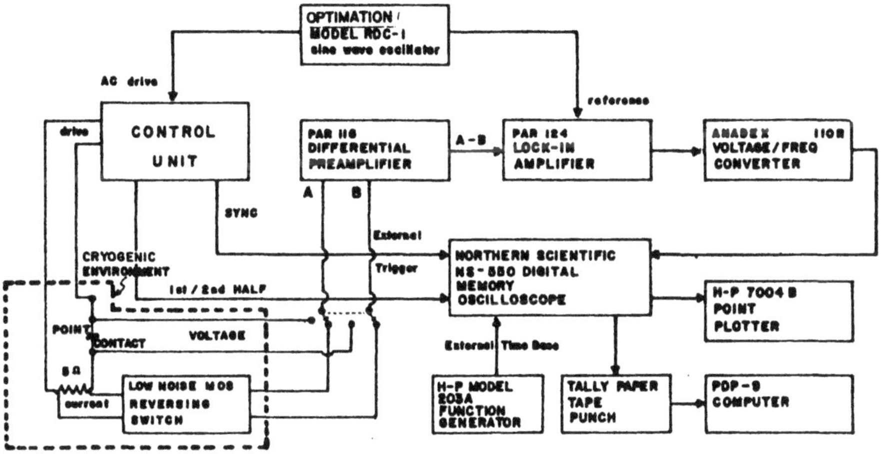

Figure 3 Block Diagram of Data Collection System

data collected in the 2048 address memory oscilloscope is output to paper tape for computer analysis.

Data from a typical run is shown in Figure 4. (A) represents the dV/dI vs I curve for a junction. This is needed

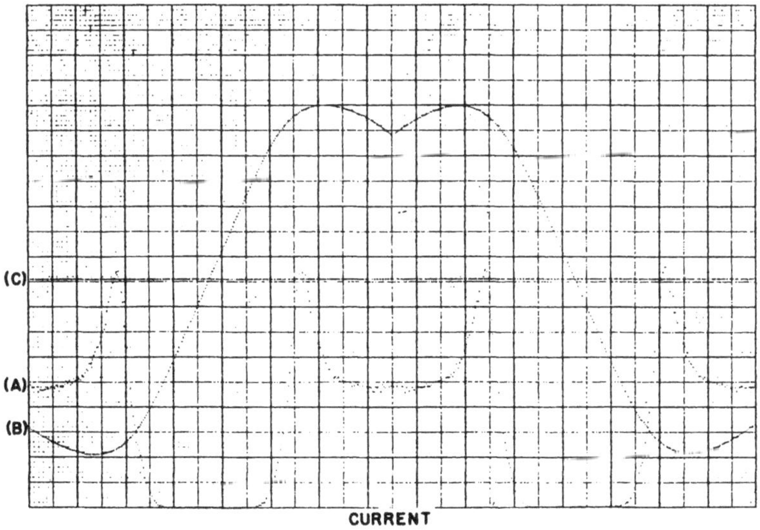

Figure 4 Typical Data Plots for a Junction (A) $\frac{dV}{dI}$, (B) Second Harmonic and (C) Second Harmonic Zero, Versus I.

since, to compute the thermoelectric constant, we must know the resistance outside the Josephson tunneling region. There are two traces, one for negative going sweep and one for positive going sweep, to eliminate the possibility of systematic effects due to the direction of current sweep. (B) represents the second harmonic measurement. To find where the second harmonic goes to zero, the ac signal is removed and the measurement repeated producing curve (C). To find the crossing point accurately the digitized data is fed to a computer which has been programmed to find the least squares fit to a straight line for the portion of the curve near the crossing and then find the intersection of the two straight lines.

RESULTS AND CONCLUSIONS

The material investigated was Nb for which the expected α is $\sim .5\ \mu V/^{o}K$. The α's found experimentally for clean samples were consistently smaller than this, and, more importantly, their signs were random with the average nearly zero.[10] The same situation holds true if we assume that the temperature dependence of α for Nb follows that demonstrated by Steele[11] to apply to some other superconducting metals, i.e.,

(4) $\alpha = \lambda T$.

The λ's have been computed, and the results are shown in Figure 5. Here the expected value is $.07\ \mu V/^{o}K$, and the average experimental value is $-.008 \pm .010\ \mu V/^{o}K$. Therefore, we must conclude that for clean Nb samples either there is no thermoelectrostatic effect or that this model is not adequate to explain it. Further, since the predicted thermoelectric power of Nb is higher than that of most other superconductors, we can reach the same conclusion about superconductors in general.

In view of this conclusion, an explanation is necessary for the many reported positive results. We tried to obtain

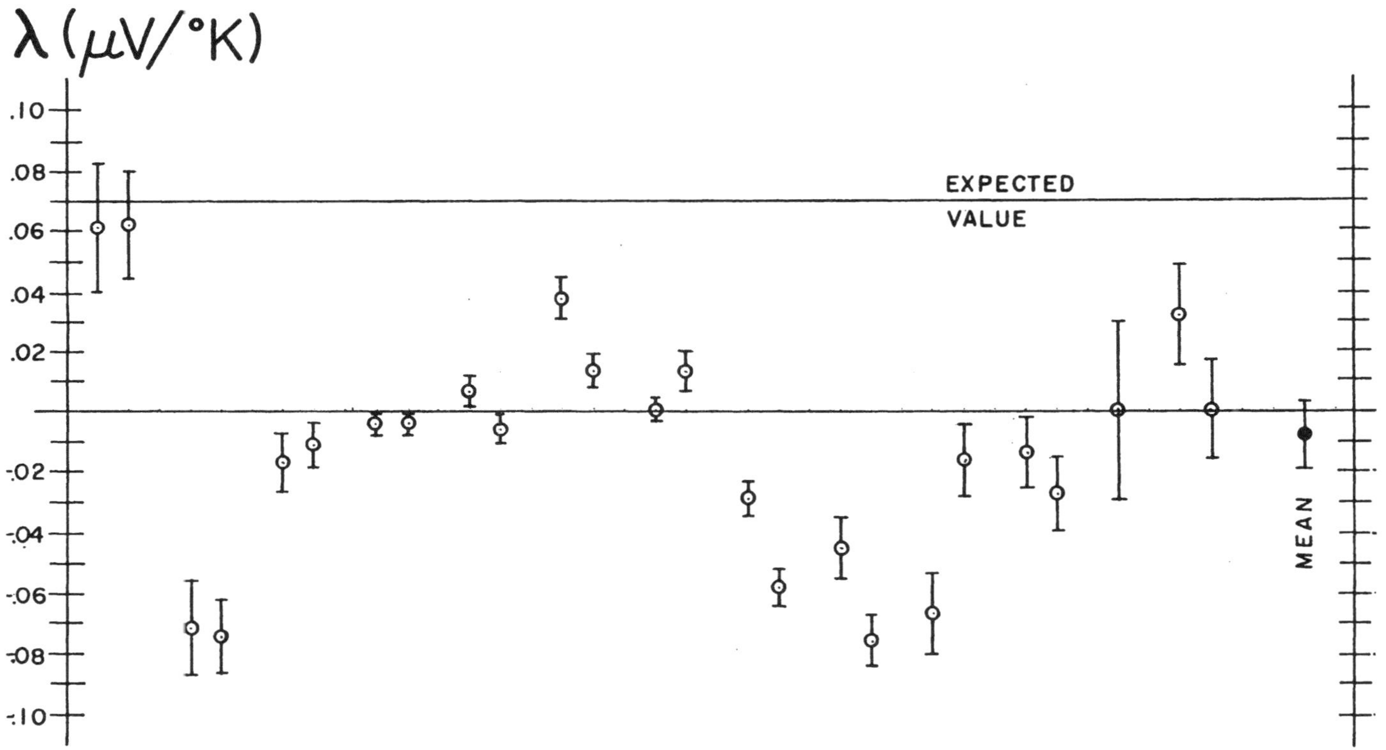

Figure 5 Values of λ for Clean Samples

such an explanation by simulating these other experimental situations by allowing our sample to age after the wire was broken but still at 10^{-10} Torr and 4.2 K so that contamination would occur in a slow and controlled fashion. We then repeated our experimental measurements after one, three, seven and nine days of aging, and the results are shown in Figure 6. A shift is now clearly observable. The magnitude varies but is of the same order of magnitude as the theory predicts with a mean value of .031 $\mu V/^{o}K$. However, it is consistently of the <u>wrong sign</u>. To explain these results we believe that a new model will have to be developed. One possibility is that the thermally generated superelectrons are moving in response to differences in super and normal electron concentrations, not in response to a thermoelectric potential α. This would be consistent with the previous point contact tunneling experiment of Clarke and Freake. Since their sample was Pb which has a negative thermoelectric coefficient α, they would observe the same polarity shift from either model. In the case of Nb the thermoelectric coefficient α is positive, and the two models would produce signals of opposite sign. It is clear from the present experiment that the model which ties the supercurrent flow to the equalization of concentrations provides for better agreement with our data. At present the model is, of course, qualitative and says nothing about the magnitude of the effect. However, any proposed model must not only explain the null result obtained with a clean surface but also must include the effect of a surface contamination layer.

We hope in the near future to further confirm this description by repeating the experiment with other materials having both positive and negative thermoelectric coefficients.

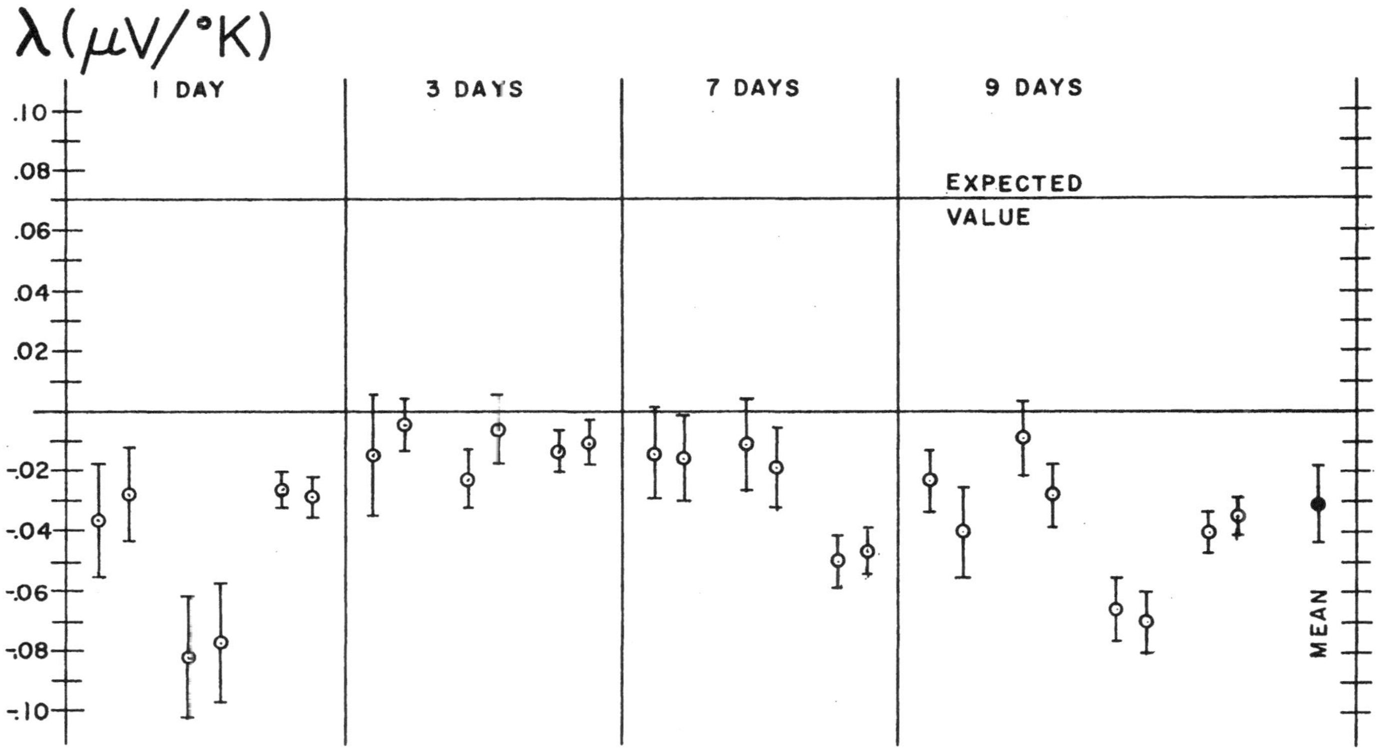

Figure 6 Effect of Aging on λ.

REFERENCES

1. Clarke, John, Goubau, Wolfgang M., Ketchen, Mark B.: A Reliable DC SQUID Made with Tunnel Junctions, IEEE Transactions on Magnetics, Vol. MAG-11, No. 2, 724-727 (1975).

2. Jackel, L.D., Clark, T.D., Buhrman, R.A.: Current-Phase Relations and Noise in rf Biased SQUIDs, IEEE Transactions on Magnetics, Vol. MAG-11, No. 2, 732-735, (1975).

3. Steiner, K., Grassmann, P.: Eine obere Grenze der Thermokraft Zwischen Supraleitern, Phys., Z. 36, 527-528 (1935).

4. Daunt, J.G., Mendelssohn, K.: An experiment on the mechanism of superconductivity, Proc. Roy. Soc. A185, 225-239 (1946).

5. Putterman S., de Bruyn Ouboter, R.: Are there Thermoelectric Effects in Superconductors?, Phys. Rev. Letters 24, 50-52 (1970).

6. London, Fritz: Superfluids (John Wiley & Sons, New York, 1950), Vol. I.

7. de Waele, A. Th. A. M., de Bruyn Ouboter, R, Pipes, P.B.: Thermoelectrostatic Effects in Superconductors, Low Temperature Physics - LT 13, K.D. Timmerhaus, W.J. O'Sullivan and E.P. Hammel, eds. (Plenum Press, New York, 1974), Vol. 3, Chap 7, 772-776.

8. Clarke, John, Freake, S.M.: Superconducting Fountain Effect, Phys. Rev. Letters 29, 588-591 (1972).

9. Pegrum, C.M., Guenault, A.M., Pickett, G.R.: An Experimental Investigation of the Thermoelectric Generation of Flux in a Superconducting Bimetallic Loop, Low Temperature Physics - LT 14, M. Krusius and M. Vuorio, eds. (North-Holland Publishing Company,

Amsterdam, 1975) Vol. 2, 513-516.

10. Welker, Nancy K., Bedard, Fernand D.: Thermoelectrostatic Effects in Superconductors, Low Temperature Physics - LT 14, M. Krusius and M. Vuorio, eds. (North-Holland Publishing Company, Amsterdam, 1975) Vol. 2, 517-520.

11. Steele, M.C.: Thermoelectromotive Force of a Superconductor Versus the Same Metal in the Non-Superconductive State, Phys. Rev. 81, 262-267 (1951).

LOW FREQUENCY APPLICATIONS

CURRENT PERFORMANCE OF SUPERCONDUCTING QUANTUM INTERFERENCE DEVICES*

John Clarke
Department of Physics, University of California, Berkeley, and Materials and Molecular Research Division, Lawrence Berkeley Laboratory, Berkeley, California 94720.

This article briefly surveys the performance of SQUIDs as low frequency detectors, and discusses how improvements in sensitivity might be achieved. The paper summarizes an extended review[1] to be published in the proceedings of the NATO Advanced Study Institute on Small Scale Superconducting Devices, Lago di Garda, Italy, September 1-10, 1976.

DC SQUID

Principles

In the dc SQUID[2], two Josephson[3] junctions are mounted on a superconducting ring of inductance L. The critical current, I_m, of the two junctions is periodic in the external magnetic flux applied to the ring, Φ_e, with a period Φ_o. The SQUID can be used to detect small changes in Φ_e ($<< \Phi_o$) by measuring the corresponding change in I_m.

An order-of-magnitude estimate of the flux resolution may be made as follows. In the limit $\beta = LI_m/\Phi_o >> 1$, the critical current modulation depth $\Delta I_m = I_m[n\Phi_o] - I_m[(n + \frac{1}{2})\Phi_o] \rightarrow \Phi_o/L$. However, most SQUIDs are operated with $\beta \approx 1$, for which value[4] $\Delta I_m \approx \Phi_o/2L$. If $L \approx 10^{-9}$ H, $\Delta I_m \approx 1\ \mu A$. We assume that each junction can be represented by a resistively shunted tunnel junction[5] with $\beta_c = 2\pi I_c R^2 C/\Phi_o \lesssim 1$, so that the current-voltage characteristic is non-hysteretic. Here I_c, C, and R are the critical current, capacitance, and shunt resistance of each junction. The current-voltage (I-V) characteristic of the SQUID is also periodic in Φ_e, provided that the Josephson frequency is not much greater than the characteristic frequency of the ring, $(2R/L)/2\pi$. If $R \approx 1\ \Omega$ and $L \approx 1$ nH, $R/\pi L$ is about 0.3 GHz, corresponding to a voltage $\Phi_o R/\pi L$ of roughly 1 μV. If the SQUID is biased with a constant current I_o near this voltage, the voltage modulation depth is $\Delta V \approx \Phi_o R/4L$. The transfer function is

$$(1) \quad \left(\frac{\partial V}{\partial \Phi_e}\right)_{I_o} \approx \frac{R/2}{L} ,$$

about 0.5 $\mu V\ \Phi_o^{-1}$. Equation (1) is a measure of the signal available from the SQUID.

The intrinsic noise of the SQUID can be approximated by the Johnson noise in the resistive shunts (for a more detailed treatment, see ref. 4). The power spectrum of the voltage noise is

$$(2) \quad S_V(f) \approx 4k_B T(R/2) ,$$

where T is the temperature of the SQUID. The power spectrum, S_Φ, of the flux noise is readily obtained from Eqs. (1) and (2):

$$(3) \quad S_\Phi \approx 8k_B TL^2/R .$$

The energy resolution is

$$(4) \quad \frac{S_\Phi}{2L} \approx \frac{2k_B T}{(R/2L)} .$$

Equation (4) indicates that the energy resolution of the SQUID per unit bandwidth is essentially $k_B T$ divided by the characteristic frequency or "sampling frequency" of the SQUID, R/2L. For $R = 1\ \Omega$, $L = 10^{-9}$ H, and $T = 4$ K, $S_\Phi^{\frac{1}{2}} \approx 10^{-5}\ \Phi_o\ \text{Hz}^{-\frac{1}{2}}$, and $S_\Phi/2L \approx 2 \times 10^{-31}\ \text{JHz}^{-1}$. A more detailed calculation[4] shows that Eq. (4) should be multiplied by a dimensionless factor that is a complicated function of I_o, R, L, and T. For the parameter values given, the detailed theory indicates that $S_\Phi^{\frac{1}{2}}$ is a factor of about 3 higher than these estimates.

For most applications, the SQUID is coupled to a superconducting flux transformer. The appropriate figure of merit in the zero frequency limit is the energy resolution per Hz referred to the input coil of inductance L_i coupled to the SQUID[1,6]:

$$(5) \quad \frac{S_\Phi}{2M_i^2/L_i} = \frac{S_\Phi}{2\alpha^2 L} ,$$

where $M_i^2 = \alpha^2 L L_i$.

Practical Device

The most sensitive dc SQUID is that of Clarke, Goubau, and Ketchen[7]. The configuration is shown in Fig. 1. The substrate is a fused quartz tube 20 mm long with an outside diameter of 3 mm. A band of Pb/In alloy (∿ 10% wt. In) about 11 mm wide and 0.3 μm thick is evaporated around the tube. A 250 μm wide, 75 nm-thick Au film is then evaporated: this film is the shunt for the tunnel junctions. Next, two 150 μm-wide, 0.3 μm-thick Nb films, separated by 1.2 mm, are dc sputtered onto the cylinder, making low resistance contacts with the Au film and, at low temperatures, superconducting contacts with the Pb/In band. The Nb is thermally oxidized, and immediately afterwards, a 0.3 μm-thick Pb/In tee is deposited. The cross bar of the tee overlaps the niobium strips to form two tunnel junctions, each with an area of about 10^{-2} mm^2. Next, the Pb/In band is scribed midway between the Nb strips. Two In pellets are pressed on as contacts, one on the base of the tee, and the other on the Pb/In band. The entire sensor is coated with a thin insulating layer of Duco cement. Finally, a 0.3 μm-thick Pb/In ground plane (not shown in Fig. 1) is evaporated over the slit in the Pb/In band and the strips that form the junctions to mini-

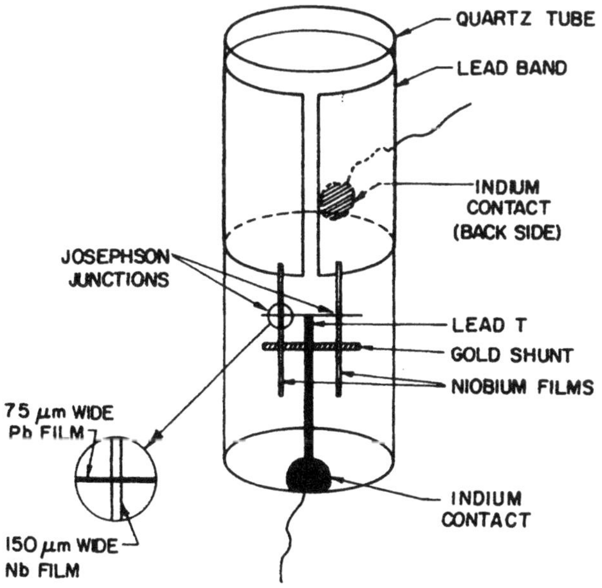

Fig. 1 Thin film tunnel junction dc SQUID

mize the parasitic inductances. A 50-nm film of Ag deposited over the ground plane protects it from oxidation.

Typical parameters for the sensor are: capacitance per junction, 200 pF; total critical current, 1 to 5 μA; parallel shunt resistance, 0.5 Ω; and total inductance, about 1.25 nH (∿ 0.75 nH for the cylinder, and ∿ 0.5 nH for the Nb and Pb/In strips). For testing, and for use in conjunction with a flux transformer, the SQUID is enclosed in a superconducting tube to screen out fluctuations in the external magnetic field. The tube acts as a ground plane that reduces the inductance of the SQUID cylinder to about 0.5 nH. The total SQUID inductance is thus about 1 nH.

The SQUID is invariably operated in a feedback mode. The SQUID is biased at a constant current I_o greater than the critical current, and a 100 kHz-modulation flux of peak-to-peak amplitude $\Phi_o/2$ is applied by means of a coil inside the SQUID. The ac voltage appearing across the SQUID is amplified with a cooled tank circuit ($Q \sim 100$) or by a cooled transformer with a gain of about 300. After further amplification with a FET preamplifier, the signal is lock-in detected, using a 100 kHz reference. The output of the lock-in is zero when $\Phi_e = n\Phi_o/2$, and a maximum or minimum when $\Phi_e = (n \pm \frac{1}{4})\Phi_o$. The output of the lock-in is further amplified, integrated, and fed back as a current into the coil in the SQUID.

Performance

For the case in which the SQUID is matched to the preamplifier with a transformer, typical performance parameters for the SQUID in the feedback mode are: dynamic range, $\pm 3 \times 10^6$ in a 1 Hz bandwith; frequency response 0 to 50 kHz; and slewing rate, $2.5 \times 10^5\ \Phi_o s^{-1}$. The noise and drift of the SQUID were measured with a 24-turn superconducting coil wound on the outside of the Pb/In cylinder. The energy resolution $S_\Phi/2\alpha^2 L$ referred to the input coil is plotted in Fig. 2, and the relevant parameters are given in Table 1. The roll-off above 200 Hz is due to filtering in the electronics. The resolution is about $7 \times 10^{-30}\ JHz^{-1}$ in the white noise region, corresponding to a rms flux noise of about $3.5 \times 10^{-5}\ \Phi_o Hz^{-\frac{1}{2}}$. We estimate $\alpha^2 \approx 0.5$, so that the energy resolution referred to the SQUID is about $3 \times 10^{-30}\ JHz^{-1}$.

The long term drift of the SQUID in a helium bath whose temperature was regulated to ±50 μK was $\lesssim 2 \times 10^{-5} \Phi_o h^{-1}$.

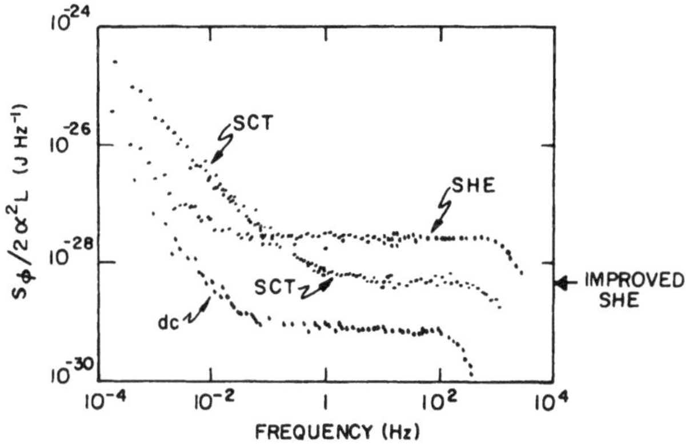

Fig. 2 Noise power spectra (plotted as energy resolution) for tunnel junction dc SQUID and S.C.T. and S.H.E. toroidal rf SQUIDs (Summer 1975). Subsequently, the white noise of the S.H.E. SQUID was improved to the level shown.

SQUID	S_Φ ($\Phi_o Hz^{-1}$)	M_i (nH)	L_i (nH)	$S_\Phi/2\alpha^2L$ ($10^{-30} JHz^{-1}$)
dc (Clarke *et al.*[7])	1.2×10^{-9}	11.5	356	7
rf (S.H.E.[16]) 19 MHz	5×10^{-9}	20	2×10^3	50
rf (S.C.T.[17]) 30 MHz	6.6×10^{-10}	3	360	50
rf (Pierce *et al.*[18]) 10 GHz	1×10^{-10}	8	500	2

Table 1 Flux noise power spectrum (S_Φ), mutual inductance with input coil (M_i), inductance of input coil (L_i), and figure of merit ($S_\Phi/2\alpha^2L$) for several SQUIDs.

Future Developments

At measurement frequencies above 2×10^{-2} Hz, the resolution of the tunnel-junction dc SQUID at 4 K is limited by its intrinsic noise. From Eq. (4), we see that at a fixed temperature, the performance can be improved by increasing the sampling frequency R/L. This increase is subject to the constraints $\beta_c = 2\pi I_c R^2 C/\Phi_o \lesssim 1$ and, as can be deduced from the more detailed theory, $\beta \sim 1$. A reduction in L improves the energy resolution only if the coupling efficiency to the input coil is not also reduced. The alternative means of improving the resolution is to increase R, simultaneously decreasing C to maintain $\beta_c \lesssim 1$. It should be possible to reduce the junction area by four orders of magnitude to $\sim 1 \times 1\,\mu m$ (a size attainable by photoresist techniques) maintaining the critical current at $\sim 1\,\mu A$ (the critical current density required, $\sim 10^2\,A\,cm^{-2}$, is easily attainable). C would be decreased by four orders of magnitude, and the shunt resistance of each junction could then be increased to $\sim 100\,\Omega$ without introducing hysteresis. If L is maintained at 10^{-9} H, this procedure should decrease $S_\Phi/2\alpha^2 L$ by two orders of magnitude to a value of 10^{-31} JHz^{-1} or less.

RF SQUID

Principles

The rf SQUID consists of a single Josephson junction mounted on a superconducting ring[8,9], with $LI_c \sim \Phi_o$. The ring is coupled to the coil of a LC-resonant circuit that is excited by a sinusoidal current, I_{rf}, at its resonant frequency (typically 30 MHz). The rf voltage, V_T, developed across the tank circuit is amplified, and detected with a diode (for example). If one plots V_T (vertically) versus I_{rf} (horizontally), one obtains a series of "steps" and "risers". On the steps, V_T is nearly independent of I_{rf}. If Φ_e is slowly changed, the voltages at which the steps appear oscillate with period Φ_o. It can be shown[1,8] that the transfer function referred to the tank circuit is

$$(6)\quad \left(\frac{\partial V_T}{\partial \Phi_e}\right)_{I_{rf}} \approx \frac{\omega L_T}{M} ,$$

where $\omega/2\pi$ is the rf frequency, L_T is the tank circuit inductance, and $M = K(LL_T)^{1/2}$. The value of M cannot be reduced indefinitely, since the influence of the SQUID on the tank circuit would become insignificant. An approximate lower limit on M is set by requiring that the dissipation in the SQUID be no smaller than the dissipation in the tank circuit. A straightforward analysis[1] shows that this requirement is equivalent to

$$(7)\quad K^2 Q \gtrsim 1 ,$$

where Q is the quality factor of the tank circuit. If $Q \sim 100$, $K \gtrsim 0.1$. Taking the values $\omega/2\pi \approx 30$ MHz, $L_T \approx 100$ nH, $L \approx 1$ nH, $K \approx 0.1$, and $M \approx K(LL_T)^{1/2} \approx 1$ nH, we find $(\partial V_T/\partial \Phi_e)_{I_{rf}} \approx 40\ \mu V \Phi_o^{-1}$.

The rf SQUID is usually operated in a feedback circuit. A flux modulation ($\Phi_o/2$ peak-to-peak and typically at 100 kHz) is applied to the SQUID; after rf detection, the 100 kHz signal is lock-in detected and the smoothed output is fed back to the tank circuit coil.

The three sources of noise in the rf SQUID, intrinsic noise, tank circuit noise, and preamplifier noise have been reviewed in detail by Jackel and Buhrman[10]. The intrinsic flux noise has a power spectrum[10,11]

$$(8) \qquad S_\Phi^{(i)} \approx \frac{\pi\varepsilon^2\Phi_o^2}{\omega},$$

where ε (assumed to be << 1) is the ratio of the voltage rise along a step in the V_T-I_{rf} characteristic to the separation in voltage of successive steps. The tank circuit Johnson noise gives rise to an equivalent flux noise with a power spectrum[1,10,12]

$$(9) \qquad S_\Phi^{(tc)} \approx \frac{4\pi\varepsilon^2 k_B T_e L}{\omega},$$

where T_e is the effective temperature of the tank circuit. We have assumed $LI_c \approx \Phi_o$. The equivalent flux noise of the preamplifier has a power spectrum found by dividing the power spectrum, $S_V^{(p)}$, of the voltage noise by $(\partial V_T/\partial \Phi_e)^2_{I_{rf}}$:

$$(10) \qquad S_\Phi^{(P)} \approx \frac{M^2 S_V^{(p)}}{\omega^2 L_T^2}.$$

This result may be written in a convenient form by defining a noise temperature, $T_N^{(p)}$, for the preamplifier through the relation

$$(11) \qquad S_V^{(p)} = 4k_B T_N^{(p)} \varepsilon R_T .$$

Here, R_T is the resistance of the tank circuit on resonance in the absence of the SQUID, and εR_T is thus the approximate dynamic resistance on a step in the presence of the SQUID. If we insert Eq. (11) in Eq. (10) and set $K^2Q = 1$ with $Q = R_T/\omega L_T$, we find

$$(12) \qquad S_\Phi^{(p)} \approx \frac{4\varepsilon k_B T_N^{(p)} L}{\omega}.$$

Notice from Eq. (11) that $T_N^{(p)} \propto 1/\omega$ if Q and ε remain constant, so that $S_\Phi^{(p)}$ is proportional to $1/\omega^2$.

The overall energy resolution is found from Eqs. (8), (9), and (12):

$$\text{(13)} \quad \frac{S_\Phi}{2L} \approx \frac{\pi\varepsilon^2\Phi_o^2}{2\omega L} + \frac{2\pi\varepsilon^2 k_B T_e}{\omega} + \frac{2\varepsilon k_B T_N^{(p)}}{\omega} .$$

This expression assumes that $LI_c \approx \Phi_o$ and $K^2Q \approx 1$. The intrinsic energy resolution is proportional to the energy available per cycle, $\sim \Phi_o^2/L$, divided by ω. The second and third terms represent the thermal energies of the tank circuit ($k_B T_e$) and preamplifier ($k_B T_N^{(p)}$) divided by ω. It should be emphasized that these expressions are approximate, and are accurate only to within factors of 2 or 3.

It is instructive to make estimates for the three contributions. If we take as typical values $\varepsilon \approx 0.2$, $L \approx 1$ nH, $\omega/2\pi \approx 30$ MHz, $T_e \approx 200$ K, and $T_N^{(p)} \approx 50$ K, we find $S_\Phi^{(i)}/2L \approx 10^{-30}\,\text{JHz}^{-1}$, $S_\Phi^{(tc)}/2L \approx 4 \times 10^{-30}\,\text{JHz}^{-1}$, $S_\Phi^{(p)}/2L \approx 2 \times 10^{-30}\,\text{JHz}^{-1}$, and $S_\Phi/2L \approx 10^{-29}\,\text{JHz}^{-1}$. The total rms flux noise is $\sim 7 \times 10^{-5}\,\Phi_o\,\text{Hz}^{-\frac{1}{2}}$. As is usually the case for the rf SQUID, the intrinsic noise is relatively insignificant.

Practical Devices

A selection of rf SQUIDs is shown in Fig. 3. These include the point contact SQUID (a) of Zimmerman <u>et al.</u>[8], the thin film SQUID (b) of Mercereau and co-workers[9], the two-hole SQUID (c)[8,13], and the toroidal SQUID (d)[14]. Another ingeneous point contact device (not shown) is the fractional-turn SQUID of Zimmerman[15]. Each device has an inductance of 1 nH or less, and a critical current $\sim \Phi_o/L$. Most devices are operated at 20 to 30 MHz. However, much higher frequencies have been used in order to improve the resolution.

Performance

In the feedback mode, typical performance parameters are: dynamic range, $\pm 10^6$ in a 1 Hz bandwidth; frequency response, 0 to a few kHz; slewing

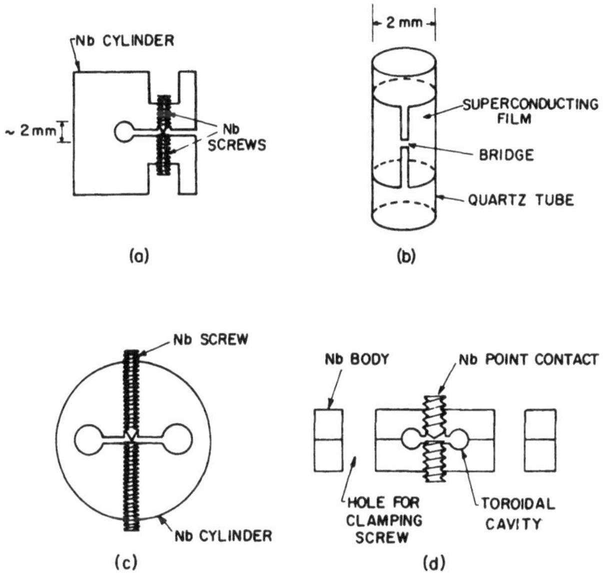

Fig. 3 Selection of rf SQUIDs: (a) point-contact rf SQUID, machined from niobium; (b) thin-film rf SQUID evaporated on quartz tube; (c) two-hole point-contact rf SQUID, machined from niobium; (d) toroidal point-contact rf SQUID, machined from niobium.

rate, 10^5 to 10^6 $\Phi_o s^{-1}$; and rms flux noise, 10^{-4} Φ_o $Hz^{-1/2}$ (in the white noise region). The flux resolution is in quite good agreement with the value calculated earlier. At Berkeley, in the Summer of 1975, we measured the noise power spectra of the toroidal rf SQUIDs of S.C.T.[16] and S.H.E.[17]. The power spectra obtained, plotted as $S_\Phi/2\alpha^2L$, are shown in Fig. 2. Since these spectra were obtained, the white noise of the S.H.E. device has been improved to about 5×10^{-29} JHz^{-1}. The values of S_Φ, M_i, L_i, and $S_\Phi/2\alpha^2L$ (in the white noise region) are presented in Table 1 for three rf SQUIDs (in the case of the S.H.E. device, the new value of S_Φ has been used).

Numerous workers have attempted to achieve better flux resolution by working at higher frequencies. For example, Pierce et al.[18] operated a thin film cylindrical SQUID at 10 GHz, and obtained a flux noise of about $10^{-5}\Phi_o$ $Hz^{-1/2}$, and an energy resolution of 2×10^{-30} JHz^{-1} at frequencies above a few kHz (see Table 1). The noise increased substantially at lower frequencies. The best flux resolution is probably that obtained by Gaerttner[19],

$7 \times 10^{-6} \Phi_o$ $Hz^{-1/2}$, using a 440 MHz point contact SQUID. Unfortunately, the parameters required to calculate $S_\Phi/2\alpha^2L$ appear to be unavailable.

Future Improvements

Jackel and Buhrman[10] have given a detailed discussion of the optimization of the flux resolution of the rf SQUID. In all present versions, the intrinsic noise is insignificant, and the preamplifier and tank circuit contributions dominate. Optimization usually involves making these two contributions equal[10]. The best 30 MHz-SQUIDs are well-optimized, and no further improvement in performance is likely at this frequency. As is evident from Eqs. (8)-(10), the intrinsic and tank circuit contributions to the energy resolution vary as $1/\omega$, while the preamplifier contribution varies as $1/\omega^2$, provided that $S_V^{(p)}$ is independent of ω; unfortunately, the voltage noise of conventional FET preamplifiers tends to increase with increasing frequency. For example, suppose that $\omega/2\pi = 10$ GHz, $\varepsilon = 0.2$ (this may be a low estimate at 10 GHz), $L = 1$ nH, $T_e = 200$ K, and $T_N^{(p)} = 500$ K. From Eq. (13), we find $S_\Phi^{(i)}/2L \approx 4 \times 10^{-33}$ JHz^{-1}, $S^{(tc)}/2L \approx 10^{-32}$ JHz^{-1}, and $S_\Phi^{(p)}/2L \approx 5 \times 10^{-32}$ JHz^{-1}. The noise is dominated by preamplifier noise. This result is an order of magnitude smaller than the experimental value of Pierce _et al._[18], indicating that their preamplifier had a noise temperature higher than 500 K, and, probably, that ε was greater than 0.2.

It is clear that high frequency preamplifiers with lower noise are necessary to reduce the preamplifier contribution to the flux noise. The most promising development appears to be the use of cooled FET preamplifiers[19], which can have lower noise temperatures than room temperature preamplifiers. A cooled preamplifier may have the additional advantage of reducing T_e substantially, thereby reducing $S_\Phi^{(tc)}$.

Jackel and Buhrman[10] have shown that a non-sinusoidal current-phase relation in the weak link gives rise to a substantially higher noise than a sinusoidal current-phase relation. This result should be borne in mind when one chooses a weak link. For example, long microbridges tend to have highly non-sinusoidal current-phase relations, and are consequently undesirable in applications where the best noise performance is required.

Buhrman and Jackel[20] have found that the ultimate intrinsic energy resolution of the rf SQUID is of order $4k_BT/\omega_{opt}$, where $\omega_{opt} = R/L$. Apart from possible numerical factors, this is essentially the same result as that for the dc SQUID, Eq. (14). If we choose $\omega/2\pi \approx 10$ GHz, and $L \approx 1$ nH, we require $R \approx 60\ \Omega$. Thus, both dc and rf SQUIDs require junctions with the highest possible resistance to achieve the best flux resolution. This requirement, and the need for a sinusoidal current-phase relation, imply that the use of small area shunted tunnel junctions in SQUIDs should receive serious consideration.

References

*Work supported by the U.S.E.R.D.A.

1. Clarke, J.: Superconducting quantum interference devices for low frequency measurements. Proc. NATO Advanced Study Institute on Small Scale Superconducting Devices, Lago di Garda, Italy, September 1-10, 1976, to be published.
2. Jaklevic, R. C., Lambe, J., Silver, A. H., and Mercereau, J. E.: Quantum interference effects in Josephson tunneling. Phys. Rev. Lett. 12, 159-160 (1964); Zimmerman, J. E., and Silver, A. H.: Macroscopic quantum interference effects through superconducting point contacts. Phys. Rev. 141, 367-375 (1966).
3. Josephson, B. D.: Possible new effects in superconducting tunneling. Phys. Lett. 1, 251 253 (1962).
4. Tesche, C. D. and Clarke, J.: A computer model for noise in the dc SQUID. Proc. 1976 Stanford Applied Superconductivity Conference, to be published.
5. Stewart, W. C.: Current-voltage characteristics of Josephson junctions. Appl. Phys. Lett. 12, 277-280 (1968); McCumber, D. E.: Effect of ac impedance on dc voltage-current characteristics of superconducting weak-link junctions. J. Appl. Phys. 39, 3113-3118 (1968).
6. Claassen, J. H.: Coupling considerations for SQUID devices. J. Appl. Phys. 46, 2268-2275 (1975).
7. Clarke, J., Goubau, W. M., and Ketchen, M. B.: Tunnel junction dc SQUID: fabrication, operation, and performance. J. Low Temp. Phys. 25, 99-144 (1976).

8. Zimmerman, J. E., Thiene, P., and Harding, J. T.: Design and operation of stable rf-biased superconducting point-contact quantum devices, and a note on the properties of perfectly clean metal contacts. J. Appl. Phys. 41, 1572-1580 (1970).

9. Mercereau, J. E.: Superconducting magnetometers. Rev. Phys. Appl. 5, 13-20 (1970); Nisenoff, M.: Superconducting magnetometers with sensitivities approaching 10^{-10} gauss. Rev. Phys. Appl. 5, 21-24 (1970).

10. Jackel, L. D., and Buhrman, R. A.: Noise in the rf SQUID. J. Low Temp. Phys. 19, 201-246 (1975).

11. Kurkijarvi, J., and Webb, W. W.: Thermal fluctuation noise in a superconducting flux detector. Proc. Appl. Superconductivity Conf. Annapolis, 581-587 (IEEE, N.Y., 1972).

12. Kurkijarvi, J.: Noise in the superconducting quantum flux detector. J. Appl. Phys. 44, 3729-3733 (1973).

13. Giffard, R. P., Webb, R. A., and Wheatley, J. C.: Principles and methods of low-frequency electric and magnetic measurements using an rf-biased point-contact superconducting device. J. Low Temp. Phys. 6, 533-610 (1972).

14. Goodman, W. L., Hesterman, V. W., Rorden, L. H., and Goree, W. S.: Superconducting instrument systems. Proc. IEEE 61, 20-27 (1973); Rifkin, R., and Deaver, B. S.: Current-phase relation and phase-dependent conductance of superconducting point contacts from rf impedance measurements. Phys. Rev. B13, 3894-3901 (1976).

15. Zimmerman, J. E.: Sensitivity enhancement of superconducting quantum interference devices through the use of fractional turn loops. J. Appl. Phys. 42, 4483-4488 (1971).

16. S.C.T., Inc., Mountain View, California.

17. S.H.E. Corp., San Diego, California.

18. Pierce, J. M., Opfer, J. E., and Rorden, L. H.: A broadband thin-film SQUID magnetometer pumped at 10 GHz. IEEE Trans. Magn. MAG-10, 599-602 (1974).

19. Gaerttner, M. R.: Low noise 440 MHz SQUID. (Reported at Int. Mag. Conf., Toronto, 1974.)

20. Private communication.

A VARIABLE TEMPERATURE HIGH SENSITIVITY SQUID MAGNETOMETER

J. A. Good
Cryogenic Consultants Ltd., Metrostore, 231 The Vale,
London, W3 7QS, England

INTRODUCTION

The Superconducting Quantum Interference Device (SQUID) is recognised as the most sensitive detector of magnetic flux available. It shows important advantages over alternative methods both in ultimate sensitivity $\sim 10^{-15}$ Tesla (1 Hz bandwidth) and in its very rapid response time. Present commercially available systems have a limited response to frequencies of 200 kHz but the sensor itself will respond at very much higher frequencies. Both small signals of magnetic flux and electrical current may be measured with a coil flux linked to the sensor. Above 100 kHz the device is probably better considered as a parametric amplifier, while below this frequency the device has a low input impedance and its principal use is as a magnetometer. A superconducting input coil system is used to transfer an external signal to the sensor. As a magnetometer its ability to follow changes in the local magnetic field, or the magnetic state of a sample, rapidly and at high sensitivity, has made possible many new experimental techniques. However, most of this work has been carried out by groups already familiar with cryogenic and superconducting equipment. It has been our intention to develop a general purpose magnetometer operating over a temperature range between 1.5K - 300K and to make it suitable for use by non specialists.

The magnetometer consists essentially of a sample chamber, thermally isolated from the helium bath so that its temperature can be varied between helium temperatures and 300K. The sample chamber is enclosed by a pair of opposed superconducting pick-up coils which are connected in an all superconducting circuit to a signal coil mounted in a two hole sensor of the usual form (1). To provide a variable local magnetic field, a small superconducting solenoid of pure niobium wire surrounds the sample chamber and pick-up coils. The SQUID is mounted external to the field of the magnetising coil. A change of magnetic flux through one only of the opposed pick-up coils is reflected by a change of flux in the signal coil as the total flux remains constant. If both coils see the same change in magnetic field, as they are of equal area and opposed no flux change occurs and there is no signal recorded at the SQUID. This form of pick-up is useful since it attenuates all external magnetic interference in the magnetising field.

APPLICATIONS

The SQUID as a detector of magnetic flux is only sensitive to changes in magnetic field rather than to absolute values so to make absolute measurements a sample must be passed through the pick-up coils of the magnetometer. The total flux change is then recorded as a positive and then negative excursion of the SQUID output signal. The peak to peak height is a measure of twice the magnetic moment of the sample. Calibration is required to allow for the sample shape, the coupling efficiency to the pick-up coil and for the efficiency of the flux transfer from the pick-up coil into the SQUID. Regrettably for these reasons it is not practical to make use of the periodic response of the SQUID to flux, based on the quantum ϕ_0 as a direct calibration, but this periodic response can be used as a point of reference and all system gains and coupling efficiency can be measured relative to it. The ultimate system linearity and system gain repeatability depend on both the usual electronic parameters, such as feedback resistor drifts, and also on some special factors affecting the superconducting circuitry and the SQUID itself.

For instance the effective inductance of both the input coils and the SQUID itself are dependent on the ambient magnetic field and on the presence of signal currents circulating within the device. As the signal current changes so does the effective loop area and inductance of the circuit, with the result that the measurements are non-linear and even hysteretic. These effects are greatly increased by the magnitude of the ambient field and also by the fact that both the SQUID and pick-up coils are usually cooled in the earth's field. Earlier workers had hoped to use the absolute value of the flux quantum as a digital counter and as it was expected to obtain linearities with gain stability of better than 10^{-8}, such units were for a time marketed commercially. But for the reasons outlined above the expected performance was not even approached and better performance is now achieved with flux locked-loop feedback operation. The linearity is limited by the usual electronic parameters such as the constancy of the feedback resistor and the stability of the mutual inductance of the feedback circuit to the SQUID. With normal feedback circuits linearity and gain stability of better than 10^{-6} are obtained. The best performance is obtained if the feedback flux is applied not to the SQUID but to the signal input circuit so that there is no net flux transfer to the SQUID, and no circulating supercurrents are set up in either the input circuit or the body of the SQUID itself.

The SQUID offers most advantage over alternative techniques in the measurement of weak magnetic phenomena at low magnetic fields. At present the most sensitive measurement of susceptibility at high fields of 8 Tesla can be made using the Faraday method, that is measuring the force on a sample in a strongly divergent field.

For the measurement of very weak magnetic moments such as rock samples for palaeomagnetism, the improvement in both signal to noise and in measuring time is many orders of magnitude over conventional techniques. As has already been pointed out such measurements are made by displacing the samples through the pick-up coils of the magnetometer and recording the flux change observed.

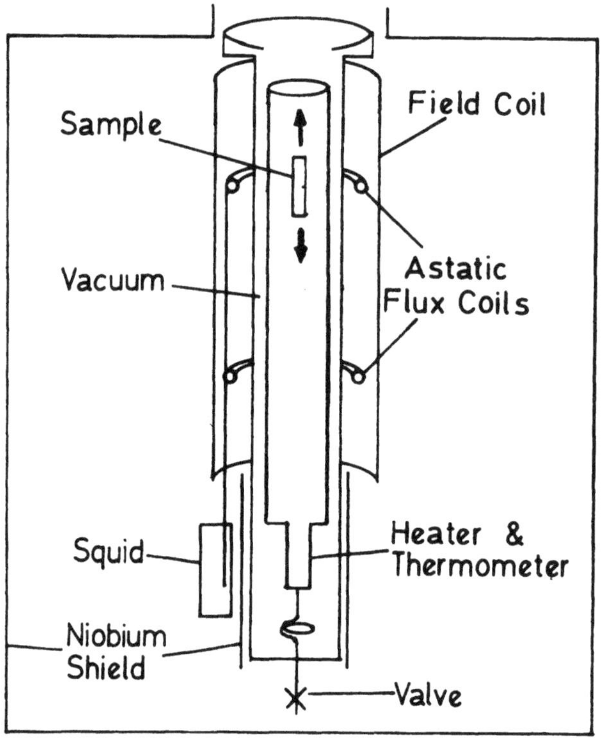

Fig. 1 Simplified diagram of the magnetometer unit

The alternative and very valuable technique is to subject a sample while mounted in a stationary position in the magnetometer to some change of intensive variable or other influence. Measurements have been very successfully made of changes in magnetic state under optical illumination and RF electromagnetic fields of varying frequency. In such experiments of nuclear magnetic resonance the SQUID is particularly useful in observing the change of magnetic state of a small number of moments where broad line width considerations make direct absorbtion NMR methods difficult. The SQUID also makes possible a whole new range of experiments in which the dynamic response of a sample to an externally imposed influence can be measured as a function of time even where the signals are very weak and the change in magnetic state is rapid. Signals of less than a fraction of a flux quantum with a response time of 1ms or less can be followed and recorded.

A third possibility is the use of an AC method either vibration or rotation of the sample or by AC modulation of the local magnetic field with a stationary sample. This technique has not yet been widely applied but offers the advantage of automatic measurement, particularly when temperature dependent properties are to be investigated. One problem of present apparatus is that of the magnetic moment of the sample support, the sample chamber, and its surrounding shields because of their larger mass, have, even after the most careful selection of materials, a much higher temperature dependent magnetic moment than typical small samples. By use of AC methods it may in the future be possible to make continuous measurements as a function of temperature.

DESIGN AND CONSTRUCTION

The most important features of and the principal problems to be resolved in the design of a variable temperature magnetometer for general purpose use are described below with the methods used to resolve them.

The principal specifications of the system are as follows:-

Sample chamber inner diameter	9mm
Diameter of SQUID pick-up coils	16mm
Flux transfer efficiency	$\sim$1%
Temperature range	1.5K-300K
Temperature stability	$\sim$10mK
Maximum field of magnetisation coil	0.14 Tesla
Field stability	$<10^{-8}$hour^{-1}
Helium capacity above the magnetometer	$\sim$7 litres
Helium hold time	60 hours
Nitrogen capacity	$\sim$10 litres
Nitrogen hold time	80 hours

The cryostat is constructed from all welded stainless steel and is both superinsulated and nitrogen shielded to reduce the helium consumption to a minimum. The sample chamber size of 9mm was chosen as being the most suitable compromise considering that most samples are of small volume, $\sim$1mm^3, but that access must be provided both for occasional larger samples and for the sample support and any RF coils or other components. To permit movement of the sample and sample replacement while the system is cold, there is direct access from the top of the cryostat to the sample chamber.

The Variable Temperature System

This operates by drawing a steady 10cc of liquid helium through a valve from the helium bath into a thermostatically controlled heat exchanger. From there the helium passes up the sample chamber tube cooling the sample and chamber to the temperature set by the heat exchanger. The temperature of the gas stream is measured by a rhodium iron resistance thermometer which has the twin advantage of excellent stability and relative linearity of resistance as a function of temperature over the whole range from 1.5K-300K. An electronic circuit is used to compare the resistance of the thermometer with a preset value and to adjust the power applied to a heating

element on the exchanger. The maximum power can be up to 10 watts, which is required to ensure rapid warming of the sample chamber. Similarly, to cool the sample chamber rapidly it is also necessary to increase the flow of helium to more than 10cc/hour.
Because the gas stream has limited thermal capacity it is important that the thermometer and the exchanger are as close as possible to the sample, but the electrical control current is sufficient to severely disturb the magnetometer pick-up coils. Complete suppression of this source of interference is achieved by placing the heat exchanger thermometer and the electrical leads within a,4K cooled superconducting shield of seam-welded niobium sheet. The shield also prevents temperature errors arising from the magneto-resistance of the thermometer which could otherwise be exposed to the fringing field of the magnetising coil.

Stability and Low Drift
These are the two most obvious requirements of the magnetometer itself. Assuming that external influences are screened out, the important sources of internal magnetic flux noise and drift are relative movement of the pick-up coils in the magnetising field and such things as the effect of temperature on the magnetic susceptibility and dimensions of the various superconducting coils in the system. Long term drifts in the magnetising field , due to imperfect persistent mode operation, are not, as might have been expected, of importance. The pick-up coils are an astatic pair so that if the field is homogeneous and the pick-up coils are perfectly balanced, a change in the magnetising field would not give rise to an error signal. In practice the coils are not perfectly balanced and changes of the magnetising field are attenuated by only a factor of 10^3. A typical flux drift for a persistent mode coil is below 10^{-8}hour^{-1}. At a field of 0.01 Tesla, and allowing for the flux transfer efficiency (10^{-2}) of the input circuit, the drift rate is estimated at

6×10^{-23} Wb sec^{-1} or $<10^{-7}$quanta sec^{-1}.
However, the effects of temperature on the flux stability of the system are not negligible. A typical expansion coefficient at 4K is $10^{-6}K^{-1}$. In a magnetic field of 0.01 Tesla and with the same assumptions about input signal transfer, a change of temperature of 50mK is required to produce a signal at the SQUID equal to one flux quantum. Since in a field of 0.01 Tesla it is possible to detect, with the SQUID, flux changes of $<10^{-3}\phi_0$ the requirement on temperature stability of the input circuit is severe, 30µK. A similar consideration applies to vibration and movements of the pick-up coils in the magnetic field of the magnetising coil. If the magnetic field uniformity of the field coil is 10^{-3} over 30mm then the displacement of a pick-up coil of 3nm will cause a flux change at the SQUID input of 10^{-3} flux quanta.
To reduce the effects of vibration and thermal expansion to a minimum, we have built the pick-up coils and the magnetising coil all into the same unit. It is possible to do this without loss of flux sensitivity only because the astatic pick-up coil pair has a very small magnetic coupling coefficient to the field coil provided that it is well balanced. All the coils are wound on the same glass-fibre and epoxy former but with a separating layer of 0.5mm thickness epoxy glass between the pick-up and magnetising coils. All the windings are epoxy bonded to the former, and because they are built as a single unit the temperature effects are minimised.
The best field stability is obtained by using a niobium single core with spot welded joints at the terminals and to the superconducting switch. The switch itself is also made of niobium as it does not have to have a high resistance in the normal state.
Typically the flux noise referred to the input of the SQUID is $10^{-3}\Phi_0$ and drifts are not observed at this sensitivity over short-term periods of 20 minutes once a steady state has been reached. As most of the experiments are of a short-term

nature the drift is unimportant and future work will concentrate on reducing the observed flux noise and on improving the flux coupling efficiency to the SQUID.

Superconducting Shielding

This is required to eliminate stray signal pick-up from the movement of magnetic objects in the laboratory as well as electrical currents in nearby apparatus. The pick-up coils are sensitive to field gradients and all laboratories contain many sources of magnetic fluctuations. We have found that welded niobium shield, 250 micron thick is vastly superior to lead-plating or 3mm lead sheet. Our magnetometer is therefore enclosed in a cylindrical niobium shielding vessel with a narrow neck at the top through which the magnetometer system is introduced. Because of the relatively high fields around the magnetometer, the shield is made very large compared to the magnetising coil so that magnetic coupling between the two units is reduced to a low level. Because the two units are mechanically separate there is always the possibility of relative motion which would appear as a noise signal of the pick-up coils. By designing the system in this way, and by paying special attention to the mounting of the magnetometer rigidly in the cryostat, so that vibrations were minimised and characteristic frequencies were high, it was possible to make the system substantially immune to vibration with field applied. Under test, the cryostat was mounted directly on the floor of the laboratory with no special precautions. The immunity to vibration was tested by the simple procedure of tapping the apparatus and stamping on the floor. This procedure had only a minimal effect and so far as normal movement of personnel in the laboratory and operation of the equipment is concerned, there is no effect due to mechanical vibrations. However, at one stage a 2mm stainless steel tube for a helium level gauge was inserted into the shielded part of the apparatus. With this tube which was not

mechanically secure inserted, a large resonant signal appeared at the magnetometer despite extensive anti-vibration mountings under the cryostat. Another source of erroneous signal was the needle valve controlling the helium supply to the magnetometer. Although this valve is mounted some 70mm above the magnetometer unit, it is still within the niobium shielding. It was observed that as the valve was shut tight a signal appeared at the magnetometer. This signal is equivalent to a DC shift in field and it is assumed to originate from magneto-strictive effects in the two stainless steel tubes which are stressed when the valve is closed.
To test the shielding due to stray magnetic fields, a bar magnet 50mm long with a field at the poles of 0.1 Tesla, was rotated close to the outer wall of the cryostat. No signal was observed at the magnetometer and by measuring the field around the bar magnet and knowing the sensitivity of the input circuit we were able to estimate the effective shielding factor of the niobium shield as being 160dB.

Radio Frequency Interference

Radio frequency interference is a problem in the use of SQUID magnetometers because the SQUID responds to low level signals at frequencies up to the Gigahertz frequency range. Even low level RF interference will obstruct correct operation of the system and it is therefore essential to suppress all sources of interference. The metal cryostat acts as a relatively good external screen. All the electrical leads which pass into the cryostat must pass through the RFI filters at the cryostat head. Suitable RFI filters are widely available even for the 10 Amp current leads to the magnetometer field coil. As it was intended to carry out RF excitation of the sample during some experiments, some RF signals are bound to be coupled into the magnetometer pick-up coils. To prevent this signal being introduced into the SQUID, the signal input coil is screened by a thin wall cupronickel tube inserted into the

SQUID. The SQUID housing and the pick-up coil are separately earthed to the cryostat. We believe that this system is superior to a shield between the pick-up coils and the sample because such a shield would be exposed to the magnetic field of the magnetometer and might be expected to show magnetic hysteresis.

Constructional Materials

All the materials used for construction inside or in close proximity to the pick-up coils will be sources of magnetic signals as the field and temperature change. It is obviously important to choose materials which are as free of magnetic impurities as possible, and where the material is as uniform as possible. Not only must individual samples of material be of uniform properties but as far as possible the material should be readily commercially available in the same uniform condition. We chose glass-fibre epoxy structures where strength and low thermal conductivity was required. For instance, the inner and outer walls of the vacuum space surrounding the sample were glass-fibre tubes of our own construction. For radiation shielding layers of aluminised mylar were used in conjunction with an additional radiation shield of high purity, five nines, 75 micron thick aluminium sheet. This latter shield is gas cooled to about 80K. Aluminium has the advantage that although it is paramagnetic and does contain considerable iron impurity, the iron is dissolved in a non-magnetic form. To prevent the circulation of electrical currents in the tube-shaped aluminium shield, it is not electrically continuous around the circumference.

MEASUREMENTS AND CONCLUSIONS

The materials used for the sample chamber were chosen on the basis of their known properties and on the assumption that they would give reliable performance from a mechanical and vacuum point of view, as well as being non-magnetic.

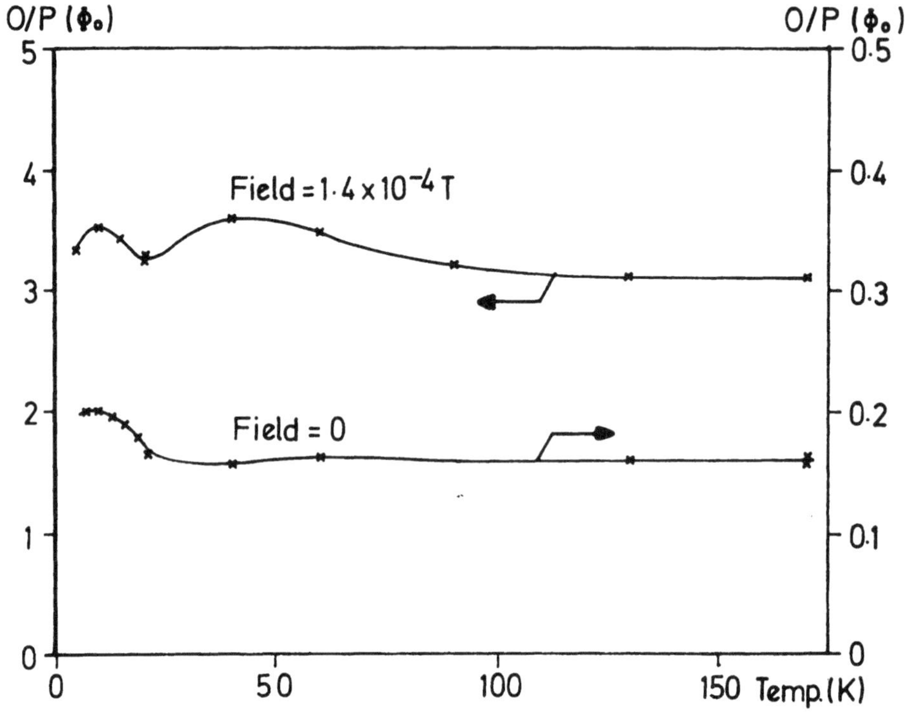

Fig. 2

One of the first measurements made with the system was of the magnetic behaviour of materials for a suitable sample holder and these clearly illustrate the problems of obtaining so-called non-magnetic material.
The material chosen was ESR quartz which is specially prepared for its non-magnetic qualities and is available in the form of rods. The signal from a long rod of 3mm diameter was measured as a function of temperature and magnetic field.

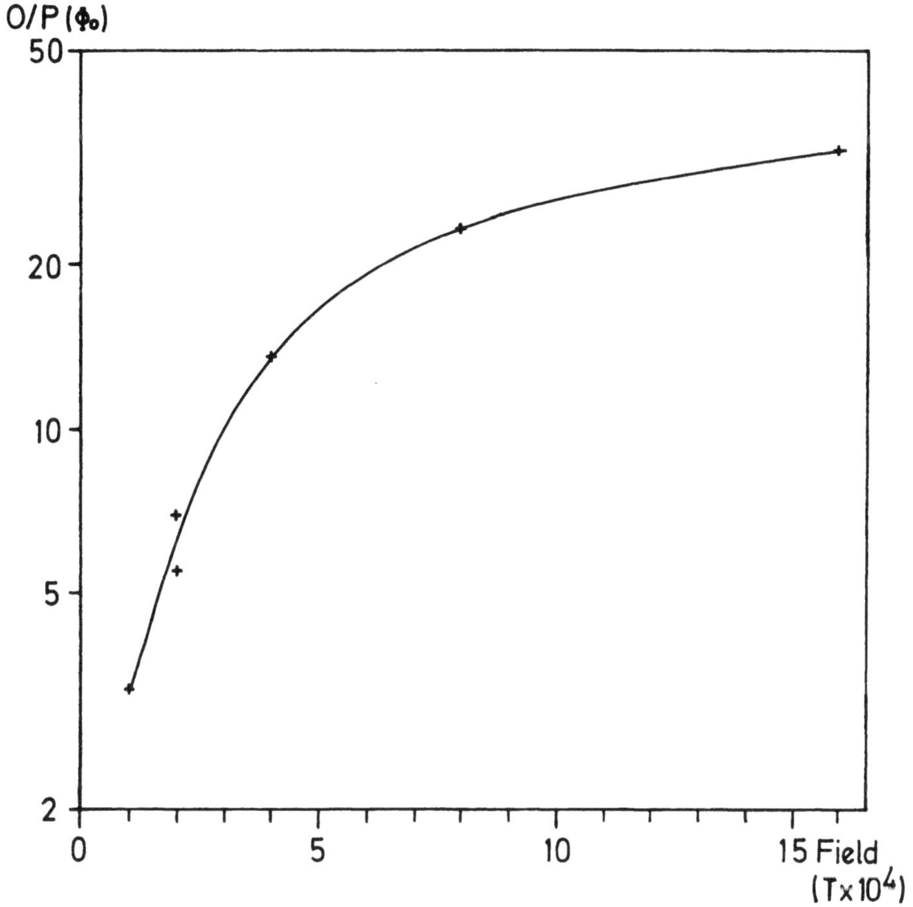

Fig. 3

Figure 2 shows the signal as a function of temperature for two different values of field and figure 3 shows the signal at a temperature of 5K as a function of field. Not only is the material quite strongly magnetic but also the behaviour is not single valued as a function of temperature. In addition the magnetic behaviour is irreversable appearing to have a ferromagnetic as well as a paramagnetic component. Since the sample is generally smaller than the support rod or tube it is clear that ESR quartz is not a suitable material for supporting the sample if full use is to be made of the sensitivity of the magnetometer.

In conclusion we believe that a magnetometer of this design meets most of the requirements that could be expected of it. Improvements can obviously be made. It is of interest to point out that with the present materials and technique we are not yet making full use of the noise performance of the 20 MHz SQUID system which gives a resolution of $\sim 10^{-4}\Phi_0$, and even at levels of $10^{-3}\Phi_0$ the magnetometer is so sensitive that there are difficulties in mounting the sample. The useful improvements therefore would include better coupling of the signal into the SQUID, better "non" magnetic construction materials which we may hope to find by using the present SQUID systems rather than improved SQUID noise figures.

Acknowledgement

The author wishes to thank Dr. H. Heidrich of the Technische Universität, Berlin for permission to make use of his experimental results on the magnetic behaviour of ESR quartz.

(1) Zimmerman, J.E., Sensitivity Enhancement of Superconducting Quantum Interference Devices through the Use of Fractional-Turn Loops. J. App. Phys. Volume 42, No. 11 4483-4486 (1971)

FLUX CREEP OF SUPERCONDUCTORS MEASURED WITH A SQUID SYSTEM

W. Ludwig and H. Gessinger
Max-Planck-Institut für Metallforschung, Institut für Physik,
Büsnauer Str. 171, D 7000 Stuttgart 80, Germany

INTRODUCTION

Magnetic aftereffects of Nb single crystals have been investigated by a SQUID magnetometer as a function of temperature and of the applied field. These measurements are a contribution concerning the question whether flux creep is a thermally activated process or not (1, 2). The time dependence of the relaxation curves turned out to be extremely sensitive to small temperature and field drifts. The influence of unstabilities is estimated and stability criteria are given using reversible thermodynamics and the temperature dependence of the thermodynamical critical field H_c.

Special stabilizing devices and shieldings reduced the flux noise amplitude to less than 1000 Φ_o ($1\Phi_o = 2\cdot 10^{-15}$ Vs). A double loop fluxtransducer and an electronic reset automatic allowed the recording of 10^7 Φ_o with a sensitivity of 47300 Φ_o/V. A special experimental procedure guaranteed reproducible and reliable results.

STABILITY CRITERIA

Fig. 1 shows the magnetic flux density B in a superconductor as a function of the applied field H_{ext} at two different temperatures. A change of the flux density may be achieved by a temperature step $\Delta T = T_2 - T_1$ at constant field (A → B') or by a field step ΔH_s at constant temperature (A → B). In both cases the new equilibrium values B_B and $B_{B'}$ are not established

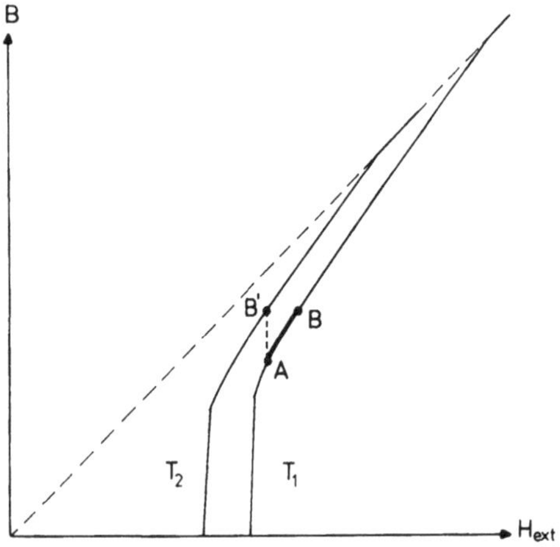

Fig.1

Flux density curve of a low κ type II superconductor at two different temperatures

spontaneously but after a considerable time period. This aftereffect measured at various fields and temperatures yields information on the kinetics of flux lines (3, 4).

A prerequisite for a successful operation of our magnetometer are extremely high stabilities of temperature and magnetic field. Instabilities may lead to a change of the flux density according to:

$$(1) \quad d\Phi/A_{sp} = (\partial B/\partial H_{ext})_T \cdot dH_{ext} + (\partial B/\partial T)_{H_{ext}} \cdot dT$$

where A_{sp} denotes the cross section of the specimen. If we accept a flux noise amplitude of $\Delta\Phi_{min}$ eq.(1) leads to a field stability at constant temperature of

$$(2) \quad \Delta H_{ext}/H_{ext} = \Delta\Phi_{min}/\left(A_{sp} \cdot (\partial B/\partial H_{ext})_T \cdot H_{ext}\right) \quad .$$

In order to get an estimate for the required temperature stability at constant field H_{ext} we make use of the relation

$$(3) \quad \begin{aligned} (\partial B/\partial T)_{H_{ext}} &= - (\partial H_{ext}/\partial T)_B \cdot (\partial B/\partial H_{ext})_T \\ &\simeq - (\partial H_c/\partial T) \cdot H_{ext}/H_c \cdot (\partial B/\partial H_{ext})_T \quad . \end{aligned}$$

From eq.(1) and (3) we obtain

$$(4) \quad \Delta T/T = (T_c^2 - T)/(2 \cdot T^2) \cdot \Delta\Phi_{min}/\left(A_{sp} \cdot (\partial B/\partial H_{ext})_T \cdot H_{ext}\right) \quad ,$$

where T_c denotes the critical temperature. By inserting eq.(2) this gives

$$(5) \qquad \Delta T/T = \Delta H_{ext}/H_{ext} \cdot (T_c^2 - T^2)/(2 \cdot T^2) \quad .$$

With $\mu_o \Delta H_s = 0.4$ mT, $(\partial B/\partial H_{ext})_T = 2\mu_o$, $A_{sp} = 10^{-5}$ m^2 we find a maximum flux change of $4 \cdot 10^6\ \Phi_o$. If the flux noise should not exceed 1000 Φ_o at $\mu_o H_{ext} = 0.2$ T and $T = 4.2$ K we need a field stability of $\Delta H_{ext}/H_{ext} = 5 \cdot 10^{-7}$ and a temperature stability of $\Delta T/T = 10^{-6}$ for $T_c = 9.2$ K. The value for $\Delta T/T$, however, seems to be overestimated because we found $\Delta T/T = 2 \cdot 10^{-5}$ to be sufficient to get reproducible results.

EXPERIMENTAL SET UP

The whole cryostat is magnetically shielded by a mumetal cylinder. In addition a cryoperm and a copper cylinder are installed in the liquid N_2 bath. The squid system consists of 3 main parts:

the field generating unit

the flux detection unit

and the temperature control device.

Field Generating Unit (Fig. 2)

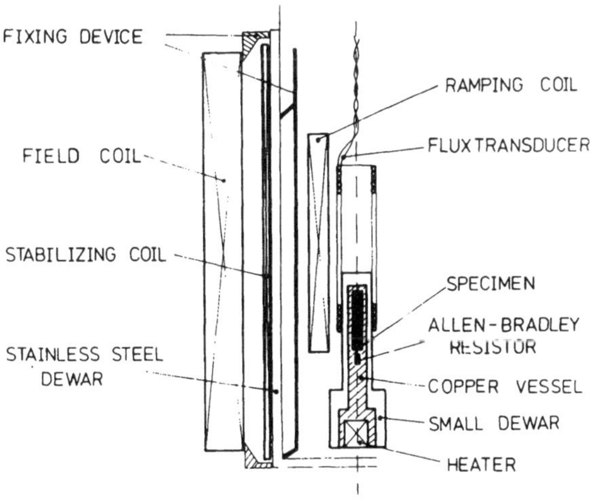

Fig. 2
Experimental set up

The field coil which is made of single core NbTi wire is able to provide a field of 1T in the persistent mode. The relative field drift follows exactly a logarithmic time law during at least one hour with a logarithmic creep rate of $d(\ell n\, t)/dt = 1.4 \cdot 10^{-4}$ at a field of 0.3 T. Under the assumption that the logarithmic time law holds one day it is necessary to wait 24 hours for a relative field drift smaller than $3 \cdot 10^{-7}$ during 3 minutes.

In order to improve the field stability we introduced a second persistent mode coil inside of the field coil. This stabilizing coil is made of NbTi wire too and is heated during the loading of the field coil. It has the effect of reversing the field drift of the field coil and suppressing it by a factor of 60. An additional stabilizing effect has the flux creep of the stabilizing coil which cancels by part the reversed field drift. After these improvements the measured field stability was better than $2 \cdot 10^{-7}$.

The ramping coil which is made of copper wire provides a maximum ramping field of 10 mT with a relative stability of 10^{-6}.

Flux Detection Unit (Fig. 3)

The fluxtransducer (FT) consists of two superconducting loops coupled by a fluxtransformer. The first loop envolving the pick up coils and the primary coil is made of NbTi wire. The pick up coils are balanced to about 1%. Therefore their flux creep cancels almost completely and does not affect the results at all. A bifilar wound heater is installed to clear the supercurrent in the FT thus preventing an overflow during changes of the applied field. The second loop of the FT consists of Nb wire and connects the fluxtransformer to the well shielded SQUID. The secondary coil of the transformer has 3 turns and is air coupled to the primary coil which has 80 turns. The flux reduction ratio is fixed to a value of 40. The centre of the flux detection unit is a commercial SQUID

Fig. 3
Flux detection unit

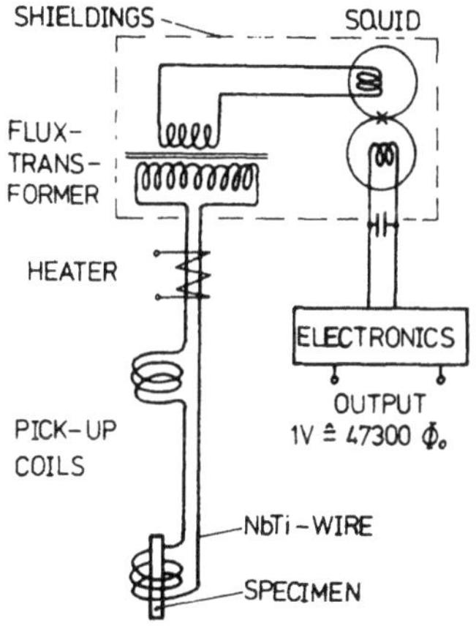

system model 202 from SHE Corp. San Diego. The electronics, however, has been slightly modified and has been equipped with an automatic reset circuit which causes the output all 11 V to rebegin at zero and allows the recording of about 10^7 Φ_o, the upper limit being given by the current carrying capacity of the FT. This value is equivalent to an output voltage of more than 200 V. An overall sensitivity of 47300 Φ_o/V allows for the resolution of flux jumps down to 1000 Φ_o with a maximum slew rate of $2 \cdot 10^8$ Φ_o/s (slew rate of the SHE system 200000 Φ_o/s).

Mechanical vibrations of the field generating unit and of the FT may cause flux noise with an amplitude of more than 10000 Φ_o. Therefore we installed two fixing devices which fixed the field coil as well as the aluminium tube where the flux detection unit is built in to the stainless steel dewar (Fig.2).

Furthermore the whole cryostat is mounted on a shock averting foundation.

Temperature Control Device (Fig. 2)

As shown above temperature variations during the measurement of flux creep are the main source of errors in the results. A stability of T = 0.1 mK is essential to get reproducible and reliable results. For example a positive or negative temperature drift of 0.1 mK/min may simulate an exponential or logarithmic time law respectively. In order to achieve the required stability, which was successful only below 4.2 K, we coupled the specimen thermally by vacuum grease to the enclosing copper vessel which is insulated against the surrounding He liquid by a small dewar of 7 mm diameter. The vessel is kept at constant temperature by means of a temperature controller regulating against an 100 Ω Allen-Bradley resistor. The heater is wound inductionfree and the heater current does not influence the output.

At 4.2 K the theoretical temperature stability should be better than 0.05 mK limited by the drift of the controller input. The highest temperature we reached during measurements on magnetite was 42 K.

EXPERIMENTAL PROCEDURE

Because it is easier to apply a field jump of well defined amplitude than to vary the temperature in a defined and sufficiently fast manner we changed the magnetic induction by an incremental change of $\mu_0 \Delta H = 0.4$ mT. This value seems to be a good compromise because smaller values result in smaller aftereffect amplitudes and larger field steps may influence the time law.

Before applying the magnetic field we made the specimen normalconducting in zero field by a short heat pulse, thus eliminating remanent flux. 3 minutes after the field rise when the relaxation effects had become smaller than 200 Φ_0/s we switched on the field step which has a rise time of 20 ms and defines the point of time zero (Fig. 4).

The flux changes due to flux flow following spontaneously the field step may surpass the critical current of the FT which is

Fig. 4

Timing of the after-effect measurements

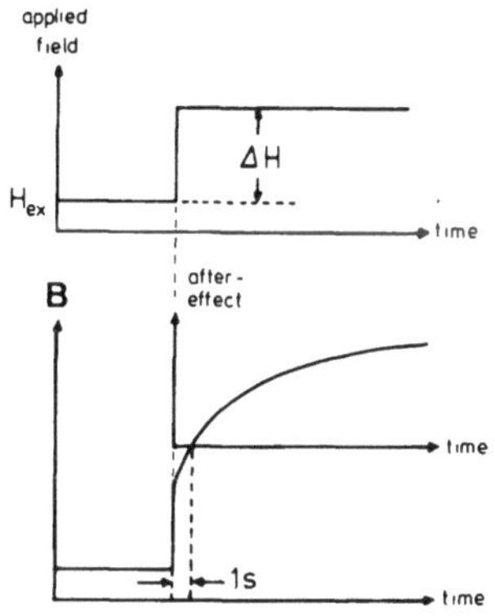

therefore heated during the first second after the field ramp. Therefore the aftereffect in our measurement is defined as that flux change which is transmitted by the FT one second after switching on the field ramp.

EXPERIMENTAL RESULTS

Fig. 5 shows some typical isothermal aftereffect curves at 4.2 K. The specimens were pure monocrystalline Nb cylinders of 30 mm length and 3 mm diameter with a resistivity ratio $R_{300}/R_{4.2} = 800$. One specimen was deformed 6% in tension. Flux jumps between 800 Φ_0 and 100000 Φ_0 could be observed only with the deformed sample and only at fields below 190 mT, whereas the undeformed specimen led to nearly flux jump free after-effect curves. The magnetization curves of these samples are shown in Fig. 6. The arrows indicate the points where the above measurements have been performed. The isochronal relaxation curves reflecting the amplitude of the aftereffect are defined by that flux change which passes through the FT between 1 s and e.g. 60 s after the field ramp. They have to be compared to the magnetization curves. The relaxation peaks at H_{c1} and H_{c2} are not yet understood. Fig. 7 demonstrates the temperature dependence of the 100 s isochrones measured with a deformed ($\Delta\ell/\ell = 29\%$) high purity Nb single crystal ($\ell = 42$ mm, $d = 3.5$ mm, $R_{300}/R_{4.2} = 2000$). The H_{c1} peak exhibits a strong temperature dependence and splits into two maxima for temperatures below 2.9 K. The temperature dependence of the H_{c2} peak seems to be neglegible. Measurements with a NbTa specimen were reported elsewhere (5).

Fig. 5
Aftereffects of a deformed and undeformed Nb sample at 4.2 K

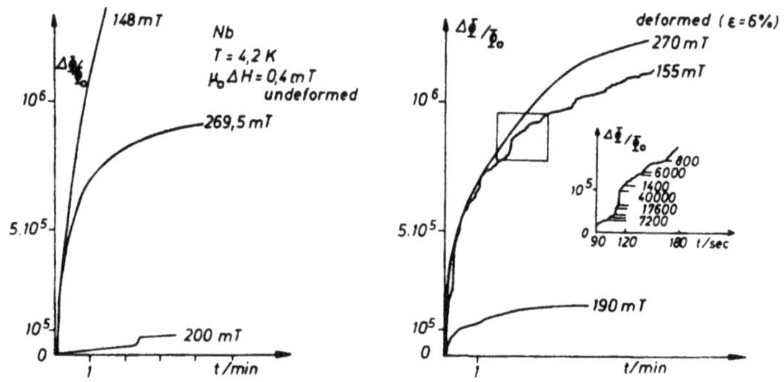

Fig. 6
60 second isochrones and the corresponding magnetization curves.
T = 4.2 K

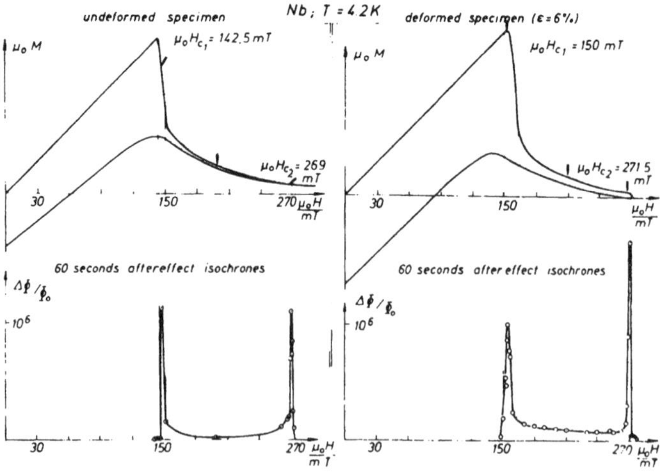

Fig. 7

Temperature dependence of the 100 second isochrones

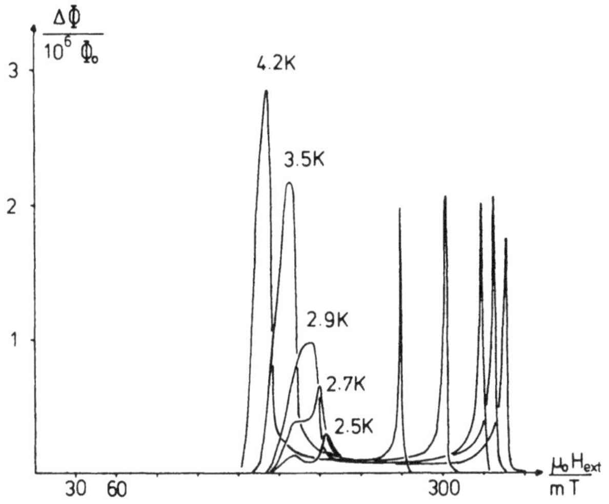

ACKNOWLEDGEMENTS

The authors wish to thank Prof. A. Seeger and Prof. H. Kronmüller for support and helpful discussions and Dr. U. Essmann for his advice and the critical review of the paper.

REFERENCES

1) Anderson, P.W.: Phys. Rev. Lett. 9, 7, 309 (1962).
2) Wade, J.M.A.: Phil. Mag. 20, 1107 (1969).
3) Kim, Y.B., Hempstead, C.F., Strnad, A.R.: Phys. Rev. 131, 6, 2486 (1963).
4) Beasley, M.R., Labusch, R., Webb, W.W.: Phys. Rev. 181, 2, 682 (1969).
5) Bodmer, A., Gessinger, H., Ludwig, W.: Proceedings of the International Discussion Meeting on Flux Pinning in Superconductors, Eds. P. Haasen and H.C. Freyhardt, Akademie der Wissenschaften in Göttingen, Göttingen 1975, 136-141 (1975).

A Magnetic Excited State in Oxy Haemoglobin A Detected with an RF-SQUID-Magnetometer

H. E. Hoenig
Physikalisches Institut der Johann Wolfgang Goethe-Universität
6000 Frankfurt am Main, Germany

K. Gersonde
Abteilung Physiologische Chemie der Rheinisch-Westfälischen
Technischen Hochschule Aachen, 5100 Aachen, Germany

ABSTRACT

The temperature dependence of the magnetic susceptibility of oxy haemoglobin A and carbonmonoxy haemoglobin A was investigated in the temperature range from 2 to 100 K in magnetic fields up to 0.07 T. A special magnetometer with a commercial two hole RF-SQUID as the sensor element was used. On the background of the diamagnetism of the sample and the paramagnetism of 4 % autoxidized haemoglobin a weak temperature dependent magnetism for oxy Hb was detected which is absent in carbonmonoxy Hb. This result indicates the existence of a partially occupied triplet spin state in oxy Hb with an excitation energy of about 17 cm^{-1}.

INTRODUCTION

Oxy haemoglobin A and carbonmonoxy haemoglobin A were considered to be purely diamagnetic compounds ever since the pioneering work done by Pauling and Corywell who measured the magnetic suceptibility at room temperature applying the Gouy method (1). On the basis of measurements of the magnetic moment (1), visible and ultraviolet absorption (2), Mössbauer effect (3) and X-ray fluorescence (4) different models for the electronic state of oxy haemoglobin have been proposed and are discussed in reference (5). From Mössbauer experiments it was concluded that a molecular orbital ground state intermediate between ferrous and ferric assignment for the iron ground state in oxy Hb could exist which should correspond to a weak temperature dependent magnetism due to the population of low lying magnetic excited states (6, 7). The high sensitivity of a SQUID and the superconducting circuitry well stabilized in superfluid Helium allowed to prove this suggestion by measuring the temperature depending magnetic susceptibility of oxy Hb.

EXPERIMENTAL

Oxy Hb buffered in 0.1 M phosphate pH 7 was prepared as described in Ref. 8. The concentration of the protein was 173 and 160 $\pm$ 1 mg per ml on the basis of M = 16114 Dalton per submit. A fraction of oxy Hb solution was taken and oxygen was replaced by carbonmonoxide. Spectrophotometric analysis and quantitative separation of oxy and CO haemoglobin from met forms by affinity chromatography on IHP-sepharose (unpublished results) were used to determine the content of met haemoglobin (see table 1). The sample tubes filled with 2 ml Hb solution were shockfrozen to 70 K in a transfer container and stored at this temperature under 1 atm of purified He gas. The samples were transferred into the precooled magnetometer without warming up for exposuring to air. The sample tubes were made of PVC with 5.8 mm diameter and 0.15 mm wall thickness.

Table 1 Met Hb content of the investigated oxy Hb and carbonmonoxy Hb samples. ρ defines the protein density and F the cross section of the frozen sample.

sample		F $[cm^2]$	ρ $[g\ cm^{-3}]$	met Hb content⁺ %	met Hb content⁺⁺ %
1a	O_2-Hb	0.25	0.173	4.6	5.0
1b	CO-Hb	0.23		6.5	6.0
2a	O_2-Hb	0.24	0.160	3.8	4.2
2b	CO-Hb	0.23		5.5	6.0

$^{+}$ determined on the basis of magnetic results

$^{++}$ determined on the basis of separation experiments on IHP-sepharose

The magnetometer is similar to that described in Ref. (9). The applied technique is outlined in Ref. (10). The principal features of the apparatus are scetched in Fig. 1. The magnetic moment of the sample (1) is measured by moving the sample tube by hydraulic means within a steady magnetic field into the upper one of two compensated pick up coils (2) which are part of a closed superconducting loop acting as a transformer for magnetic flux. The transformer integrates the flux change in coil (2) caused by the sample movement and transmits a selected fraction of this flux to the SQUID (3). The rigidly designed cold box is kept at 2 K the temperature being stabilized within one mK. The magnetic field is supplied by a superconducting solenoid (10) operating in the persistent mode.

Perfect magnetic shielding of the experimental area and the detection chamber is obtained by a superconducting coating (9) of the heavy brass construction. In order ot reduce the effects of vibrations the cold box is mounted rather loosely.

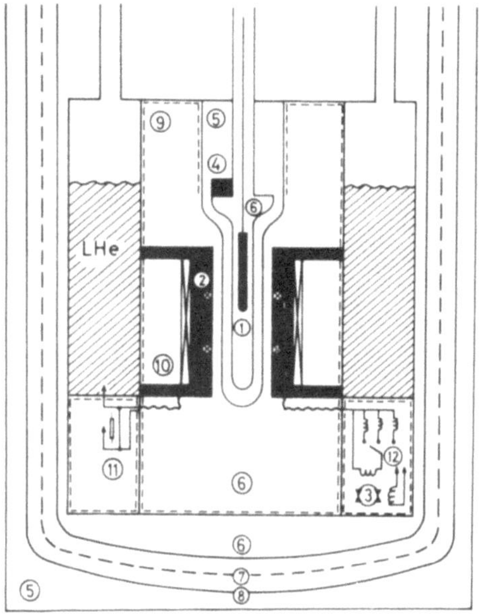

Fig. 1 SQUID-Magnetometer-Cryostat
(1) sample, (2) flux transformer pick up coils, (3) SQUID, (4) Ge and GaAs thermometers, (5) vacuum jacket, (6) He exchange gas, (7) lN_2-shield, (8) Mu metal shield, (9) superconducting shield, (10) solenoid, (11) solenoid heat switch, (12) mechanical flux switch (Ta)

The temperature of the sample can be varied between 2 and 200 K. Thermal equilibrium within the sample area is achieved with a slotted 5N copper tube. A radiation shield inside the vacuum spacing around the sample finger is also made of 5N copper and thermally anchored at the bath.

The absolute calibration of the signal U which is obtained by moving the "long" ($\gtrsim$ 4 cm) uniform sample from infinity into one of the two pick up coils in terms of μ' [Gcm^2], which is the magnetic moment per cm of sample length was performed with superconducting wire. For such a sample μ' is given by $\mu' = -FH/(4\pi) \propto U$, where H is defined as the magnetic field and F as the cross section of the sample. The magnetic susceptibility χ_{Fe} per mole of Fe is given by μ'/HA. A is the number of moles of protein subunits per cm of sample length and can be calculated from $A = F\rho/16144$ g. ρ [gcm^{-3}] defines the protein density and F [cm^2] the cross section of the frozen sample.

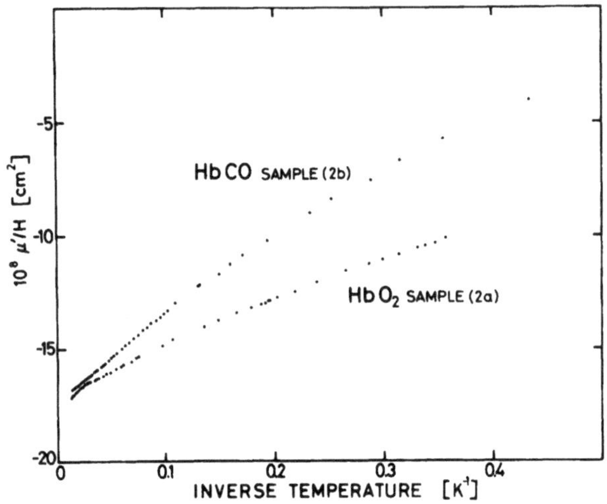

Fig. 2 Magnetic moment per cm sample length per Oerstedt as function of the inverse temperature for oxy Hb sample 2a and for carbonmonoxy Hb sample 2b

RESULTS AND DISCUSSION

For oxy Hb sample 2a and carbonmonoxy Hb sample 2b the measured magnetic moment per cm sample length devided by the magnetic field is plotted in Fig. 2 versus the reciprocal temperature. The samples stay diamagnetic down to 2 K. The temperature dependence of both samples is mainly due to the unavoidable contamination of paramagnetic met Hb: A calculation of the met Hb concentrations (e.g. Ref. 11, with met Hb high spin-low spin ratio of 1.7 for our freezing conditions, as confirmed independently) gives the results collected in table 1, which correspond with the met Hb concentrations found with spectrophotometric analysis. Since the correction for met Hb in the oxy Hb samples is rather critical we use the corresponding carbonmonoxy Hb samples (same preparation) as background reference: It is assumed that i) the susceptibility of pure carbonmonoxy Hb is temperature independent (1) and ii) the groundstate of oxy Hb is nonmagnetic leading to a temperature independent susceptibility of pure oxy Hb below 4 K. The temperature dependent part of the susceptibility of the carbonmonoxy Hb samples is scaled as to fit the slope of the susceptibility of the oxy Hb samples below 4 K and is subtracted therefrom.

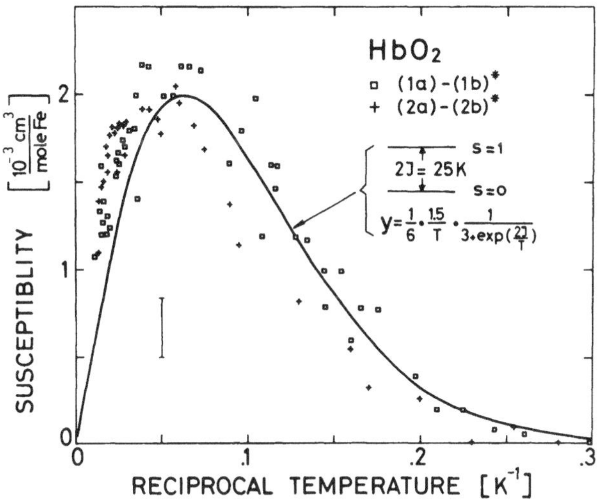

Fig. 3 Susceptibility of oxy Hb samples 1a and 2a corrected for met Hb paramagnetism as derived from carbonmonoxy Hb samples 1b and 2b respectively plotted as function of the reciprocal temperature. The full line is the calculated molar susceptibility for a spin singlet-triplet level sequence, scaled by a factor 1/6.

On the basis of our assumptions the remaining weak temperature dependent susceptibility is due to oxy Hb. The results for two independent sets of data (samples 1a/b and 2a/b) are plotted in Fig. 3. For comparision the susceptibility of a system with two spins of $S = 1/2$, which couple antiferromagnetically with an exchange constant $2J = 25$ K, was calculated, applying a scale factor of 1/6. If the susceptibility of Fig. 3 is due to a charge transfer state (Fe^{3+} - O_2^-) in oxy Hb, as proposed by Ref. (6) and (7), this state is characterized by a weigth of about $1/\sqrt{6}$ and thus is of importance for the room temperature properties of oxy haemoglobin A.

REFERENCES

(1) L. Pauling, C. D. Coryell, Proc. Natl. Acad. Sci. 22, 210 (1936)
(2) D. W. Smith, R. J. P. Williams, in "Structure and Bonding" Vol. 7, Springer-Verlag, Berlin p. 1 (1970)
(3) J. E. Maling, M. Weissbluth, in "Solid State Biophysics" (ed. by S. J. Wyard), Mc Graw-Hill, New York, p. 327 (1969)
(4) A. S. Koster, J. Chem. Phys. 56, 3161 (1972)
(5) H. B. Gray in "Bioinorganic Chemistry", Advances in Chemistry Series, no. 100, American Chemical Society, Washington, D. C. (1971)
(6) J. Peisach, W. E. Blumberg, B. A. Wittenberg, J. B. Wittenberg, J. Biol. Chem. 243, 1871 (1968)
(7) J. B. Wittenberg, B. A. Wittenberg, J. Peisach, W. E. Blumberg, Proc. Natl. Acad. Sci. 67, 1846 (1970)
(8) E. Trittelvitz, K. Gersonde, K. H. Winterhalter, Eur. J. Biochem. 51, 33 (1975)
(9) J. W. Dawson, H. B. Gray, H. E. Hoenig, R. G. Rossmann, J. M. Schredder, R. H. Wang, Biochemistry 11, 461 (1972)
(10) R. P. Giffard, R. A. Webb, J. C. Wheatley, J. Low Temp. Phys. 6, 533 (1972)
(11) N. Nakano, J. Otsuka, A. Tasaki, Biochim. Biophys. Acta 236, 222 (1971)

High Temperature Application of SQUID Voltmeter

Masayuki Konishi, Toshizo Fujita, and Taiichiro Ohtsuka
Department of Physics, Faculty of Science, Tohoku University, Sendai 980, Japan

Abstract

The use of SQUID voltmeter has been extended to high temperature measurements in order to investigate the transport phenomena in metals and alloys over a wide temperature range. The details of the cryostat and the SQUID voltmeter system are described, and the performance as well as possible applications of the voltmeter are discussed from a practical point of view.

1 Introduction

The use of Josephson junction device in measurement of very small voltage has been reported by several authors.[1)-5)] Fujita et al.[5)] described a SQUID voltmeter system for the precise measurement of the transport coefficients of metals and alloys at low temperatures. The voltmeter system was found to be of practical use with sufficient reliability. Their applications, however, were restricted to the temperature region below about 10 K.

Recently there has been a new interest in the effects of anisotropies and magnetic fields on the transport coefficients, especially on the thermoelectric power of single crystals, where anisotropic relaxation times of the electron-phonon interaction and detailed structures of the Fermi surface seem to play important roles.[6),7)] The study of the effects requires experimental data over a wide range of temperatures around several tens of Kelvin where such effects often make their appearances. The resolution of the voltmeter is mainly limited by two sources of noise.[5),8)] One is the inherent noise of the device, and the other is Johnson noise generated in the resistive elements of the input circuit. The noise voltage equivalent to the device noise increases proportionally with the resistance of the input circuit, while the Johnson noise increases with both temperature and resistance. Therefore the SQUID voltmeter has been considered to lose its advantage over competing conventional techniques for high resistance signal source at high temperature. We believe, however, that the employment of SQUID voltmeter retains some merits from a practical point of view.

With an intention mainly for application in thermoelectric power measurements, we tried to extend the application of the SQUID voltmeter to higher temperatures. In the following, the details of the cryostat and the SQUID voltmeter system are described, and also the performance as well as possible applications of our system are discussed.

2 Apparatus

Figure 1 is a block diagram of the SQUID voltmeter system.

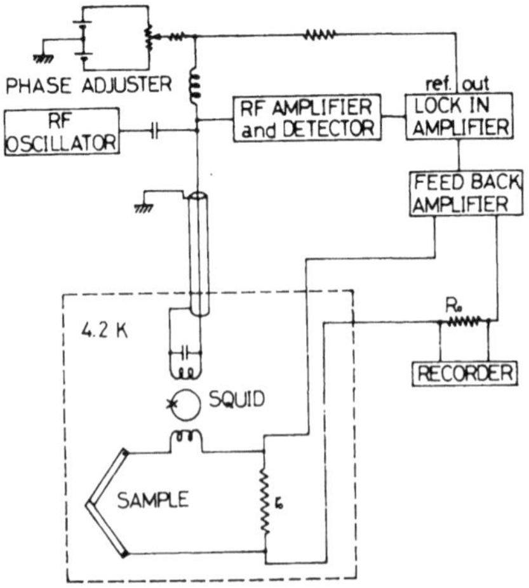

Fig.1 Block diagram of the SQUID voltmeter system.

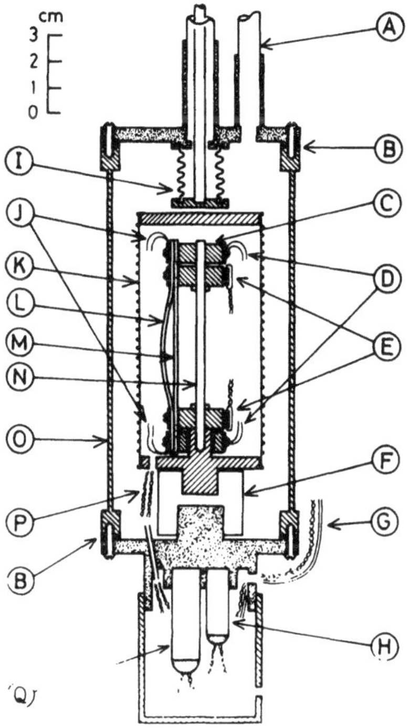

Fig.2 Main part of the cryostat. A: evacuation line, B: indium O-ring, C: copper block, D: thermocouples to measure the temperatures of the hot and cold ends of a sample, E: carbon heaters, F: nylon block, G: various leads to room temperature section, H: standard feedback resistor, I: mechanical heat switch, J: thermocouple to measure the temperature difference between the hot and cold ends, K: manganese heater non-inductively wound on the radiation shield of copper cylinder, L: sample, M: Pb reference, N: bakelite support, O: copper vacuum jacket, P: voltage leads of normal silver, Q: encapsulated SQUID sensor.

We employed a symmetric rf SQUID and our instrumental arrangement is almost the same as that previously reported.[5] The emphasis is layed on new designs of the sample chamber and the input circuit including a sample. The sample chamber is illustrated in Fig.2. The encapsulated SQUID sensor and the standard feedback resistor are installed at the bottom of the cryostat and are kept at the fixed temperature of 4.2 K. The sample chamber is evacuated to below 10^{-7} torr. The copper blocks with the sample mounted are thermally insulated from the liquid helium bath by a nylon block, and are surrounded by a radiation shield made of thin copper cylinder. The sample is cooled down to the temperature of liquid helium bath with a mechanical heat switch on to the copper cap at the upper end of the radiation shield. The manganese wire of 1 kΩ is uniformly wound on the radiation shield to heat the whole assembly of the sample and the holder. The two carbon heaters of resistance of 680 Ω are also attached to the cold and hot end of the sample. The heater at the cold end is used to make a fine control of the temperature of the sample, and the heater at the hot end is used to establish a temperature difference. Currents are supplied to the manganese heater from an external current source with high stability, and to carbon heaters from manual current controllers. Three pairs of Au + 0.07 at % Fe vs. chromel thermocouple, which are calibrated between 4.2 K and 273.17 K, are employed to measure the absolute temperatures of the hot and cold ends, and the temperature difference between them. All electrical leads to the sample chamber pass from the helium bath through a stycast seal.

For measurements below 10 K, superconducting wires can be employed for all leads in the potentiometric loop except for the sample and the feedback resistor.[5)] In the present system normal conducting "normal silver" wires are employed instead for the part of potential leads in the sample chamber where the temperature is raised above 4.2 K. As can be seen from the arrangement shown in Fig.3, the thermal e.m.f.'s generated in a pair of voltage leads cancell each other. All the leads and thermocouples are twisted tightly to prevent pick-up of electromagnetic noise, and are thermally anchored to various points to prevent direct heat flows to the sample. The total resistance of the potentiometric loop is typically ∿0.1 Ω. The standard resistor is a thin copper plate with a resistance

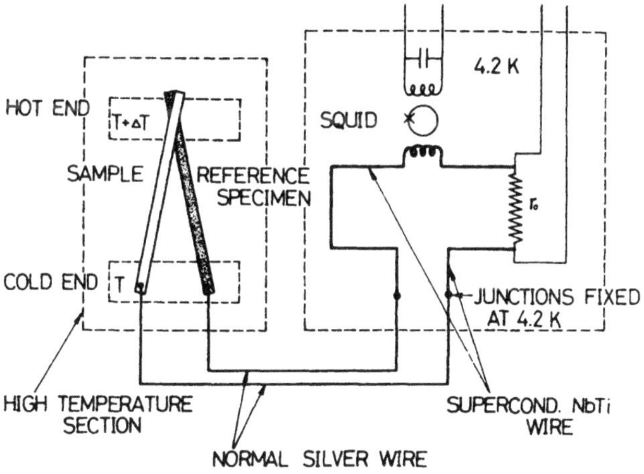

Fig.3 Detail of the input circuit including the sample thermocouple.

of $r_o = 3.9911 \times 10^{-3}$ Ω at 4.2 K, through which the current is fed back to balance the voltage. The value of the feedback current can be read from the voltage drop across a 1 kΩ standard resistor in series at room temperature.

All the samples prepared were well annealed before they were mounted on the sample holder. The electrically insulated junctions of the differential thermometer are sandwitched at both ends between the sample of interest and the reference specimen, which are electrically insulated from each other except for at the hot end and are tightly mounted on the holder as illustrated in Fig.3.

3 Measurement

The SQUID driven by the rf bias current of 18 MHz is used as a sensitive galvanometer to detect the small change in voltage across the sample in a potentiometric arrangement. Figure 4 shows a typical pair of the recorder traces, which indicate the voltage change (A) across the 1 kΩ standard resistor R_o, and the output (B) of the differential thermocouple due to a

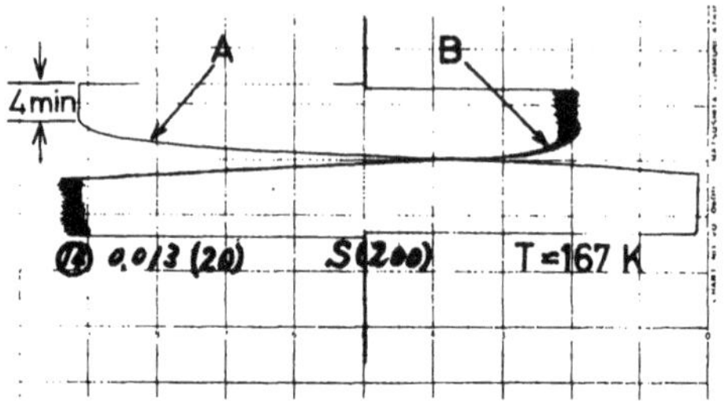

Fig.4 A pair of recorder traces indicating outputs of A the SQUID voltmeter (full scale: 0.8 μ V) and B a conventional amplifier voltmeter (full scale: 20 μ V) with a sensitivity of 10^{-8} V. The former gives the thermal e.m.f. of the sample and the latter the e.m.f. of the Au + 0.07 % Fe vs. chromel thermocouple for temperature measurement.

temperature gradient. Let ΔV_A and ΔV_B be the net change in the output A and B induced by the temperature change at the hot end from T_1 to T_2 respectively. The temperature of the cold end is regulated to remain constant by decreasing the current supplied to the carbon heater at the cold end. The values of the temperature at the hot and cold ends are monitored with a digital voltmeter. We define the mean temperature of the sample by $T = (T_1 + T_2) / 2$. The change ΔV in thermal e.m.f.'s produced in the sample thermocouple of interest is calculated from the relation

$$\Delta V = \Delta V_A / \left(\frac{R_o}{r_o} \right)$$

where $\left(\frac{R_o}{r_o} \right) = 2.5056 \times 10^5$ is the voltage gain in our voltmeter system. The temperature difference ΔT between both ends of the sample is given by

$$\Delta T = \Delta V_B / S_B (T)$$

where $S_B(T)$ is the temperature derivative of the thermal e.m.f. of the Au + 0.07 at % Fe vs. chromel thermocouple. Thus the absolute thermoelectric power of the sample S_{sample} (T) at the temperature T can be determined by

$$S_{sample}(T) = S_{ref}(T) + \Delta V / \Delta T$$

where $S_{ref}(T)$ is the absolute thermoelectric power of a reference specimen. If pure lead is adopted as the reference specimen, the values for $S_{ref}(T)$ have been given as a function of T by J. W. Christian et al.[9] In order to calculate the value for $S_B(T)$ and $S_{ref}(T)$ at a given temperature, we used the interpolation formulae

$$S_B(T) = dE / dT \quad \text{where} \quad E(T) = \sum_{n=0}^{5} a_n T^n ,$$

$$\text{and} \quad S_{ref}(T) = \sum_{n=0}^{5} C_n T^n$$

which are computer-fitted to the calibration data on the thermal e.m.f. for the Au + 0.07 at % Fe vs. chromel thermocouple and the published data on the absolute thermoelectric power for Pb, respectively.

4 Performance

The SQUID voltmeter system is regarded as a dc amplifier, the first stage element of which is the SQUID sensor coupled with the tank circuit. The voltage gain is given by $\mu = \frac{R_o}{r_o} \approx 2.5 \times 10^5$. We carried out preliminary experiments on the thermoelectric power of some samples relative to a lead reference specimen. In practice it is easy to raise the temperature of the sample from 4.2 K to 200 K without much loss of liquid helium and also to maintain the temperature at a desired value in-between; typically the sample is raised to 200 K by supplying currents of 25 mA and 5 mA to the manganese heater and the cold and carbon heater, respectively. The recorder traces shown in Fig.4 were obtained at T = 167 K. The longitudinal direction shows the time passage. The running time of ∿7 min below 100 K, ∿12 min between 100 K and 150 K, and ∿20 min above 150 K was required to get one plot of datum. This considerably long time necessary for the thermoelectric power measurement is spent to establish a temperature difference, whereas the SQUID voltmeter responds very rapidly. The temperature difference along the sample was varied with the mean temperature, typically as $\Delta T = 50$ mK near the liquid helium temperature and $\Delta T = 1$ K near 200 K.

From a direct comparison between these two traces the noise in the SQUID voltmeter is found to be negligibly small. The Johnson noise generated in a resistive element R_{IN} is expressed as

$$< \Delta V_{JN}^2 >^{1/2} = \sqrt{4kTR_{IN}B} \quad .$$

According to a simple analysis with the detection band width of $B = 1$ Hz, the Johnson noise at 200 K with the input resistance of $R_{IN} = 0.1\ \Omega$ is 3×10^{-11} V. On the other hand, the input voltage equivalent to the device noise is given by[4)5)8)]

$$< \Delta V_{DN}^2 >^{1/2} = (\phi_o/M)\ R_{IN}\ \varepsilon B^{1/2} \ .$$

where M is the mutual inductance and $\varepsilon B^{1/2}$ is the flux resolution in unit of the flux quantum ϕ_o with the detection band width B. For $R_{IN} = 0.1\ \Omega$, $M \simeq 10^{-9}$H, $\varepsilon \simeq 10^{-4}$, and $B = 1$ Hz $< \Delta V_{DN}^2 >^{1/2}$ is calculated to be $\sim 10^{-11}$V. Therefore the ultimate resolution of our SQUID voltmeter system remains roughly constant and, for samples with resistance lower than 0.1 Ω, is estimated to be no worse than $\sim 4 \times 10^{-11}$ V / $\sqrt{Hz}$ over this wide temperature range.

As examples, experimental data on the temperature dependence of the absolute thermoelectric power of Au + 2.1 at % Co, polycrystalline indium, and Al + 0.013 at % Mn are shown in Fig.5(a)-(c) after the values for lead are subtracted. These results obtained were compared for check performance with the previous data reported by W. H. Keesom and C. J. Matthijs,[10)] and G. Borelius et al.[11)] for Au + 2.1 at % Co, B. Bosacchi and R. P. Huebener for indium,[12)] J. R. Cooper et al. for Al-Mn alloys,[13)] and R. P. Huebener for pure Al.[14)] Reasonable agreement can be found between these data.

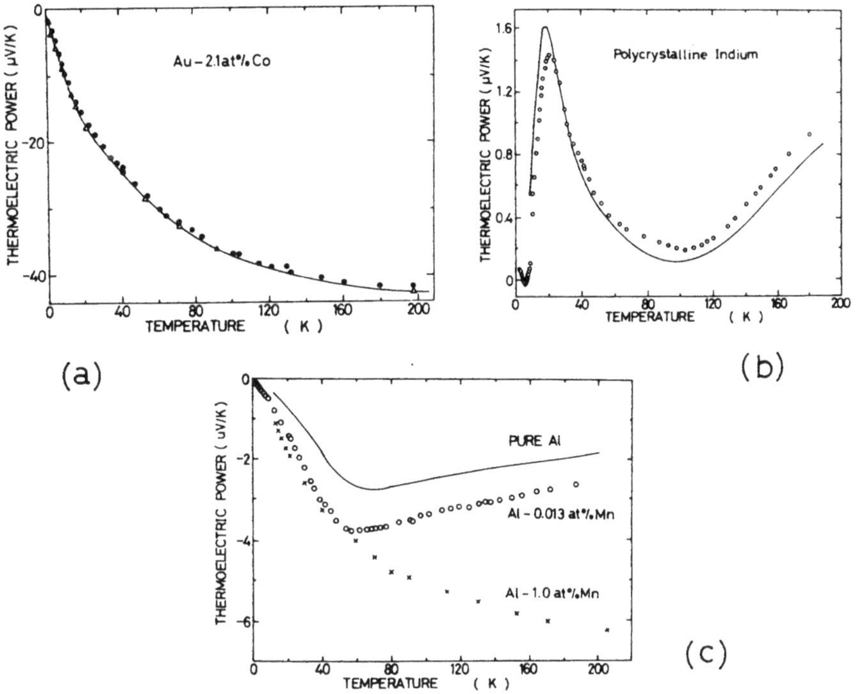

Fig.5 The absolute thermoelectric power vs. temperature. (a): Au + 2.1 at % Co. The data denoted by —△— are those reported by W. H. Keesom and C. J. C. J. Matthijs, and G. Borelius et al. (b): Polycrystalline indium. The solid curve was obtained by B. Bosacchi and R. P. Huebener. (c): Al + 0.013 at % Mn. The solid curve for pure Al was reported by R. P. Huebener, and the data denoted by x are those by J. R. Cooper et al. for 1.0 at % Mn.

The experimental errors may arise from various sources. In particular the calibration of the thermometer, the standard feedback resistor, and the reference specimen lead to errors in the determination of the absolute thermoelectric power which amounts to 1 ∿ 2 %, 0.1 %, and 1 ∿ 2 %, respectively. The total error is estimated to be less than 3 ∿ 4 %.

5 Discussion

A variety of works have been reported on the improvement of

the technique to measure thermoelectric powers using conventional types of voltmeters. B. Bosacchi and R. P. Huebener[12] used a microvolt-potentiometer at room temperature for experiments between 4.2 K and 300 K and attained a resolution of 10 nV, while J. R. Cooper et al.[13] obtained a resolution as high as 1 nV between 2 K and 440 K with a nanovoltmeter also at room temperature. The present SQUID voltmeter was found to operate with high resolution of 4×10^{-11} V / $\sqrt{Hz}$ for the source resistance lower than 0.1 Ω at temperature below 200 K. In practice a voltmeter which has such a high resolution and can cover such a wide temperature range will be useful for experimental investigations on the transport phenomena. Another practical merit is that no special care is required against stray e.m.f.'s because the signal is processed in low temperature environment.

Due to its high sensitivity, our SQUID voltmeter system promises to be a powerful tool for investigating very small effects of, e.g., point defect, dislocation, dilute impurity, magnetic field, and various kinds of anisotropy on the thermoelectric power by employing difference methods using a reference material appropriate to the investigation in question. Furthermore our voltmeter has a large flexibility for a variety of experiments involving small voltage measurements, such as electrical resistivity studies.

Finally it is noted that the essential points for successful use of our system to measure transport coefficients are how to thermally insulate the sample assembly from the liquid helium bath, and how to achieve good thermal contact between the thermometer and the sample.

References

1) J. Clarke: A Superconducting Galvanometer Employing Josephson Tunnelling, Phil. Mag. 13, 115-127 (1966).

2) J. W. McWane, J. E. Neighbor and R. S. Newbower: Ultrasensitive Potentiometer for Use at Liquid Helium

Temperatures, Rev. Sci. Instrum. 37, 1602-1603 (1966).

3) D. A. Zych: Measurement of Small dc Voltages Using Phase Sensitive Detection Techniques, Rev. Sci. Instrum. 39, 1058-1059 (1968).

4) R. P. Gifford, R. A. Webb and J. C. Wheatley: Principles and Methods of Low-Frequency Electric and Magnetic Measurements Using an rf-Biased Point-Contact Superconducting Device, J. Low Temp. Phys. 6, 533-610 (1972).

5) T. Fujita and T. Ohtsuka: Practical SQUID Technique for the Precise Measurements of Transport Coefficients, Japan. J. appl. Phys. 15, 881-886 (1976).

6) A. D. Caplin, C. K. Chiang, J. Tracy, and P. A. Schroeder: Some Thermopower Measurements on Aluminium and Indium in Zero and Non-Zero Magnetic Fields, Phys. stat. sol. (a) 26, 497-506 (1974).

7) F. J. Blatt, A. D. Caplin, C. K. Chiang, and P. A. Schroeder: PHONON DRAG THERMOPOWER OF NOBLE METALS IN A HIGH MAGNETIC FIELD, Solid State Commun. 15, 411-414 (1974).

8) O. V. Lounasmaa: Experimental Principles and Methods Below 1 K, Academic Press, 1974.

9) J. W. Christian, J. P. Jan, W. B. Pearson and I. M. Templeton: Thermo-electricity at low temperatures VI. A redetermination of the absolute scale of thermoelectric power of lead, Proc. Roy. Soc. (London) A 245, 213-221 (1958).

10) W. H. Keesom and C. J. Matthijs: Measurements on thermoelectric forces of some alloys at temperatures from 2.5 K to 17.5 K, Comm. Kamerlingh Onnes Lab. Leiden 238, 1-9 (1935).

11) G. Borelius, W. H. Keesom, C. H. Johansson and J. O. Linde: Measurements on thermo-electric forces down to temperatures obtainable with liquid or solid hydrogen, Comm. Kamerlingh Onnes Lab. Leiden 217, 22-33 (1931).

12) B. Bosacchi and R. P. Huebener: Absolute thermoelectric power of polycrystalline indium, J. Phys. F: Metal Phys.

1, L27-29 (1971).
13) J. R. Cooper, Z. Vucic and E. Babic: The thermoelectric power of AlMn alloys, J. Phys. F: Metal Phys. 4, 1489-1500 (1974).
14) R. P. Huebener: Thermoelectric Power of Aluminum and Dilute Aluminum Alloys, Phys. Rev. 171, 634-641 (1968).

USE OF PERSISTENT SUPERCURRENTS IN SQUID CURRENT STABILISERS AND THEIR APPLICATION TO A RESISTIVITY MEASUREMENT ON NIOBIUM AND A PORTABLE VOLTAGE STANDARD.

J.C. Gallop
National Physical Laboratory
Teddington, Middlesex, TW11 OLW

1 INTRODUCTION

The quantum periodicity of the SQUID response to applied magnetic flux, together with its very high sensitivity has clear applications to the control and stabilisation of direct currents. Systems using these properties have already been described (Sullivan 1972). This paper describes a system which includes a persistent supercurrent as a reference for the SQUID, leading to improved long term stability, less sensitivity to cryostat temperature changes and the ability to stabilise large currents.

2 LIMITATIONS OF CONVENTIONAL SQUID CURRENT STABILISERS

The basic SQUID current stabiliser consists of a superconducting ring containing one or two Josephson junctions with an input coil carrying the current to be stabilised. The SQUID detection system gives an output which is periodic in the current so that digital measurement together with analogue interpolation is possible. The analogue output is fed back to the input coil to keep the flux coupled into the SQUID constant (see fig. 1a). Such a system works very well for short term stabilisation (less than one hour, say) of currents of about 10mA or less. The drawbacks of such a system are that the current periodicity of the device depends slightly on the effective area of the Josephson junction and the current distribution in it. Most SQUID junctions are of the point contact or thin film microbridge type. For the first the point contact area is prone to small, but non-negligible, changes, presumably due to mechanical vibrations, whereas for the second type, the critical current i_c and its distribution are very temperature dependent. Thus in both cases the current reference scale becomes uncertain and this effect becomes more important the higher the current (see section 4). A second related problem is that the finite area

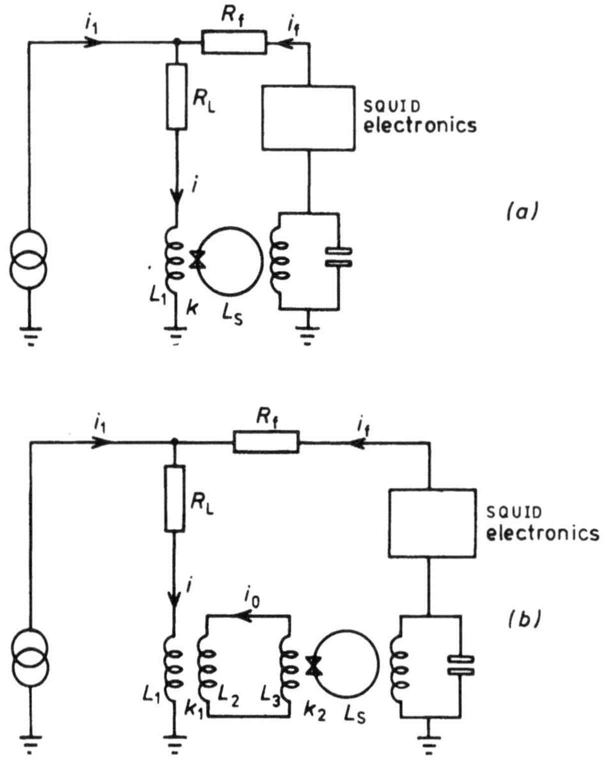

Fig. 1 Schematic circuit diagram of a) a conventional SQUID current stabiliser and b) a stabiliser using a persistent supercurrent reference.

of the weak link leads to a Fraunhofer type 'diffraction' envelope being superimposed on the usual interference pattern of the SQUID. Thus as the input current i is increased the SQUID response has nodes and the envelope is attenuated roughly as i^{-1} for high currents which sets an upper limit on the current which may be stabilised. A third problem which is equally important is that the inductance L_s of the SQUID loop may itself be temperature dependent.

3 USE OF A PERSISTENT CURRENT COIL

All of the above limitations may be avoided by operating the SQUID sensor in the near zero flux condition. This has been achieved by coupling to the SQUID via an intermediate superconducting loop carrying a circulating supercurrent of the appropriate magnitude and direction to cancel out the flux at the SQUID resulting from the current to be stabilised. Use of the SQUID in this near zero flux condition means that the current reference scale is no longer dependent on the SQUID parameters so that changes in critical current or L_s with time or temperature are very much less important. It is, of course, essential that the new feature introduced to the system, the intermediate persistent current coil, should be very stable, both mechanically, thermally and magnetically. Consider the current stabiliser system shown in fig. 1b. A current i_1 from a d.c. source is passed through a coil coupled via an intermediate superconducting coil to the SQUID, with overall mutual inductance M. The output from the SQUID system (transfer function $A = di_f/d\phi_s$) is fed back to the input coil, so that

$$i = i_1 + i_f$$

where

$$i_f = A(Mi - n\phi_0 - M_2 i_0)$$

and i_0 is the circulating supercurrent coupled to the SQUID with mutual inductance M_2. The term $n\phi_0$ (n an integer) takes care of the periodic response of the device, the periodicity being given by the flux quantum $\phi_0 = h/2e$). If the various coil self inductances are as shown in the figure and the coupling coefficients are k_1 and k_2 as shown then the mutual inductances may be written

$$M = \frac{k_1 k_2 (L_1 L_2 L_3 L_s)^{\frac{1}{2}}}{(L_2 + L_3)}$$

The effect on i of a change in the SQUID loop inductance L_s becomes

$$(di/dL_s) = -(i_0 M_2 - iM)A/2L_s(1-AM)$$

If $i_0=0$ then (di/dL_s) is proportional to Mi, that is to the amount of flux coupled into the SQUID. However if we choose $i_0 = iM/M_2$ then the stabilised current becomes insensitive to changes in L_s. A persistent current of the correct magnitude and sign is easily set up in the loop by cooling the whole system slowly through the superconducting transition temperature with the current to be stabilised already flowing in the input coil. Then the flux linking the coil remains constant during the superconducting transition so that the circulating supercurrent has the correct value to make $(di/dL_s) = 0$.

The most important of the three effects limiting conventional SQUID stabilisers is probably the temperature dependent loop inductance. It is suggested here that the effect arises from the change with temperature of the superconducting penetration depth $\lambda(T)$. Typically $\lambda(T) = \lambda(0) \times (1 - (T/T_c)^4)^{-\frac{1}{2}}$. The finite penetration depth of magnetic field into the superconductors gives rise to a surface reactance effect which must be added to the simple geometric inductance of the SQUID loop. This additional contribution is temperature dependent. The size of the contribution depends on the width to length ratio of the superconducting circuit in the direction of current flow. It might be expected that the intermediate coil would suffer from the same temperature dependence as the SQUID, but although the effect should exist it is too small to be significant since the width to length ratio is of the order of 5×10^{-5} for the coils used, but of the order of 1 for the SQUID itself.

4 CONSTRUCTION AND OPERATION OF PROTOTYPE CURRENT STABILISERS

So far two systems have been constructed and tested. One using a type of d.c. SQUID is suitable for high currents (~1 A) and the other using a commercial 20 MHz point-contact SQUID for currents in the range 1-100 mA. The construction of the primary of the transformer was similar in both cases, being about 200 turns of insulated Niobium wire (0.05 mm dia.) wound in a single layer on a tufnol former (10 mm dia.). The external current coil was of insulated copper wire wound directly on top of the niobium in a similar single layer. The ends of the superconducting coil were argon arc welded. In the first system a solder drop SQUID (Clarke 1966) was formed around the niobium wire well away from the coil, and glued to the same former. For the low current system the niobium coil was also wound into a 200 turn input coil for the r.f. SQUID, giving a superconducting flux transformer arrangement. Two different detection systems were used with the d.c. SQUID, one using a photocell galvanometer amplifier as a low noise, low input impedance amplifier (Gallop 1974) and the other a cryogenic, high Q (~ 300) resonant circuit to impedance match the d.c. SQUID to a conventional low noise transistor amplifier (Clarke 1974). Both systems achieved a flux resolution of around $2 \times 10^{-5}\ \phi_0 Hz^{-\frac{1}{2}}$. The second of these techniques has the advantage of a broader bandwidth and more rapid recovery from overloads. The detection system for the r.f. SQUID was of the conventional 20 MHz type, giving a limiting flux noise of $\sim 5 \times 10^{-4}\ \phi_0 Hz^{-\frac{1}{2}}$. (For both d.c. and r.f. systems the long term stability of the SQUID itself was of the order of $10^{-4}\phi_0$/(hour) typically.) The stability of the external current when servoed by the SQUID was measured by passing it through a stable resistor in a temperature controlled oil-bath and comparing the volt-drop across it with a saturated standard cell also temperature controlled to better than 1 mK.

5 RESULTS

The results obtained with the two systems at 4.2K are compared in table 1. The measured stabilities are probably limited by drifts in the standard cell emf's or standard resistors used for the voltage comparison. The very significant improvement achieved in the temperature coefficient by operating the SQUID near zero applied flux should be apparent.

Table 1	d.c. SLUG	r.f. SQUID
Current stabilised	100 mA - 1A	10 mA
Stability	$<2.5 \times 10^{-8}$/hr.	$<2 \times 10^{-9}$/hr.
Noise	$2\times10^{-5}\phi_0/(Hz)^{\frac{1}{2}}$ $4\times10^{-10}A/(Hz)^{\frac{1}{2}}$	$5\times10^{-4}\phi_0/(Hz)^{\frac{1}{2}}$ $5\times10^{-11}A/(Hz)^{\frac{1}{2}}$
Current periodicity	$20\ \mu A/\phi_0$	$0.1\ \mu A/\phi_0$
(di/dT)/i with persistent current	$\lesssim 4\times10^{-8}$/mK	$\lesssim 2\times10^{-8}$/mK
(di/dT)/i without perisistent current	5×10^{-6}/mK	2×10^{-7}/mK

The technique outlined here shows how long term, high current SQUID current stabilisers may be made. An interesting additional result from this work is that it has allowed a new upper limit to be set for the resistivity of superconducting niobium at 4.2K. This upper limit has been set by observing the stability of the persistent supercurrent flowing in the additional coil. If the supercurrent flowing in this coil decayed with time the SQUID servo-loop would set up an external current to compensate for the decay. The absence of any observable decay together with the self-inductance of the coil and the wire length and diameter allow an upper limit for ρ to be set from the expression

$$\rho = \pi r^2 (L_2 + L_3)(di_0/dt)/i_0 l$$

where r is the radius and l the total length of the wire. Putting into this expression the appropriate values yields a result

$$\rho < 10^{-26}\Omega m \quad \text{at } 4.2K$$

(Since the supercurrent flows only in a surface layer of order λ thick, the area of the conductor should perhaps be taken as $2\pi r\lambda$ rather than πr^2, allowing the upper limit set on ρ to be reduced to $\lesssim 2 \times 10^{-29}\Omega m$.) This is believed to be the lowest experimental value so far obtained, and illustrates the very high sensitivity attainable with SQUIDs.

6 PORTABLE VOLTAGE STANDARD BASED ON A NULL FLUX SQUID

The high stability of direct currents stabilised using a SQUID and a persistent current carrying coil suggests that such devices may have wider applications in the field of precise electrical measurements. The proven stability of cryogenic low temperature coefficient resistors (Gallop and Petley) and of the SQUID stabilised current sources described above together with the superconducting direct current comparator principle developed by Harvey (1973) and Hahlbohm et al. (1974) suggested the possibility of constructing a portable transfer voltage standard. In this device, which is undergoing tests in prototype form, there is a closed superconducting coil which carries a persistent supercurrent. This provides a reference and is coupled to a multihole SQUID (Zimmerman) via a superconducting current comparator shield of the 'snake swallowing its tail' variety. This shield also contains a winding which carries a current from an external source (see fig. 2). The current comparator has a very useful property of making the mutual inductance per turn for the

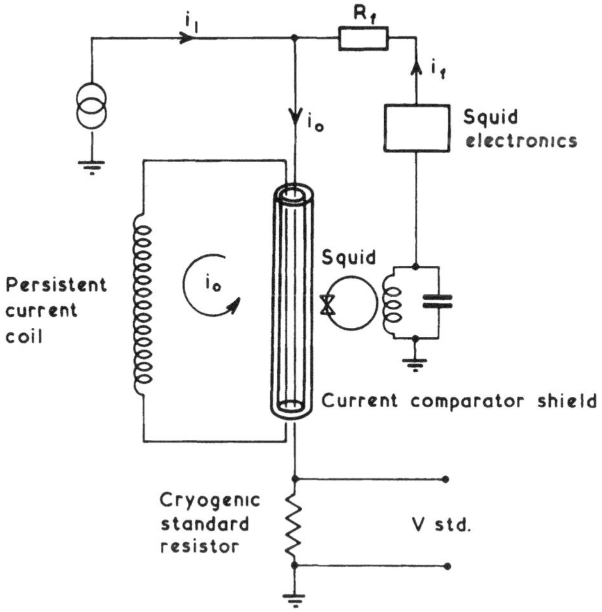

Fig. 2 Schematic circuit diagram of cryogenic transfer voltage standard described in text.

two windings coupled to the SQUID exactly equal, independent of the positions of the individual windings within the superconducting shield. The output from the SQUID can be used to servo the external current supply to keep it exactly equal and opposite to the persistent supercurrent, which provides a current reference standard. By passing the servoed current through a stable low temperature coefficient phosphor bronze resistor in the cryostat, the voltage drop across the resistor should be highly stable, in view of the stability of the current (already shown to be at least as good as 2×10^{-9}/hour) and the stability of the resistor value (ageing rate at 77K or lower $< 1 \times 10^{-10}$/hour, temperature coefficient at 4.2K of 4×10^{-6}/K.)

Ideally the coil carrying the persistent current should be perfectly shielded from external magnetic influences. However since it must be coupled to the external current carrying coil, total isolation, such as might in principle be attained by totally enclosing the coil in a superconducting box, cannot be realised. There will always be a finite mutual inductance M_{12} between the persistent and external current carrying coils. This means that any change in i will change the persistent reference current according to the relationship

$$\phi_t = n\phi_0 = i_0 L_0 + i M_{12}$$

where ϕ_t is the total flux linking the superconducting coil and is constant in time being equal to an integral number n of flux quanta. Thus the change induced in i_0 by a change in i is

$$\frac{di_0}{di} = -\frac{M_{12}}{L_0}$$

This parameter should be made as small as possible in order to maximise the stabilisation factor of the device. The superconducting current comparator geometry ensures that M_{12} is small. If the shield has a length l and an inner diameter D and the windings (of n_1 and n_2 turns), are of wire of diameter d then an upper limit on the inter-winding mutual inductance is set by

$$M_{12} < \mu_0 l \log (D/d)\, n_1 n_2 / 2\pi$$

Taking realistic values for the parameters in this expression suggests $M_{12} < 2 \times 10^{-7}$H. The persistent current carrying coil is a large multi-turn toroidal coil which in the prototype system has an inductance of around 2mH so that the ratio M/L_0 should be less than 10^{-4}. This already small figure could be very much reduced in future designs by using a current comparator shield winding with two separate axial holes through it. One hole would carry the supercurrent winding and the other the external current winding. Provided that the holes emerge far from each other on the scale of the hole size the inter-winding mutual inductance can be made vanishingly small.

The complete volt-transfer system is housed inside a lead superconducting shield in a small commercially available liquid helium cryostat, with a height of 0.6m and a weight of 10 kg. A liquid helium capacity of 1.4 l gives a hold time between fillings in excess of 50 hours at 4.2K. The system has been designed to achieve a transfer accuracy of better than 1 in 10^8. It is hoped that it will be possible to use it to transfer standard voltages between primary and secondary standards laboratories, with a precision and reliability greater than can be achieved at present using standard cells.

REFERENCES

Clarke, J.: A superconducting galvanometer employing Josephson tunnelling. Phil. Mag. 13, 115-127 (1966).

Clarke, J., Gouban, W.M. and Ketchen, M.B.: Thin film d.c. SQUID with low noise and drift. IEEE Trans. Mag. 11, 724-7 (1975).

Gallop, J.C.: Noise limitations on measurements made with SLUGs. J. Phys. E. 7, 855-9 (1974).

Gallop, J.C.: Use of persistent supercurrents in SQUID current stabilisers. J. Phys. D. 9, 2111-6 (1976).

Gallop, J.C. and Petley, B.W.: Operational experience with a cryogenic volt monitoring system. IEEE Trans. Instr. Meas. 23, 267-71 (1974).

Grohmann, K., Hahlbohm, H.D., Lubbig, H. and Ramin, H.: Ironless cryogenic current comparators for a.c. and d.c. applications. IEEE Trans. Instr. Meas. 23, 261-263 (1974).

Harvey, I.K.: Precise superconducting d.c. ratio transformer. J. Phys. E. 6, 812-813 (1973).

Sullivan, D.B.: Superconducting quantum interference devices; an operational guide. NBS Technical Note 629 (1972).

Sullivan, D.B. and Dziuba, R.F.: Low temperature direct current comparators. R.S.I. 45, 517-9 (1974).

THE USE OF RESISTIVE SQUIDS FOR CALORIMETRY

J.G. Park
Dept. of Physics, Blackett Laboratory, Imperial College, London SW7 2BZ, UK

INTRODUCTION

A resistive SQUID (RSQUID) may be driven by the same system that would be used for a SQUID of the same shape and size. The system usually used at excitation frequencies of order 20 MHz is indicated in figure 1. It has the property that if a potential difference V is established across the normal metal region (N in the figure) the output voltage of the detector (v_o) oscillates at a frequency $V/(\alpha \phi_0)$ where ϕ_o is the flux quantum(1). The value of the coefficient α depends on the geometry of the RSQUID.

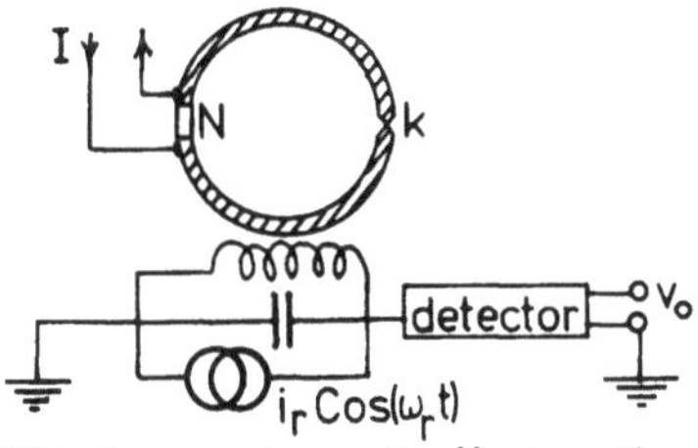

Figure 1 A resistive SQUID shown schematically together with the system used to drive it and so detect the amplitude of oscillation in the tuned circuit. k is the weak link in a ring of superconductor (shaded) a part of which (unshaded) has been replaced by normal metal (N.). The tuned circuit, coupled to the ring by mutual inductance, is driven at a frequency f_r at or near to its frequency of resonance.

Usually $\alpha \sim 1$ and we shall take $\alpha = 1$ in what follows. α is only equal to 1 when the RSQUID is toroidal. If an external current I is passed through the region of normal metal (Fig. 1) v_o will oscillate at a frequency $f = IR/\phi_o$ provided $I \gg I_a$. I_a is a current $\sim 0.1\ \mu A$ which may be regarded as the 'critical current' of the weak link in the sense that $V = 0$ when $|I| < I_a$. Thus v_o does not oscillate; the current I passes as a supercurrent through the weak link. I_a is smaller than the true critical current of the weak link that would be measured in the absence of RF excitation (3) and its precise value varies with the amplitude of the RF excitation current i_r. When the external current I is comparable with, but not smaller than I_a, f varies with I in a non-linear fashion. When $|I| < I_a$, v_o, instead of oscillating, takes up a value which is constant in time and is a function of I (3,4,5). v_o fluctuates about this value as a result of Johnson noise. In this paper we explore the idea of making measurements on the thermal properties of materials such as their specific heat by using the sensitivity of the RSQUID to a heat current $\dot{Q}$ rather than to an electric current I. Suppose that the normal metal has a thermopower S and carries a heat current $\dot{Q}$ which flows from one normal-superconducting (NS) interface to the other but that the electric current $I = 0$. If the thermal resistance of the normal metal N between the two interfaces is W the potential difference between them $V = S\dot{Q}W$. Thus v_o will oscillate at a frequency

(1) $f = S\dot{Q}W/\phi_o$ (as long as $S\dot{Q}W/R \gg I_a$)

(If $S\dot{Q}W/R < I_a$ a current $S\dot{Q}W/R$ flows around the ring in such a direction as to reduce the potential difference across N to zero). The effect of a heat current in producing oscillations in v_o is observed in most experiments on RSQUIDs and is generally regarded as a nuisance. Here we examine the possibility of putting the effect to good use.

In figure 2 we indicate the construction of a form of toroidal RSQUID on which we have made a number of experiments[2-5]. We give it here as an example of a toroidal RSQUID which is easy to visualise. Moreover it has a geometry which is likely to be suitable for heat current measurement. This RSQUID was made by pressing a copper ring N between cylindrical niobium discs P and D. The weak link was made by the point of a niobium screw which must be adjusted at low temperature. The ring had previously

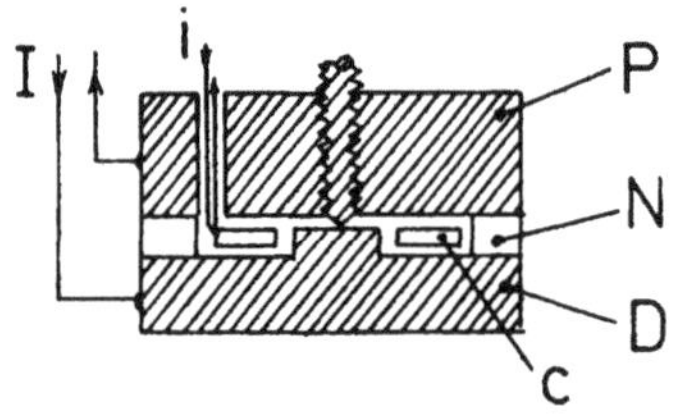

Figure 2 Section through a toroidal RSQUID. The current in the toroidal coil c may include flux modulation current as well as RF current. Superconducting parts are shaded. The drawing is approximately to scale. In our experiments the RSQUID was excited at 20 MHz. The outer diameter was about 1.3 cm.

been tinned with soft solder (superconducting below about 7 K) on its top and bottom faces. The electrical contact resistance could in this way be made exceedingly small. It was probably less than a tenth of the resistance of N in our experiments; i.e. less than $10^{-9}\,\Omega$.

THE RSQUID AS WATTMETER

If this RSQUID is surrounded by a vacuum (for thermal isolation) and power Q supplied to a heater attached to the lower disc D while the upper disc P is held at a constant temperature by contact with a thermal reservoir, a heat current will pass through N and cause v_o to oscillate. Used in this way it should be possible to use the RSQUID as a sensitive wattmeter of low thermal resistance and high dynamic range.

The sensitivity of such a device will be determined by the resistance of N (R $\mu\Omega$) and its thermopower S. If we define a parameter σ by writing $S = \sigma T$ $\mu V\ K^{-1}$ and assume that the Wiedemann-Franz law (WF) holds for the alloy from which N is made (in which case $W = 41\ R/T\ KW^{-1}$) we find for the sensitivity $f/\dot{Q} = 2000\ \sigma R\ erg^{-1}$. If $\sigma \sim 2$ and $R \sim 10^{-2}$, which are typical values, $f/\dot{Q} = 40\ ergs^{-1}$. Such high sensitivity is combined with a low contribution of the material itself to the thermal resistance. One can see this directly by noting that the temperature difference across the normal metal is given by $\Delta T = f\ \phi_o / S$. Even at the highest frequency ($\sim$ 100 kHz) at which one is likely to use a conventional 20 MHz system ΔT is only about 20 μK even if $S \sim 10$ $\mu V\ K^{-1}$; and this corresponds to substantial heat current when $R \sim 10^{-2} \mu\Omega$ of about 2 mW. The thermal resistance of the device acting as a wattmeter will usually be determined by the resistance of the interfaces rather than the normal metal itself.

We have neglected the heat current through the weak link. We shall now discuss why it is reasonable to neglect it. If it was not likely to be negligible the idea of using an RSQUID as a wattmeter would be much less promising than at present it appears to be. If the weak link in the normal state has electrical resistance R_k and obeys the WF law it will have a thermal resistance of $4\ 10^7 R_k / T\ KW^{-1}$ i.e. about $10^8\ KW^{-1}$ at 4 K if $R_k \sim 10\Omega$, a typical value. Even with pressure contacts at the NS interfaces, as in the RSQUID described above, the thermal resistance of the path through N is unlikely to exceed 100 KW^{-1} at a compression force of about 50 kg wt (8). This resistance could be reduced by an improved design in which the NS junctions are made by soldering with a solder that is not superconducting at the working temperature. But even under unfavourable conditions using pressure contacts the proportion of the heat current flowing through the weak link is not likely to exceed about 10^{-6}. The WF law can be badly in error for pressure contacts between metal, but not usually by very much when most of the heat is transported by electrons rather than by phonons. We do not know for certain what proportion is transported by electrons in point contact weak links but we believe that it will be large enough,and thus the WF law closely enough obeyed,that the heat conducted through the weak link may be neglected. Our conclusion should be tested by experiment nevertheless.

Note that we are limited, if the method of construction outlined here is used, to temperatures which lie below the superconducting transition temperature of lead (7.2 K). With more exotic materials such as Nb_3Sn this limit might conceivably be raised to a temperature approaching 18 K.

HEAT CAPACITY MEASUREMENT

As an example of a novel kind of measurement that might be made using an RSQUID wattmeter consider how one might use it to measure heat capacity at low temperatures. If the frequency of oscillation f is proportional to the heat current $\dot{Q}$ through N, then the number (n) of cycles of oscillation that take place in a given interval of time is proportional to the amount of heat Q that has flowed through N in the same interval. If it can be arranged that this is the heat that flows into (or out of) a specimen during the time that its temperature changes from T to T+ δ T then the heat capacity of the specimen may be deduced. It will be given by

(2) $$CV = Q/\delta T = (\phi_o n)/(SW\delta T)$$

where V is the volume of the specimen and C is its specific heat per unit volume. Note that W here is the thermal resistance of N itself and does not include the resistance of the NS interfaces.

In figure 3 we have sketched a version of the RSQUID described earlier (Fig. 2) which has been modified to make it more suitable for this kind of measurement. The RF excitation and detection system would not need to be altered but the ring N and the lower disc D - to which the

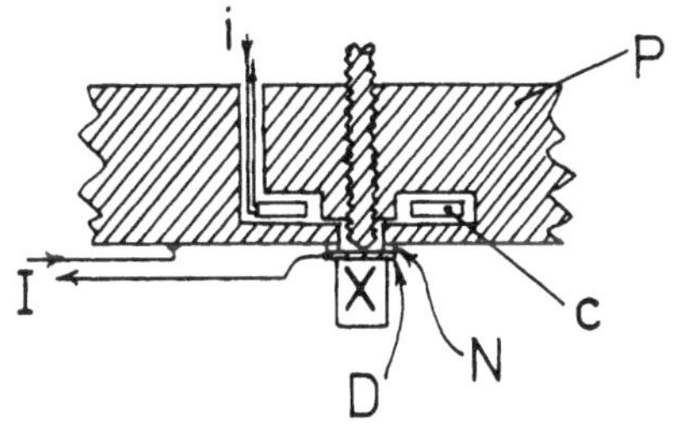

Figure 3 The RSQUID of figure 2 modified for heat capacity measurement. It is imagined to be surrounded by a vacuum. X is the specimen

specimen X is attached - have been considerably reduced in size so that the correction for their heat capacity may be made as small as possible. The upper disc P has now been expanded to become part of a larger block in close thermal contact with a thermal bath. It is presumed that the heat conducted by any electrical leads or clamping device can be neglected and that the whole device is surrounded by a high vacuum. When the temperature of P changes from T to T+ δ T there fore, a count of the number of cycles n enables the heat capacity of X to be determined.

To get some idea of magnitudes and sizes consider a possible special case. Suppose we use CuFe (0.1 at.%) for the normal metal. Below 2 K the specific heat is predominantly electronic, and is therefore proportional to temperature, with a coefficient $\gamma \sim 3\ 10^{-4}\ JK^{-2}cm^{-3}$(11). The resistivity $\rho \sim 1\ \mu\Omega$ cm; σ varies from 2.9 at 1 K to 2.0 at 6 K (6). If this material is formed into a ring 3 10^{-3} cm thick, 1 cm in diameter, 1 mm wide, it will have a volume of about 10^{-3} cm^3, a heat capacity at 1 K of 3 ergs K^{-1} and a resistance R of $10^{-2} \mu\Omega$. The sensitivity, just as it was in the example considered earlier, will again be $f/\dot{Q} = n/Q \sim 40$ $ergs^{-1}$. If X was a 1 gm specimen of the same material (volume $\sim 10^{-1}$ cm^3) it would have a heat capacity at 1 K of 300 ergs K^{-1} corresponding to a count per degree $n/\delta T \sim 15000\ K^{-1}$. The correction for the heat capacity of the ring N would be less than 1%. (It is less certain how small the correction for D could be made). Within wide limits the magnitude of the thermal resistance due to the NS interfaces should not affect the accuracy of the method; merely the time for the system to come to thermal equilibrium after a change in temperature of the thermal reservoir.

The maximum value of f that can be permitted during a measurement is likely to correspond to a lower frequency than either an ultimate limit ($\sim f_r/50$) determined principally by the frequency of excitation f_r(5) or a practical limit set by the filters in the electronic system. It will have a typical value of about 10 kHz, set by the bandwidth of the system following the detector which must be small enough for the signal to noise ratio of the v_o oscillations to be sufficient to allow accurate counting. According to Soulen et al (7) a ratio of at least 14 dB is necessary. This at any rate is what they find they need for their RSQUID noise thermometer. Whether this limit can be raised when the 'up and down' counting method (outlined below) is used, remains to be seen.

One might be afraid that reliable results could only be obtained with this method if great care was taken to make the distribution of heat current density q throughout the normal metal extremely uniform. If, for a given heat current, V varied according to the distribution of q over the cross-section of N it would be necessary to attach the specimen in such a way that q was always uniform; otherwise the calibration would vary with the manner in which the specimen was attached, and might, in unfavourable circumstances, even vary with temperature differently for different specimens. This might conceivably happen if the variations in thermal conductivity with temperature caused the distribution of q in the normal metal to vary with temperature.

It is probable however that q does not have to be extremely uniform in order for $V/\dot{Q}$ to be a reproducible function of temperature. This may be seen by considering the following simple system. Suppose we have a rectangular strip of unit width along whose length the thickness d and the temperature difference ΔT vary. The potential difference V must be the same at all points if the two faces of the strip are everywhere in intimate contact with a superconductor, as we assume them to be. At each point V is the sum of two terms: V_e associated with electric current of density j and V_t associated with heat current of density q. We will assume the strip to be thin enough, and the variation of ΔT and d sufficiently gradual that we may treat the currents j and q as though they flow normal to the surfaces. Then $V_e = j\rho d$, where ρ is the resistivity of the strip: $V_t = Sqd/K$ where K is the thermal conductivity of the strip and S its thermopower. Since the total electric current must be zero, $\int_0^D j\,dx = 0$ where x is the distance measured along the strip from one end and D is the length of the strip. Thus $\int_0^D (V_e/d)dx = 0 = \int_0^D [(V-V_t)/d]\, dx$ and hence $V = [S\dot{Q}/(KD)] / (\overline{1/d})^{-1}$ which is independent of $\Delta T(x)$. Thus V has the same value that it would have had if ΔT was uniform and the strip had a uniform thickness $d_e = (\overline{1/d})^{-1}$. This simple analysis will break down if there are variations of composition as well as thickness. The more non-uniform the distribution of q is over the cross-section of N the more uniform the composition on which ρ and S depend must be.

The ultimate limit to the resolution of this method is determined by Johnson noise. If a frequency measurement is made with a gate time τ, the RMS uncertainty $\delta f = (t_n \tau)^{-\frac{1}{2}} + (12\,\tau^2)^{-\frac{1}{2}}$ where the characteristic time $t_n = \phi_o^2/(2kTR)$. If $R \sim 10^{-8}\Omega$, $\sigma \sim 2$ and T = 1 K, the uncertainty in a measurement of Q made in a time of 10 secs say would be $2\,10^{-2}$ ergs. Careful attention will have to be paid to heat leaks if resolution of this order is to be achieved.

METHODS OF COUNTING

If $\dot{Q}$ can be relied upon not to change sign during a measurement the cycles of oscillation may be counted directly from the number of times v_o passes through zero with a slope dv_o/dt of a given sign. The non-linearity of the function $f(\dot{Q})$ when $\dot{Q}$ is small is potentially a source of error however in this direct counting method. If $I_a \sim 0.1\ \mu A$ for example and $R \sim 10^{-8}\Omega$ there will be significant non-linearity when $f \sim I_a R/\phi_o \sim 2$ Hz. An approximate estimate for the error can be obtained by using the relationship derived by Sullivan et al (9) for the frequency of oscillation of a voltage-biased Josephson junction when the capacitance across the junction is negligible. We merely replace I by the equivalent quantity $SW\dot{Q}/R$. If $\dot{Q}_o$ is the heat current at which $SW\dot{Q} = I_a R$ their expression can be written

(3) $\quad f = (SW\dot{Q}/\phi_o)(1 - \dot{Q}_o^2/\dot{Q}^2)^{\frac{1}{2}}$

For the particular ring and sample taken as an example above $\dot{Q}_o = 0.02$ ergs s^{-1} and corresponds to a rate of change of temperature $(dT/dt)_m = 0.14$ mK s^{-1}. Equation (3) is not strictly accurate for our RSQUID but is probably accurate enough for making a rough estimate of the non-linearity error.

It should not be difficult in measurements on alloy samples of this sort to ensure that the temperature T always changes at a rate fast enough that $f \gg I_a R/\phi_o$ but slow enough that thermal equilibrium is sufficiently well maintained throughout the sample. If the temperature of the sample changes at a rate of $20(dT/dt)_m$ the linearity error estimated from equation (4) is about 0.1% and the temperature inhomogeneity along the length of the sample (ℓ) given approximately by $20(dT/dt)_m(\ell^2/D_t)$. For the CuFe specimen taken as an example above $D_t \sim 10^2 cm^2 s^{-1}$. If 0.1% non-linearity is not to be exceeded dT/dt must be greater than 3 mK s^{-1}: this corresponds to an inhomogeneity in temperature if $\ell = 1$ mm of about 5 μK.

It is possible to count the cycles in another way (10), referred to above as the 'up and down' method, which discriminates between positive and negative heat currents. AC flux modulation is applied at a frequency f_m much higher than the maximum value that can be allowed for f during a measurement of n. The oscillating signal v_o is fed to two phase-sensitive detectors (PSDs) to which different reference signals are applied, one at a frequency f_m, the other at $2f_m$. By converting the output of these PSDs into logic states '0' when the output is negative, '1' when it is positive, one can apply the two signals thus obtained as inputs to a logic circuit which counts quarter cycles of oscillation, adding to the total when $\dot{Q}$ is positive, subtracting from it when $\dot{Q}$ is negative. If it is important to reduce the non-linearity error, this method of counting has the advantage that, even when the discrimination between positive and negative is not important, the AC flux modulation reduces the effective value of I_a. That it will do so may be seen by considering a special case (5). Suppose that the weak link is one for which the mean current I_k (averaged over the cycle around which the RSQUID is driven by the RF excitation current) varies sinusoidally with the mean phase difference Θ_j across the weak link, i.e. suppose that $I_k = I_a \sin \Theta_j$. If this is modulated by flux alternating as a triangular function of time with a peak to peak amplitude $2\Delta\phi_m$, it can be shown (5) that $\Theta_j(t)$ is triangular also with peak to peak amplitude $2\Delta\Theta_j = 4\pi\Delta\phi_m/\phi_o$ and that the mean current I_k, averaged now over a period of the modulating flux, is given by $\overline{I_k}(\Theta_j) = I_k(\Theta_j) \sin \Delta\Theta_j/\Delta\Theta_j$. In effect I_a is multiplied by the factor $\sin \Delta\Theta_j/\Delta\Theta_j$. In general then (not only in this special case) the effective value of I_a is reduced by the presence of modulation current. There will be a limit however to the amount by which it can be reduced. It cannot be made zero without at the same time making the amplitude of the required signal at f_m

zero also. The oscillations in the output of the PSD whose reference signal is at f_m will be reduced (relative to $\Delta\phi_m$) by approximately the same factor as the effective value of I_a. Thus if the maximum value of f is limited by the signal to noise ratio in the oscillating signal v_o, a reduction of a factor of 10 (say) in the minimum value of f before non-linearity becomes appreciable will be accompanied by a reduction by a factor 10 in the maximum value of f. However this technique could be a useful one in calorimetry for reducing the non-linearity error.

It is possible to measure a feedback current proportional to n rather than count n directly. One merely has to apply AC flux modulation current and to take a feedback current to the coil just as one might do in a conventional SQUID magnetometer. This current changes at a rate proportional to V : hence this method does not suffer from any non-linearity error due to the intrinsic properties of the RSQUID. On the other hand the possible resolution is probably not as great as one might hope to achieve using counting methods, a factor which might be important when making corrections for the heat capacity of addenda like N and D; moreover one is subject to errors introduced by amplifier drifts.

A programme of experimental work is now underway to try out some of these ideas. One possible application of the method proposed above for measuring heat capacity has already been suggested. Dr. J.A. Lee and his colleagues at A.E.R.E., Harwell, are interested in extending measurements of the heat capacity of the actinide elements down to lower temperatures. The range of conventional methods is limited by the self-heating of samples as a result of their radio-activity. It should be possible to extend the range down to lower temperatures by making use of the method described here, since the specimen can be connected to the thermal reservoir by a very small thermal resistance. The heat current due to self-heating will then produce a very small rise in temperature of the sample and can be balanced, as far as its effect on the RSQUID goes, by a constant external (electric) current I.

REFERENCES

1. Zimmerman, J.E., Silver, A.H., : A high sensitivity superconducting detector J. Appl. Phys. 39, 2679-82 (1968)
2. Park, J.G., Farrell, D.E., Kendall, J.P. : The response of a toroidal resistive SQUID to modulation by magnetic flux. J. Phys. F :Metal Phys. 3, 2169-73 (1973)
3. Park, J.G. : The physics of resistive SQUIDs modulated at low frequency. J. Phys. F : Metal Phys. 4, 2239-63 (1974)
4. Park, J.G. :Resistive SQUIDs in the absence of modulation current. Int. Conf. on Low Temp. Phys. (LT14) Helsinki, 4, 164-7 (1975)
5. Park, J.G. : Resistive devices. NATO Institute on small scale superconducting devices (Gardone Riviera, Italy, September 1976); to be published
6. Kjekshus, A., Pearson, W.B. : Thermoelectricity and electrical resistivity of dilute copper alloys of chromium, manganese, iron and cobalt at low temperatures. Can. J. Phys., 40, 98-112 (1962)
7. Soulen, R.J., Finnegan, T.F. : A microwave resistive SQUID for noise thermometry. Rev. de Phys. Appl., 9, 305-7 (1974)
8. Berman, R., Mate, C.F. : Thermal contact at low temperatures. Nature, 182, 1661 - 3 (1956)
9. Sullivan, D.B., Peterson, R.L., Kose, V.E., Zimmerman, J.E. : Generation of harmonics and subharmonics of the Josephson oscillation. J. Appl. Phys. 41, 4865-73 (1970)
10. Nisenoff, M : Superconducting magnetometers with sensitivity approaching 10^{-10} gauss. Rev. de Phys. Appl., 5, 21-24 (1970)
11. Franck, J.P., Manchester, F.D., Martin, D.L., : The specific heat of some dilute copper iron alloys showing a minimum in the electrical resistance at low temperatures. Proc. Roy. Soc., A263, 494-507 (1961)
12. Kamper, R.A. : Metrology and standards. NATO Institute on small scale superconducting devices (Gardone Riviera, Italy, Sept. 1976): to be published.

POSSIBLE CRYOCOOLERS FOR SQUID MAGNETOMETERS*

J. E. Zimmerman, R. Radebaugh and J. D. Siegwarth
Cryogenics Division, National Bureau of Standards,
Boulder, Colorado 80302 USA.

ABSTRACT

In a study to determine if a cyclic cryogenic refrigerator is suitable for cooling SQUID magnetometers and similar instruments, we have used a specially designed Stirling machine with a three-stage displacer to achieve and maintain a temperature of 13 K. The displacer and cylinder were made of low-susceptibility, nonconducting plastic to minimize magnetic interference. For the same reason, and also to minimize mechanical noise, the machine operates at low displacement and low speed, and it uses only about 10 W of mechanical power, requiring 50 W input to a typical small electric motor. Since the temperature achieved is within the range of NbN and Nb_3Sn SQUIDs, it can, in principle, be used to cool a SQUID. The estimated magnetic signal from the reciprocating displacer may not seriously affect the magnetometer sensitivity, but the signal due to the cyclic pressure-induced geometric distortion of the SQUID mounting in the earth's field may be very difficult to eliminate.

INTRODUCTION

In this paper we address a subject that is always implicit, but not usually expressed in detail, in any application of superconductivity, namely, the cryostat. For small devices, such as SQUIDs, the universal practice is to use liquid-helium cryostats with a variety of shapes, sizes, and special features.(1) It need hardly be pointed out that the cost and inconvenience of procuring and handling liquid helium, the attendant restrictions on how the instrument can be mounted and handled, and the inherent intermittent nature of the operation, all contribute to the obvious disadvantage of superconducting devices relative to competing devices operating at or near ambient temperature.

The alternative to liquid-helium cooling is to use a closed-cycle cryogenic refrigerator (hereafter referred to as a cryocooler). The use of a cryocooler replaces the disadvantageous features noted above with a whole set of new ones. Limited reliability is perhaps the most often mentioned; others are maintenance requirements, cost, and, as we shall see, magnetic and mechanical interference due to moving parts. With regard to reliability, there seem to be two schools of thought, corresponding roughly with the high reliability expounded by cryocooler manufacturers on the one hand, and the low reliability expected by potential users on the other.(2) The fact is that very few cryocoolers are in use today to operate small superconducting devices, so there is no substantial body of experience regarding reliability or anything else.

There are several intermediate solutions to the cryogenic cooling problem. One of these is a liquid-helium cryostat incorporating one or more radiation shields maintained at rather low temperatures by an auxiliary cryocooler, which may greatly reduce the helium evaporation rate.(3) Such a system combines some of the best, and some of the worst, features of both worlds. One of the good features is that it permits one to get practical experience in the use of cryocoolers without being fully committed, in the sense that if the cryocooler turns out to be no good one still has a cryostat. Another useful feature is that the cryocooler can be turned off for periods of several hours of critical measurements, where the magnetic and mechanical noise of the cryocooler might be intolerable, with only a slight increase in liquid-helium consumption.

There are some other types of cryostats which may be useful for particular applications. For example, a cascaded Joule-Thomson system operating from cylinders of compressed gases may be suitable for intermittent short-term operation with quick cooldown. It is beyond the scope of this paper to discuss all possible cryostatic systems, even by way of introduction, and we will hereafter restrict the discussion to the use of liquid-helium and closed-cycle refrigerators (cryocoolers) using helium gas as the working fluid.

In fact, there is only one known type of practical, small, self-contained system which can maintain superconducting temperatures in a 300 K environment. This is the type using helium gas as a working fluid in a cycle involving isothermal compression at room temperature and isothermal expansion, or a combination of isothermal expansion and other processes, at one or more low temperatures. The particular point to be emphasized here is that compression and expansion require moving parts, which may be very difficult to operate in a way that will not interfere with the operation of highly sensitive magnetometers.

We have chosen the Stirling machine as an analytical and experimental model to investigate the problems associated with operating SQUID magnetometers with a small closed-cycle helium-gas cryocooler. It is a very simple machine, shown schematically in Figure 1, which has only two essential moving parts, a piston and a displacer.(4) The displacer fits loosely in its cylinder so that gas can move freely past it. Thus the pressure P is nearly the same throughout the total volume V of

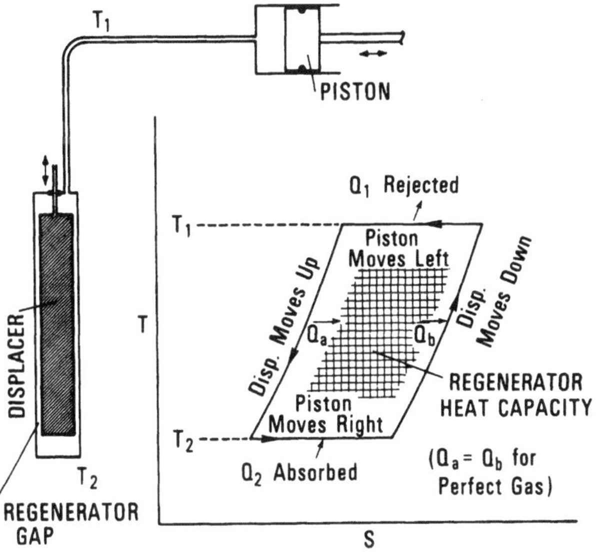

Figure 1 Schematic representation of operation of split Stirling machine and idealized thermodynamic cycle on temperature-entropy diagram.

the system. Work $W = \int PdV$ may be done on the system as the piston moves back and forth, but no work is done by or on the displacer since displacer motion does not change the volume. Referring to Fig. 1, when the displacer is in its lowest position, the piston is moved to the left, compressing the working fluid and producing a heat of compression Q_1 which is rejected at ambient temperature T_1, typically 300 K. Next, the displacer is moved to the top of its cyliner, so the working fluid moves (is "displaced") from the top to the bottom of the cylinder. If we may assume that the machine is already in steady-state operation, with the bottom end at a low temperature T_2, and a stable temperature gradient already established along the displacer, then the fluid being displaced gives up heat Q_a to the displacer and cylinder walls along the annular gap, and arrives at the bottom end at temperature T_2. Then the piston is moved to the right, expanding the fluid and producing a flow of heat Q_2 into the fluid at temperature T_2 at the bottom of the cylinder. Finally, the displacer is moved to its lowest position, displacing the remainder of the fluid back to the top of the cylinder and completing the cycle. The fluid picks up heat Q_b from the walls of the annular gap as it travels up from T_2 to T_1.

In steady state operation it is required that $Q_a = Q_b$, otherwise the temperature gradient along the displacer and cylinder will change with time. With real gases, such as helium at 20 K and below, the requirement $Q_a = Q_b$ is incompatible with the assumption of isothermal heat exchange Q_1 and Q_2 at T_1 and T_2. (i.e., The enthalpy change Q_a or Q_b between two given temperatures T_1 and T_2 depends on pressure, so the requirement $Q_a = Q_b$ means that the expansion cannot be strictly isothermal. This is a limitation, or at least an analytical complication, in the operation of the Stirling cycle.

A more serious limitation, which we will have more to say about, is that the heat capacity of the walls at the cold end becomes insufficient, relative to that of the fluid, to efficiently provide the necessary heat exchange Q_a and Q_b. Considerable research has been devoted to finding materials with favorable properties (large heat capacity and thermal conductivity) to use in regenerative heat exchangers at low temperatures. For use with SQUID magnetometers we would like to impose additional restrictions on the material, namely, that it also have favorable properties relative to generation of electromagnetic interference and noise in the SQUID.

EXPERIMENTS WITH PLASTIC DISPLACER-REGENERATORS

Our experiments are intended only to demonstrate feasibility of specific materials in a specific refrigerating mechanism which may have a possibility of satisfying the stringent requirements of a SQUID magnetometer.

The first experiment was to build a singlestage displacer-regenerator (Fig. 2) of the type invented by Stirling in 1816, but using modern materials. The displacer was a nylon rod 9.6 mm diameter by 300 mm long. It operated inside a spun-glass epoxy cylinder (designated G-10 by the National Electronic Manufacturers Association) with 2.4 mm wall thickness. The radial clearance between displacer and cylinder was about .07 .07 mm, and the length of the cylinder was 313 mm, allowing a 12.7 mm stroke for the displacer, which was connected to a thin brass push rod running through a sliding rubber seal at one end of the cylinder. The push rod was operated by a scotch yoke at the end of a shaft which was turned at 1 Hz by a small electric motor. A pair of cams, 90° out of phase, on the same shaft, operated the inlet and exhaust valves.(5) The working fluid was helium gas, supplied at 14 atmospheres from a storage cylinder and exhausted to the atmosphere when expanded. Twenty-five layers of aluminized plastic sheet were wrapped around the displacer cylinder, which was mounted in an evacuated metal cylinder. Operated at 1 Hz, the machine attained a temperature in the neighborhood of 50 K three or four hours after starting, and maintained the temperature for about two days on a cylinder of helium.

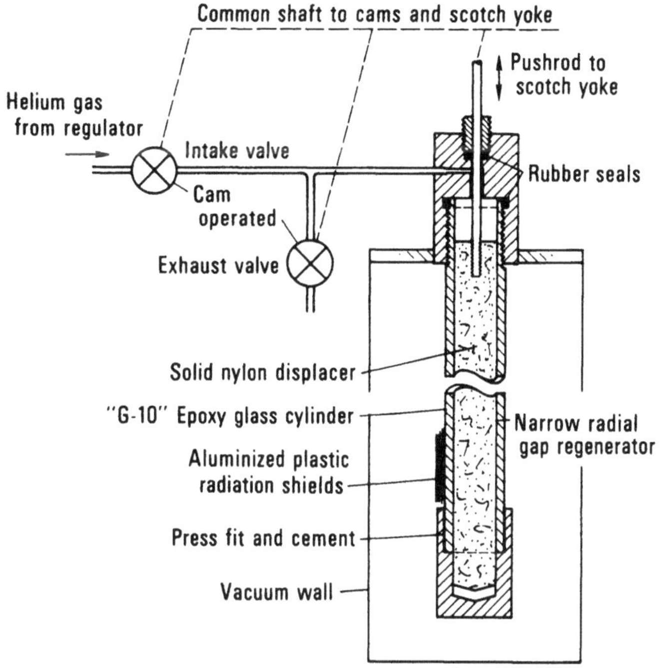

Figure 2 Details of simple cryocooler operating from compressed helium gas cylinder. Radial gap around displacer is too narrow (∿0.05 mm) to be shown in this drawing. Displacer was 9.5 mm diameter by 300 mm long, with 12.7 mm stroke.

The success of this simple apparatus in producing rather low temperatures encouraged us to construct a 3-stage, closed-cycle, split Stirling machine,(6) using the same materials and techniques (Fig. 3). Dimensions of the stages were as follows: The first stage displacer-regenerator was 245 mm long and 19 mm diameter, with 0.05 mm radial clearance inside the spun-glass epoxy cylinder. The second stage was 143 mm long and 9.45 mm diameter with 0.07 mm radial clearance. The third stage was 144 mm long and 4.7 mm diameter with 0.04 mm radial clearance. The cylinders all had a 2.4 mm wall thickness and were epoxied together using a commonly available epoxy glue. A brass cap was epoxied on the bottom or cold end. Cylindrical aluminum radiation shields were fitted over the lower end of the first and second stages. Several layers of aluminized plastic sheet were wrapped within the shields and around the outside of the assembly for additional radiation shielding. A polished stainless steel cylinder containing a 35.6 mm diameter plastic piston formed the compressor. The various dimensions of these machines were dictated by intuition and by immediate availability of materials. In subsequent work we hope to optimize the size and shape on a more rational basis. The piston was driven through a 38 mm stroke by a

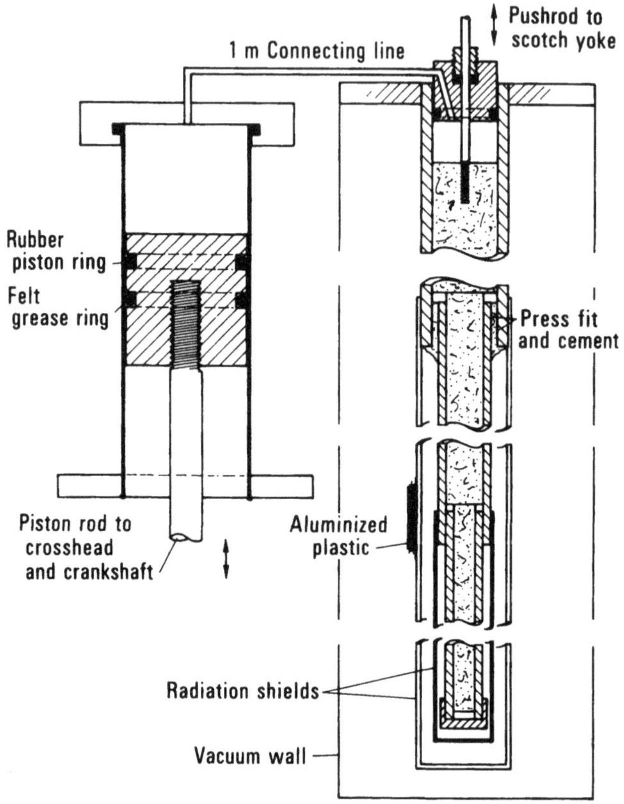

Figure 3 Three-stage Stirling cryocooler. Displacer sections were 19 mm, 9.5 mm, and 4.7 mm diameter, and 245 mm, 143 mm, and 144 mm long, with 12.7 mm stroke. Piston was 35.6 mm diameter with 38 mm stroke.

crankshaft and crosshead mechanism adapted from a small commercial freon compressor. Operating at 1 Hz with a displacer stroke of 12.7 mm and a pressure excursion of about 3 to 13 Pa, this machine maintained a temperature below 16 K.

This machine was also operated with the same valved cycle used for the first cryocooler, and reached almost the same minimum temperature of 16 K. The performance was improved by reducing the dead space around the displacer. We found that the nylon displacer had a larger coefficient of expansion than the spun-glass epoxy cylinder, so that there were several millimeters of "dead space" at the cold end after steady-state operation was achieved. We therefore carefully measured the differential contraction and machined the displacer accordingly, so that the displacer and cylinder fitted together precisely in steady-state operation, rather than at room temperature. After initial cooldown, the dead space was eliminated by readjusting the length of the push rod. When this was done, and the phase angle between piston and displacer was empirically optimized, the cryocooler maintained a temperature around 13 K or less, for continuous periods as long as 5 weeks. The limit on non-stop operation appeared to be due to air being entrapped by grease on the piston ring and the pushrod seal and contaminating the helium gas. Such contamination would eventually build up, reduce the radial clearance of the displacer, and cause friction. We eliminated most of the contamination by maintaining an ambient helium atmosphere on the back side of the piston, but have not yet done so for the pushrod seal.

This three-stage displacer has been operated a total of nearly 5000 hours with no obvious signs of wear. It will maintain a temperature between 12-1/2 and 13 K at the cold end with no heat load. With a heat load of 11 mW the temperature rises to 14.5 K. The outer radiation shield is maintained at 120 K and the inner radiation shield at 40 K. Operated at f = 1Hz with a pressure excursion as noted above, the mechanical power required for the compression stroke, assuming an isothermal process, is $P_1 V_1 f(\ln P_2/P_1) = (3 \times 10^5)(38 \times 10^{-6}) \ln(13/3)$ watts $\cong$ 15 W, where $V_1 = 38\ cm^3$ is the piston displacement. Since a considerable fraction of the work is returned on the expansion stroke, the net power required was estimated to be of the order 10 W or less. This estimate was confirmed by operating a similar machine with a 50 watt motor whose efficiency was less than 25%.

We have recently made fairly extensive measurements and calculations of the various components of intrinsic heat load in this machine, namely, shuttle heat, conduction, radiation, and regenerator inefficiency, for the purpose of optimizing the design. In this connection, we may wish to complicate the design slightly by incorporating a high-heat-capacity matrix at the cold end of the displacer cylinder, to improve the regenerative heat exchange and achieve lower temperature. The results of our measurements and the optimization will be submitted for publication subsequently.

CONSIDERATION OF COMPATIBILITY WITH A SQUID MAGNETOMETER

Having demonstrated that superconducting temperatures (high-T_c) can be achieved and maintained for long times with a very low-power cryocooler made of plastics with favorable magnetic properties, it remains to be seen whether such a machine can be used with a highly sensitive SQUID magnetometer. Ignoring for the present the problem of the compressor and its drive mechanism, consider the magnetic signals generated by just those parts that would be near the SQUID, namely the bottom end of the displacer and its cylinder. In our experimental machine, the end of the displacer is 4.7 mm diameter and moves up and down about 13 mm. In the field of the earth, 0.5 x 10^{-4} T, this gives a peak-to-peak magnetic moment at 1 Hz of about 0.3 cm^3 x 0.5 x 10^{-4}T x 1.1 x 10^{-5} = 0.16 x 10^{-9} T cm^3. The susceptibility of nylon, 1.1 x 10^{-5}, was measured for us by F. Fickett.

If a SQUID were mounted 5 cm (for example) from the end of the displacer, the peak-to-peak field at 1 Hz would be about 10^{-12} T. A coherent oscillation of this magnitude could be subtracted from the output with an accuracy of perhaps 10^{-14} T, which is comparable to present SQUID sensitivities in a 1 Hz bandwidth. Thus, one may be able to cope with the magnetic signal from a small reciprocating displacer of nylon or comparable material.

On the other hand, a peak-to-peak pressure excursion of 10^6 Pa (10^5 Pa $\cong$ 1 atmosphere) in the displacer cylinder will cause it to bend owing to non-uniform wall thickness and other defects. If the SQUID were mounted on the end of the cylinder for mechanical support, bending of the cylinder in the earth's field may cause a flux change in the SQUID far too large to be subtracted from the output with sufficient accuracy. There is no need to use precise numbers to show this. An elastic modulus of 10^{11} Pa and a pressure change of 10^6 Pa in a cylinder whose ratio of length to diameter is 10 gives a bend of the order of 10 x 10^6x 10^{-11} x 10^{-2} = 10^{-6} radians, if the lateral inhomogeniety (different wall thickness, for example) is of order 10^{-2}. In the earth field this would appear as an oscillation of the order of 10^{-10} T. Two alternative approaches suggest themselves for reducing this signal. One is to measure the bending and empirically correct it by grinding or scraping one side of the cylinder. The other is to mount the SQUID on an independent support with only a flexible thermal connection to the cold end of the cylinder. In any case, it may be quite difficult to reduce or compensate the bending signal to a level of 10^{-14} T.

COMPARISON OF THE CRYOCOOLER WITH CONVENTIONAL LIQUID-HELIUM CRYOSTATS

We have shown that the lowpower cryocooler can maintain temperatures below 15 K with ten or twenty milliwatts heat load. Bias power requirements for SQUIDs are many orders of magnitude smaller than this, and the only significant heat load imposed by the SQUID is conduction down the electrical leads. Our experience with conventional liquid-helium cryostats indicates that, with good design, the conduction heat load can be reduced to less than ten milliwatts. Actually, designing or specifying cryostats on the basis of refrigeration capacity at the coldest point is quite misleading. What is important is to distribute the refrigeration capacity, using a multi-stage machine, so that the bulk of the heat input (conduction, radiation, etc.) is pumped at relatively high temperatures, leaving very little to be pumped at the lowest temperature. The distribution of refrigeration capacity vs temperature is another aspect of design optimization which has received little attention in the literature. For comparison, a typical small liquid-helium cryostat with an evaporation rate of one liquid liter per day (corresponding to about 700 liters of gas at room temperature) is supporting a total heat load, intrinsic plus imposed, of 29 mW at 4.2 K, and has the capability of absorbing an additional 7.6 m W/K due to the heat capacity of the vapor, which is approxmately independent of temperature. These figures might be used as a guide for the necessary refrigeration capacity of a small cryocooler. Actually, the figures should be quite conservative because (1) a cryocooler would not have the large cold reservoir and (2) optimum design would probably not require the temperatureindependent heat absorption capacity, but rather would concentrate the refrigeration capacity at higher temperatures.

FURTHER PERTINENT COMMENTS

The work we have reported here is unique in two respects, first in the choice of materials and construction methods for the displacer, and second in building a machine with at least an order of magnitude lower input power requirements than previous comparable machines. Both of these features are, of course, related to our particular aim of minimizing magnetic interference, but they will have considerable effect on cost and portability as well. It appears from our experiments that the cost of a cryocooler could easily become comparable to the cost of a SQUID magnetometer or similar superconducting instrument. Such systems might well be justified economically, in view of the fact that the cost of liquid helium will increase as the supply diminishes, whereas the cost of cryocoolers should decrease substantially if they are produced in sufficient numbers to replace present liquid-helium cryostats.

This work completes what might be called a preliminary study of the feasibility of using a small mechanical cryocooler for SQUID magnetometers and other small superconducting devices. We conclude that a second phase of development is justified. We believe this should include (1) optimizing the design as noted above, (2) incorporating a high-T_c SQUID with which to measure and evaluate the level of magnetic interference from the various moving parts, and (3) evaluating other potentially useful materials and refrigeration processes which may improve the performance.

ACKNOWLEDGEMENT

We are grateful to several individuals for helpful comments and criticism, particularly Sam Collins, Dave Daney, T. R. Strobridge, and R. O. Voth. This work is supported jointly by the National Bureau of Standards, the Office of Naval Resarch and the Naval Electronics Systems Command.

REFERENCES

1. For example, we have built 1.7-liter spun-glass-epoxy cryostats with hold times of nearly three days on one filling of liquid helium. At the opposite extreme, a 180-liter cryostat has been built with a hold time of 102 days. See R. J. Dinger, J. R. Davis and M. Nisenoff, NRL Memorandum Report 3256, Naval Research Laboratory, Washington, DC (March, 1976).

2. Some contrasting comments on reliability by different people (not directed specifically at each other) may be found in Proc. Cryogenic Cooler Conf., AFFDL-TR-73-149, Air Force Flight Dynamics Laboratory, Wright-Patterson Air Force Base, Ohio 45433 (Dec. 1973). This report also contains several papers on recent developments in miniature cryocoolers for the 10-20 K range.

3. R. C. Longsworth, p. 390, Proc. Fifth International Cryogenic Engineering Conf., IPC Science and Technology Press (1974), has shown that liquid-helium hold times can be increased by a factor of 4 or more by maintaining a shield at 20 K rather than at 77 K. He estimates that one might achieve hold times of the order of a year with a 25-liter cryostat. His results should be quite conservative as applied to cryostats for lightweight low-energy devices like SQUIDs.

4. Much historical and recent work on Stirling and related machines is described by G. Walker, Stirling-Cycle Machines. Oxford University Press (1973).

5. Operation of a Stirling-type displacer-regenerator with a valved helium supply was developed by H. O. McMahon and W. E. Gifford, p. 354, Advances in Cryogenic Engineering, Vol. 5, Plenum Press (1960).

6. The stepped-displacer idea used here was invented by Fokker and Köhler. The equivalance of a stepped-displacer machine to a multi-stage machine is explained in some detail by G. Prast, Philips Tech. Rev. 26, 1 (1965). The term "split" refers to the fact that the compressor cylinder and displacer cylinder are geometrically separated from each other.

ON LIMITING MAGNETIC FIELD SENSITIVITY OF SUPERCONDUCTING QUANTUM INTERFEROMETER WITH TWO JOSEPHSON JUNCTIONS

S.I.Bondarenko
Physico-Technical Institute of Low Temperatures,
UkrSSR Academy of Sciences, Kharkov, U.S.S.R.

Unlike single-junction SQUIDs, superconducting quantum interferometer (SQI) with two Josephson junctions has the advantage of (1) achieving much higher voltages just in the interferometer contour; (2) stable (without hysteresis) operation with much higher relations $L_0 J_c/\Phi_0$ (L_0 is the SQUID or SQI contour inductance, J_c the critical current of SQI junctions, Φ_0 the magnetic flux quantum equalling 2.07×10^{-15} Weber) amounting to hundreds and thousands, while in single-junction SQUIDs multiquantum transitions occur even at $L_0 J_c/\Phi_0 = 10 - 20$; (3) avoiding magnetic noises induced in single-junction SQUIDs by temperature fluctuations of the remagnetization loop in the hysteresis regime; (4) using simpler circuits to separate the effective signal. Thus, one may achieve not worse magnetic field sensitivity than in single-junction SQUIDs. Let us analyse the two-junction SQI voltages, transformation coefficients, magnetic and electric noises, using the following assumptions: 1) the current - voltage characteristics (CVC) of the SQI junctions are governed by the Aslamazov-Larkin equation [1] in the whole range of operating temperatures ($T = 0 - T_c$) and are similar for both junctions; 2) SQI is perfectly symmetric, i.e. has the same normal resistance and critical currents for the junctions ($R_1 = R_2 = R$; $I_{c1} = I_{c2} = I_c$) and equal inductions of the SQI arms ($L_1 = L_2$); 3) the junctions of $R \leqslant 10$ Ohm are possible; 4) the frequency band Δf of magnetic and electric processes equals unit ($\Delta f = 1$); 5) possible application of superconductors and temperatures permitting the energy gap (Δ) of several millivolts; 6) the maximum permissible SQI induction is 10^{-7} Henry at $T = 4^{\circ}$ K; 7) Josephson junctions are possible whose intrinsic induction is not less than 10^{-13} Henry; 8) in terms of the reasonable cryostat size and SQI capabilities we assume the SQI quantization area S to be within $10^{-2} - 10^{-12} m^2$; 9) the external magnetic field (H) in SQI is homogeneous; 10) the limiting power $P = I_c V_o$ ($V_o = RI_c$) in the junction must not exceed the possible heat dissipation (10^{-6}W); 11) the critical current I_c is related to the junction normal resistance R, temperature T and energy gap Δ by

$$(1) \qquad I_c = [\pi \Delta(T)/2eR]\, th\, [\Delta(T)/2kT]$$

where e is the electron charge, 1.6×10^{-19} C, $K = 1.6 \times 10$ J/deg, Δ in J, R in Ohm.

If $L_o I_c \ll \Phi_c$ in the SQI resistive state the maximum amplitude of quantum oscillations V(H) is $\Delta V_m = RI_c$. The transformation coefficients are $A_\Phi = \frac{\Delta V_m}{\Phi_o/2} = \frac{2I_c R}{\Phi_c}$ for the magnetic flux and $A_H = \frac{\Delta V_m}{\Delta H/2} = \frac{\Delta V_m}{\Phi_o/2S_c} = 2RI_cS_o/\Phi_c = A_\Phi S_c$ for the magnetic field. Thus, for $T \to 0^{\circ}$ K $\Delta V_m \approx [\frac{\Delta(0)}{e}]$ V;

$A_\Phi \simeq [\Delta(0)/\Phi_c e]$ V/Weber ;for $T \to T_c$ $\Delta V_m \simeq [\Delta^2(T)/ekT]$ V

$A_\Phi \simeq [\Delta^2(T)/ekT]$ V/Weber . At $L_o I_c \gg \Phi_c$ one may obtain

$\Delta V_m \approx R\sqrt{I_c \Phi_o / L_o}$. With Eq.(1) involved, we have

$\Delta V \backsim [R\Delta(0)/L_c]^{1/2}$ for $T \to 0^{\circ}$ K and $\Delta V_m \backsim [R/L_o T]^{1/2}$

for $T \to T_c$. The appropriate transformation coefficients are $A_\Phi = 2[R\Delta(0)/eL_c\Phi_c]^{1/2}$, $A_H = 2S_o[R\Delta(0)/eL_c\Phi_o]^{1/2}$ at $T \to 0^\circ K$ and $A_\Phi = 2[R\Delta^2(T)/L_c kTe\Phi_o]^{1/2}$, $A_H = 2S_o[R\Delta^2(T)/L_c kTe\Phi_o]^{1/2}$ at $T \to T_c$.

Similar voltages can be obtained in the circuit with a single junction SQUID when a high-Q resonance circuit is achieved. We shall consider noise effects: 1) intrinsic magnetic noises of SQI; and 2) intrinsic electric noises of SQI. Using the earlier derived formulas, we can find the maximum permissible sensitivity and parameters of two-junction SQI.

The magnetic noise induced in the resistive state of SQI may be represented as a noise flow $\delta\Phi_1$: $<\delta\Phi_1> = L_o(2kT\Delta f/R)^{1/2}$. The electric noise by the Niquist formula is $\delta V_1 = (2kT \cdot R\Delta f)^{1/2}$. Comparing $<\delta H_1> = <\delta\Phi_1>/S_o$ with $\Delta H/2 = \Phi_o/2S_o$ and $<\delta V_1>$ with ΔV_m, we shall find what part of $\Delta H/2$ and ΔV_m can be detected. Since ΔV and ΔH are related by curve $V(H)$, then the highest of the relations $\delta H_1/\Delta H/2$ or $\delta V_1/\Delta V_m$ governs the limiting SQI sensitivity, i.e. the minimum detectable share of the period in the magnetic flux (or field). The governing factor depends on the relationship between $L_o I_c$ and Φ_o. The dependences $\delta\Phi_1$, L_c; $<\delta H_1>$, $<\delta V_1>$, ΔV_m, $2<\delta H_1>/\Delta H$, $<\delta V_1>/\Delta V_m$, $P = RI_c^2$ on R for various $L_o I_c/\Phi_o$ are plotted in the figure at a) $T \to 0^\circ K$ $\Delta(0) = 10^{-22}$J, and b) $T \to T_c$, $\Delta(T) = 10^{-24}$J. To calculate S_o the formula proposed for the cylindric SQI: $L_c = \mu_o S_o/\ell$ where S_o is in m^2, $\mu_o = 4\pi \times 10^{-7}$ Henry, ℓ is the cylinder length in meters. Putting $\ell = 10^{-2}$ M gives $S_o \simeq L_o \times 10^{+4} m^2$. The calculation gives $T = 4^\circ K$. The plots at $T \to 0^\circ K$ show that a)the optimum properties of SQI are realizable only at $R \geq 1$ Ohm, since at $R < 1$ Ohm $P_o > 10^{-6}$W. At

$L_oI_c < \Phi_0$ the magnetic flux resolution in fractions of a flux quantum Φ_0 is determined by electric noises in SQI and amounts to $0.5 \times 10^{-7}\,\Phi_0$ at $R = 10^2$ Ohm and $L_oI_c = 1/10$ (Curve IX) and $0.5 \times 10^{-8}\,\Phi_0$ at $R = 1$ Ohm. In field units this is $(4\cdot10^{-10} - 4\cdot10^{-9})$ A/m, respectively. Otherwise, at $L_oI_c > \Phi_0$ magnetic noises predominate in SQI. As a result, the maximum resolution of SQI at $L_oI_c = 100\,\Phi_0$ is $10^{-6}\Phi_0$ at $R = 1$ Ohm and $10^{-5}\,\Phi_0$ at $R = 10^2$ Ohm, which corresponds to $(8\cdot10^{-10} - 8\cdot10^{-11})$ A/m in the field units, respectively. Note that for the latter case $L_o = 2\cdot10^{-8}$ Henry, $S_o = 3\cdot10^{-4}$ m^2, $I_c = 10^{-5}$ A.

The analysis of the plots at T T_c shows that at all values of R P_o suffers no limitation, but an increase in $L_oI_c/\Phi_0 > 1$ sets limits due to L_o exceeding 10^{-7} Henry. Thus, regarding intrinsic noises, at $L_oI_c = 10\,\Phi_0$ SQIs with $R \lesssim 1$ Ohm are preferable. Similarly to the case at $T \to 0$K, as the quantum share resolution Φ_0 is regarded, at $L_oI_c < \Phi_0$ electric noises predominate and at $L_oI_c > \Phi_0$ magnetic ones. The former case permits $2.5\cdot10^{-4}\,\Phi_0$ for $R = 10^2$ Ohm, which corresponds to the field $8\cdot10^{-11}$ A/m and $2.5\cdot10^{-5}\,\Phi_0$ at $R = 1$ Ohm ($8\cdot10^{-10}$ A/m). If $L_oI_c = 100\,\Phi_0$, then R 10^{-1} Ohm is permissible, at $R = 10^{-2}$ Ohm the resolution is $10^{-3}\Phi_0$ ($8\cdot10^{-9}$ A/m), in this case $L_o = 10^{-8}$ Henry, $S_o = 10^{-4}$m^2 , $I_c = 2\cdot10^{-5}$A. The figure at $T \to 0$ and $T \to T_c$ gives information on the other parameters of optimum two-junction SQI.

REFERENCES

1. Aslamazov, G., Larkin, A.I.: The Josephson effect in point superconducting contacts. Pisma v ZhETF 9, vyp.2, 150 (1969)

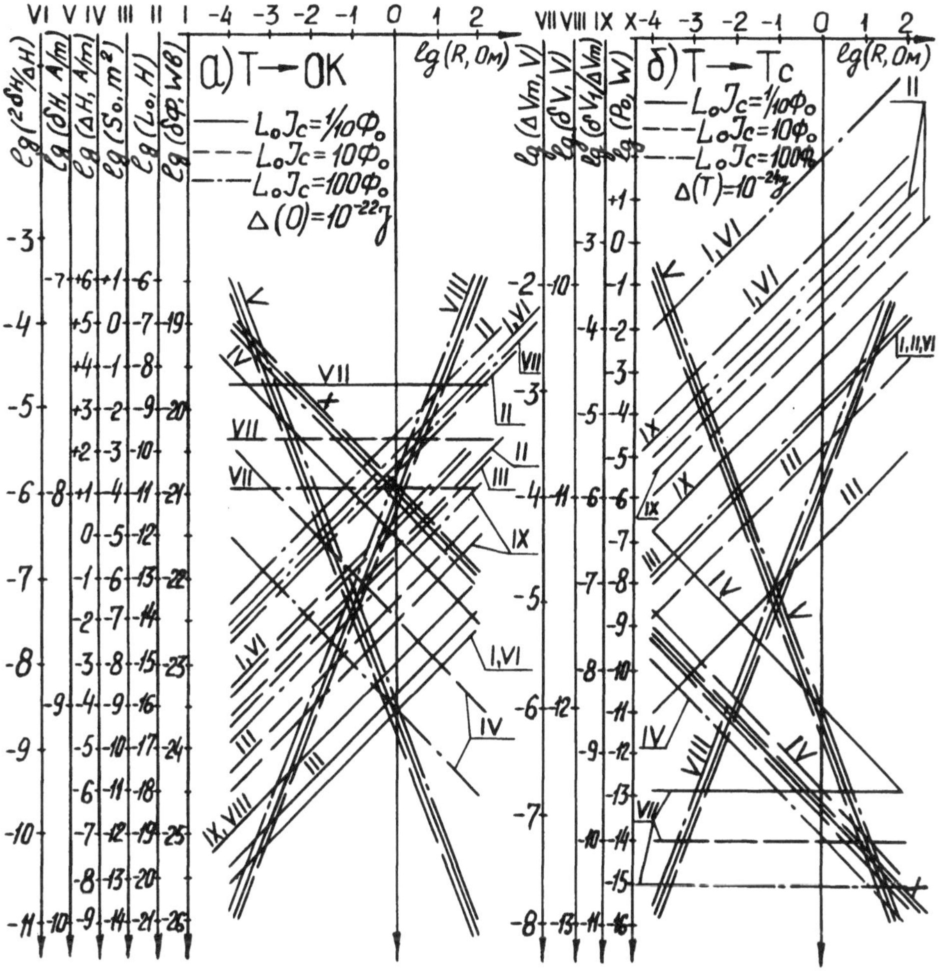

Figure

Nomograph for calculation of magnetometer principal parameters

AN INTEGRATED THIN FILM GRADIOMETER BASED ON A DC SQUID

Gordon B. Donaldson
Department of Applied Physics, University of Strathclyde, Glasgow G4 ONG, Scotland
Mark B. Ketchen, John Clarke and Wolfgang Goubau
Department of Physics, University of California and Materials and Molecular Research Division, Lawrence Berkeley Laboratory, Berkeley, California 94720, USA

ABSTRACT

We describe a magnetic gradiometer which is completely formed from thin films deposited on a glass substrate 89 x 21 x 12mm in dimensions. The active element is a DC SQUID. The as-fabricated device is balanced to the point at which it will operate in a normal environment without low frequency magnetic shielding. It measures the off-diagonal gradient $\partial H_z/\partial x$ with a resolution of 2×10^{-10} gauss-cm^{-1}-Hz$^{-\frac{1}{2}}$ in the range 0.05 Hz to 100 Hz, and has a parallel rejection factor of 10^{-4}. Using movable niobium films which are coplanar with the gradiometer, we have achieved a balance (rejection of uniform H_z field) of 1 part in 10^5.

Possibilities for improving the resolution are discussed.

INTRODUCTION

Several workers[1, 2] have shown how the SQUID magnetometer, with its extreme sensitivity, can be rendered much more versatile if it is produced in a gradiometer form. In such a structure, the single superconducting pick-up coil of a magnetometer flux transformer is replaced by two counterwound parallel and coaxial coils separated by a suitable distance. These coils are connected in series with the input coil to the SQUID. Provided the system is properly balanced, there is no net transformer current, and therefore no input flux, if the coils are placed either in a uniform field parallel to their axis, or in any field parallel to their planes. A non-uniform field parallel to the axis, however, causes different fluxes to link the coils, and hence there is a net flux at the SQUID proportional to the field gradient.

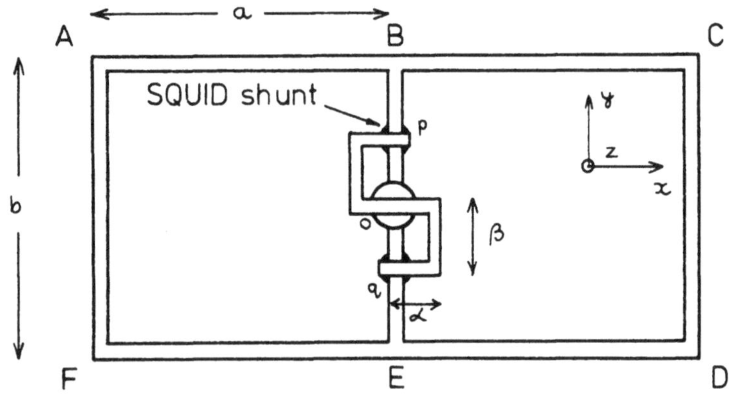

Fig. 1 Principle of DC SQUID gradiometer. All films except the shunts are superconducting. There are Josephson tunnel junctions at p and q, but an insulated crossover at 0.

The most sensitive system so far described has been that of Wynn et al[2] who, with a coil separation of 30cm, achieved a resolution of 3 x 10^{-12} gauss-cm^{-1}-$Hz^{-\frac{1}{2}}$ in the gradient $\partial H_z/\partial z$. Fields perpendicular to the axis, and uniform fields parallel to it, were rejected by using small adjustable superconducting tabs to 'tune' the coils. The best parallel and perpendicular rejection factors achieved were better than 10^{-7}, these factors being defined in each case as the change of SQUID signal produced by the field divided by the change which would be obtained if the same field were applied perpendicular to one only of the pick-up loops.

The recent advent of thin film DC SQUIDS[3] made it logical to explore the possibilities of producing a complete gradiometer, deposited on a single substrate in thin film form. The principle of the device we have produced is a little different from that of the conventional gradiometer, and is illustrated in Figure 1. Superconducting films are used to form the rectangular structure ABCDEF on a substrate in the xy plane. In magnetic fields H_x, H_y parallel to the plane, the structure has no response. In a uniform field H_z perpendicular to the plane, a screening current circulates round the outer loop, but there is no current in the cross-strip BE. Only if the left hand and right hand loops link different magnetic fluxes will a net current flow down BE, and this is given, to first order, by

$$i_{BE} = \left(\frac{\partial H_z}{\partial x}\right) \left(\frac{a^2 b}{L_1 + M_{12}}\right) \qquad (1)$$

a and b being the length and width respectively of one rectangle, while L_1 is the self inductance of one loop (ABEF or CBED) and M_{12} is the mutual inductance between these loops.

We may now imagine that i_{BE} couples flux into a double junction SQUID loop pOq formed by evaporating an S shaped loop over BE in such a way that Josephson junctions are formed at p and q, but there is an insulated cross-over at O. The net flux in the SQUID can then be regarded as the sum of the flux coupled in by i_{BE}, and that directly applied by the field, and is

$$\Phi_s = i_{BE} M_\alpha + \frac{\partial H_z}{\partial x} \beta \alpha^2 \qquad (2)$$

where M_α is the mutual inductance between BE and the SQUID loop, while β and α are SQUID dimensions. Thus Φ_s, which can in principle be measured using standard SQUID instrumentation, is proportional to the off-diagonal field gradient $\partial H_z/\partial x$.

CONSTRUCTION, OPERATION AND PERFORMANCE

The details of construction of the practical gradiometer are as follows. The substrate (see Figure 2) is an 89 x 21 x 12mm pyrex sheet, and on it is deposited the outer rectangle ACDF (dimensions 48 x 16mm) in the form of a 150 μm wide, 500nm thick Pb - In (5%wt) film. Gold SQUID shunts[3)] and gold electrical contact pads are deposited next, and are followed by the strip BE which is a sputtered Nb film 300 nm thick and 100μm wide. The niobium is then thermally oxidised,[3, 4)] and tunnel junctions completed at p and q by evaporating 50μm wide lead strips across the strip BE and the gold shunts. The SQUID loop is completed, first by depositing an insulating disc consisting of 100 nm of SiO followed by 500 nm of Ge over the midpoint of the Nb, and then by evaporating the rest of the S-shaped loop using Pb/In. Finally a thin Duco cement layer is applied to coat the structure and protect it from moisture.

Note that no ground plane is deposited, in contrast to the original cylindrical thin film DC SQUID[3)].

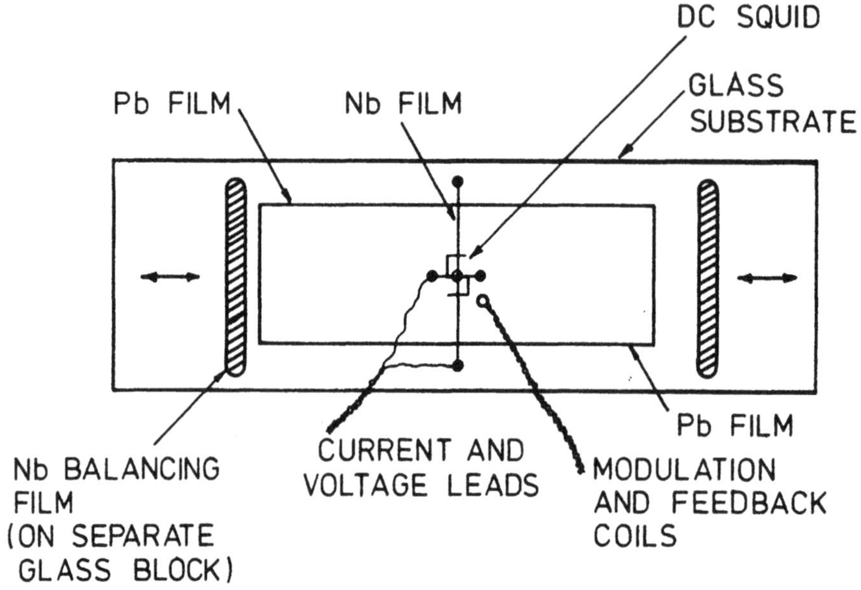

Fig. 2 Fabrication of gradiometer

In operation, the gradiometer is mounted in the cryostat without any magnetic shielding (although RF screening is usually necessary outside the laboratory). Leads are connected to the opposite sides of the SQUID loop, as shown, and nearby a small 100 kHz modulation and feedback coil is glued to the substrate. The device is operated in a flux-locked mode as described elsewhere for the DC SQUID,[3] current biased to a DC voltage of ~1μV, and coupled to a preamplifier through an impedance matching LC circuit resonant at 100 kHz with Q ~ 100. The inductance has to be wound from copper: a superconducting coil would represent a large diamagnetic anomaly at low frequencies, and so upset the balance of the unshielded gradiometer.

The inherent SQUID noise was dominated by 1/f noise below 0.05 Hz, but was white at higher frequencies, up to the 100 Hz roll-off frequency of the electronics, with an rms value of $8 \times 10^{-5}\ \Phi_0 - Hz^{-\frac{1}{2}}$, which is only about 2.5 times higher than that of the best cylindrical DC SQUID[3].

Gradiometers were tested for their response to uniform fields, by using a 2m diameter Helmholtz coil pair driven by a sinusoidal current in the

0.5 - 100 Hz range. As fabricated they had parallel rejection factors of 10^{-4}, and were balanced (perpendicular rejection) to 1 or 2%.

To obtain more precise balances we devised a tuning arrangement involving superconducting elements which were coplanar with the gradiometer (Figure 2): these are 300nm thick niobium films sputtered on to glass blocks which are pressed against the gradiometer substrate and moved along it by a differential screw arrangement. Depending on their position relative to the ends of the gradiometer, these films divert more or less H_z flux through the individual pick up loops, and so control the balance; H_x and H_y field components, on the other hand, are in principle not distorted by the tuners, which thus have no effect on the parallel rejection.

Obviously increasing the tuner width will increase the amount of H_z flux which can be 'steered', and so enhance the range of tuning possible. However it is desirable to make the width to thickness ratio of the tuner films no larger than that of the gradiometer films, in order to ensure that their demagnetising factor does not make them any more susceptible to entering the mixed state in weak perpendicular fields[5]. This implies that the tuner films should be quite thick. However, given that the tuners can never be exactly coplanar with the gradiometer films, such thick films could increase the risk of significant distortion of parallel fields into the z-direction, and so degrade the parallel rejection of the device. Fortunately, it is possible to show that tuner film dimensions can be chosen in such a way as to give a total variation in balance of several per cent, while not compromising the demagnetising factor condition, and still adding only about 10^{-6} to the parallel rejection factor. Using tuner films of such design we were able to improve the balance of our gradiometers to 1 part in 10^5, while still maintaining a parallel rejection of $\leqslant 10^{-4}$.

With our gradiometer balanced we applied a known gradient to it using a current in a long wire parallel to the y axis, but 0.5m from the gradiometer. The measured response implies that the limiting $\partial H_z/\partial x$ sensitivity of the device (corresponding to the white SQUID noise) is 2×10^{-10} gauss - cm^{-1}-$Hz^{-\frac{1}{2}}$.

Neither the balance nor the resolution were frequency dependent in the range 0.5 - 100 Hz.

DISCUSSION

The ease of fabrication and the inherent natural balance of this prototype gradiometer suggest that it is capable of further development. The parallel field rejection is currently thought to be limited by the imperfect flatness of the glass substrate and can obviously be improved. An extension of the baseline to 30 cm, and a doubling of the width should improve the resolution by a factor of 10. Further improvement with this basic design would be difficult essentially because of the poor coupling of the pick-up coils to the SQUID itself. (A total of 109 Φ_o has to be applied to a large SQUID loop to produce a single Φ_o in the SQUID.) However improvements in the flux resolution of the SQUID, such as may be expected with smaller tunnel junction dimensions and higher resistances, should give some improvements in the gradiometer noise.

We finally note that since the as-fabricated gradiometers are fairly closely, and stably, balanced before tuning, it may be possible to achieve high, permanently adjusted, balance and parallel rejection by mechanically trimming tuner films which are permanently deposited on the substrate.

ACKNOWLEDGMENTS

This work was supported by the U.S.E.R.D.A. and by the Geological Survey, Department of the Interior, under U.S.G.S. Grant No. 14-08-001-G-224.

One of us (GBD) acknowledges the hospitality and support of the University of California, and the assistance of the Fulbright Commission.

A slightly fuller account of this work was presented at the Applied Superconductivity Conference, Stanford, 1976.

REFERENCES

1. J.E. Zimmerman and N.V. Frederick, "Miniature and Ultrasensitive Superconducting Magnetic Gradiometer and its Use in Cardiography and Other Applications". Appl. Phys. Lett., Vol. 19, pp. 16-19, 1971.

2. W.M. Wynn, C.P. Frahm, P.J. Carroll, R.H. Clark, J. Wellhoner, and M.J. Wynn, "Advanced Superconducting Gradiometer/Magnetometer Arrays and a Novel Signal Processing Technique", IEEE Trans. Magn., Vol. MAG-11, pp. 701-707, 1975.

3. J. Clarke, W.M. Goubau, and M.B. Ketchen, "Thin Film dc SQUID with Low Noise and Drift", Appl. Phys. Lett., Vol. 27, pp. 155-156, 1975; "Tunnel Junction dc SQUID: Fabrication, Operation, and Performance", J. Low Temp. Phys., Vol. 25. pp. 99-144, 1976.

4. P.K. Hansma, "A New Method of Fabricating Niobium Oxide Barrier Josephson Junctions", J. Appl. Phys. 45, 1472-1473, 1974.

5. G. Lasher, "Mixed State of Type I Superconducting Films in a Perpendicular Magnetic Field", Phys. Rev. 154, 345-348, 1967.

A CRYODEVICE FOR INDUCTION MONITORING OF DC ELECTRON OR ION BEAMS WITH NANO-AMPERE RESOLUTION

K. Grohmann, D. Hechtfischer and J. Jakschik
Physikalisch-Technische Bundesanstalt, Institut Berlin,
Abbestr. 2-12 , D-1 Berlin 10, Germany

INTRODUCTION

Charged particle beams are a wellknown important appliance in nuclear and solid state physics. Exact beam current measurements are necessary to investigate nuclear reactions, cross-sectionsor ion implantation effects for example.

In principle there are two different ways to measure charged particle beam currents. First a direct measurement, in which the particles must interact with some material, for example the gas in a beam-through ionization chamber or the walls of a Faraday cage, and second an electromagnetic measurement via the magnetic field generated by the beam.

The accuracy of the direct measurement is limited by numerous secondary effects [1]. These must be known as a function of all possible parameters to correct the results. In the case of electron beams such effects are backscattering or production of photo and Compton electrons, and other effects due to bremsstrahlung. In the case of ion beams the measurements will be influenced, for example, by secondary electrons or ions or by space-charge neutralized particles in the vacuum system of the target chamber. The second method, where the beam's magnetic field is measured by means of a toroidal transformer surrounding the beam, has none of these drawbacks. It allows absolute beam monitoring with a negligible feedback to the beam. Unfortunately, this method needs a pulsed beam for high sensitivity. In connection with dc-beams the sensitivity of best current transformers with high permeability iron cores and second harmonic detection is limited to some μ-Amperes. It is the intention of this contribution to demonstrate that the electromagnetic method can be extended to the n-Ampere region for dc-beams as well, if a SQUID is used as sensor.

APPARATUS

In principle, it would be sufficient to surround the beam with a toroidal superconducting detector coil which forms a flux-transformer in connection with the signal coil of the SQUID. However, this simple configuration would have two essential drawbacks: the signal would be a function of the beam's location and of the coil's geometry and furthermore time-dependent magnetic stray fields would also affect the measurements preventing any SQUID operation.

These two drawbacks can be overcome by a special system of superconducting shields, which has been investigated in connection with Cryo-Current-Comparators (CCCs). The underlying principle has been described elsewhere [2], therefore we shall restrict here to a brief summary.

Current Comparators are devices by which currents can be compared with very high accuracy. Whereas conventional current comparators have cores of high permeability for this purpose, the Cryo Current Comparator has only superconducting screens normally made of lead. The principle is shown in fig. 1: the superconducting screen protects the detection area - where the detector coil of a flux transformer coupled to the SQUID is located - against the direct influence of the magnetic field of the current windings. (Only one of these windings is shown in the figure.) The necessary coupling between the inner and the outer region is maintained by surface currents on the screens. Hence the screens must not be closed - they must have a gap, either a cylindrical one or a ring-like one, as shown in fig. 1, or a combination of both. These gaps behave like filters: the only information which can pass the gap without

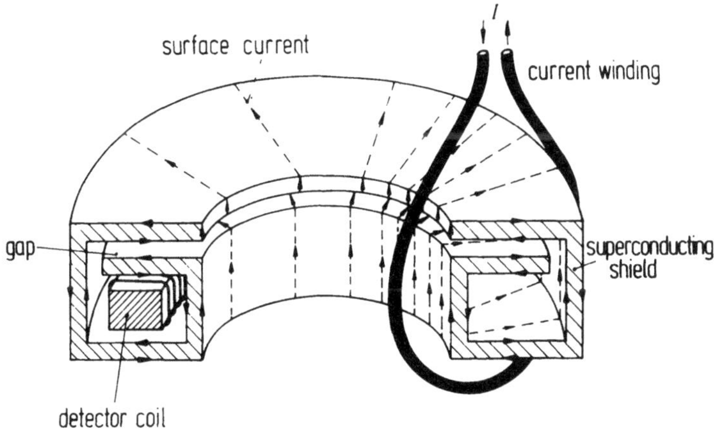

Fig. 1
Shielding system of a Cryo Current Comparator

attenuation is the information about the net current linkage, that is the total current flowing through the bore of the comparator. The information concerning the special position of the field generating current is attenuated on intrusion of the gap. It is true that the higher the multipolarity of the corresponding magnetic field, the stronger is its attenuation. For a cylindrical gap the attenuation is exponential. The lowest attenuation is about one order of magnitude for each gap length of 1.15 diameters. It is associated with fields of dipolar symmetry. The same attenuation laws hold for any external field. Therefore, in a properly designed CCC no external field can reach the detection coil, if it is not strong enough to destroy the superconductivity of the screens. This means that the two tasks mentioned above are fulfilled by the shielding system of a CCC: on the one hand we get full coupling between the beam current penetrating the comparator and the detector coil of the flux-transformer independent of the beam's location. On the other hand we get full screening of the detector coil against magnetic stray fields.

The complete device is shown in fig. 2. The beam is fed through the room temperature bore of the cryostat. The comparator is of toroidal form and surrounds the beam. On its inner perimeter it has two cylindrical gaps, connected in series, which magnetic fields must penetrate to reach the detector coil. The attenuation by these two gaps for magnetic stray fields is calculated to be at least eight orders of magnitude. The device can easily be calibrated by means of a single auxiliary turn, fed through the bore of the comparator. If this auxiliary turn is used for feedback the comparator works in the real sense of its name. The SQUID is used only as a zero indicator in this case. Otherwise, that is without feedback, the comparator is used as a dc current transformer with the SQUID acting as current indicator.

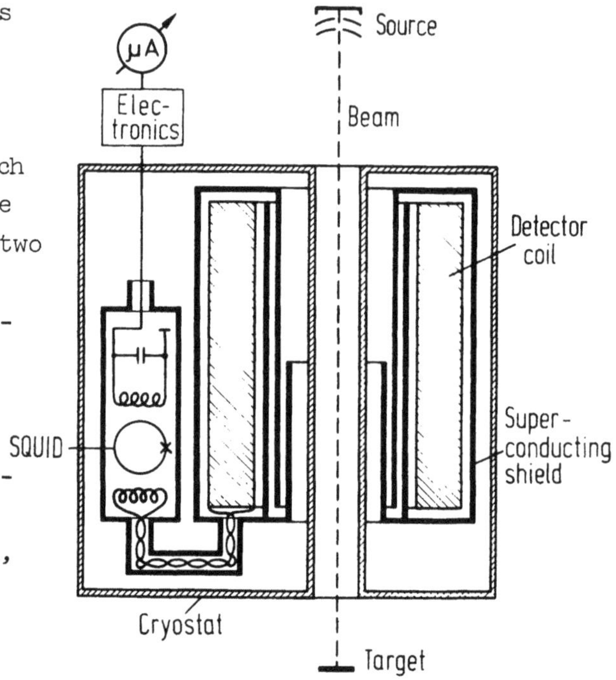

Fig. 2 Principle of dc beam current measurement

RESULTS

For given SQUID parameters the current sensitivity of the device will be the higher, the greater the volume of the detector coil in the inner of the shields is. We have chosen the following coil dimensions: outer diameter 190 mm, inner diameter 110 mm and height 100 mm. The coil has ten distributed turns. The measured current sensitivity for these parameters is 4.16 μA for a shift of one flux-quantum (Φ_0) in the SQUID. A peak-to-peak noise of $10^{-3}\,\Phi_0$ has been found with a bandwidth of 1 Hz for this device, which limits the current resolution to about 5 nA.

Fig. 3 shows the chart record of an electron beam current of 108 nA generated by a van de Graaff-accelerator at 2.5 MeV. The enhancement of noise after switching on the beam is caused by beam flickering, which can be resolved by observing the same signal with greater bandwidth and higher time resolution on an oscilloscope.

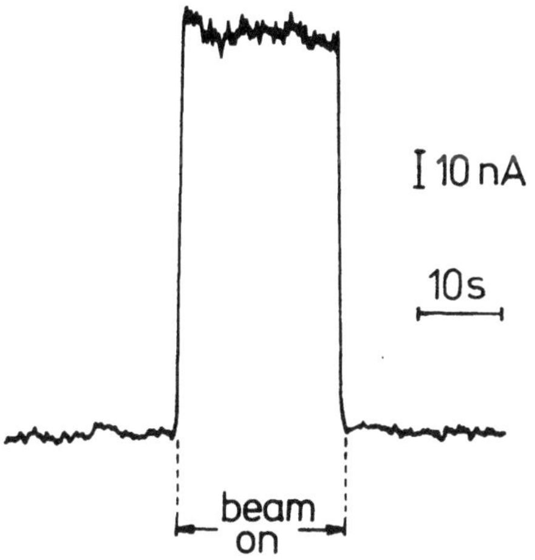

Fig. 3 SQUID response to an electron beam current of 108 nA (mean value). Bandwidth 1 Hz

SUMMARY

A device consisting of a rf-SQUID in connecting with a Cryo Current Comparator is described, which allows dc charged particle beams to be inductively monitored with a resolution of some n-Amperes, known hitherto only for pulsed beams.

REFERENCES

[1] S. Amelinckx et al.: Ion Implantation, North-Holland Publ. Comp., Amsterdam (1973)

[2] K. Grohmann, H.-D. Hahlbohm, D. Hechtfischer and H. Lübbig: Field attenuation as the underlying principle of cryo current comparators, part I, Cryogenics 16, 423 - 429 (1976); part II: Ring cavity elements, Cryogenics 16, 601 - 605 (1976)

A SQUID COMPARATOR BRIDGE FOR LOW-TC-CRYORESISTOR MEASUREMENTS

D. Andreone, E. Arri, G. Boella, G.C. Marullo
Istituto Elettrotecnico Nazionale Galileo Ferraris (IEN)
Corso M. d'Azeglio 42, 10125 Torino (Italy)

E.P. Balsamo
Laboratori Nazionali del CNEN, 00044 Frascati (Italy)

U. Michi
Istituto di Fisica Università di Genova
Viale Benedetto XV 5, 16132 Genova (Italy)

1 INTRODUCTION

The most recent and improved Josephson standards for the maintenance of the dc voltage unit use cryogenic voltage comparators, i.e. transfer the critical elements of the comparison circuit between the Josephson voltage and the reference voltage to the low temperature of the junction.

In setting up a cryogenic voltage standard at IEN, a novel circuit configuration for this comparison has been recently proposed [1] [2]. It is based on the conversion of the ratio between the two voltages to be compared (10 mV and 1 V) into an equal 1/100 ratio between two dc currents, measured with a superconducting dc ratio standard and related SQUID null detector. The conversion is obtained via two cryoresistors of equal nominal value (100 Ω).

This voltage cryocomparator uses the irradiated Josephson junction as an ideal voltage generator and, unlike the usual cryocomparators of resistive type, requires no calibration of resistance ratios. Referring to the 100 Ω cryoresistors, fed by two currents of 0.1 and 10 mA respectively, only one verification is necessary. In fact the final uncertainty in voltage comparison is of the same order than the variation in resistance ratio from the nominal 1/1 value due to the temperature and power coefficients of the cryoresistors.

Variations of the order of 10^{-9} or less are possible using some now available alloys [3] [4], which permit the construction of 100 Ω cryoresistors with power and temperature coefficients of the order of 10^{-8}/W and 10^{-7}/K respectively. Therefore, one of the main problems in realizing the mentioned cryocomparator consists of the measurement of very low temperature and power coefficients of cryoresistors with adequate resolution.

2 THE CRYOBRIDGE SYSTEM

These measurements are possible with a system completely at low temperature by using a simple cryobridge (fig. 1) [5], which retains some elements of the cryocomparator, i.e. a superconducting dc ratio standard with ratios $N_2/N_1 = 1/1$ and $N_2/N_1 = 1/10$ and a SQUID detector. The system both permits further studies toward the definitive implementation of these elements and seems competitive with respect to more conventional methods based on bridges at ambient temperature.

The system requires the comparison of two cryoresistors R_1 and R_2 supplied with the same voltage. For compensating the deviations of the bridge a third winding N_3 of the comparator is used.

2.1 Temperature Coefficient Measurement

For the determination of the temperature coefficient the two cryoresistors are of equal nominal value $R_2 = R_1$ and the comparator is in the $N_2/N_1 = 1/1$ ratio. The cryoresistors R_1 is us-

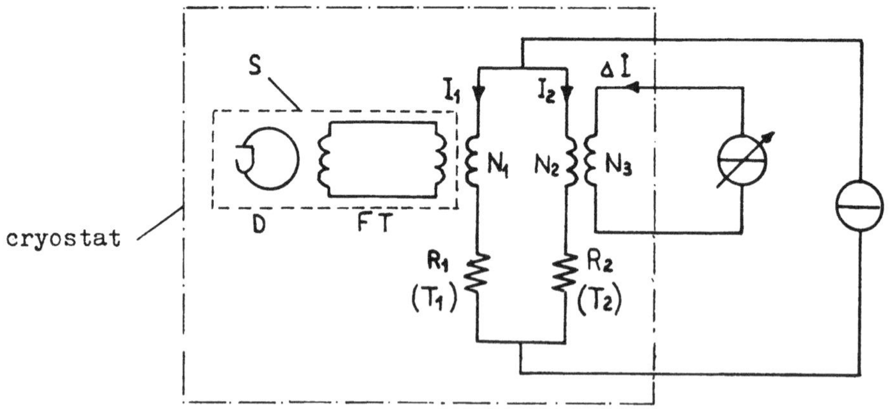

Fig. 1 Simplified circuit diagram of the cryobridge for measurement of temperature and power coefficients of cryoresistors: R_1 and R_2 - cryoresistors at temperatures T_1 and T_2; N_2/N_1 - superconducting current comparator; N_3 - compensation winding; D - SQUID detector; FT - flux transformer; S - r.f. shield.

ed as a reference resistor and kept in the helium bath at constant temperature T_1; the temperature T_2 of R_2 is varied, for example between 1.5 K and 20 K, by means of a thermostatic environment and the ratio R_2/R_1 is measured.

2.2 Power Coefficient Measurement

The power coefficient is determined by using two cryoresistors with $R_2/R_1 = 1/10$, inserted in the comparator circuit with $N_2/N_1 = 1/10$ ratio and kept at the same constant temperature $T_1 = T_2$ in the above thermostatic environment. By comparing the two resistors at different values of currents I_1 and I_2, and by extrapolating the results, it is possible to compute the power coefficient.

In fact, if the resistors are built with the same wire (same material, same diameter, same heat transmission coefficient), their power coefficients are inversely proportional to the resistance values(*). In this case, let us the relationships for the two resistors R_1 and R_2 in the cryobridge

$$R_1 = R_{01} (1 + \gamma_1 P_1)$$
$$R_2 = R_{02} (1 + \gamma_2 P_2),$$

where R_{01} and R_{02} are their resistances at zero power, γ_1 and γ_2 their power coefficients, P_1 and P_2 their powers: if $R_2 = 10 R_1$,

$$P_1 = 10 P_2 \qquad \gamma_1 = 10 \gamma_2$$

and then the simple relationship holds:

$$\frac{R_1}{R_2} = \frac{R_{01}}{R_{02}} \frac{1 + 100\gamma_2 P_2}{1 + \gamma_2 P_2} \cong \frac{R_{01}}{R_{02}} (1 + 100\gamma_2 P_2) .$$

(*) For cylindrical resistors of circular crosssection, at thermal equilibrium the relationship

$$\gamma = (4\alpha\rho)/(\pi^2 k d^3 R)$$

holds, where γ, α, k, d, ρ and R are respectively: power, temperature and heat transmission coefficients; diameter; resistivity and resistance of the resistor.

3 FIRST IMPLEMENTATION

The first comparison system for the cryobridge of fig. 1 is of the same type than the one used in the preliminary implementation of the cryocomparator [2]. It consists (fig. 2) of a superconducting current comparator $N_1 - N_2$ of the overlapped-tube type [6], which is coupled with a two-hole point-contact SQUID magnetometer [7] via an r.f. shielded superconducting flux transformer. A lead shield avoids coupling with slowly varying

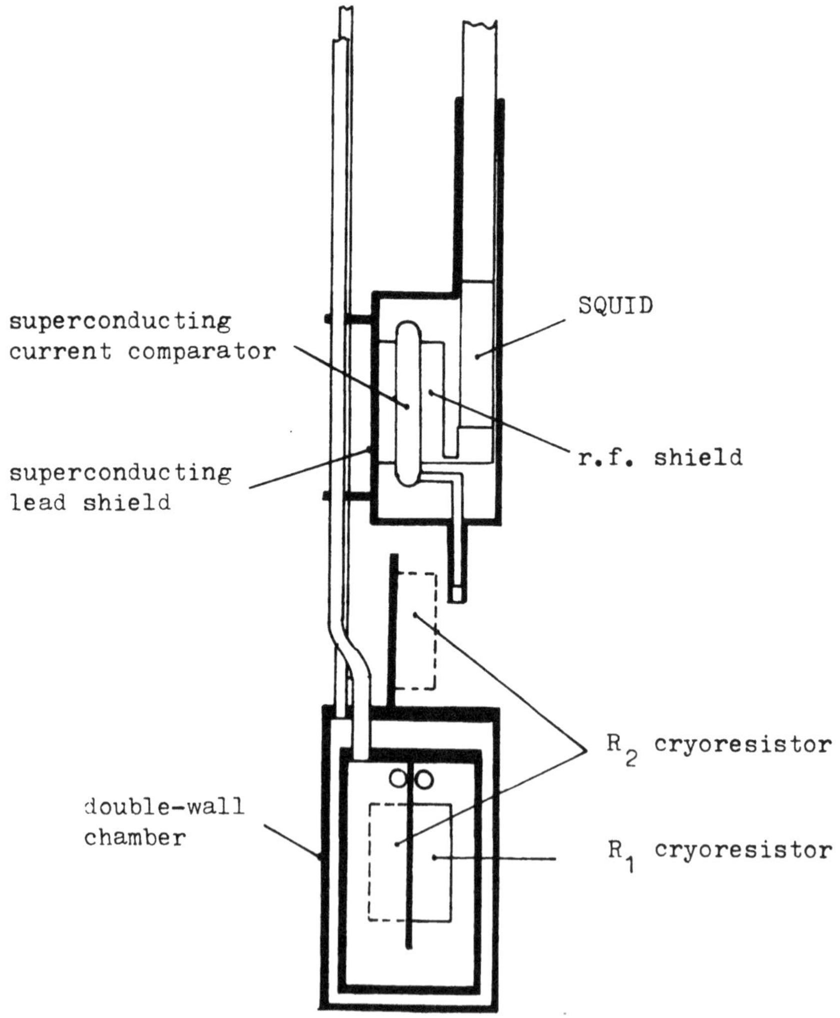

Fig. 2 Practical implementation of the cryobridge

external magnetic fields, which reduce the system resolution. The current comparator has six windings with 3 x 100 turns, 2 x x 200 turns and 1 x 400 turns. The third winding has 100 turns. All the connections of the cryobridge are made with superconducting material.

The thermostatic environment for the cryoresistors consists of a double-wall chamber, thermally insulated from the helium bath, which is suitable to contain two cryoresistors and whose temperature can be varied in the range 1.5 - 20 K approximately.

The system sensitivity is about 2 μA • turn/ϕ_0, better than the one of the previous realization [2] owing to a better coupling between comparator and SQUID. The noise level is about $10^{-2}\phi_0$; therefore, the system resolution is of the order of 10^{-8} A • • turn. In these conditions the limit on the uncertainty of the current 1/10 ratio is 1 • 10^{-9}. This value is also valid for the intermediate ratios, which can be realized with different arrangements of the windings.

4 PRELIMINARY RESULTS

Preliminary measurements of temperature coefficient was carried out on 10 Ω cryoresistors realized with 0.1 mm diameter AlMg5 alloy wire noninductively wound on a cylindrical aluminum core. The current comparator was in the arrangement 500 against 500 turns. With a current of 10 mA in both cryoresistors, a relative resolution of the order of 10^{-8} was obtained. A typical resistance vs. temperature diagram of a 10 Ω cryoresistor in the range 1.5 - 4.2 K is shown in fig. 3.

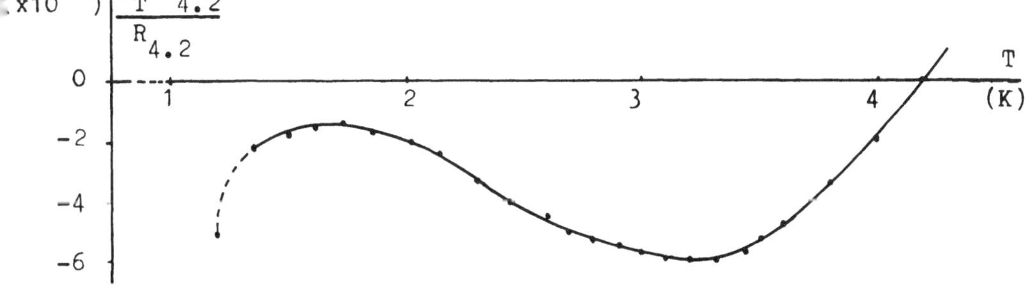

Fig. 3 Typical resistance versus temperature curve of a cryoresistor realized with an AlMg5 alloy wire.

Measurements of power coefficient are in progress using 10 Ω and 100 Ω cryoresistors, fed with currents of 10mA and 1 mA respectively. The comparator is in the arrangement 100 against 1000 turns. The relative resolution is reduced to $1 \cdot 10^{-7}$.

5 CONCLUSIONS

A cryobridge system, based on a superconducting current comparator and a SQUID detector, is under investigation at IEN for measuring very low temperature and power coefficients of cryoresistors. The preliminary results obtained with a first implementation of the system look very promising.

It can be estimated that the performance of the system will improve by more than one order by using both a SQUID of the highest available sensitivity and better r.f. shielding techniques.

REFERENCES

[1] Arri, E.: Proposal for a cryogenic voltage comparator for 2e/h measurement. Alta Frequenza 45, 53E-55E (1976).

[2] Andreone, D., Arri, E., Boella, G., Marullo, G.C.: A novel cryogenic voltage standard. IEEE Trans. Instrum. Meas. IM-25, 303-306 (1976).

[3] Cimberle, M.R., Michi, U., Mori, F., Rizzuto, C., Siri, A., Vaccarone, R.: A search for low temperature standard resistance alloys. Proc. ICEC-6, (Grenoble, France), 190-193 (1976).

[4] Kose, V., Fuhrmann, B., Warnecke, P., Melchert, F.: Cryogenic voltage standard at PTB. Proc. AMCO-5, (Paris, France), 478-484 (1975).

[5] Harvey, I.K., Collins, H.C.: Precise resistance ratio measurements using a superconducting dc ratio transformer. Rev. Sci. Instrum. 44, 1700-1702 (1973).

[6] Dziuba, R.F., Sullivan, D.B.: Cryogenic direct current comparator and their applications. IEEE Trans. Magn. MAG-11, 716-719 (1975).

[7] Balsamo, E.P., Paternò, G., Luciani, G., Pace, S., Romani, G.L.: A RF superconducting magnetometer. Internal Report (in Italian), LNF-73/75, (CNEN, Frascati, Italy) (1973).

HIGH FREQUENCY APPLICATIONS

SUPERCONDUCTING DEVICES FOR MILLIMETER AND SUBMILLIMETER WAVELENGTHS[†]

P. L. Richards
Department of Physics, University of California, Berkeley, California 94720

Josephson junctions are highly nonlinear and respond up to infrared frequencies. A wide variety of high frequency applications have been explored. In addition, several high frequency superconducting devices are being developed which do not use the Josephson effect. These include a superconducting version of the Schottky barrier diode for use as a heterodyne mixer and a video detector, and also a composite superconducting bolometer. In This review, we will discuss the state of development of superconducting mixers, detectors, and parametric amplifiers at microwave, millimeter, and submillimeter wavelengths. Emphasis will be placed on developments which have taken place since 1974 when a previous review was circulated[1].

Super-Schottky Mixer

In an ordinary Schottky barrier diode, thermally excited carriers give an I-V curve of the form $I = I_o \exp(SV)$, where $S = e/kT$ at room temperature. When such diodes are cooled, the S parameter usually saturates at a few times the 300 K value, and non-thermal noise is seen. Although not understood in detail, these effects are usually ascribed to leakage currents. A version of the Schottky diode has been developed[2] to obtain large S-values at low temperatures. It operates in the tunneling dominated region and is nonlinear because of the sharp energy gap structure on the superconducting density of states near the Fermi surface. The effect is essentially the same as quasi-particle tunneling in a superconductor-insulator-normal metal junction.

The best super-Schottky performance demonstrated thus far has been with 5 μm diameter Pb spots on high conductivity p-type GaAs. Values of $S \simeq 10^4\ V^{-1}$ have been obtained which are essentially equal to e/kT for the ambient pumped ^4He temperature. When care is taken to minimize leakage currents, the range of voltage ΔV over which S is constant can be sufficiently wide to obtain an intrinsic conversion loss $L_o < 5$ dB. Parasitic loss due to the junction capacitance and the spreading resistance into the semiconductor contributes additional conversion loss. In experiments at 9 GHz, this additional loss L_p was ≃ 3 dB. Since L_p will increase as ω^2, it is a serious problem for higher frequency operation. It can be reduced by increasing the conductivity of the semiconductor, or by changing the conduction geometry. Improvement is possible by using a number of smaller diodes connected in parallel. Alternatively, a very thin semiconducting barrier, such as the etched single crystal Si barrier[3], could be used.

Experiments at 9 GHz on a 1.1 K super-Schottky have given a single sideband mixer noise temperature T_m = 6 K and a conversion loss of 7.5 dB. This is the lowest microwave mixer noise temperature ever reported. It demonstrates the high potential of these devices for low noise applications. The saturation level of this mixer has not been reported, but it can be estimated from given values of $S\Delta V = 6.5$ and $S = 10^4$. A mixer with R_D = 400 Ohms should saturate severely at $P_S = P_{LO} \simeq 10^{-9}$ W. This corresponds to a dynamic range of $< 10^4$ for T_m = 6 K and B = 1 GHz.

Super-Schottky Video Detector

Because of its high S-value at low temperature, the super-Schottky is an extremely quiet video detector[2]. In this application, the range of voltages over which S is constant is of importance only to the saturation level. Experiments at 9 GHz have shown an NEP of 5.4×10^{-16} W/$\sqrt{\text{Hz}}$ at a temperature of 1.06 K. This is the best microwave video NEP yet reported. Parasitic loss contributes more than a factor 2 to this NEP and would be increasingly important at higher frequencies, as is the case for the mixer application.

Josephson Effect Mixer With External Local Oscillator

A Josephson junction can be operated as an efficient mixer with a very small amount of local oscillator power (P_{LO}). The Josephson mixer therefore resembles the large S-value mixer discussed above. The details of its operation, however, are entirely different. If P_{LO} at a frequency $\omega_{LO}/2\pi$ is coupled to a Josephson junction, constant voltage steps with zero differential resistance appear which are separated by multiples of the voltage $V = \hbar\omega_{LO}/2e$. The steps are zero frequency beats between the AC Josephson frequency and the nth harmonic of ω_{LO}. The most useful type of Josephson mixer is biased with a constant current to a region between the $n = 0$ and $n = 1$ steps. In order to do this, it is necessary to use a type of Josephson junction such as a point contact, Dayem bridge, or small area tunnel junction, which has no hysteresis on the static I-V curve. The RF impedance of a Josephson junction biased in this way is real and nonlinear. It will serve as an efficient heterodyne down-converter for small RF signals.

The operation of a Josephson mixer can be most simply understood as follows: The total RF current, $I_{RF} = I_S + I_{LO}$ is amplitude modulated at ω_{IF}. If $I_{RF} \simeq I_o/2$, the height I_o of the zero voltage ($n = 0$) step is suppressed essentially linearly by I_{RF}, so is modulated at ω_{RF} by an amount proportional to dI_o/dI_{RF}. If the junction is biased with a constant direct current to a point between the $n = 0$ and $n = 1$ steps, the bias voltage swings at ω_{RF} by an amount proportional to $R_D(dI_o/dI_{RF})$. The conversion efficiency L^{-1} is thus proportional to $R_D(dI_o/dI_{RF})^2$.

The detailed performance of this mixer has been calculated using the resistively shunted junction (RSJ) model. At values of the normalized frequency $\Omega_{RF} = \omega_{RF}/2eI_cR << 1$, the conversion efficiency into a single sideband can be considerably greater than unity. The nonlinearity is of high order so that efficient harmonic mixing is possible.

If the frequency at which the Josephson mixer is operated corresponds to $\Omega > 1$, the conversion loss increases as Ω^2. Even for $\Omega >> 1$, the Josephson mixer retains an extraordinary capability as a harmonic mixer. Down-conversion from 891 GHz with an $\sim$ 1 GHz local oscillator has been observed[4]. The performance of Josephson effect fundamental mixers has been evaluated at least semi-quantitatively[5,6,7] at 36, 135, and 300 GHz using Nb point contact junctions with less than the full theoretical values of I_cR. In most of this work, $\Omega = 1$ corresponded to $\leq$ 300 GHz. The most complete results are at 36 GHz[5] where the measured conversion efficiency of 1.4 and the noise T_m = 54 K were in good agreement with analog computer calculations based on the RSJ model including thermal noise in the shunt resistance. Based on this model, it is possible to estimate the performance of Josephson effect mixers if we assume that non-hysteretic Nb junctions with the full theoretical value of I_cR will be available, so that $\Omega = 1$ at $\sim$ 1 THz. The mixer contribution to the system noise temperature should be $T_m \approx 0.05$ to 0.1 times νT where ν is the frequency in GHz and T is the ambient temperature in Kelvin[8].

There are practical lower limits to the value of T_m/T which can now be reached for small Ω. The predicted proportionality $T_m/T \propto \nu$ occurs only if the RF source impedance is optimized. The optimum source impedance is $\ll R$ for $\Omega \ll 1$. Since it is difficult to transform the impedance of a free space wave below a few Ohms, optimum coupling to single junctions is hard to achieve for $\Omega \lesssim 0.3$. Consequently, the experimental values of T_m/T should be ≈ 30 throughout the millimeter region. We will discuss below the problem of avoiding hysteresis under the operating condition required for best performance. A series array of junctions would in principle solve most of these problems and allow the predicted values of T_m/T to be approached[8]. This prediction has not yet been tested experimentally.

Although the performance of the Josephson mixer cannot be understood in terms of a curvature parameter S, it shares some features of large S value mixers, including a low saturation power level. The mixer is badly saturated when the signal power is comparable to $10^{-2}\left(I_c^2R\right)$ compared with a local oscillator power $\simeq I_c^2R$. It seems plausible that the saturation level would be increased by the use of a high impedance series array of junction since the power per junction would be decreased.

Josephson Effect Mixer with Internal Local Oscillator

The AC Josephson currents in a point contact junction which is biased at a finite voltage have been used as the local oscillator for a millimeter wave heterodyne down-converter[9]. Since the low frequency noise in the junction frequency modulates this local oscillator over many hundreds of MHz, coherent down-conversion of monochromatic signals is not obtained. When used as a receiver for broad band thermal sources, temperature sensitivities $\Delta T \leqslant 0.3$ K were measured in 1 sec from 40 to 260 GHz. This corresponds to a receiver noise temperature $T_R \lesssim 6 \times 10^3$ K. It is thought that this number could be significantly reduced with better coupling of the junction to the RF signal. Although the usefulness of this device is limited by its wide LO bandwidth, the sensitivity is high enough to be interesting, and the absence of any requirement for a local oscillator leads to great convenience of operation.

Josephson Effect Video Detector

The maximum zero voltage current of a Josephson junction decreases as I_{RF}^2 for $I_{RF} << I_c$. Consequently, if no local oscillator power is applied, a non-hysteretic junction which is biased in the resistive region with a constant bias current will act as a square law detector for small signals[10]. As was the case with the mixer, the complexity of the Josephson effect makes this device quite different from the classical diode rectifier. For signal frequencies ω_s which are low compared with the Josephson frequency set by the bias voltage $\omega_J = 2eV_{bias}/\hbar$, there is classical video detection which is proportional to d^2V_o/dI_o^2. There is a dispersive enhancement and a change of phase when $\omega_s = \omega_J$, and the response is proportional to $R_D = dV_o/dI_o$ for $\omega_s > \omega_J$.

Relatively low impedance junctions, in which the rounding of the I-V curves by fluctuations is not severe, must be resonantly coupled to the RF radiation. Higher impedance junctions are easier to couple to free space radiation, but are dominated by fluctuations[11]. The experiments in both cases are in reasonable agreement with the RSJ model theory and give NEP $\simeq$ 3 to 5×10^{-15} W/$\sqrt{Hz}$ at 90 to 120 GHz. This NEP should scale as Ω^2 at higher frequencies. Some frequency selectivity can be obtained if the junction is biased with a modulated voltage centered around $\hbar\omega_{RF}/2e$. Since the response is dispersive in form, it is only useful for narrow-band signals. The observed NEP for junctions operated in this mode is considerably higher than the values quoted above[12].

Superconducting Bolometer

Thermal (bolometric) detectors have historically been the most sensitive wide band receivers of submillimeter wave radiation. These consist of a radiation absorbing element, a thermometer, and a thermal link to the heat sink. To keep thermal noise and heat capacity to a minimum, sensitive bolometers are operated at pumped ^{4}He (or even ^{3}He) temperatures. In contrast to diode video detectors which are in practice limited to a single spatial mode of electromagnetic radiation, the area of a bolometer can be adjusted to couple to any number of modes. The best submillimeter wave bolometers available[13] use an ∿ 200 Ohms/□ metal film on a transparent dielectric substrate as an absorbing element, and a superconducting Al film operated near its critical temperature of 1.4 K as the thermometer. Such bolometers have shown sensitivities of NEP = 2×10^{-15} W/$\sqrt{\mathrm{Hz}}$ with areas of 0.08 cm^2 and response times of τ = 50 ms. This NEP can, in principle, be reduced by a factor ∿ 3 by decreasing the substrate thickness or increasing its Debye temperature, and a factor in excess of 10 by operating at ^{3}He temperatures. The numbers achieved represent a major recent improvement in performance at submillimeter wavelengths compared to that previously available from monolithic semiconductor bolometers. The primary reasons for the improvement are the increased radiation absorption efficiency, and the low heat capacity.

Parametric Amplifiers

Under a variety of bias conditions, Josephson junction exhibits a nonlinear inductive reactance. If this reactance is pumped from a suitable source, parametric gain can be obtained. Gain is generally observed only if some care is taken to control dissipation at parasitic (unwanted) combination frequencies. If the junction resistance is made large compared with the termination impedance of the parasitic, then the idler is said to be shorted out. In this case no power is dissipated due to current flow at the parasitic frequency. Since junction impedances rarely exceed 20 Ohms, this condition is difficult to meet at microwave frequencies. Alternatively the circuit impedances can be made large compared with the junction resistance. Although this termination is frequently used, it does not prevent dissipation at the parasitic frequency, since currents can still flow and dissipate power in the quasiparticle shunt which is internal to the junction.

Josephson effect paramps can be pumped either externally with a microwave oscillator, or internally by means of the AC Josephson currents. Two classes of parametric amplifiers have been developed which are based on these two approaches. We consider first the externally pumped, and then the internally pumped paramps.

Externally Pumped Paramp

The best performance of an externally pumped Josephson effect parametric amplifier has been obtained with a doubly degenerate negative resistance mode of operation in which the junction has zero DC (current and voltage) bias. The physical symmetry of a junction which is biased in this manner implies that the nonlinear inductance will depend only on even powers of the current. The amplifier is therefore operated with $2\,\omega_p = \omega_s + \omega_i$ and $\omega_p \approx \omega_s \approx \omega_i$. The lowest parasitic frequencies are at $\approx 3\,\omega_p$ and $\approx 5\,\omega_p$.

Amplifiers of this type have been operated at 10 and 33 GHz using a series array of Dayem bridges[14], at 9 GHz using a series array of tunnel junction[15], and at 36 GHz with a single point contact[16]. Gain in excess of 10 dB is generally obtained if the bandwidth is sufficiently narrow to avoid saturation on external noise[17]. Since the external power which can be carried without saturation increases with the number of junctions employed, the gain bandwidth parameter $G^{\frac{1}{2}}\Delta\omega/\omega$ is larger for the array paramps. Values of 12-15 dB gain with 10 percent bandwidths have been observed at 10 and 33 GHz using arrays of $\sim$ 100 bridges. The dynamic range of these amplifiers is relatively small, and is influenced by parametric oscillations (or amplification of junction noise) in a way which is not well understood. It appears not to be a sensitive function of the number of junctions employed.

The noise measured in this type of paramp has generally been dominated by noise from the (room temperature) circulator and the heterodyne receiver which follows it. In most cases it has only been possible to set an upper limit to the noise expected from an amplifier operated with a cooled circulator. These upper limits typically correspond to $T_N \simeq 20$ to 50 K.

A theoretical model has been used to interpret the array paramp which assumes that all important idlers are shorted[14]. Since each individual junction sees a high impedance RF source, and thus becomes unstable for pump currents in excess of I_c, it appears that the open circuit idler theory[16] may be more appropriate. Based on this theory, the maximum normalized frequency at which such an amplifier can be operated is $\Omega = 0.3$. Although heating can still be important, some forms of hysteresis on the DC I-V curve do not directly affect this device, since they disappear under the operating conditions in which the zero voltage current is suppressed essentially to zero by the pump.

Attempts have been made to operate an externally pumped Josephson paramp in the singly degenerate mode in which $\omega_s \simeq \omega_i \simeq \omega_p/2$. Parametric oscillations have been seen in tunnel junctions and point contacts, but not gain[16,18].

Internally Pumped Paramps

Gain has been observed[19,20] at 30 MHz and 9 GHz in an internally pumped singly degenerate parametric amplifier with $\omega_s + \omega_i = \omega_p$ and $\omega_s \simeq \omega_i$. The point contact junction was voltage biased in a low inductance (resistive SQUID) configuration so as to provide a voltage-clamped pump at $\omega_p = 2eV/\hbar$. The circuit was resonant at $\omega_s \simeq \omega_i$ and short circuited other important combination frequencies. This amplifier closely resembles a conventional type of varactor paramp with a single important idler frequency. It is difficult to provide the required circuit properties for this amplifier at microwave frequencies because of the $\lesssim$ 10 Ohm impedance of typical point contacts. A series array of junctions might alleviate this problem, but it would have to be operated in a synchronized, voltage biased mode which has not yet been observed at microwave frequencies.

A more promising type of internally pumped amplifier for high frequency applications has been developed independently by two groups[20,21]. Although the devices appear qualitatively similar, the discussion of them is quite different. One was analyzed as a resistive SQUID without reference to the I-V curve of the contact, and the other was analyzed in terms of the step induced on the measured I-V curve by a cavity which was resonant at ω_s. In both cases, $\omega_p < \omega_s$ and the important idler frequencies are at $\omega_s - \omega_p$ and at $2\omega_p - \omega_s$. The measured value of $G^{1/2}\Delta\omega/\omega$ was only 4×10^{-3} due to coupling limitations for the low impedance junctions employed[21]. The maximum theoretical values are ≤ 0.4 for normalized frequency $\Omega \simeq 1$ and $\leq (2\Omega)^2$ for $\Omega > 1$. It therefore appears promising for operation at frequencies above 100 GHz. The noise temperature of this amplifier reaches a minimum value of 42T at $\Omega = 1$. Because of its large noise, this amplifier will be competitive only at relatively high frequencies.

The two idler parametric amplifier discussed above will up-convert signals from the idler at $\omega_s - \omega_p$ to the signal frequency ω_s. When resonated at $\omega_s - \omega_p = 115$ MHz, it served[20] as a parametric up-converter to 9 GHz with a low noise temperature $\lesssim 3.6$ T and a power gain of 25. The relatively large noise in the two idler paramp can be understood in terms of the parametric up-conversion of ambient temperature noise from the low frequency idler at $\omega_s - \omega_p$.

This parametric up-converter can in principle be coupled with a low noise down-converter such as the super-Schottky to provide low noise single frequency amplification. This combination has been proposed as a low noise IF amplifier for a heterodyne receiver based on the super-Schottky heterodyne mixer[22].

Properties of Real Josephson Junctions

The resistively shunted junction (RSJ) model with the full (energy gap limited) value of $I_cR = \pi\Delta/2e$, and thermal noise in the shunt resistor, has generally been used to predict the ultimate performance of high frequency Josephson effect devices. The actual performance of such devices has usually been limited by the (frequently) less favorable properties of real junctions.

For high frequency operation, a large value of the I_cR product is nearly always desirable. This can be obtained with high T_c superconductors under ideal conditions. Clean Nb point contacts have the full theoretical I_cR (except at high impedances) but, unfortunately, are rather unstable. The more stable oxidized Nb point contacts tend to have smaller values of I_cR. Full values of I_cR are generally found in tunnel junctions and variable thickness bridges, where the phase variations are confined to a small region, and not in Dayem or proximity effect bridges.

Devices which must be biased at finite voltage suffer from the general problem that hysteresis appears on the I-V curve of most well-shielded junctions when they are operated at low temperatures and with large values of

I_cR. According to well known models, hysteresis can be caused by internal capacitance, by heating, and by a frequency dependent external impedance. Some progress has been made in understanding the importance of these sources of hysteresis in various types of junctions.

Internal junction capacitance does not generally contribute to hysteresis on the I-V curves of point contacts and thin film bridges, although Josephson plasma resonances in the submillimeter wavelength range are certainly possible. Capacitance certainly dominates the performance of the usual types of evaporated-film, oxide barrier tunnel junctions. Recent work[23] with small area high current density tunnel junctions has shown, however, that their properties are essentially similar to these of point contacts and variable thickness bridges.

Heating appears to be a severe limitation to the performance of the classical Dayem bridge which has uniform film thickness. The characteristic times for excited quasiparticles to relax to the phonons, and for the phonons to escape to the substrate or He bath correspond to low microwave frequencies[24].

Heating effects are reduced if the junction region looks into a large enough solid angle of (superconducting) metal that the excited quasiparticles can diffuse away and be replaced by equilibrium quasiparticles, rather than relax in the junction region. This condition can be met in point contacts, in tunnel junctions with thick metal films, and in variable thickness bridges with thick "banks". Recent experiments with these last structures have shown RF induced steps out to 3.7 mV[25], where the temperature at the center of the bridge has been computed to be $\sim$ 20 K. Steps have been seen out to 17 mV in Nb point contacts[26] which corresponds to a temperature at the center of $\sim$ 100 K. These estimates based on the Wiedemann-Franz law[27] strongly suggest that tunneling through a normal layer plays an important role at high frequencies and/or high voltages, even in three-dimensional point contacts and variable thickness bridges.

An additional source of hysteresis can arise from the effects of the external circuit impedance. If this impedance is very high at all frequencies, then the I-V curve obtained from the RSJ model (without noise or RF bias) is hyperbolic and shows no hysteresis. A reduction of the external impedance (at a given frequency) leads to a depression of the static I-V curve at the corresponding voltage. If the current at the bias point is depressed below I_c, there will be negative resistance and hysteresis. This effect is easily understood in the limit of zero external impedance at the bias frequency. The pair current then varies sinusoidally in time so that it makes no time average contribution to the total (DC) current. A junction which is used in a high frequency device will usually be tightly coupled to a finite RF source resistance. The static I-V curve will therefore be depressed in the neighborhood of the RF frequency, and negative resistance will occur in the absence of fluctuations ($T = 0$).

The effects of current fluctuations, which arise from external noise or Johnson noise in the shunt resistance, is to smooth out hysteresis. The noise parameter $\Gamma = 2ekT/I_c\hbar$ is the ratio of an effective fluctuation current to the critical current. As Γ approaches unity it becomes possible to DC bias at any point on the I-V curve. Since the fluctuations also contribute directly to device noise, it is desirable to minimize them by operating in a shielded environment and at low temperatures. There is a tendency for devices which require a stable bias point at finite voltage to have a minimum practical operating temperature, below which hysteresis appears.

It is difficult in practice to distinguish between different sources of hysteresis. Hysteresis from any source grows rapidly as the temperature decreases, since I_c grows and the fluctuation current decreases. For a junction with a constant I_cR product, operating voltage, and temperature, the noise parameter Γ increases in proportion to R. At the same time R grows relative to the external circuit impedance, so that both the hysteresis caused by the external circuit, and the fluctuations which mask it, tend to increase with R. Hysteresis from heating will depend primarily on V (in the three-dimensional Wiedemann-Franz model[28]) or on the dissipated power V^2/R, so will decrease in importance as R, and therefore Γ, increases.

The use of series arrays of Josephson junctions has often been mentioned as a way to improve device performance. Several important advantages can be obtained in this way. The array impedance can be several hundreds of Ohms, so that junction matching is simplified. Each junction in a series array looks out into the high impedance of the other junctions, so that hysteresis arising from the matching circuit should not be a problem in arrays containing many junctions.

The primary difficulty with arrays of junctions is that the spread in junction properties often makes it difficult to obtain the required bias condition. Arrays have been successfully used only in the case of the externally pumped parametric amplifier which requires no DC bias. If series arrays can be made in which most of the junctions have critical currents which differ by only a few percent, they would be useful for such devices as the video detector or the heterodyne mixer with external LO. These require a DC bias, but do not depend critically on the value of the Josephson frequency. Unsynchronized series arrays will not be useful for devices such as internally pumped parametric amplifiers or mixers which require a narrow width of the Josephson oscillations.

Experiments at relatively low frequencies have shown a tendency for closely spaced junctions to synchronize and act coherently[28]. It is speculated that the injection of non-equilibrium densities of pairs and quasiparticles from one junction into its neighbors may cause this effect. If such synchronization could be extended to the microwave frequency range, it would permit the use of arrays in several of the internally pumped devices. Such arrays would couple useful amounts of microwave power into an external circuit. At frequencies low enough (or with arrays small enough) to avoid internal resonances, synchronized arrays might be useful as wide band voltage controlled microwave oscillators.

Conclusions

The future of Josephson effect high frequency devices depends critically on the detailed properties of available junctions. Idealized calculations based on the RSJ model predict useful performance at millimeter and submillimeter wavelengths for many of the devices discussed here. Although some measurements at millimeter wavelengths approach the theoretical performance, many do not.

Difficulties in obtaining large values of I_cR without hysteresis are often mentioned. In some cases these are due to heating and improvement is possible in principle. The fundamental limits to the performance of single junction devices which arise from the coupling to the RF source, however, have not generally been given adequate attention. Although arrays of junctions will in principle alleviate these problems, a major effort will be required to produce arrays with adequate uniformity.

Very similar performance is now being obtained from the best point contacts, variable thickness bridges, and small area tunnel junctions. The properties of all of these structures deviate from the RSJ model in ways which are understandable from the full microscopic theory. The effect on high frequency device performance of such phenomena as subharmonic gap structure, etc, are not generally known. They may be very important, especially for junctions operated with samll Γ. Careful measurements of the high frequency impedance[29] of appropriate types of junctions may be the best way to obtain general answers to these questions.

In the enthusiasm over the Josephson effect, there has been relatively little recent work on non-Josephson superconducting devices. The excellent prospects of the super-Schottky diode and the superconducting composite bolometer show that these possibilities should not be neglected.

The author gratefully acknowledges helpful discussions with many persons including R. Chiao, J. H. Claassen, K. K. Likharev, Y. Taur, and M. Tinkham.

References

†Work supported by the U.S. Office of Naval Research.

1. Richards, P. L.: The Josephson junction as a detector of microwave and far infrared radiation. "Semiconductors and Semimetals", Willardson, R. K., and Beer, A. C. Eds. (Academic Press, New York, to be published) V. 12.
2. McColl, M., Pedersen, R. J., Bottjer, M. F., Millea, M. F., Silver, A. H., and Vernon, F. L. Jr.: The super-Schottky diode microwave mixer. Appl. Phys. Letters 28, 159-162 (1976); McColl, M., Millea, M. F., Silver, A. H., Bottjer, M. F., Pedersen, R. J., and Vernon, F. L. Jr.: The super-Schottky microwave mixer (Proceedings of the 1976 Stanford Applied Superconductivity Conference, to be published).
3. Van Duzer, T., and Huang, C. L.: Schottky diodes and other devices on thin silicon membranes. IEEE Trans. ED-23, 579-583 (1976).
4. Blaney, T., and Knight, D. J. E.: Direct 825th harmonic mixing of a 1 GHz source with an HCN laser in a Josephson junction. J. Phys. D6, 936-952 (1973).

5. Taur. Y., Claassen, J. H., and Richards, P. L.: Conversion gain and noise in a Josephson effect mixer. Rev. Phys. Appl. 9, 263-268 (1974); Conversion gain in a Josephson effect mixer. Appl. Phys. Letters 24, 101-103 (1974); Noise in Josephson point contacts with and without rf bias. Appl. Phys. Letters 25, 759-761 (1974).

6. Claassen, J. H., and Richards, P. L. (to be published).

7. Edrich, J., Sullivan, D. B., and McDonald, D. G.: A low noise Josephson mixer for the 1 mm wavelength range (to be published).

8. Richards, P. L., Claassen, J. H., and Taur, Y.: Josephson effect mm-wave receivers. "Low Temperature Physics - LT14", Krusius, M., and Vuorio, M., Eds. (American Elsevier, New York, 1975) V. 4, pp. 238-241.

9. Avakjan, R. S., Vystavkin, A. N., Gubankov, V. N., Migulin, V. V., Shtykov, V. D.: Millimeter waveband signal conversion in S-c-S Josephson junctions with self-pumping. IEEE Trans. MAG-11, 838-840 (1975).

10. Kanter, H., and Vernon, F. L., Jr.: High frequency response of Josephson point contacts. J. Appl. Phys. 43, 3174-3183 (1972); Response of superconducting point contacts to high frequency radiation. Phys. Letters 35A, 349-350 (1971).

11. Tolner, H., and Andriesse, C. D.: High impedance point contact Josephson junctions. IEEE Trans. MAG-11, 866-869 (1975); Tolner, H., Andriesse, C. D., and Schaeffer, H. H. A.: Wide-band detection with high impedance Josephson junctions. (Proceedings of the 1975 Zürich International Conference on Infrared Physics, to be published).

12. Kulikov, V. A., and Likharev, K. K.: Microwave detection, selfselective mode of operation at 30-40 GHz with Nb point contacts. Izv. VUZOV-Radiofiz. 19 (1976).

13. Clarke, J., Hoffer, G. I., Richards, P. L., and Yeh, N.-H.: A superconducting transition edge bolometer. "Low Temperature Physics - LT14", Krusius, M., and Vuorio, M., Eds. (American Elsevier, New York, 1975) V. 4, pp. 226-229; (and to be published).

14. Parrish, P. T., and Chiao, R. Y.: Amplification of microwaves by superconducting microbridges in a four-wave parametric mode. Appl. Phys. Letters 25, 627-629 (1974); Chiao, R. Y., and Parrish, P. T.: Parametric amplification by unbiased Josephson junctions. J. Appl. Phys. 46, 4031-4042 (1975); Feldman, M. J., Parrish, P. T., and Chiao, R.

Y.: Operation of the Suparamp at 33 GHz. J. Appl. Phys. 47, 2639 (1976).

15. Claeson, T. (private communication).
16. Taur, Y., and Richards, P. L.: Parametric amplification and oscillation at 36 GHz using a point contact Josephson junction. J. Appl. Phys. (to be published); A Josephson effect parametric amplifier at 36 GHz. (Proceedings of the 1976 Stanford Applied Superconductivity Conference, to be published).
17. Feldman, M. J.: The thermally saturated suparamp. (to be published).
18. Mygind, J., Pedersen, N. F., and Sorensen, O. H.: Direct detection of the parametrically generated half-harmonic voltage in a Josephson tunnel junction. (to be published).
19. Kanter, H., and Silver, A. H.: Self-pumped Josephson parametric amplification. Appl. Phys. Letters 19, 515-517 (1971).
20. Kanter, H.: Parametric amplification with self-pumped Josephson junctions. IEEE Trans. MAG-11, 789-793 (1975); Two idler parametric amplification with Josephson junctions. J. Appl. Phys. 46, 4018-4025 (1975).
21. Vystavkin, A. N., Gubankov, V. N., Kuzmin, L. S., Likharev, K. K., Migulin, V. V., and Senenov, V. K.: One frequency parametric amplifier using self-pumped Josephson junction. (Proceedings of the 1976 Stanford Applied Superconductivity Conference, to be published).
22. Silver, A. (private communication).
23. McDonald, D. G., Johnson, E. G., and Harris, R. E.: Modeling Josephson junctions. Phys. Rev. B13, 1028-1031 (1976).
24. Kaplan, S. B., Chi, C. C., Langenberg, D. N., Chang, J. J., Jafarey, S., and Scalapino, D. J.: Quasiparticle and phonon lifetimes in superconductors. (to be published).
25. Octavio, M., Skocpol, W. J., and Tinkham, M.: Improved performance of tin variable-thickness superconducting microbridges. (Proceedings of the 1976 Stanford Applied Superconductivity Conference, to be published).
26. McDonald, D. G., Kose, V. F., Evenson, K. M., Wells, J. S., and Cupp, J. D.: Harmonic generation and submillimeter wave mixing with the Josephson effect. Appl. Phys. Letters 15, 121-122 (1969).
27. Tinkham, M. (private communication).
28. Palmer, D. W., and Mercereau, J. E.: Coherent effects in series arrays

of proximity effect superconducting bridges. IEEE Trans. MAG-11, 667-670 (1975).

29. Claridge, D. E., Giffard, R. P., Michelson, P. F., and Fairbank, W. M.: Nine gigahertz impedance measurements on Ta and Nb point contacts. (Proceedings of the 1976 Stanford Applied Superconductivity Conference, to be published).

A Closed Loop Broadband Microwave Operated SQUID

H. Rogalla and C. Heiden[+]
Institut für Angewandte Physik der Universität Münster,
Roxeler Str. 7o-72, 44oo Münster, Germany

INTRODUCTION

Magnetization measurements at low temperatures by means of electronic integration of induced voltages from pick-up coils around the specimen under test usually suffer from drift effects due to thermal emfs. Small changes in magnetization - whatever their origin may be - cannot be distinguished from the drift if slow sweeps of magnetic field are used. It therefore appears attractive to employ a SQUID for this purpose. Sudden spontaneous magnetization changes like flux jumps in superconductors, even if small,however pose a serious problem, if the slew rate of the SQUID is not sufficient. In this paper we report on experiences made with a microwave operated SQUID system which makes use of a comparatively fast phase sensitive detector. Also preliminary magnetization measurements performed with this system on different materials are reported.

THE SYSTEM

The system uses the principle of the RF-biased SQUID[1,2]. The flux sensitive part (see block diagram fig. 1) is a reflective microwave SQUID (1) similar to a device described in the literature[3]. Microwave power at 9.4 GHz is supplied by a Gunn oscillator (2) via directional couplers (3,6), attenuator (4),

+ present address: Institut für Angewandte Physik der Universität Giessen, Heinrich Buff-Ring 14-16, 63oo Giessen, Germany

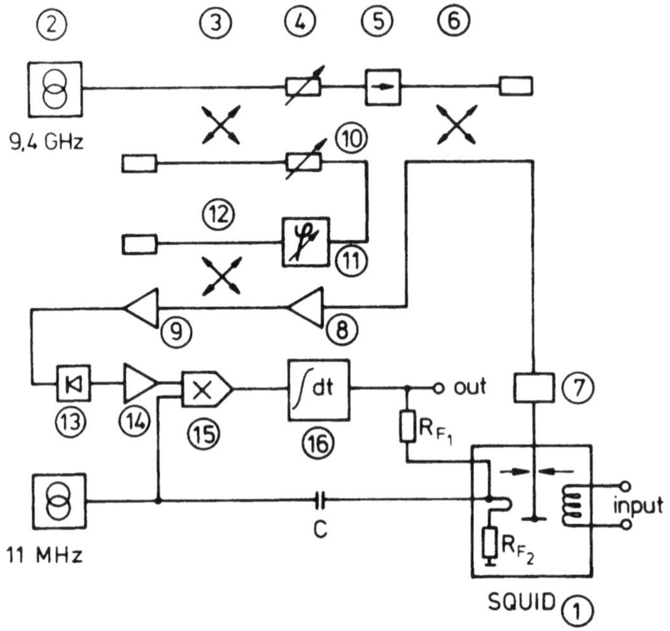

Fig. 1 Block diagram of SQUID electronics

isolator (5) and impedance transformer (7), a near optimum taper[4] with 1% reflection. Microwaves reflected from the SQUID are amplified (36dB) with tunnel diode amplifiers (8,9) and then rectified by a Schottky-diode (13). In order to reduce system noise a compensation is introduced which feeds additional microwave power from the Gunn oscillator (2) to the input of the second tunnel diode amplifier (9) via directional couplers (3,12), attenuator (lo) and phase shifter (11). Amplitude and phase of the compensating network are adjusted to obtain near loo% amplitude modulation at the rectifier input. The rectified signal is further amplified (14) by a low noise video amplifier (3dB points: 325 Hz and 16 MHz). Signal to noise ratio at its output is 26dB under optimum conditions. For fast operation in the lock-in mode, the frequency of the modulating signal should be as high as possible. In the realization presented here, 11 MHz were chosen in order to use standard components. The output of the phase sensitive detector (15) is applied to an integrator (16) with 3.3 μs time constant and then coupled to the SQUID via feedback resistors R_{f1} and R_{f2}. After first using a fast small signal transconductance multiplier for phase sensitive detection, improved performance was achieved with a double balanced mixer. Data presented in this paper were obtained with the latter one. The output signal (at port OUT) has a range of ± 7oo $\emptyset_o$. Input to the SQUID is provided by a small toroidal coil with mutual inductance of about 5oo pH.

SYSTEM PERFORMANCE

For a sinusoidal input signal, the maximum possible output voltage (peak to peak value) was measured as a function of frequency. The result obtained under normal laboratory conditions (simple lead shield, no particular precautions against vibrations) is shown in fig. 2.

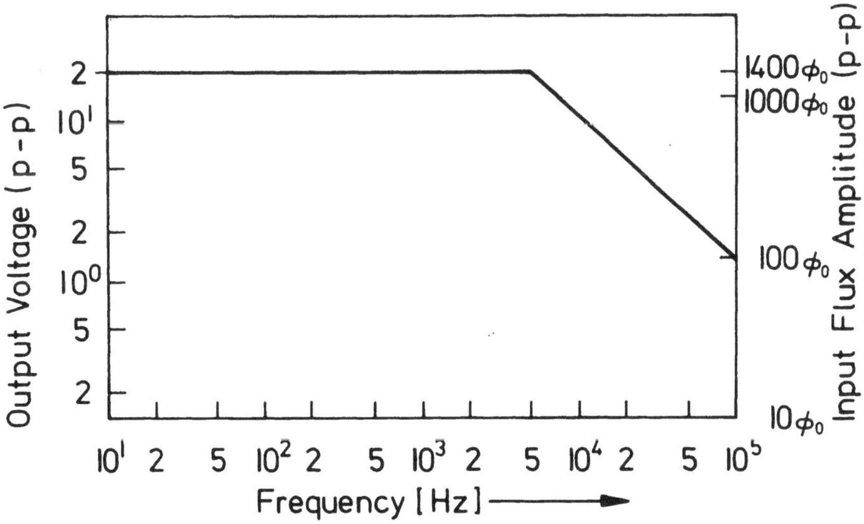

Fig. 2 Frequency dependence of maximum output voltage

Full output (2o Vpp) is available up to 5kHz, which corresponds to a maximum slew rate of ca. $2 \cdot lo^{7}$ $\emptyset_o$/s. A f^{-1}-behavior for the maximum output voltage is observed for higher frequencies. A safe operation of the SQUID in the lock-in mode is still possible at loo kHz with a sinusoidal flux modulation of loo $\emptyset_o$ peak to peak. (fig.3).

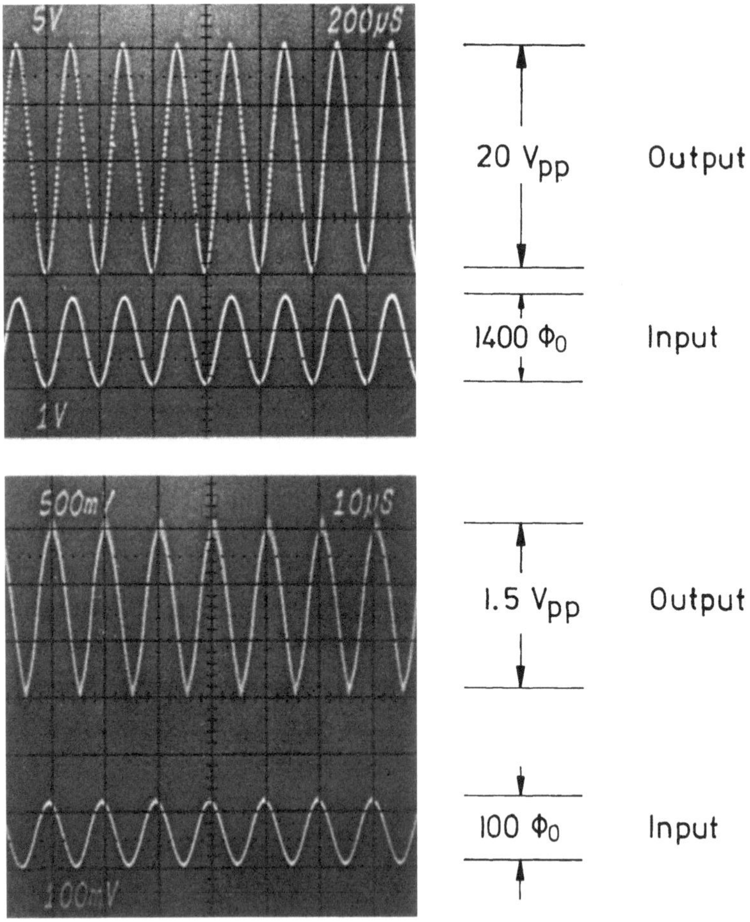

Fig. 3 Maximum system output voltage for 5 kHz and loo kHz

System noise was analysed in the frequency range between lo^{-2} Hz and lo^{5} Hz, the input being terminated with o.2 Ω.

Microwave input power to the SQUID was 10^{-9} W. Except for larger contributions at low frequencies due to hum and vibrations the observed noise was white with a value of $3\cdot 10^{-4}$ $\emptyset_o/\sqrt{Hz}$. If the whole system is used with a band from dc to 100 kHz, a signal to noise ratio of ca. 83dB is obtained, if the maximum flux modulation of 1400 $\emptyset_o$ is employed.

The spectrum of system noise in the range between 3 Hz and 10^5 Hz is shown in fig. 4. The peak at 5 Hz is due to a mechanical resonance of the system, the maxima between 50 Hz and 500 Hz are due to hum (50 Hz and harmonics). No particular precautions against vibrations and hum were taken.

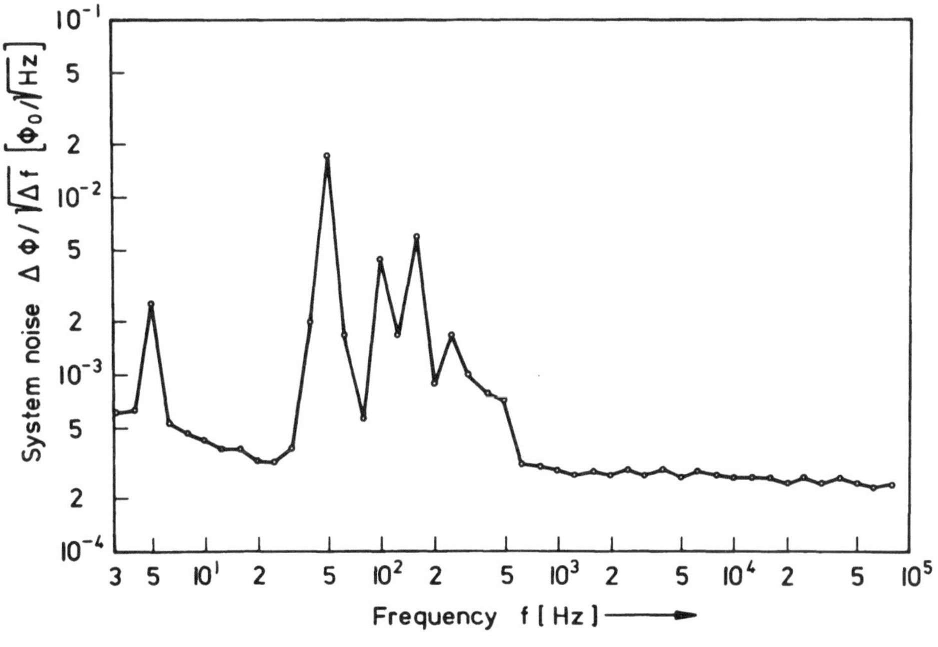

Fig. 4 Effective flux noise density, $\Delta\emptyset/\sqrt{\Delta f}$, as function of frequency.

MAGNETIZATION MEASUREMENTS

The described SQUID-system is being used to study the magnetization behavior of superconducting materials with very low drift. Either the magnetic induction B or the magnetization $M = B-\mu_o H_a$, where H_a is the magnetic field applied to the specimen by a superconducting solenoid, can be measured using the flux transformer system shown in fig. 5. It consists of two pick-up coils, one of which is wound around the specimen while the other senses the solenoid field only. Each of these coils forms a superconducting loop with a secondary coil. Proper compensation can be adjusted by moving a further superconducting coil between these secondaries which couples a net flux corresponding to B or M to the SQUID-input. The flux transformer system is wound with single-core Nb-Ti-wire and is shielded with one lead-foil.

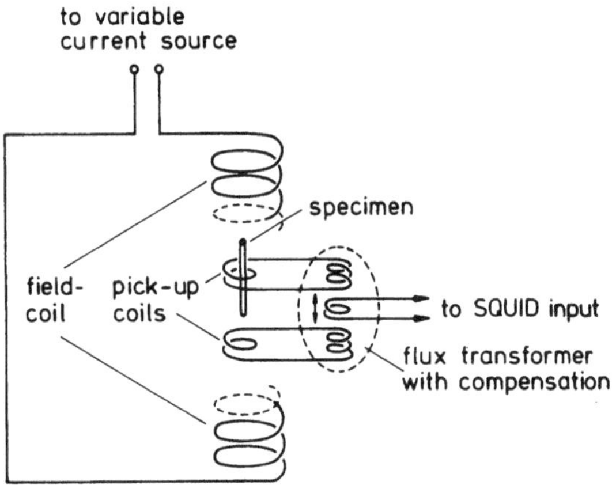

Fig.5 Flux transformer system for magnetization and induction measurements.

Fig.6 shows magnetization curves of a Vanadium sample, fig.7 induction curves of a $Pb_{o.98}$ $In_{o.o2}$ specimen. For each branch the field was swept from zero to maximum field or in reverse direction using a time of 2o minutes.Three hours were needed to obtain the whole plot in fig. 6, within resolution of the used x-y recorder no drift was observed.

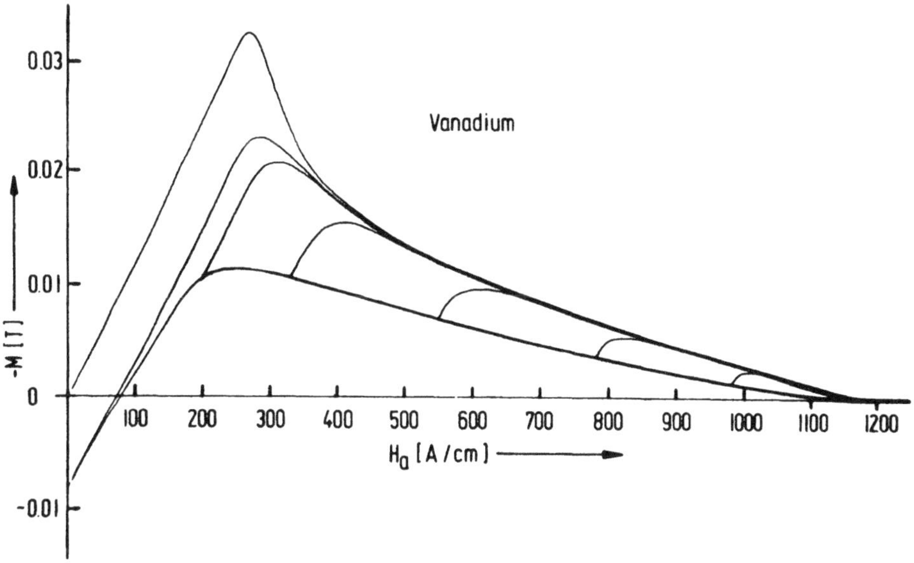

Fig.6 Initial, complete secondary and minor secondary magnetization curves of a Vanadium sample (circular cylinder of o,9 mm diameter and 28 mm length). After obtaining these data, the sample was removed from the system and the field swept back to zero value. In this way, the line corresponding to zero magnetization was obtained.

Little flux jumps can be seen in fig. 7 for the case of decreasing field. They did not produce any stability problems in the lock-in mode of the SQUID-system. This system stability also is demonstrated in fig. 8 which shows a secondary magnetization loop of the Vanadium sample swept back and forth with a frequency of 4o Hz. The results presented in this paper are preliminary. Improvements appear to be possible both with respect to the noise performance and to the bandwidth of the system.

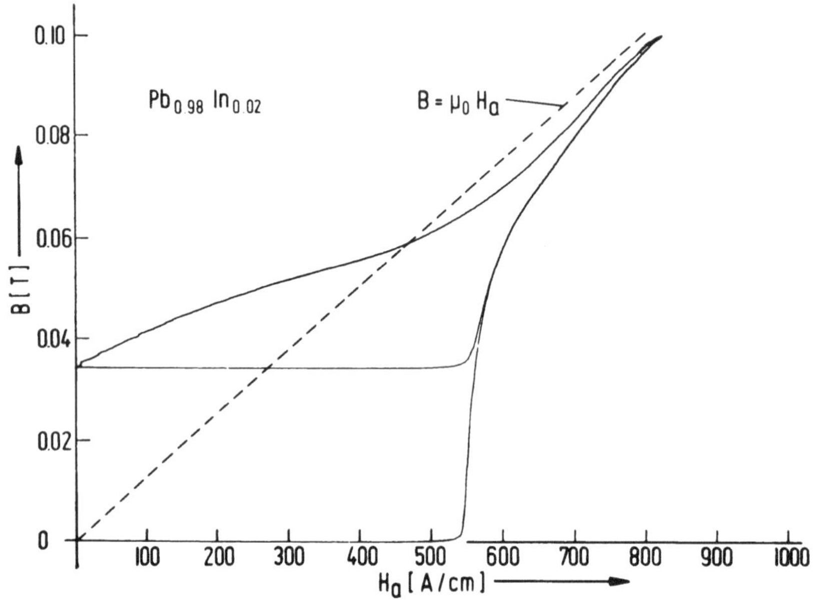

Fig.7 Initial and secondary induction curves of a $Pb_{0.98}$-$In_{0.02}$-sample (circular cylinder of 1,3 mm diameter and 6o mm length). Duration of field sweeps for each branch was 16 min.

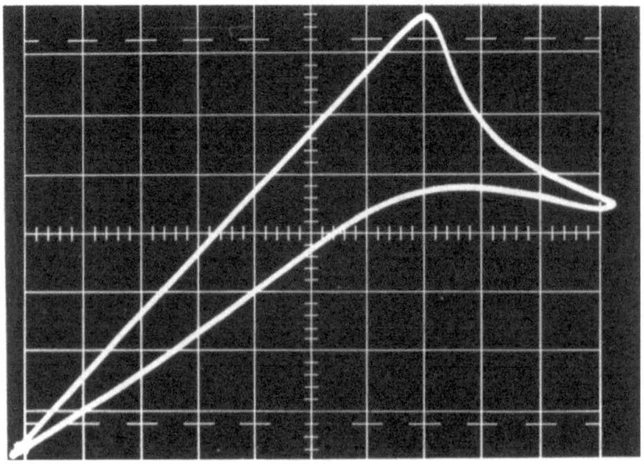

Fig.8 Oscilloscope display of secondary hysteresis loop of the Vanadium sample (see fig.6). The loop was swept with a frequency of 4o Hz.

REFERENCES

1. J.E. Zimmermann, P. Thiene and J.T. Harding, J. Appl. Phys. 41, 572 (197o)

2. R.A. Kamper and M.B. Simmonds, Appl.Phys.Letters 2o, 27o (1972)

3. R.A. Kamper, M.B. Simmonds, R.T. Adler and C.A. Hoer, Proc. Intern. Conf. on Applied Superconductivity, Anapolis, (1972)

4. R.P. Hecken, IEEE Transactions on Microwave Theory and Techniques, MTT 2o, 734 (1972)

A NUMERICAL TREATMENT OF THE LUMPED MODEL OF AN RF BIASED SQUID BY MEANS OF A HYBRID COMPUTER

S.N. Erné and H. Luther
Physikalisch-Technische Bundesanstalt, Institut Berlin
Abbestr. 2-12, D1 Berlin 10, Germany

Over a wide range of frequency an RF SQUID can be described by a lumped model. The SQUID itself is represented by a weak link which interrupts a superconducting ring magnetically coupled with a tank circuit. By using the resistively shunted model one obtains the well-known lumped circuit as shown in fig. 1. On the left hand side we see the SQUID loop containing the current paths of the weak link and the ring inductance L_S. C_J denotes the capacitance of the junction and, of course, parasitic capacitances and $I_0 \sin\phi$ the typical Josephson expression for the pair current. The third term $\frac{1 + \alpha \cos\phi}{R_q}$ describes the quasiparticle current and the quasiparticle pair interference term, also known as the $\cos\phi$-term. R_q is the quasiparticle resistance and α the coefficient of the $\cos\phi$-term. ϕ denotes the quantum mechanical phase difference. The right hand part in fig. 1 is a conventional tank circuit with inductance L_T and capacitance C_T. R_T is the equivalent resistance for the losses in the tank circuit and k is the coupling coefficient. $I_{RF} \sin\omega t$ is the radiofrequency bias current with amplitude I_{RF} and angular frequency ω. I_S and I_T denote the currents in the SQUID loop

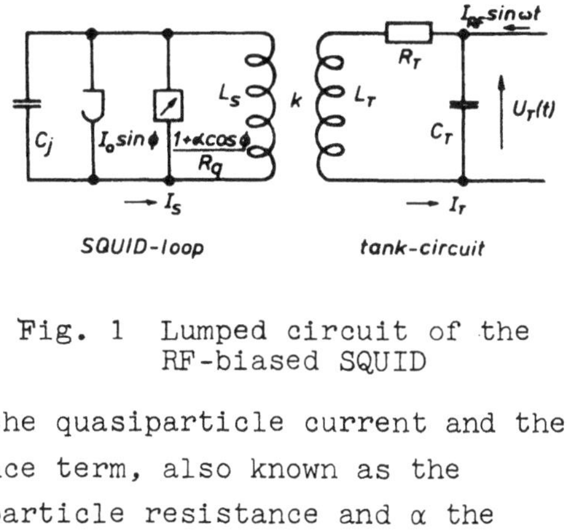

Fig. 1 Lumped circuit of the RF-biased SQUID

and in the inductance of the tank circuit respectively.

A numerical treatment of the lumped model is based on the following system of differential equations:

$$(1.)\qquad L_S C_j \ddot{\phi} + \frac{L_S}{R_q}\dot{\phi}(1+\alpha\cos\phi) + \frac{2e}{\hbar}L_S I_o \sin\phi + \phi = \frac{2e}{\hbar}k\sqrt{L_S L_T}\, I_T + \frac{2e}{\hbar}\Phi_{dc}$$

$$(2)\qquad (1-k^2)\ddot{I}_T + \frac{R_T}{L_T}\dot{I}_T + \frac{1}{L_T C_T}I_T = -\frac{\hbar}{2e}\frac{k}{\sqrt{L_T L_S}}\ddot{\phi} - \frac{1}{L_T C_T}I_{RF}\sin\omega t$$

Equation (1) describes the behaviour of the SQUID loop if a dc-flux Φ_{dc} is applied and if it is driven by a flux generated by the current I_T flowing in the inductance of the resonator. Equation (2) describes the behaviour of the tank circuit and determines the current I_T with regard to the bias current $I_{RF}\sin\omega t$ and the SQUID-loop-feedback.

With respect to SQUID-magnetometer applications we are mainly interested in the dependence of the voltage U_T across the tank circuit on the dc-flux Φ_{dc}

$$(3)\qquad U_T(t) = \frac{1}{C_T}\int_0^t [I_{RF}\sin\omega t' - I_T(t')]\,dt'$$

Up to now this system of equations could be solved analytically only for the non-hysteretic case $\frac{2e}{\hbar}L_S I_o < 1$ assuming voltage independent coefficients I_o, R_q and α [1]. Numerical solutions using digital computers are not feasible because of the extremely long computation time, due to the high non-linearity of eq. (1) in connection with a large difference between the characteristic times of the equations (1) and (2): The flux transitions in the SQUID-loop occur in periods of the order of ps; the characteristic time of the tank circuit however is in the order of ns. An analog computer is suitable for solving the SQUID equation (1) but not for equation (2) and (3) because of stability problems which cannot be overcome. For these reasons we have used a hybrid computer system.

Therein the digital processor solves the equations (2) and (3) and supports the analog computer in solving equation (1). Fig. 2 demonstrates the main parts of the program's flow. The digital processor transforms the physical system parameters into dimensionless coefficients suitable for analog computation. The normalization factors are floating and depend on the parameter configuration to allow optimal application of the analog computer. Using a Runge-Kutta-Method in the main routine the digital processor solves equation (2) and feeds the calculated value of I_T to the analog computer. Correspondingly, the analog computer provides $\ddot{\phi}$ as feedback to the digital processor. The schedule time of this routine is 70 ms corresponding to a real-time-resolution of 0.7 ps.

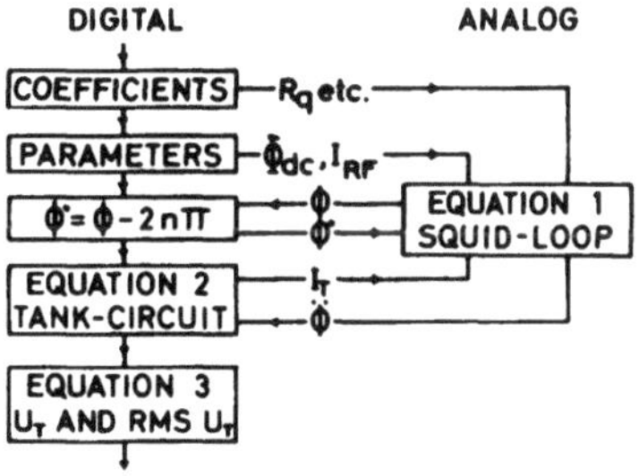

Fig. 2 Main part of the hybrid program

During the same routine the voltage at the tank circuit and its RMS-value are calculated. Besides this the digital computer has to support the analog computer in reducing the phase difference to the main argument $\phi^* \leq 2\pi$ for appropriate driving of the analog function generators for sin ϕ and cos ϕ .

Running the program we at first were interested in a survey of essential properties of the solution. For the hysteretic case, that is $\frac{2e}{\hbar} L_S I_o > 1$, we can distinguish between two different types of behaviour:

- stationary solution for the tank voltage or
- appearance of relaxation oscillations in the tank voltage amplitude

We found that the first case is the common behaviour at high frequencies. Relaxation oscillations are possible in special cases, but we could not find reliable conditions for their existence. In fig. 3 is shown the time-dependence of $U_T(t)$ for two different values of the applied dc flux Φ_{dc}. The beginning of each curve is governed by the transient behaviour of the system. The stationary state is shown in the remaining parts: In the right hand curve (b) a constant amplitude of the tank voltage, in curve (a) relaxation oscillations. The amplitude of the relaxation oscillations is small compared with the total voltage amplitude and with respect to the voltage changes due to a changing of the applied dc-flux too. In consequence, the relaxation oscillations do not strongly affect the amplitude or RMS-value of the tank voltage and so the sensitivity of the SQUID-magnetometer to dc-flux (modulation depth $U_M = U_{TRMS}(\Phi_{dc}=0) - U_{TRMS}(\Phi_{dc}=\frac{\Phi_0}{2})$).

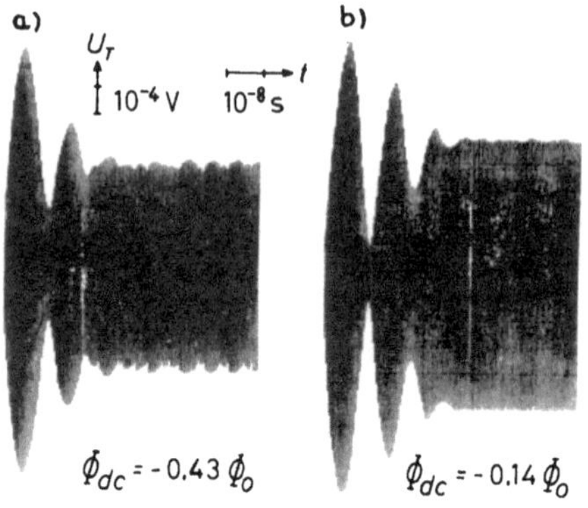

Fig. 3 Time-dependence of the tank-circuit-voltage $C_S = 0.1$ pF, $I_0 = 20\ \mu A$, $R_q = 5\ \Omega$, $L_S = 0.1$ nH, $I_{RF} = 2{,}8\ \mu A$, $\omega/2\pi = 2{,}03$ GHz, tank-circuit resonance frequency and Q are 2,00 GHz and 35 resp., $\Phi_0 = \frac{h}{2e}$.

A further aim of our investigation was to calculate characteristics of the SQUID-magnetometer corresponding to those which are commonly observed in experiments. Although the influence of relaxation oscillations on the RMS-value of the tank-voltage is in general of secondary order we intended to avoid any relaxation oscillations of the solution by choosing a high working frequency in connection with a suitable set of SQUID-parameters.

In fig. 4 the RMS-value of the tank-voltage U_{TRMS} versus the bias-current-amplitude I_{RF} is shown for a working frequency of 10 GHz. The structure of the curves is characterized by monotonic increasing of $U_{TRMS}(I_{RF})$ similar to the non-hysteretic case. This is a consequence of the continuous (Bessel-like) dependence of the amplitude of the current in the SQUID-loop on the amplitude of the exciting flux.

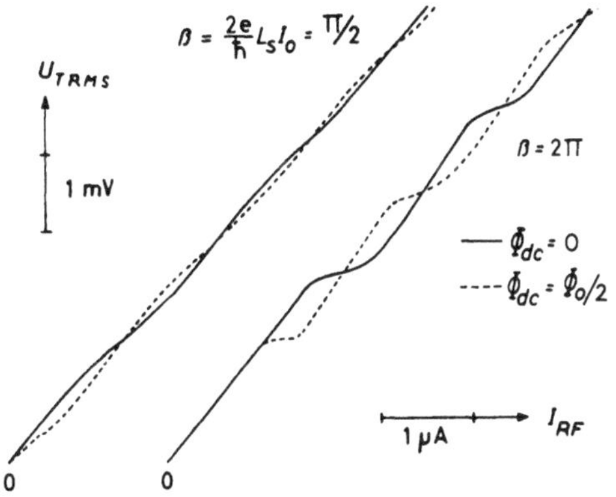

Fig. 4 Calculated staircase-curves

On the left hand side of fig. 4 curves for a value of the parameter $\beta = \frac{2e}{\hbar} L_S I_o$ not much larger than one ($\beta = \pi/2$) are shown. The staircase-like structure is strongly reduced; with decreasing β we found a continuous approach to the behaviour of the SQUID-magnetometer in the non-hysteretic case. However we wish to point out that the signal generation in the hysteretic case is in principle different from that in the non-hysteretic case. This follows from the results in fig. 5 where the dependence of the modulation depth is shown as a function of a normalized off-tune-parameter $\delta = 2\,Q\,\frac{\omega - \omega_R}{\omega_R}$ for $\beta = \pi/2$ (ω_R denotes the resonance angular frequency of the tank circuit with coupled SQUID, but with $I_o = 0$ and $\alpha = 0$, Q is the Q-factor of the tank circuit). Even for β almost unity, we get in the hysteretic mode a shape of the curve like that of a resonator response curve. This is quite different from equivalent curves in the non-hysteretic case, where the SQUID acts reactively, and where we get a second lobe of negative values of the modulation depth for positive detuning. The general feature of the SQUID, to work dissipatively, was confirmed too

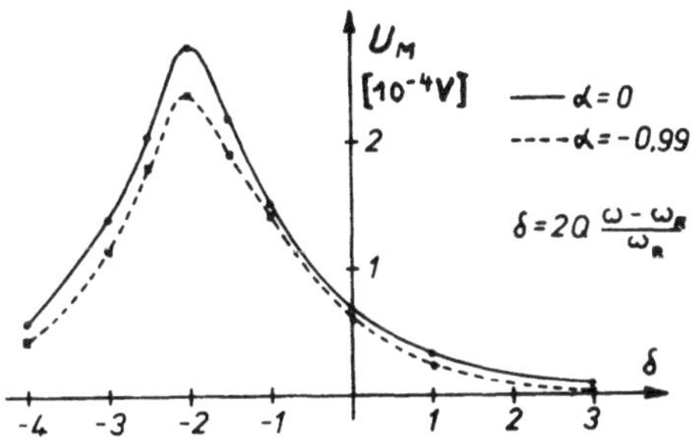

Fig. 5 Dependence of the signal-voltage on the off-tune-parameter

by studies of the influence of the cos ϕ-term on the modulation depth. The small reduction of the signal level even at large values of α as it can be seen from fig. 5 indicates that the dissipation in connection with the cos ϕ-term is small in comparison with that in the quasiparticle path. Fig. 6 shows the Bessel-function-like dependence of the modulation depth U_M on the bias amplitude I_{RF} and (by comparing the upper and lower graph) the well known dependence of the maximum modulation depth on the SQUID-parameter ß. Again we recognize the small influence of the cos ϕ -term in decreasing the amplitude of U_M and shifting its zero crossing point.

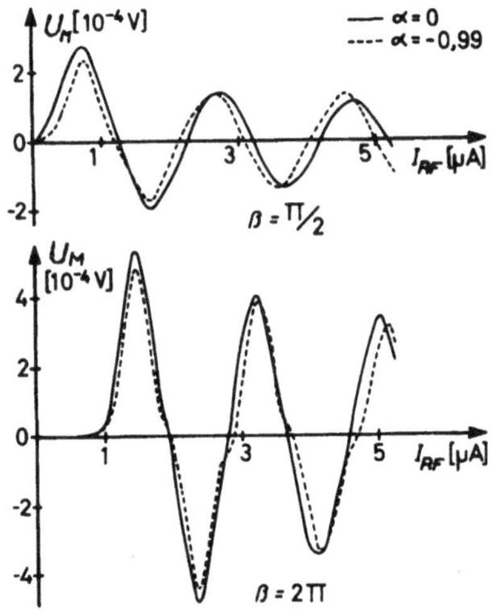

Fig. 6 Dependence of the modulation depth on the bias-current-amplitude

REFERENCE

[1] Ernė, S.N., Hahlbohm, H.-D., Lübbig, H.: Theory of the rf-biased superconducting quantum interference device for the nonhysteretic regime. J. Appl. Phys. 47, 5440, (1976)

RF POWER MEASUREMENTS USING QUANTUM INTERFERENCE IN SUPERCONDUCTORS†

D. B. Sullivan, N. V. Frederick and R. T. Adair
National Bureau of Standards, Institute for Basic Standards, Boulder, Colorado 80302 USA

ABSTRACT

A SQUID system for high sensitivity and broad dynamic range measurement of rf power at 30 MHz has been developed. The system is a modification of earlier concepts and consists of two SQUIDs, each sensing power in adjacent but slightly overlapping ranges, providing an overall dynamic range of 105 dB. A dc calibration of one of the SQUIDs provides a power uncertainty of less than 0.1 dB for the system at power levels as low as -131 dBm ($\sim 8 \times 10^{-17}$ watts).

INTRODUCTION

The application of the microwave-biased SQUID to the measurement of rf attenuation has been demonstrated in two separate laboratories (1-4). Several reports have also discussed the measurement of power with the same device (5,6). At levels of 10^{-9} watts (-60 dBm), the SQUID was shown to provide an excellent absolute measure of power, but below this level a substitution technique was used. The present paper discusses a different technique for measuring power, which utilizes two SQUIDs in cascade to achieve higher sensitivity without the substitution process. The pump frequency of 1 GHz for these SQUIDs limits the measurements to an upper frequency of about 100 MHz but the concept can be extended to higher frequency.

The static response of an rf-SQUID to applied magnetic flux can be made very nearly sinusoidal by operation in the non-hysteretic mode (7) or by operation in the overdamped mode (8). The precise measurement of rf quantities with the SQUID is, in fact, dependent on the SQUID's sinusoidal response. The principal of measurement of power and attenuation can be simply illustrated as follows. The SQUID response is assumed to have the sinusoidal form

$$V(\phi) = V\cos(2\pi\phi/\phi_o) \qquad (1)$$

where $\phi_o = h/2e$ is the flux quantum. It is assumed that this response prevails at the measurement frequency, ω_m. This is equivalent to saying that the readout (or pump) frequency, ω_o, is much larger than ω_m. If a signal, $\phi = \phi_m \sin(\omega_m t)$ is applied to the SQUID then the average value of the detected SQUID output is

$$\langle V(\phi) \rangle = VJ_o(2\pi\phi_m/\phi_o), \tag{2}$$

where J_o is the Bessel function of zero order. The Bessel function has zeros at well known values of its argument and these zeros thus provide an absolute measure of the quantity ϕ_m/ϕ_o. If the flux ϕ_m is generated by a current through some fixed coil, then the Bessel function argument can be expressed as $2\pi I_m/I_o$ where I_o is the current required to drive one flux quantum through the SQUID. One immediately notes that the system can measure rf current, I_m, in units of I_o. Furthermore, if the coil does not perturb the line impedance at the measurement frequency, then a dc measure of I_o provides a calibration of this rf-ammeter. Implicit in this last statement is the assumption that penetration-depth effects are negligible, that is, that the instantaneous field generated by the rf signal in the coil is the same as that generated by an equal direct current. The question of coil impedance will be discussed later. If the line is terminated (after the SQUID) with a resistance, R_o, which is equal to its characteristic impedance, then the power delivered to the load is $I_m{}^2R_o/2$ where the factor of two is needed because I_m is the current amplitude as opposed to an rms current. It is thus clear that both the terminating impedance and current sensitivity must be determined in order to measure power.

As explained in previous work (1-6), attenuators can be calibrated by noting the ratio of currents at the zeros of the Bessel function. In practice, attenuation can be measured in this way with an rms uncertainty of a few thousandths of a dB. This same uncertainty is potentially available for power measurement, but the uncertainty in connector reflections and line losses limit the value of power measurement to ∿0.1 dB uncertainty. At this level of uncertainty the process of locating the Bessel function zeroes becomes a trivial matter and we refer the reader to the previous work on attenuation (1-6) for details.

A significant portion of our recent effort has been spent on the development of a permanent point contact for the SQUID. The purpose is to simplify the operation so that the systems can be integrated into routine calibration service with no extraordinary 'junction twiddling' skill required by the operating personnel. A brief description of the permanent junction which is contained in a replaceable cartridge has recently been given and we refer the reader to that work (3) for details as the same junctions are used in this work.

ABSOLUTE POWER AT MODERATE SENSITIVITY

Previous systems for broadband measurement of rf power (5,6) provided 0.1 dB resolution at -60 dBm. Because of its utility, we decided to try to improve the sensitivity of a broadband system. Figure 1 shows the SQUID used for this measurement. The pump frequency is 1 GHz and the signal frequency is limited by the system bandwidth to 100 MHz. The question then is how sensitive can this SQUID be to rf power and retain a 100 MHz bandwidth. Although the 1 GHz readout circuit is tuned, the Q which is approximately 5 does not limit the readout bandwidth. A coil of 17 turns of 0.25 mm diameter insulated wire was placed in the toroidal SQUID cavity (2mm x 2 mm channel). This forms a short section of helical transmission line which is inductively coupled to the SQUID. The characteristic impedance, $\sqrt{L/C}$, of this section of helical transmission line was measured to be $\sim 200\Omega$. The length of this line is much less than $\lambda/4$ at 100 MHz and the overall signal line impedance as measured at the room temperature end is 50Ω from dc to 100 MHz. With this coil the power sensitivity at the first zero of the Bessel function is -79.0 dBm (1.26×10^{-11} watts) as determined by the dc calibration. This calibration is referenced directly to the SQUID, consequently, account must be taken of line losses and reflections if a measurement is to be made at a room temperature port. At 30 MHz we estimate the sum of these corrections

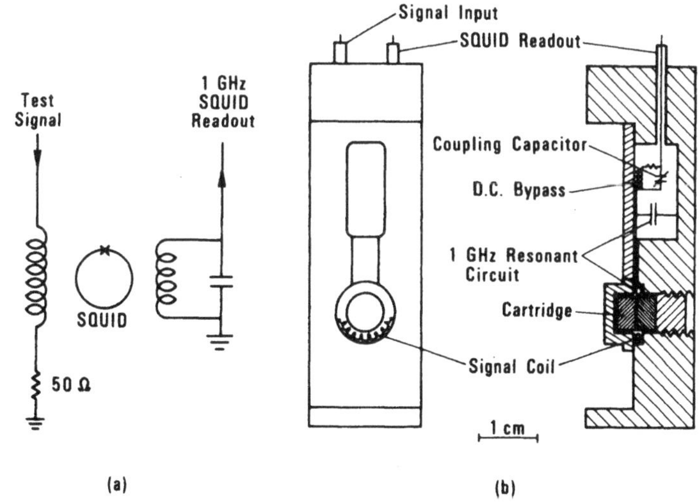

Figure 1 Broadband SQUID for rf power measurement. The impedance looking into the test signal line (a) is constant at 50Ω from dc to 100 MHz. The toroidal SQUID configuration is shown in (b). The permanent point contact junction is contained in the cartridge which is replaceable.

to be ∿0.3 dB for the stainless coax to the SQUID, the connectors, and a short length of flexible coax at room temperature. This reduces the sensitivity at the room temperature port to -78.7 dBm. An attenuator and bolometric power meter were utilized in a substitution method to provide a 30 MHz signal at this same power level and the results were found to agree within 0.1 dB with the dc calibration. We performed a second test at 500 kHz where the corrections mentioned above were significantly less and again the agreement with the dc calibration was within 0.1 dB. Although it was not directly verified, the impedance measurements suggest that the response should be flat to 100 MHz.

In the construction of the SQUID coil only minor consideration was given to impedance perturbation problems and it would appear to be quite simple to extend the concept to yet higher sensitivity. The characteristic impedance of the helical line can be reduced to 50Ω by increasing the capacitance of the line to the toroidal cavity or by decreasing the inductance. One could thus achieve higher current sensitivity by adding turns to the coil and increasing the coupling to the SQUID without introducing significant line perturbation. It would be necessary, of course, to provide some suitable transition from the coaxial to the helical line. Since power sensitivity improves as the square of current sensitivity, a 10-fold improvement in current sensitivity would yield a power sensitivity of -100 dBm (10^{-13} watts) at the first zero. This appears to be quite feasible.

The foregoing arguments should make the frequency limitations clear. As frequency increases, the sensitivity of a SQUID power meter will have to be decreased if one wants to maintain the broadband character of the response. While one can achieve higher sensitivity in a smaller bandwidth (as described in the next section), it is nonetheless desirable to push the limits for broadband detection, since such instrumentation would be more versatile.

A TUNABLE NARROW BAND SYSTEM FOR HIGHER SENSITIVITY

The concept for a higher sensitivity system is shown in Fig. 2. The upper SQUID provides a broadband measure of current and is similar to that described in the previous section. The second SQUID is similar to the first except that the coil is tuned with capacitor C_1 and matched to the line with capacitor C_2. The sensitivity of this tuned SQUID to current is established by comparison with the broadband SQUID which is in turn calibrated with a direct current. The increased sensitivity of the second SQUID is derived from multiplication of the coil current by the Q of the circuit. The sensitivities of the two SQUIDs are selected so that the first Bessel function zero for the broadband SQUID is at a level which yields a high order zero

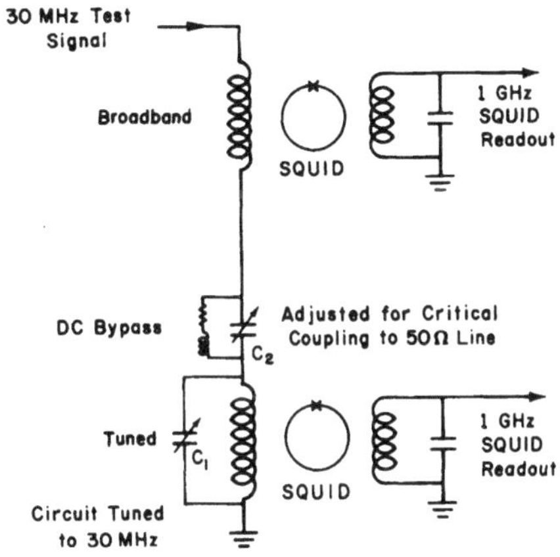

Figure 2 Dual SQUID for high sensitivity and broad-dynamic-range measurement of rf power. The sensitivity of the broadband SQUID is determined by a direct current calibration. The sensitivity of the tuned SQUID is determined by comparison with the broadband SQUID since the ranges of the two overlap. Both the coupling capacitor and the tuning capacitor can be adjusted from outside the cryostat. Both SQUIDs are of the toroidal configuration shown in Figure 1B.

for the tuned SQUID. Because the spacing between zeros beyond the 100th zero is less than 0.1 dB, it is a simple matter to interrelate the two sensitivities. A bypass of the coupling capacitor provides a dc path to ground so that the broadband SQUID can be calibrated with a direct current.

The power sensitivities (at the first zero of J_o) for the broadband and tuned SQUIDs are -76.4 dBm and -111.2 dBm, respectively. The coil in the tuned SQUID is made up of only 10 turns and hence, with more turns an improvement in sensitivity can be achieved. As will be noted in the next section, the useable sensitivity extends well below the location of the first zero.

The use of two SQUIDs in cascade requires some caution. It was necessary to maintain the 1 GHz bias on the tuned SQUID while measuring power with the broadband SQUID in order to avoid a strong nonlinearity in the terminating impedance. Errors as large as 1.5 dB were found when this practice was not followed. The nonlinear impedance is undoubtedly related to impedance variations of the junction. The tuned circuit was matched to the line at a high power level and the error due to impedance mismatch was found to appear at low power levels. With the microwave bias applied to both SQUIDs, these variations of the terminating impedance could be made to affect the power measurements at less than the 0.1 dB level, but it is nonetheless clear that this type of termination is much less reliable than a simple resistive termination. This is amply demonstrated by the fact that the system resolution and precision was no better than a few hundreths of a dB when it was used to calibrate attenuators.

EXTENDING MEASUREMENTS BELOW THE FIRST ZERO

The zero crossings of the Bessel function provide a convenient set of calibration markers, but this is not the only way to utilize the system. One can use the functional form of the Bessel function response to definite advantage in extending power measurements below the first zero. The output of the SQUID at zero measuring signal amplitude provides a reference for such measurement. Figure 3 depicts the function, J_o, and it is obvious that the system has little sensitivity for very small values of the argument $2\pi I_m/I_o$. However, as shown in Figure 3 for $2\pi I_m/I_o = 0.3$, $J_o = 0.9776$ and the magnitude of the slope is only four times less than the slope at the first zero. The argument of J_o should be derivable from a measurement

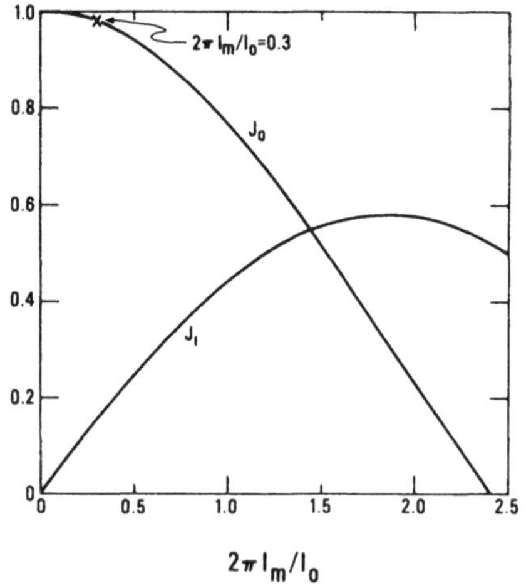

Figure 3 Bessel functions, J_o and J_1. Power measurements can be extended from the first zero of J_o down to an argument of ∿0.3. Below this level J_o provides insufficient resolution. Further sensitivity can be achieved by using J_1 as discussed in the text.

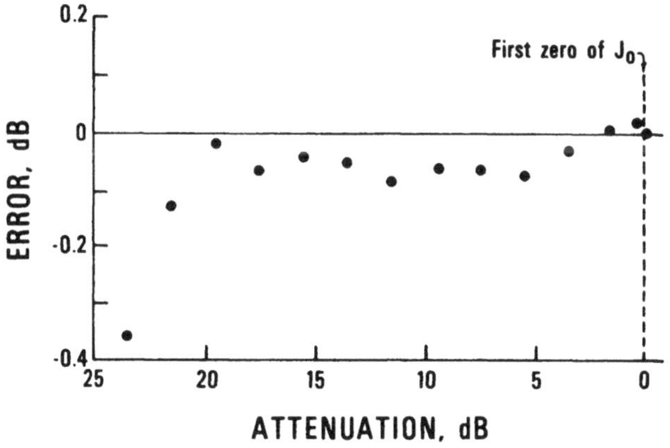

Figure 4 Extension of measurements below the first zero of J_o. Attenuation is determined experimentally with a piston attenuator and the value thus obtained is compared with that obtained from a table of Bessel functions and a measurement of J_o amplitude. The measurements are referred to the first zero of J_o. Since the power sensitivity of the zero crossings can be determined from dc measurements, this represents an extension of power measurement below the first zero.

of J_o amplitude at least to this level which is ∿18 dB below the first zero crossing. An improvement of this magnitude is well worth studying and we have therefore measured the system response in this interval below the first zero.

The results are displayed in Fig. 4 as measurements of attenuation. These can be clearly translated to an extension of the power scale. As expected, the error increases as power decreases, but the error does not exceed 0.1 dB until the signal is ∿20 dB below the first zero. With this extension of range the single broadband SQUID discussed in the second section of this paper has a sensitivity of -99 dBm (0.1 dB uncertainty) while the dual SQUID system discussed in the previous section is sensitive to a power level of -131 dBm (0.1 dB uncertainty). This also extends the dynamic range of the latter system to 105 dB.

Better sensitivity can yet be achieved if a strong reference signal is available at the signal frequency. With this signal as reference, the first signal sideband about the center (pump) frequency can be demodulated. The amplitude of this sideband (3) goes as $J_1(2\pi I_m/I_o)$, a quantity which has a large slope at zero signal power as shown in Figure 3. This could extend the measurement capability another 20 dB or more below the limits achieved to date. However, this technique is more complex and requires the reference signal for demodulation. It is likely to be of value only in certain limited situations.

SUMMARY

The concept of broadband measurement (dc → 100 MHz) of rf power has been extended to a sensitivity of -99 dBm with a 0.1 dB uncertainty. A narrow band system of extended dynamic range (> 100 db) was found to determine power with the same uncertainty at a level of -131 dBm ($\sim 8 \times 10^{-17}$ watts). This latter system provides a terminating power measurement with no attenuation substitution required. The simple nature of the dc calibration of these instruments offers a significant advantage in standards measurement procedures. Further improvement in sensitivity can be expected to accompany refinements in the design of the coil which couples the signal into the SQUID. It would be especially interesting to determine the ultimate limit of sensitivity while maintaining 0.1 dB uncertainty for the broadband device.

REFERENCES

† Work partially supported by the U. S. Department of Defense, Calibration Coordination Group. Contribution of the National Bureau of Standards, not subject to copyright.

1. R. A. Kamper, M. B. Simmonds, R. T. Adair, and C. A. Hoer: A new technique for RF measurements using superconductors. Proc. IEEE 61, 121-122 (1973).

2. R. T. Adair, M. B. Simmonds, R. A. Kamper and C. A. Hoer: RF attenuation measurements using quantum interference in superconductors. IEEE Trans. Instrum. Meas. IM-23, 375-381 (1974).

3. N. V. Frederick, D. B. Sullivan and R. T. Adair: Advances in the use of SQUIDs for RF attenuation measurement. Proc. of the 1976 Applied Superconductivity Conf., to be published in IEEE Trans. Mag.

4. B. W. Petley, K. Morris, R. W. Yell, and R. N. Clarke: Moulded microwave SQUID for RF-attenuator calibration. Electronics Letters, 12, 237-238 (1976).

5. R. A. Kamper, M. B. Simmonds, C. A. Hoer and R. T. Adair: Measurement of RF power and attenuation using superconducting quantum interference devices. Nat. Bur. Stand. Tech. Note 643, Aug. 1973.

6. R. A. Kamper, M. B. Simmonds, R. T. Adair and C. A. Hoer: Advances in the measurement of RF power and attenuation using SQUIDs. Nat. Bur. Stand. Tech. Note 661, Sept. 1974.

7. J. E. Zimmerman, P. Thiene and J. T. Harding: Design and operation of stable rf-biased superconducting point-contact quantum devices, and a note on the properties of perfectly clean metal contacts. J. Appl. Phys., 41, 1572-1580 (1970).

8. M. B. Simmonds, private communication.

AN ANALYSIS OF THE LOW INDUCTANCE ac-SQUID OPERATED AS A dc-MAGNETOMETER IN A MIXING MODE

O. H. Soerensen
Physics Laboratory I, The Technical University of Denmark,
DK-2800 Lyngby, Denmark

INTRODUCTION

The conventional ac-SQUID magnetometer in the low inductance limit has been studied experimentally in great detail (1, 2) and good agreement with theoretical results has been obtained (2, 3, 4). A recent analysis (5) has shown that an upper limit to the useful range of bias frequencies exists related to the cut-off frequency of the SQUID loop. At bias frequencies above cut-off the relative modulation of the output voltage by a slowly varying magnetic field approaches zero even though the peak-to-peak modulation may stay finite. The magnetic field dependent part of the output is generated by the non-linear weak link and more favorable conditions for the detection of that part of the signal may be realized if the response is separated in frequency from the (comparatively) large excitation. One such scheme is to apply simultaneously ac-bias at two different frequencies and take as the output the mixing product at, e.g., the intermediate (IF)-frequency (6). This mode of operation will be analyzed below under some simplifying assumptions:

1. The single junction SQUID loop is in the low inductance limit, i.e., $L_R I_o \lesssim \Phi_o/2\pi$. Here, L_R is the loop inductance, I_o is the critical current of the weak link, and $\Phi_o = h/2e$ is the flux quantum.

2. The weak link may be represented by the equivalent circuit model consisting of the parallel combination of the ideal Josephson element carrying current $I_o \sin\varphi$; a conductance, $G(1 + \varepsilon\cos\varphi)$, and a capacitance, C_J, as shown in Fig. 1

3. The quantities I_o, G, and ε are constants. Physically, that means that the bias frequencies and the SQUID loop cut-off frequency, $1/GL_R$, must be less than the energy gap frequency of the superconducting material.

4. The SQUID loop is ac biased by the applied magnetic flux $\Phi_1\sin(\omega_1 t - \theta_1) + \Phi_2\sin(\omega_2 t - \theta_2)$ and is coupled magnetically to a resonant circuit tuned to the IF-frequency, $\omega = \omega_2 - \omega_1$. Here, it is assumed that the dimensions of the SQUID loop is smaller than the wavelengths of the bias signals.

5. The current induced in the resonator or tank circuit contains only one frequency component, $I\cos\omega t$. With a mutual inductance, M, this current contributes $MI\cos\omega t$ to the applied flux. Including the quasi-static flux, Φ_x, which is to be measured, the externally applied flux may be written

$$\Phi_{ext} = \Phi_x + MI\cos\omega t + \Phi_1\sin(\omega_1 t-\theta_1) + \Phi_2\sin(\omega_2 t-\theta_2) \tag{1}$$

6. Due to the current circulating in the SQUID loop the enclosed flux differs from Φ_{ext}. Here, it is assumed that the circulating screening current contains the same frequency components as in eq. (1), i.e.,

$$\Phi_{int} = \frac{\Phi_o}{2\pi}\,(\beta_o+\beta\sin(\omega t-\psi)+\beta_1\sin(\omega_1 t-\psi_1)+\beta_2\sin(\omega_2 t-\psi_2)) \tag{2}$$

This assumption permits a consistent solution of the circuit equations to be obtained to second order in the normalized amplitudes, β_1, $\beta_2 \lesssim 1$.

CIRCUIT ANALYSIS

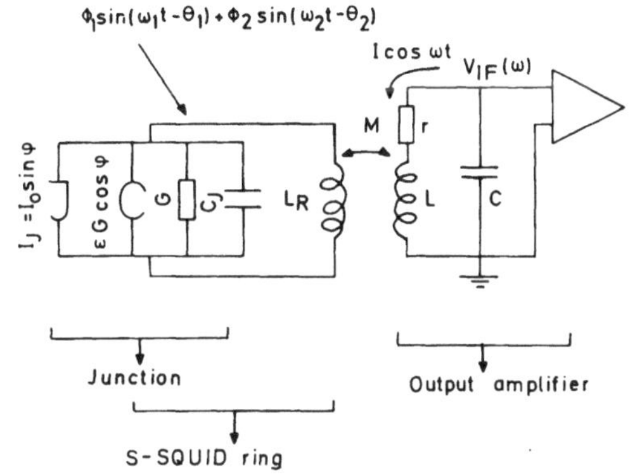

Fig. 1 The mixing magnetometer lumped circuit model.

The magnetometer circuit is shown in Fig. 1. The SQUID loop is represented by the equivalent circuit of the weak link further shunted by the loop inductance, L_R.

The current circulating in the SQUID loop is

(3) $$I_c = I_o\sin\varphi + G(1+\varepsilon\cos\varphi)V_J + C_J\dot{V}_J$$

where the phase, φ , and the junction voltage, V_J, are related to the enclosed flux, Φ_{int}, by

(4) $$\varphi = -2\pi\,\Phi_{int}/\Phi_o \quad \text{and} \quad V_J = -\dot{\Phi}_{int}$$

Finally, the screening equation relates the applied to the enclosed flux

(5) $$\Phi_{ext} = \Phi_{int} - L_R I_c$$

Below the normalized notation introduced in table I will be used. Then the SQUID loop equation may be written

External flux	Enclosed flux	Squid parameters	Tank circuit parameters
$\alpha_{ext} = 2\pi\Phi_{ext}/\Phi_o$	$\beta_{int} = 2\pi\Phi_{int}/\Phi_o$	$\gamma = 2\pi L_R I_o/\Phi_o$	$\omega_o = (LC)^{-\frac{1}{2}}$
$\alpha_1 = 2\pi\Phi_1/\Phi_o$	β_1	$\tau = GL_R$	$Q_o = \omega_o L/r$
$\alpha_2 = 2\pi\Phi_2/\Phi_o$	β_2	$\tau_c^2 = C_J L_R$	$Z_o = (L/C)^{\frac{1}{2}}$
$\alpha = 2\pi MI/\Phi_o$	β	ε	
$\alpha_o = 2\pi\Phi_x/\Phi_o$	β_o	$k^2 = M^2/LL_R$	

Table I Definition of notation

(6) $$\tau_c^2\, \ddot{\beta}_{int} + \tau(1+\varepsilon\cos\beta_{int})\dot{\beta}_{int} + \beta_{int} + \gamma\sin\beta_{int} = \alpha_{ext}$$

Under the assumption that only the component at ω is significant in the tank circuit the equation

(7) $$Q_o\ddot{\alpha}(t) + \omega_o\dot{\alpha}(t) + \omega_o^2 Q_o\alpha(t) = \omega^2 k^2 Q_o\big(\beta(t) - \alpha(t)\big)$$

results. Here, $\alpha(t) = \alpha\cos\omega t$ and $\beta(t) = \beta\sin(\omega t - \psi)$. From eq. (7) it follows that at resonance, $\omega = \omega_o/\sqrt{1-k^2}$,

(8) $$\psi = 0 \quad \text{and} \quad \alpha = -k^2 Q_o\beta/\sqrt{1-k^2}$$

With an externally applied flux of the form (1) eq. (6) may be solved to second order by eq. (2) using the method of the harmonic balance. The term $\sin\beta_{int}$ is expanded by repeated use of the addition formulae and the Fourier-Bessel expansions and the term $\cos\beta_{int}\cdot\dot{\beta}_{int}$ is found by differentiation. Even though the expansions are only made to second order in β, β_1, β_2, the calculations are tedious and will not be accounted for in detail. All terms of the expansion contain products of three Bessel functions, $J_n(x)$, of arguments $x = \beta, \beta_1, \beta_2$. Since no combinations of the form $J_o\ J_o\ J_2$ results only Bessel functions of order $n = 0, 1$ are maintained.

The result expresses the external flux amplitudes in terms of the internal amplitudes. In an abbreviated form the result is

$$\alpha_1 = \beta_1\left[\left(R_1(\beta,\beta_2,\beta_o) + \beta_2^2 P_1(\beta_1,\beta_2,\beta_o)\right)^2 + \left(Q_1(\beta,\beta_2,\beta_o) - \beta_2^2 S_1(\beta_1,\beta_2,\beta_o)\right)^2\right]^{\frac{1}{2}} \tag{9}$$

$$\alpha_2 = \beta_2\left[\left(R_2(\beta,\beta_1,\beta_o) - \beta_1^2 P_2(\beta_1,\beta_2,\beta_o)\right)^2 + \left(Q_2(\beta,\beta_1,\beta_o) - \beta_1^2 S_2(\beta_1,\beta_2,\beta_o)\right)^2\right]^{\frac{1}{2}} \tag{10}$$

$$\beta = \frac{(\gamma^2 + (\varepsilon\omega\tau)^2)^{\frac{1}{2}} J_o(\beta)\beta_1\beta_2 \sin\beta_o}{2\left[R(\beta_1,\beta_2,\beta_o)^2 + Q(\beta_1,\beta_2,\beta_o)^2\right]^{\frac{1}{2}}} \tag{11}$$

$$\alpha_o = \beta_o + \gamma\, T(\beta,\beta_1,\beta_2)\sin\beta_o \tag{12}$$

The current in the tank circuit is related to the mixing product, β, by eq. (8). The functions R, Q, P, and S are defined in Appendix I. Apart from the dependence explicitly displayed above they depend also on the circuit parameters and the frequencies.

In order to determine the phases it should be noted, firstly, that the phase of the tank current ($I(t) = I \cos\omega t$) has been specified and must be used as reference in the calculation; secondly, that the sign of β follows the sign of $\sin\beta_o$. Then it follows that the phase difference, $\psi_1 - \theta_1$, is determined by

$$\psi_1 - \theta_1 = \tan^{-1}\left\{\frac{Q_1(\beta,\beta_2,\beta_o) - \beta_2^2 S_1(\beta_1,\beta_2,\beta_o)}{R_1(\beta,\beta_2,\beta_o) + \beta_2^2 P_1(\beta_1,\beta_2,\beta_o)}\right\} \tag{13}$$

and similarly

$$\psi_2 - \theta_2 = \tan^{-1}\left\{\frac{Q_2(\beta,\beta_1,\beta_o) - \beta_1^2 S_2(\beta_1,\beta_2,\beta_o)}{R_2(\beta,\beta_1,\beta_o) - \beta_1^2 P_2(\beta_1,\beta_2,\beta_o)}\right\} \tag{14}$$

where finally,

$$\psi_2 - \psi_1 = \tan^{-1}\left\{\frac{\gamma R(\beta_1,\beta_2,\beta_o) + \varepsilon\omega\tau Q(\beta_1,\beta_2,\beta_o)}{\gamma Q(\beta_1,\beta_2,\beta_o) - \varepsilon\omega\tau R(\beta_1,\beta_2,\beta_o)}\right\} \tag{15}$$

In each case the numerator and the denominator of the argument in $\tan^{-1}$ has the same sign as sine and cosine, respectively, of the corresponding phase difference. Due to the particular initial condition chosen in this calculation the difference, $\Delta\theta = \theta_2 - \theta_1$, between the phases of the ac bias flux components is dependent on the quasi-static flux β_o (or α_o). Noting that

$$(16) \qquad \Delta\theta = (\psi_1-\theta_1) - (\psi_2-\theta_2) + (\psi_2-\psi_1)$$

this dependence is determined by eq.s (13) to (15).

The tank voltage, $V_{IF}(t)$, is determined by

$$(17) \qquad V_{IF}(t) = -\omega L(1-k^2)\,\frac{\Phi_o}{2\pi M}\,\alpha\,\sin\omega t = \omega k Q_o\sqrt{\frac{L(1-k^2)}{L_R}}\,\frac{\Phi_o}{2\pi}\beta\sin\omega t$$

where the latter form is found eliminating M and using eq.(8).

NUMERICAL EVALUATION

The amplitude of the IF-voltage is according to eq. (17) proportional to β. In order to determine the dependence on the quasi-static magnetic flux eq.s (9) to (12) must be solved in terms of the external flux amplitudes. The equations form a set of coupled non-linear algebraic equations and numerical methods must be used for their evaluation.

The calculation of the α_o-dependence of V_{IF} is made in five steps for fixed values of the circuit parameters:

1. The ac bias amplitudes, α_1 and α_2, are chosen such that the internal amplitudes, β_1 and β_2, for no value of β_o exceeds one. It is assumed that the second order result is a reasonable approximation even for $\beta_1, \beta_2 \simeq 1$ and further that β_1 and β_2 does not depend significantly on β.

2. Once α_1 and α_2 are chosen the coupled equation (9) and (10)

are solved for β_1 and β_2 as functions of β_o.

3. For each value of β_o the solutions $\beta_1(\beta_o,\alpha_1,\alpha_2)$ and $\beta_2(\beta_o,\alpha_1,\alpha_2)$ are inserted in eq. (11) and β is determined.

4. Finally, the corresponding value of α_o is found from eq. (12). The α_o vs. β_o relation is certainly single valued if $\gamma < 1$ and if β_1, β_2 are both single valued functions of α_1, α_2.

5. The IF-voltage is calculated from eq. (17) and plotted as a function of α_o (or Φ_x/Φ_o).

When β, β_1, and β_2 are determined as functions of β_o (or α_o) the phase difference $\Delta\theta$ may be calculated from eq.s (13) to (16).

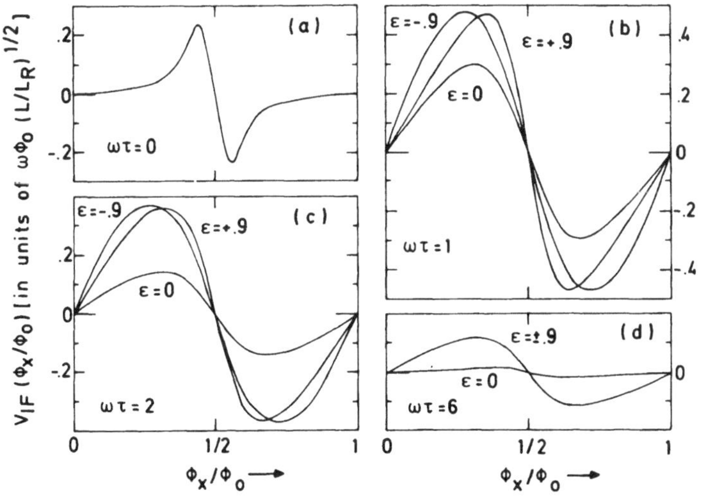

Fig. 2 IF-Voltage as a function of quasi-static external flux. Parameter values not given in the figure: $\gamma = 0.75$, $\tau_c/\tau = 1$, $Q_o = 100$, $k^2Q_o = 1$, $\omega_1/\omega = 20$.

The IF-voltage is shown in Fig. 2 as a function of Φ_x/Φ_o with $\omega\tau$ and ε as parameters. The voltage is an almost sinusoidal function of Φ_x for $\omega\tau > 0$, slightly skewed due to the screening effect. The voltage does depend on the cosine phi amplitude, ε, but not strongly on its sign, and finally the voltage is seen to decrease as $\omega\tau$ increases. It is important to notice, however, that V_{IF} in Fig. 2 is measured in units of ω. The strong distortion in the limiting case, $\omega\tau \to 0$ (Fig. 2a), reflects the dependence of β_1 and β_2 on Φ_x, i.e., is caused by a strong ac screening. If $\omega_1/\omega = 20$ and $\tau_c/\tau = 1$ as in Fig. 2 this phenomenon becomes unimportant if just $\omega\tau \gtrsim 0.2$.

The dependence of the phase difference, $\Delta\theta$, on Φ_x/Φ_o is shown in Fig. 3 with the same parameter values as in Fig. 2b. It is evident that $\Delta\theta$ is virtually independent of Φ_x in this case (the peak-to-peak modulation in Fig. 3c is less than one tenth of a degree). This result is important and is basic to the

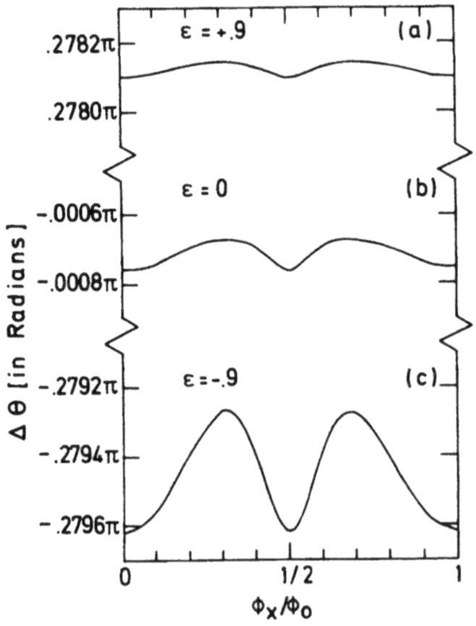

Fig. 3 Phase difference $\Delta\theta = \theta_2 - \theta_1$ as a function of quasi-static external flux. Parameters as in Fig. 2b.

flux-locked mixing magnetometer suggested in the following section.

The maximum IF-voltage amplitude is a function of $\omega\tau$ as shown in Fig. 4. When $V_{IF,max}$ is measured in units of ω the amplitude is seen to approach zero as $\omega\tau$ is increased, slowly if $\varepsilon \neq 0$, and more rapidly if $\varepsilon = 0$. At low frequencies (the dotted curves in Fig. 4) the behaviour is complicated. If $\omega_1\tau_c$, $\omega_2\tau_c \lesssim 1$ the V_{IF} vs. Φ_x dependence becomes hysteretic due to the existence of multiple solutions to the eq.s (9) and (10). This has been consistently the case for a variety of circuit parameters (7). Hence, only ac bias frequencies above the $L_R C_J$ resonance frequency of the SQUID loop should be used implying that the shunt capacitance, C_J, is an indispensable

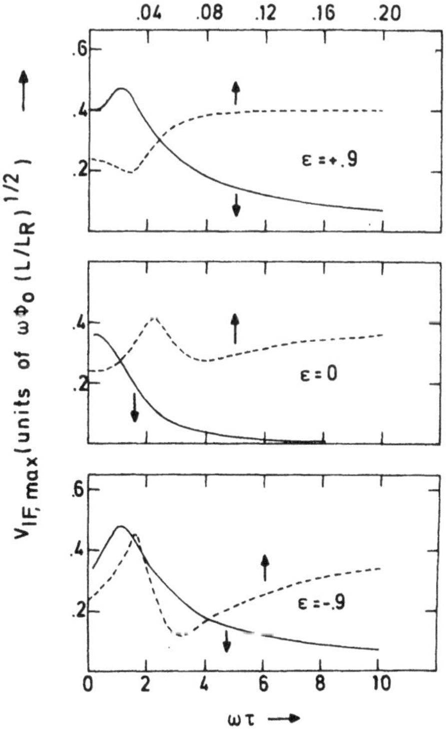

Fig. 4 Maximum IF-voltage as a function of $\omega\tau$. Full curves: lower scale; dotted curves: upper scale. Parameter values as in Fig. 2.

component in a well behaved mixing magnetometer.

Due to the presence of a shunt capacitance the ac screening in the loop becomes more and more efficient as the bias frequencies are increased. The high frequency current components by-pass the non-linear elements. Accordingly, it is expected that larger ac bias amplitudes are needed in order to preserve a significant response. That this is the case is shown in Fig. 5, where the ac bias amplitudes necessary to maintain $\beta_1, \beta_2 \lesssim 1$ are plotted as functions of $\omega\tau$. It is not possible on the scale of this figure to distinguish Φ_1 from Φ_2 and their dependence on ε can also not be seen. Part of the large ac screening currents will dissipate energy in the shunt conductance and this will eventually set an upper limit to the useful frequency range. It may also happen, however, that the shunted junction model breaks down before this limit is reached.

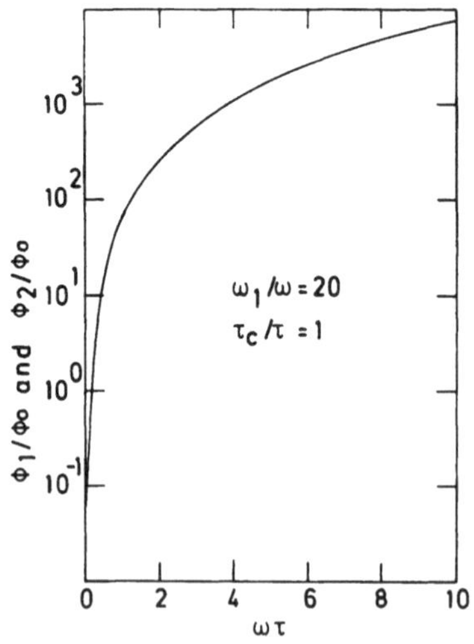

Fig. 5 Externally applied ac bias flux necessary to maintain $\beta_1, \beta_2 \lesssim 1$ as a function of $\omega\tau$. It is not possible to distinguish Φ_1 from Φ_2 on the scale of this figure.

DISCUSSION AND CONCLUSION

It was already pointed out that the shunt capacitance of the weak link is necessary in order to avoid a hysteretic response of the magnetometer. The sandwich tunnel junction is then a favourable weak link structure. The choice of a tunnel junction may be advantageous also in another respect, since a device with a low critical current is fairly easily realized even at temperatures well below the transition temperature. The thermal fluctuations may be the major cause to low frequency noise (8). The device is, however, becoming less sensitive to temperature variations as the reduced temperature is decreased and hence the device noise is minimized.

The presence of the shunt capacitance does not deteriorate the high frequency response provided ε is non-zero. This follows from eq.s (9) to (11). Letting $\omega\tau \to \infty$ with $\omega_1, \omega_2 > \omega$ it is found that β_1 and β_2 are virtually independent of β_o and the condition $\beta_1, \beta_2 \simeq 1$ may be satisfied if just α_1 and α_2 are increased as shown in Fig. 5.
Then

(18) $$\beta \simeq \tfrac{1}{2}|\varepsilon| \frac{\tau}{\omega\tau_c^2} \sin\beta_o \qquad \text{for} \quad \omega^2\tau_c^2 \gg \omega\tau > 1$$

The limiting amplitude of the tank voltage, V_{IF}, is from eq. (17)

(19) $$V_{IF} \simeq \tfrac{1}{2}|\varepsilon| \left(\frac{\tau}{\tau_c}\right)^2 \sqrt{\frac{L(1-k^2)}{L_R}}\, kQ_o \frac{\Phi_o}{2\pi\tau} \sin\beta_o,$$

which is independent of ω. Considered as a function of quasi-static flux β_o (or α_o) the peak-to-peak modulation is

(20) $$\Delta V_{IF} \simeq |\varepsilon| \left(\frac{\tau}{\tau_c}\right)^2 \sqrt{\frac{L(1-k^2)}{L_R}}\, kQ_o \frac{\Phi_o}{2\pi\tau}$$

This result is identical to the expression for the limiting modulation depth which may be derived for the conventional ac-SQUID operated in the impedance mode (5) under similar conditions: $\omega\tau \to \infty$, enclosed ac bias flux $\Phi_1 \simeq \Phi_o/2\pi$, and the tank circuit biased at resonance (the shunt capacitance is readily included in the calculation of ref. (5)). The relative modulation depth of the conventional ac-SQUID is, however,

$$(21)\qquad \left|\frac{\Delta V_T}{V_T}\right| \simeq 2|\varepsilon|\frac{\tau}{\tau_c}\,\frac{k^2 Q_o}{(\omega\tau_c)^3} \quad .$$

Whereas the relative modulation depth eq. (21) of the impedance SQUID clearly goes to zero as ω is increased it is not straight forward to estimate the limit of eq. (20). Although not explicitly dependent on frequency both the L and the Q_o of the tank circuit may need readjustment in order to realize a frequency change. Consequently, it must be specified which quantities are to be kept constant under a frequency variation. Specifying, e.g., that Q_o and the $Z_o = \omega L$ are constants the result is that $\Delta V_{IF} \propto \omega^{-\frac{1}{2}}$. However, even though the response under such conditions may approach zero the modulation depth in the mixing mode is always 100%.

In conclusion, a flux-locked operation of the mixing magnetometer (MIXMAG) is suggested. The basic idea is shown schematic in Fig. 6. The output power from the two ac sources are

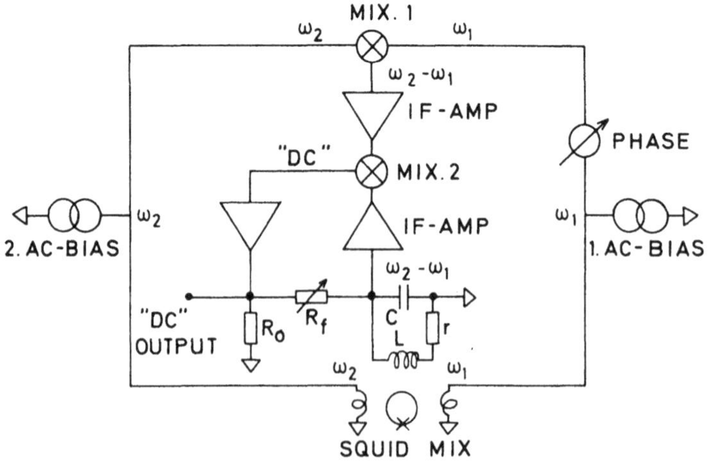

Fig. 6 Flux-locked mixing magnetometer. Lumped circuit model.

both devided between two arms, a signal arm and a reference arm. According to the results in Fig. 3 the IF-frequency components generated in the SQUID and in the mixer 1 are phase coherent independent of the quasi-static flux at the SQUID. The two IF-signals are then mixed to "dc" in the mixer 2 and the output may be maximized by means of the phase shifter in the high frequency arm. By closing the feed back loop via a resistor, R_f, the magnetometer will lock to one of the points, $\Phi_x = \frac{n}{2}\Phi_o$, since the SQUID output has a phaseshift of π at these points. An output voltage proportional to the quasi-static external flux is located across the resistor, R_o. The MIXMAG is sensitive to changes in the external flux at frequencies ranging from dc to an upper cut-off determined by the Q_o of the tank circuit. If the resonance frequency is sufficiently high (e.g., 500 MHz) and the Q_o is moderate ($\sim$ 100) the upper cut-off will be in the MHz range.

APPENDIX I

Definition of the functions T, R, Q, S, P used in the eq.s (9) to (15) in the text.

(A1) $R(\beta_1, \beta_2, \beta_o) = 1-\omega^2\tau_c^2+\gamma J_o(\beta_1)J_o(\beta_2)\cos\beta_o$

(A2) $Q(\beta_1, \beta_2, \beta_o) = \omega\tau(1+\varepsilon J_o(\beta_1)J_o(\beta_2)\cos\beta_o)+\frac{k^2Q_o}{\sqrt{1-k^2}}$

(A3) $R_1(\beta, \beta_2, \beta_o) = 1-\omega_1^2\tau_c^2+\gamma J_o(\beta)J_o(\beta_2)\cos\beta_o$

(A4) $Q_1(\beta, \beta_2, \beta_o) = \omega_1\tau(1+\varepsilon J_o(\beta)J_o(\beta_2)\cos\beta_o)$

(A5) $R_2(\beta, \beta_1, \beta_o) = 1 -\omega_2^2\tau_c^2+\gamma J_o(\beta)J_o(\beta_1)\cos\beta_o$

(A6) $Q_2(\beta, \beta_1, \beta_o) = \omega_2\tau(1+\varepsilon J_o(\beta)J_o(\beta_1)\cos\beta_o)$

(A7) $$P_1(\beta_1, \beta_2, \beta_o) = \frac{\varepsilon\omega_1\tau(\gamma Q-\varepsilon\omega\tau R)-\gamma(\gamma R+\varepsilon\omega\tau Q)}{4(R^2 + Q^2)} \sin^2\beta_o$$

(A8) $$S_1(\beta_1, \beta_2, \beta_o) = \frac{\varepsilon\omega_1\tau(\gamma R+\varepsilon\omega\tau Q)+\gamma(\gamma Q-\varepsilon\omega\tau R)}{4(R^2 + Q^2)} \sin^2\beta_o$$

(A9) $$P_2(\beta_1, \beta_2, \beta_o) = \frac{\varepsilon\omega_2\tau(\gamma Q-\varepsilon\omega\tau R)+\gamma(\gamma R+\varepsilon\omega\tau Q)}{4(R^2 + Q^2)} \sin^2\beta_o$$

(A10) $$S_2(\beta_1, \beta_2, \beta_o) = \frac{\varepsilon\omega_2\tau(\gamma R+\varepsilon\omega\tau Q)-\gamma(\gamma Q-\varepsilon\omega\tau R)}{4(R^2 + Q^2)} \sin^2\beta_o$$

(A11) $$T(\beta, \beta_1, \beta_2) = J_o(\beta)J_o(\beta_1)J_o(\beta_2)$$

In eq.s (A7) to (A10) the quantities R and Q are the functions given by eq.s (A1) and (A2) respectively. The zero order Bessel functions should be understood as their second order expansions: $J_o(x)\sim 1 - \frac{1}{4}x^2$. Numerical evaluations have shown that if $\beta_1, \beta_2 \lesssim 1$ then $\beta << 1$, or, $J_o(\beta) \sim 1$. Consequently, the mixing amplitude, β, does not appear in eq.s (9) and (10) of the text. Once β_1 and β_2 are determined from these eq.s, β is given directly by eq. (11).

REFERENCES

(1) Nisenoff, M., Wolf, S.: Observation of a cosφ term in the current-phase relation for "Dayem"-type weak link contained in an rf-biased superconducting quantum interference device. Phys. Rev. B 12, 1712-1714 (1975)

(2) Rifkin, R., Vincent, D.A., Deaver, B.S., Hansma, P.K.: rf SQUID's in the nonhysteretic mode: Detailed comparison of theory and experiment. J. Appl. Phys. 47, 2645-2650 (1976)

(3) Hansma, P.K.: Observability of Josephson pair - quasiparticle interference in superconducting interferometers. Phys. Rev. B 12, 1707-1711 (1975)

(4) Erné, S.N., Hahlbohm, H.-D., Lübbig, H.: Theory of the rf-biased superconducting quantum interference device for the nonhysteretic regime. (Unpublished).

(5) Soerensen, O.H.: An analysis of the ac-SQUID with low inductance and low critical current. J. Appl. Phys. (to be published).

(6) Zimmerman, J.E.: Heterodyne detection with superconducting point contacts and enhanced heterodyne signals from tightly coupled contacts. J. Appl. Phys. 41, 1589-1593 (1970).

(7) It should be noted that a detailed investigation of the hysteretic regions is not possible within the second order theory presented here.

(8) Clarke, J., Goubau, W.M., Ketchen, M.B.: An improved DC-SQUID. Proceedings of LT 14 vol. 4, 214-217 (1975) ed. by M. Krusius and M. Vuorio, North-Holland Publ.Co.

HIGH FREQUENCY PROPERTIES OF STABLE Nb-Nb OXIDE-Pb JOSEPHSON TUNNEL JUNCTIONS*

T. F. Finnegan, J. Wilson†, and J. Toots
Institute for Basic Standards, National Bureau of Standards, Washington, D.C. 20234, USA

INTRODUCTION

An important factor in the widespread application of Josephson junctions is the development of stable reproducible devices with well-characterized device parameters. A number of techniques for fabricating "stable" Nb-Nb oxide-Pb tunnel junctions have been reported in the literature (1-8). DC tunnelling measurements were used to infer certain properties of the junctions and to monitor device stability, but no significant effort was made to characterize the microwave properties of these devices. Published results also suggest that the stability of Nb-Nb oxide-Pb junctions depends strongly on the detailed method of fabrication. The major source of variation in junction properties is the rich diversity of Nb oxides and suboxides which can exist in the junction barrier region.

In this paper, we describe a relatively simple, straightforward method for fabricating Nb-Nb oxide-Pb devices in which all stages of junction preparation are carried out without opening the vacuum chamber. Properties of the rf sputtered Nb films, such as the low temperature resistivity ratio, and superconducting transition temperature, are presented. DC tunnelling measurements were used to infer certain properties of the junctions and also to monitor device stability over periods of more than two years. Various microwave properties of our Nb-Nb oxide-Pb junctions were measured at microwave frequencies (4 - 10 GHz) and are compared with similar results for Pb-Pb oxide-Pb junctions. Our paper concludes with a discussion of the future prospects for using Nb-Nb oxide-Pb junctions as a replacement or alternative for Pb-Pb oxide-Pb junctions in various high frequency applications.

DEVICE DESIGN AND FABRICATION

Microwave Circuit Design Considerations

A major factor in the design of thin-film Josephson junctions for microwave-related applications is the coupling characteristics of the device. The use of microstrip techniques to couple to tunnel junctions has proved practical for Pb-Pb oxide-Pb junctions (9,10). These techniques can be applied to Nb-Nb oxide-Pb junctions; however, additional considerations (such as the limited resolution with which a film can be sputter-deposited through a mask, the need to deposit the Nb film as the lower electrode, and the large dielectric constant of the predominant niobium oxide Nb_2O_5) strongly influence the choice of junction geometry. Internally-resonant Nb-Nb oxide-Pb junction structures are about three times smaller than the corresponding Pb-Pb oxide-Pb junction structures. Consequently, cross-type junction geometries were used exclusively in this work.

The principal geometry which we investigated is shown in Fig. 1. (The dimensions of the device substrate are 2.54 x 2.54 x 0.12 cm.) For Nb-Pb devices, the microstrip coupling film was of Pb and is the vertical strip at the top of Fig. 1. The narrower film, colinear with the Pb coupling

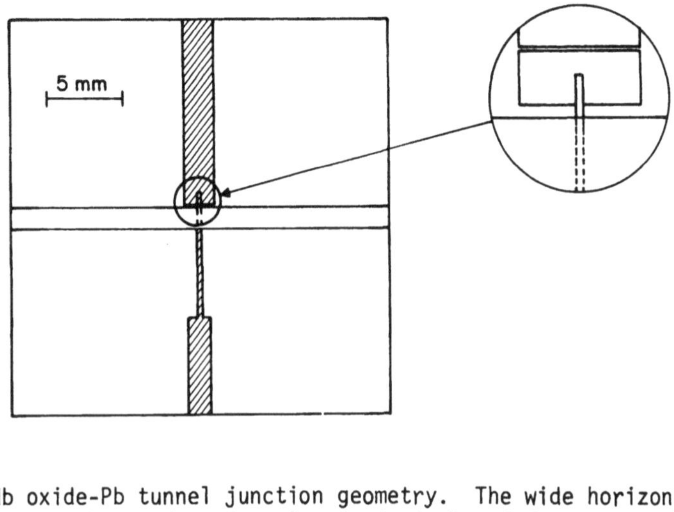

Fig. 1 Nb-Nb oxide-Pb tunnel junction geometry. The wide horizontal cross-strip is upper Pb electrode. Junction is dc isolated from external strip-line by narrow scratch (~10 μm wide) as indicated in the insert.

strip, was niobium and its width was typically 0.37 mm at the junction. The upper junction electrode (the horizontal cross-strip in Fig. 1) was Pb, and its width determined the fundamental resonance frequency ν_R of the junction. This width varied between 0.75 and 1.85 mm. As discussed elsewhere (10), the device was designed so that the Pb coupling strip formed a 50 Ω microstrip transmission line when the substrate was mounted in a microwave-integrated-circuit package. The microwave properties of the junction device were readily measured via well known techniques. Some limited measurements were also made with a second junction geometry, having smaller dimensions, which is discussed in a later section.

Nb-Nb oxide-Pb Junctions

The thin-film Nb-Pb junctions were fabricated in a commercial oil-diffusion-pumped high vacuum system. As many as six glass substrates (Corning 7059) could be processed in a pump-down cycle. Sequential sputtering and evaporating were accomplished by positioning the substrate above the desired source via a high-vacuum rotary feedthrough. The geometry of each deposited film was determined by a mask located directly above the respective source. During deposition, the substrate was maintained in nominal contact with the mask. The Nb films were rf sputtered at a rate of about 18 nm/min in a flowing, high-purity Ar atmosphere. Chamber pressure during sputtering was maintained at 1 Pa (7×10^{-3} torr). The rf power level required was about 700 watts. The substrate was at nominal ambient temperature at the start of the deposition; however, no effort was made to control substrate temperature during sputtering. Table I contains a representative list of resistivity ratios measured between 295 K and 10 K, ρ_{295}/ρ_{10}, and superconducting transition temperatures T_c, for films with thickness between 60 and 2000 nm. Specimen 4 was an actual junction electrode. Our Nb film electrodes were typically about 350 nm thick; these had low-temperature mean free paths ℓ (deduced from resistivity measurements) of about 5 nm.

Each Nb electrode was subsequently oxidized in the vacuum chamber via a dc glow discharge in a flowing, high-purity O_2 atmosphere maintained at a steady-state pressure of 10 Pa (7×10^{-2} torr). During oxidation, the Nb film electrode was positioned about 15 cm above the glow discharge

TABLE I Properties of Nb films rf sputtered through Be-Cu masks.

Sample	ρ_{295}/ρ_{10}	t(nm)	T_c(K)
1	1.94	60	7.80
2	2.83	100	8.80
3	2.81	280	8.76
4	2.64	350	8.60
5	3.24	1000	9.04
6	4.11	2000	9.10

assembly, which was similar to one described by Payley et al. (11). The applied dc voltage was 750 V. The length of oxidation time determined the Josephson current density. Junction oxidation times were between 1 and 8 min. Nb-Pb junctions were fabricated with current densities J_1 between 10^{-2} and 3×10^2 A/cm^2. The chamber was then evacuated to a background pressure below 10^{-5} Pa (10^{-7} torr) and the upper Pb electrodes (about 200 nm thick) were evaporated. Additional thin-film structures (such as microstrip coupling structures) were usually deposited after completing the junction. All junction fabrication steps were thus accomplished without exposing the device to the atmosphere. Furthermore, no sophisticated photolithographic processing was necessary. Immediately after backfilling the vacuum system with N_2 gas, each device was overcoated with a thin layer of (KPR) photoresist several hundred nanometers thick. This crucial step not only protects the Nb and Pb films from the atmosphere but also facilitates the connection of dc bias leads to the device with indium based solders. The size of the cross-type junctions made via this technique is limited by the resolution achievable in sputtering Nb through a mask. For Nb films, a film width of about 0.2 mm was the lower practical limit. For Pb films, the corresponding resolution is at least ten times better.

Pb-Pb oxide-Pb Junctions

Thin-film Pb-Pb junctions were also fabricated in a commercial oil-diffusion-pumped high vacuum system. The Pb electrodes (~200 nm thick) were thermally evaporated at a rate of about 20 nm/sec. Precision bimetallic masks, prepared in our laboratory via photoresist techniques, were used to define the film geometries. The lower Pb electrode was thermally oxidized in a static atmosphere of high purity O_2 at temperatures near 50°C. Junction oxidation times were between 15 min. and 60 min. Pb-Pb junctions with J_1 between 10^{-2} and 10^2 A/cm^2 were readily achieved. Completed junctions were overcoated with photoresist and routinely stored in liquid N_2. The devices could be temperature cycled many times to room temperature for short periods of time without degradation of junction properties.

DC TUNNELLING CHARACTERISTICS

Conductance Measurements

A typical current-voltage (I-V) characteristic for one of our Nb-Pb junctions is shown in Fig. 2. The dimensions of this junction were 0.77 x 0.38 mm. The reduced dc Josephson current resulted from self-shielding effects associated with λ_J (the Josephson penetration depth which was significantly less than the junction length L.) Below 5 K, small but distinct anomalies (which are characteristic of Nb-Pb junctions) were observed in the I-V data; namely, an excess quasiparticle current above the Pb energy gap Δ_{Pb} and an extra conductance minimum just above the

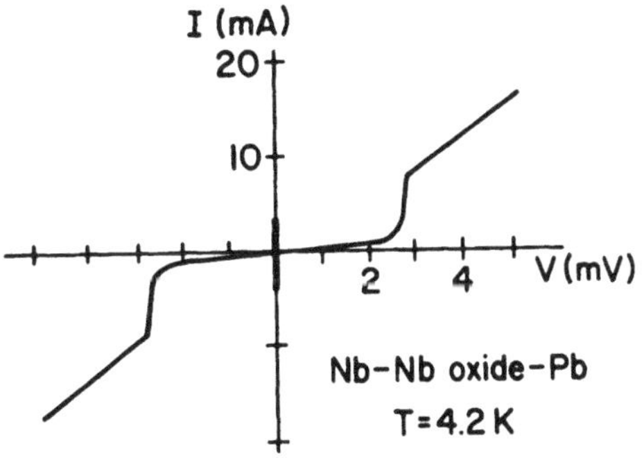

Fig. 2 Current-voltage characteristic of typical Nb-Nb oxide-Pb tunnel junction (L/λ_J = 2.8).

total gap voltage $\Delta_{Pb} + \Delta_{Nb}$. The latter feature first appeared near 5K and became more pronounced as the temperature was lowered. However, no negative resistance region was observed even for temperatures as low as 1.6 K. This additional conductance minimum or knee effect was not observed above about 5 K. The explanation for this anomalous tunnelling behavior is usually assumed to be a normal or gapless layer at the Nb-Nb oxide interface.

The dependence of I_M, the maximum dc Josephson current, on an externally applied magnetic field H is shown in Fig. 3. The field was applied perpendicular to the junction width W which was approximately equal to λ_J. The calculated amplitudes of the first four maxima were within about 10% of the fitted mean for I_1. The London penetration depths λ_L for the Pb and Nb film electrodes were deduced from I_M vs H measurements of Pb-Pb and Nb-Pb junctions. We found $\lambda_L = 43 \pm 3$ nm for the Pb films and 90 ± 6 nm for the Nb films. The large λ_L for Nb was due to the short mean free path ℓ of the films. When values of λ_L and ξ_0 for <u>pure</u> Nb (12) were combined with our measured values of ℓ to estimate $\lambda_L(\ell)$, it was in excellent agreement with the λ_L observed for the films.

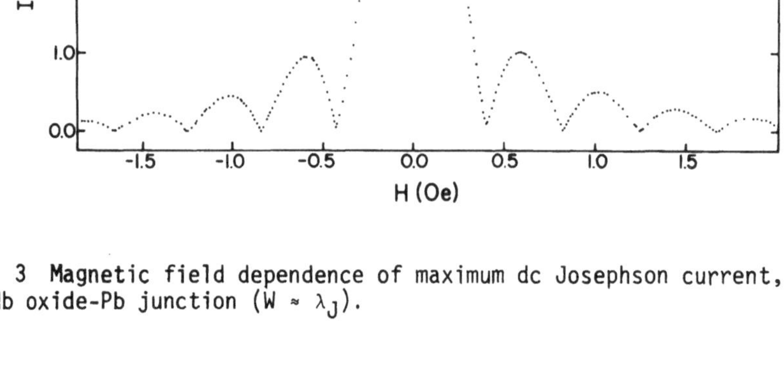

Fig. 3 Magnetic field dependence of maximum dc Josephson current, I_M, for Nb-Nb oxide-Pb junction ($W \approx \lambda_J$).

AGING CHARACTERISTICS

Nb-Pb devices have been stored routinely at room temperature (in the atmosphere) for periods of more than two years. For example, twelve devices with J_1 in the range 1-100 A/cm^2 were tested several days after fabrication and again after a two year interval. The normal state resistance R_N in each case was the same within the resolution of the measurements (about 10%). The devices could be repeatedly cycled between room temperature and cryogenic temperatures with no noticeable change in properties. In contrast, our Pb-Pb junctions had to be stored at low temperatures (usually in boiling liquid N_2)and cycled to room temperature for only short periods of time. With careful handling, individual Pb-Pb devices could be readily stored, shipped, and reused for periods of longer than two years.

MICROWAVE PROPERTIES

Junction Self-Resonances

A preliminary study of the high frequency properties of geometrically self-resonant Nb-Pb junctions was made. In particular, the temperature dependence of ν_R (the fundamental resonance frequency of the junction) and the resonator Q (quality factor) were experimentally determined for several devices by measuring the response of the microstripline-coupled junction to externally applied microwave fields. A resonance curve was measured at each of a series of temperatures by using the junction itself as a microwave detector (the I-V curve was monitored via an ac swept bias supply) and by applying the microwave power via an externally-levelled backward wave oscillator (at 4-8 GHz or 8-12.4 GHz) in series with a calibrated attenuator. The observed temperature-dependence of ν_R for one of the junctions is shown in Fig. 4. The theoretical curve was fitted to the data by assuming the explicit temperature dependence of the penetration depth predicted by the two fluid model. (The observed T_c's of the Nb and Pb film electrodes were 8.6 K and 7.15 K respectively.) Although, overall agreement is very good, significant deviations from ideal behavior are clearly discernible in the region just below 5 K. The data were also used to determine $\lambda_{eff}(T)$ for the Nb film since the functional

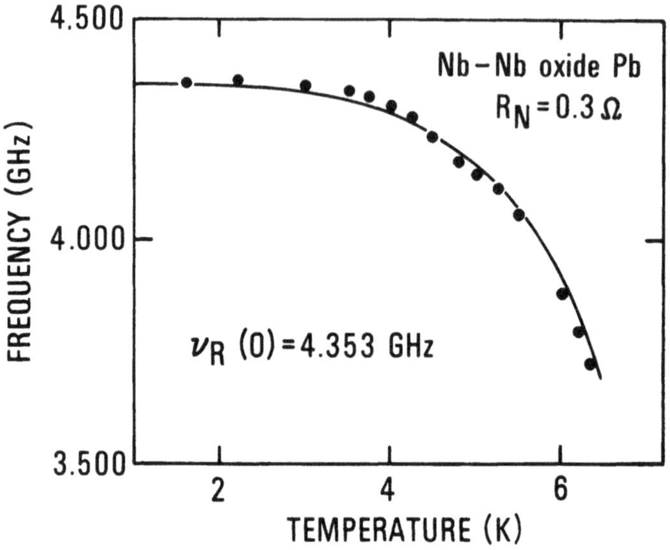

Fig. 4 Temperature dependence of fundamental junction cavity resonance frequency ν_R. (L · W = 0.77 x 0.38 mm).

dependence of ν_R on the two electrode penetration depths is known and $\lambda_L(T)$ for our Pb films fits the Gorter-Casimir phenomenological theory very well in the range of interest. A comparison of $\lambda_{eff}(T)$ with the expected $\lambda_L(T)$ of the Nb film revealed a small but significant change between 5.5 and 6.0 K. As the temperature was lowered, the ratio λ_{eff}/λ_L exhibited an abrupt decrease at 5.5 K. This decrease is most readily explained as the onset of superconductivity in a boundary layer of normal metal (e.g. NbO) at the Nb-Nb oxide interface. In the normal state, this layer gives negligible shielding of the microwave fields, but when it becomes superconducting, the additional shielding results in a smaller effective λ. A detailed analysis of the temperature dependence of λ_{eff} in such a regime is beyond the scope of this paper. However, simple models for λ_{eff} can be readily constructed having the proximity-layer thickness t and penetration depth λ_0 as parameters. Our initial results suggest that if the low temperature value of λ_0 is comparable to λ for the Nb film, then t is about 8 nm for the particular junction whose data is shown plotted in Fig. 4. This type of model can thus be confirmed and evaluated quantitatively in experiments with junctions of the form superconductor-insulator-normal metal-superconductor where the normal layer is a very thin deposited film of Cu or Al, and it can probably be applied to Nb-Nb junction structures as well.

The ratio of oxide dielectric constant to dielectric thickness ε_r/ℓ_{ox} was determined from data for a large number of junctions with fundamental resonance frequencies ν_R between about 2 and 10 GHz and current densities J_1 between about 2 and 15 A/cm^2. The result was ε_r/ℓ_{ox} = 15.4 $\pm$ 1.2 nm^{-1}. A value for ε_r was deduced from this result by means of the ellipsometrically determined oxide thickness data of Basavaiah and Greiner (13). We obtain ε_r = 46 $\pm$ 5 for the Nb oxide barriers which were formed by a dc oxygen glow discharge technique.

The temperature dependence of the junction resonator Q for one of the Nb-Pb junctions is shown in Fig. 5. (These data correspond to the resonance frequency data shown in Fig. 4.) The data were fitted with a function of the form $Q^{-1} = Q_1^{-1}(T) + Q_2^{-1}(T)$. Q_1 (T) was the calculated surface

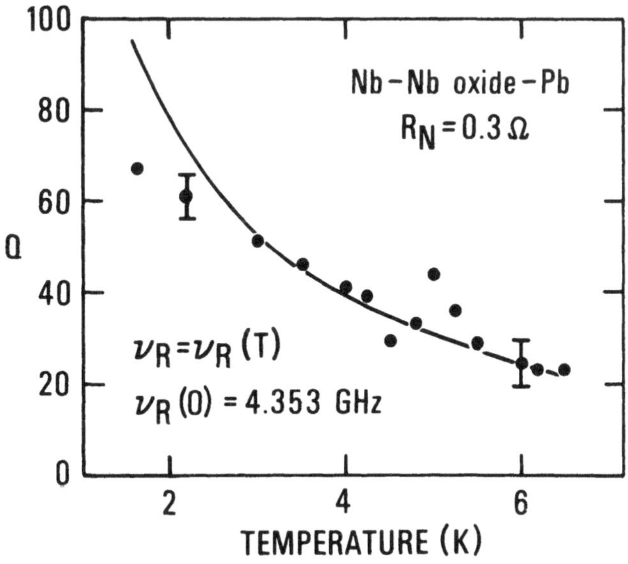

Fig. 5 Temperature dependence of quality factor of junction cavity resonator.

impedance loss factor for the Pb film, and $Q_2(T)$ was assumed to have a temperature dependence of the form $e^{\Delta/kT}$. A gap Δ of about 0.15 meV gave the fit shown in Fig. 5. The losses in the Nb-Pb junctions at these frequencies were substantially higher than those observed in comparable Pb-Pb junctions resonant at the same frequency. (Near 5 GHz, the Q's differed by a factor of three.) The Q data in Fig. 5 shows an anomaly near 5 K at about the same temperature that the additional conductance minimum or knee effect appears. The existance of substantial NbO at the Nb interface is rather certain. NbO in bulk form is a known superconductor with a T_c of about 1.3 K and a corresponding gap Δ of about 0.2 meV. The magnitude of the gap implied by our Q data agrees well with the value for bulk NbO. The Q's of Nb-Pb junctions resonantly excited at higher frequencies exhibited higher Q's. The microwave losses at a superconductor/proximity-layer/oxide interface are not well understood, and further work is needed.

Radiation Emission

The microwave radiation emitted by Nb-Pb junctions (biased on self-induced steps in an applied magnetic field) was observed with a sensitive super-heterodyne microwave receiver. In microstrip geometries of the type shown in Fig. 1, radiated powers up to 5 x 10^{-12} W were detected from junctions with resonance frequencies between 1.6 and 10 GHz. The power emitted by the Nb-Pb devices was consistently less than that emitted by comparable Pb-Pb junctions and was due primarily to the lower characteristic impedances of the Nb-Pb junctions. Microwave measurements were also made with microstrip-coupled devices having the basic geometry shown in Fig. 6. The cross-type junction dimensions were typically 0.34 x 0.2 mm with $\nu_R \sim 10$ GHz. The radiated power detected from one of these junctions ($R_N = 1\Omega$) coupled via microstrip and an external semi-rigid cable is shown in Fig. 7. Two sets of data, taken on different days, are shown. In each case, the junction was dc biased on the n = 1 self-induced step in an applied field of about 0.20 Oe. The Q of the junction resonator inferred from the junction response to applied microwave fields was about 65.

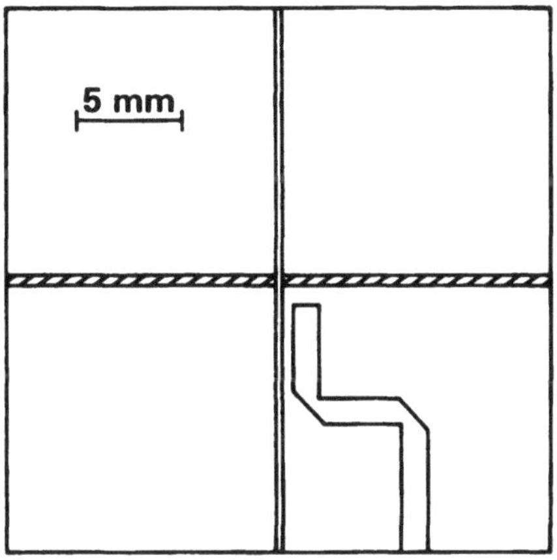

Fig. 6 Simple cross-type junction geometry with external microstrip coupling line.

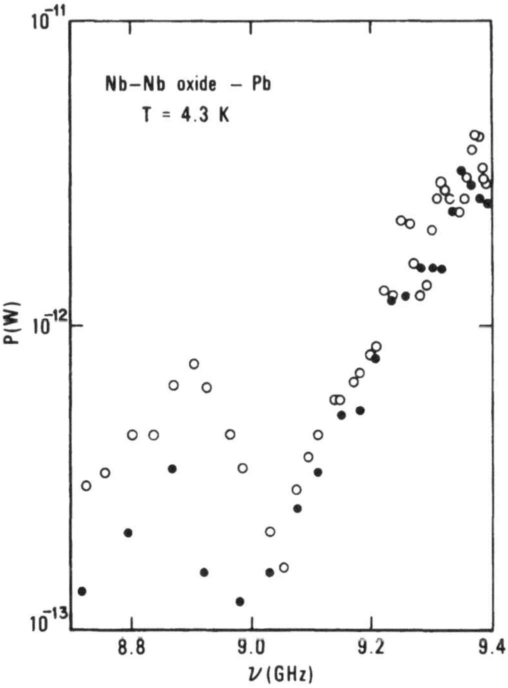

Fig. 7 Radiated power output from cross-type junction (L·W = 0.34 x 0.23 mm) biased on n = 1 self-induced Fiske step.

FUTURE PROSPECTS FOR Nb-Pb JUNCTIONS

Josephson tunnel junctions are presently used for such diverse applications as digital computer elements, high frequency parametric amplifiers, and precise frequency-to-voltage transducers for maintaining and defining the unit of voltage. In these applications, Pb or Pb alloy junctions have already been successfully used. The long-term device stability of Nb-Pb or Nb-Pb alloy junctions makes them an attractive replacement for Pb-Pb devices. However, the microwave properties of Nb-Pb junctions are significantly different from Pb-Pb junctions. Table II summarizes typical device parameters for both our Pb-Pb and Nb-Pb tunnel junctions.

TABLE II Measured junction parameters for Pb and Nb (ν_R = 10 GHz)

Parameter	Pb-Pb oxide-Pb	Nb-Nb oxide-Pb
$\lambda_L(\ell)$[nm]	43	90
ε_r	8	45
L(mm)	1.0	0.33
Q(T = 4K)	100	70
Q(T = 2K)	≥400	100

The parameter L is the junction length required to have a fundamental resonance frequency occur at 10 GHz. The large difference in L is indicative of the large difference in device capacitance for the two cases. In general, the larger capacitance of Nb-Pb junctions results in poorer microwave coupling to external transmission lines and a larger RC time constant for junctions of equal area. However, a more serious disadvantage of the Nb-Pb junctions for certain microwave-related applications may be the proximity layer at the Nb-Nb oxide interface. For example, in 2e/h voltage standard applications, precise phase coherence is required between the two superconductors forming the junction. A continuous proximity layer between the superconductors, with a low critical magnetic field H_c, could give rise to non-vertical 2e/h steps if the applied microwave field exceeds H_c. If the effective proximity layer is thick, it may seriously restrict the maximum dc voltage achievable to much smaller values than currently achievable with microstrip coupled Pb-Pb junctions (14). Similar considerations would of course apply to Nb-Nb junctions.

ACKNOWLEDGMENT

We wish to thank L.B. Holdeman for useful discussions concerning sputtered Nb films and valuable assistance with portions of the microwave measurements.

* Work supported in part by the United States Office of Naval Research.

†Present address: The BDM Corp., Vienna, Va. 22180.

REFERENCES

1. Nordman, J. E. : Thin-Film Josephson Junctions Using Getter-Sputtered Niobium. J. Appl. Phys. 40, 2111-2115 (1969).

2. Mullen, L. O. and Sullivan, D. B. : Fabrication of Tunnel Junctions of Niobium Films. J. Appl. Phys. 40, 2115-2117 (1969).

3. Graeffe, R. and Wiik, T. : Preparation of Josephson Junctions by Plasma Oxidation of Nb. J. Appl. Phys. 42, 2146-2147 (1971).

4. Greiner, J. H. : Josephson Tunneling Barriers by rf Sputter Etching in an Oxygen Plasma. J. Appl. Phys. 42, 5151-5155 (1971).

5. Schwidtal, K. : dc and ac Josephson Effect in Sputtered Nb-Nb Ox-Pb Junctions. J. Appl. Phys. 43, 202-208 (1972).

6. Hansma, P. K. : A New Method for Fabricating Niobium Oxide Barrier Josephson Junctions. J. Appl. Phys. 45, 1472-1473 (1974).

7. Rissman, P. and Palholmen, T. : Preparation and Lifetest of Niobium Josephson Junction Tunnel Diodes and Arrays. Solid-State Electron. 17, 611-615 (1974).

8. Owen, S. and Nordman, J. E. : Application of Integrated Circuit Technology to the Fabrication of Large Numbers of Niobium Based Josephson Junctions. IEEE Trans. Magn. MAG-11, 774-777 (1975).

9. Finnegan, T. F. , Wilson, J. , and Toots, J. : Interactions in Small Systems of Josephson Junctions at Microwave Frequencies. Revue Phys. Appl. 9, 199-205 (1974).

10. Finnegan, T. F., Wilson, J., and Toots, J. : Coupling Between Josephson Junctions and Microstriplines. IEEE Trans. Magn. MAG-11, 821-824 (1975).

11. Paley, S., Wilson, J., and Gayley, R. I. : Studies of tin Josephson junctions of anomalously high supercurrent. Phys. Rev. B 12, 157-162 (1975).

12. Varmazis C. and Strongin, M. : Inductive transition of niobium and tantalum in the 10 MHz range. I. Zero-field superconducting peretration depth. Phys. Rev. B 10, 1885-1895 (1974).

13. Basaviah, S. and Greiner, J. H. : Capacitance and ellipsometrically determined oxide thickness of Nb-oxide-Pb Josephson tunnel junctions. J. Appl. Phys. 47, 4201-4202 (1976).

14. Dziuba, R. F., Field, D. F., and Finnegan, T. F. : Cryogenic Voltage Comparator System for 2e/h Measurements. IEEE Trans. Instrum. Meas. IM-23, 264-267 (1974).

NOISE

NOISE LIMITATIONS OF rf SQUIDs

R. A. Buhrman
School of Applied and Engineering Physics, and The Materials Science Center, Cornell University, Ithaca, NY 14853, USA

I INTRODUCTION

In this paper a discussion is given of the noise sources and sensitivity limits of the rf biased SQUID. Both the low critical current "inductive" and the higher critical current "hysteretic" devices are considered. The intrinsic SQUID noise is treated by considering the internal flux response of a SQUID ring to an applied rf flux. While approximate, this approach yields the most direct picture of the nature of the intrinsic noise. Circuit optimization procedures for maximizing SQUID performance under various operating conditions are also outlined. Finally the effects of non-ideal weak link behavior upon SQUID operation are treated.

II SQUID RING RESPONSE

A basic understanding of the origins of the intrinsic noise of an rf SQUID can be obtained through study of the general nature of the response of an isolated SQUID ring to quasi-static and time varying applied magnetic fields. In this section we briefly sketch the essential details of this response, emphasizing those features which ultimately affect SQUID sensitivity. More complete discussions can be found elsewhere in the literature.[1-7] We begin by considering the case of very low frequency or dc applied flux. The general case of SQUID ring response to a combined rf and dc applied flux is then treated.

II.1 Quasi-Static Applied Flux

A standard schematic representation of an rf SQUID ring is shown in Fig. 1. A superconducting ring of inductance L_s is closed by a superconducting Josephson junction or weak link. The usual shunted junction model of the weak link is assumed, where a superconducting element with a current-phase relation $i_c f(\theta)$ is shunted by a resistor R_s and capacitor C_s. For proper SQUID operation it is necessary that the weak link be overdamped which requires $C_s < (\phi_o/2\pi i_c)/R_s^2$. In the discussion that follows we assume this is the case and that hence C_s can be neglected when determining SQUID ring response. Ideally the current-phase relation is sinusoidal although in practice this is often not the case.[7] Here we will assume $f(\theta) = \sin\theta$; the effect of non-sinusoidal $f(\theta)$ will be discussed later.

When an external flux ϕ_x is applied to the ring, the response of the internal flux ϕ is given by

$$1) \quad L_s\dot{\phi}/R_s + \phi = \phi_x - L_s i_c \sin(2\pi\phi/\phi_o) \quad .$$

As discussed by Kurkijärvi[8] Eq. (1) describes the motion of ϕ in the potential

$$2) \quad U(\phi,\phi_x) = (\phi-\phi_x)^2/2L_s - (\phi_o/2\pi) i_c \cos(2\pi\phi/\phi_o) \quad .$$

In the absence of thermal fluctuations and in the limit $\dot{\phi}_x/\phi_o \ll R_s/L_s$ the total current i flowing in the ring is the equilibrium supercurrent i_s. Under these conditions Eq. (1) reduces to

$$3) \quad \phi = \phi_x - L_s i_c \sin(2\pi\phi/\phi_o) \quad .$$

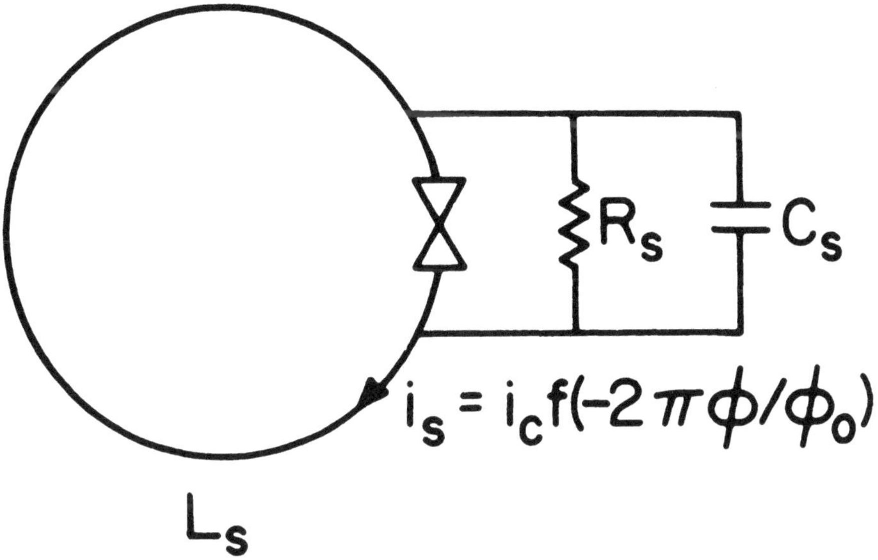

FIGURE 1 A schematic representation of an rf SQUID ring. Ideally the current-phase relation $f(-2\pi\phi/\phi_o)$ is sinusoidal and C_s sufficiently small that the weak link is very overdamped.

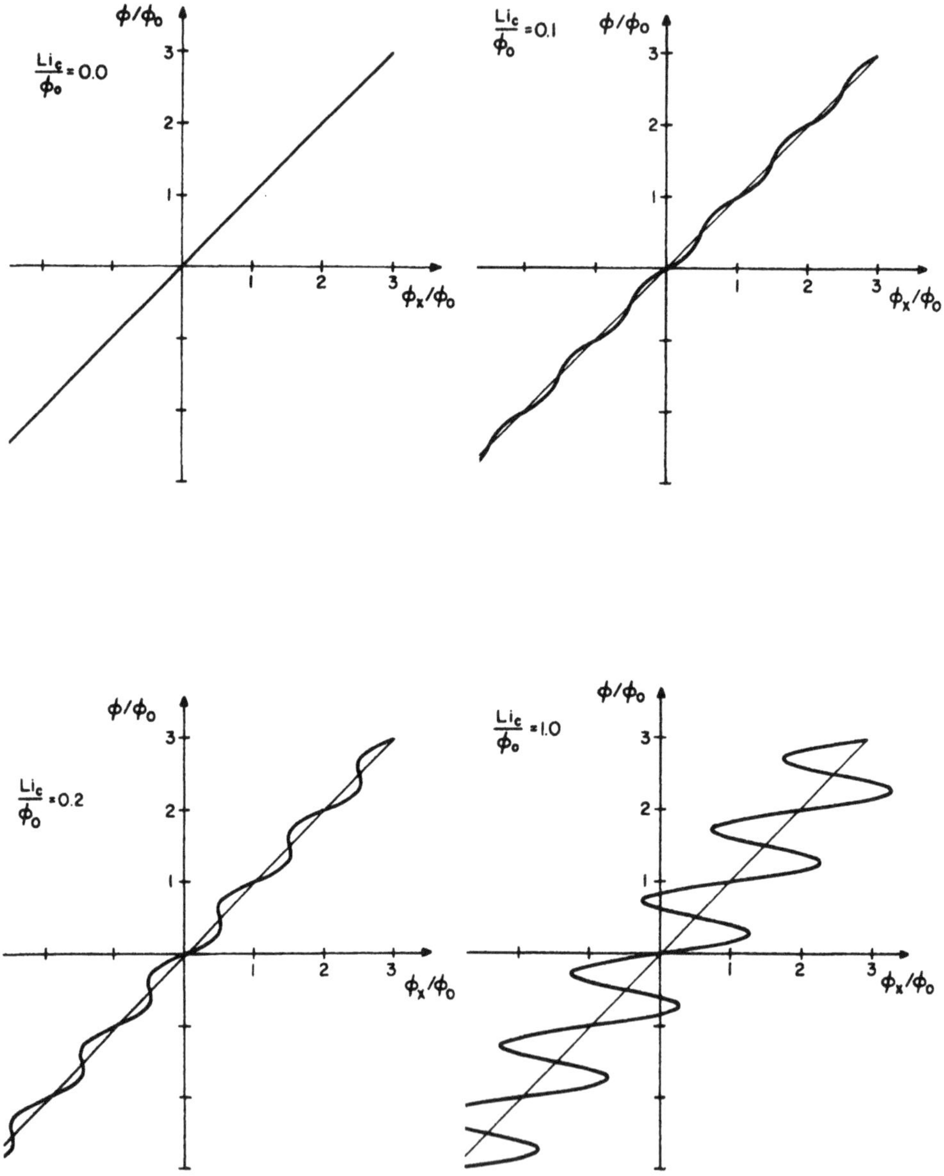

FIGURE 2 Quasi-static internal flux versus applied flux response curves for various values of critical currents.

As illustrated in Fig. 2, Eq. (3) in general describes a snake-like curve in the ϕ vs ϕ_x plane which becomes reentrant for $Li_c \gtrsim \phi_o/2\pi$. For $Li_c < \phi_o/2\pi$ the system has only one potential minimum and ϕ is a continuous function of ϕ_x. For $Li_c \gtrsim \phi_o/2\pi$ the system has multiple fluxoid quantum states. Segments of ϕ vs ϕ_x curves such as in Fig. 2 with positive slope correspond to regions of local potential minima, while those segments with negative slope correspond to local potential maxima. This is shown most clearly in Fig. 3 where the potential $U(\phi,\phi_x)$ is plotted for the case $Li_c = 2.0\phi_o$. If, for example, we initially set $\phi_x = 0$ and $\phi = -1\phi_o$ as in Fig. 3 and then increase ϕ_x, in the absence of thermal fluctuations ϕ will follow the potential minimum along the curve given by Eq. (3) until the potential barrier vanishes and the system makes a rapid transition in time $\sim L_s/R_s$ to the next lower energy state. The system will remain in this new fluxoid quantum state until ϕ_x is varied to the point where a new lower energy state is accessible. If an oscillatory flux is applied to the ring with amplitude $\gtrsim \phi_{xc}$, where $\phi_{xc} \approx Li_c + \phi_o/4$ and is the critical applied flux, the system will follow a discontinuous, hysteretic path in the ϕ vs ϕ_x plane as indicated in Fig. 4.

II.2 rf Applied Flux

In general when the SQUID ring is to be operated as a flux detector, a high frequency biasing flux $\phi_{xm}\sin\omega t$ and a "dc" signal flux $\phi_x^{(dc)}$ is applied to the SQUID ring. It is the non-linear response of the SQUID ring to this applied flux

4) $$\phi_x = \phi_{xm}\sin\omega t + \phi_x^{(dc)} ,$$

which ultimately yields the SQUID signal.

This response can be determined by solving the equation of motion given by Eq. (3). An approximate solution can be

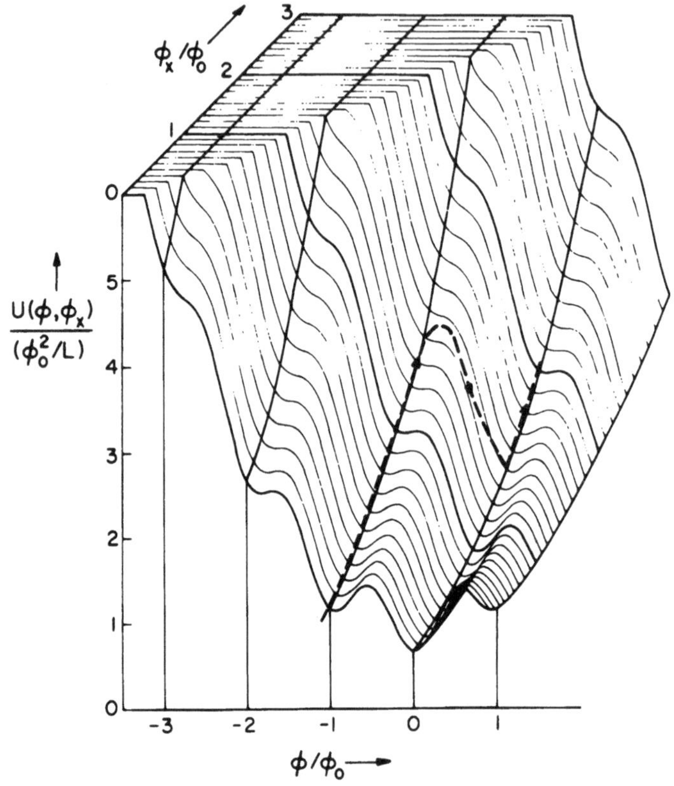

FIGURE 3 Motion of the applied flux ϕ in the potential $U(\phi,\phi_x)$ for a slowly increasing ϕ_x and $L_s i_c = 2\phi_o$. Initially $\phi = -\phi_o$, $\phi_x = 0$.

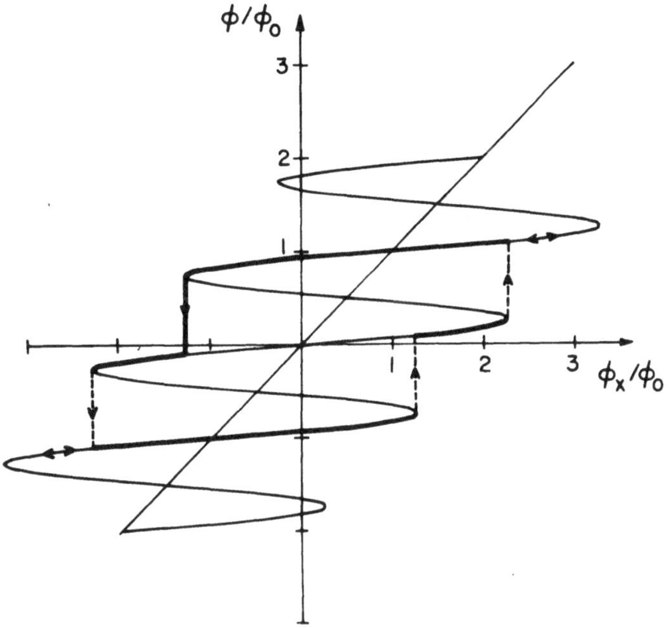

FIGURE 4 An example of the hysteretic path followed by ϕ in the ϕ-ϕ_x plane for a slowly varying periodic applied flux. In this case the critical current $L_s i_c = 2\phi_o$.

obtained in a straightforward manner by assuming an internal flux response

$$5)\quad \phi = \phi_m \sin(\omega t-\theta) + \phi^{(dc)} \quad .$$

While this approximation neglects higher order terms in ω it retains the essential features of SQUID ring behavior. Assuming this response, three coupled equations are obtained:

$$6a)\quad \phi_m[\mathrm{ctn}\theta + \omega L_s/R_s] = \phi_{xm}/\sin\theta - 2\mathrm{ctn}\theta L_s i_c \times J_1(2\pi\phi_m/\phi_o)\cos(2\pi\phi^{(dc)}/\phi_o)$$

$$6b)\quad \mathrm{ctn}\theta = R_s/\omega L_s[1 + 2L_s i_c \times J_1(2\pi\phi_m/\phi_o)\cos(2\pi\phi^{(dc)}/\phi_o)]$$

$$6c)\quad \phi^{(dc)} = \phi_x^{(dc)} - L_s i_c \times J_o(2\pi\phi_m/\phi_o)\sin(2\pi\phi^{(dc)}/\phi_o) \quad .$$

These equations do not, in general, have simple analytical solutions. It is however instructive to examine Eqs. (6) in various limits and to compare the results to complete digital solutions of Eq. (1).

In the usual case of SQUID operation the bias frequency $\omega << R_s/L_s$. This paper will only consider this low frequency situation. The case where $\omega \gtrsim R_s/L_s$ is discussed elsewhere.[9] When $\omega << R_s/L_s$ the internal flux response to $\phi_{xm}\sin\omega t$ follows the quasi-static ϕ vs ϕ_x response curve as given by Eq. (3) and as shown, for example, in Fig. 2. The effect of the dc flux $\phi_x^{(dc)}$ is simply to shift the equilibrium point about which the internal flux is modulated at frequency ω. In discussing this response, three separate critical current regimes can be considered:

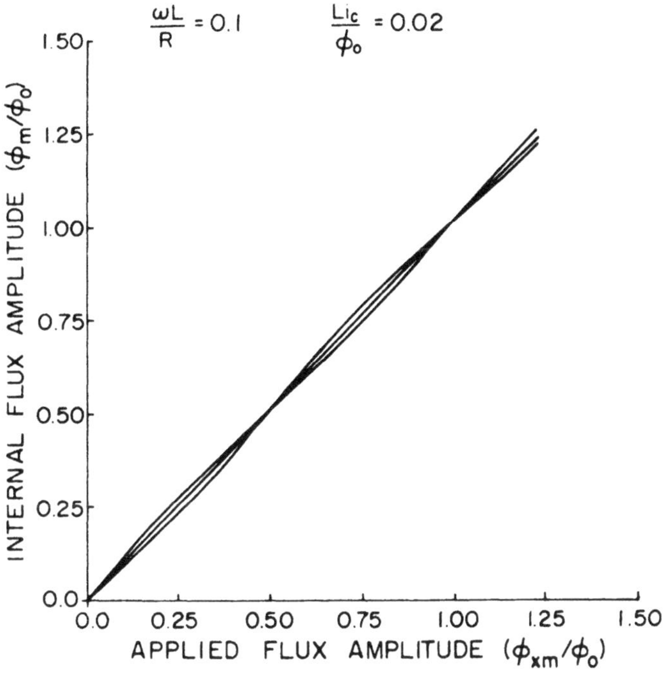

FIGURE 5 The peak amplitude of the internal flux response ϕ_m to an applied rf flux $\phi_x = \phi_{xm}\sin\omega t$. The response function is plotted for $\phi_x^{(dc)} = 0$, $\frac{1}{4}\phi_o$ and $\frac{1}{2}\phi_o$.

1. $L_s i_c << \phi_o/2\pi$

In this very low critical current limit Eqs. (6) yield the solutions

7a) $$\theta \approx 0$$

7b) $$\phi^{(dc)} \approx \phi_x^{(dc)}$$

7c) $$\phi_m \approx \phi_{xm} - 2L_s i_c J_1(2\pi\phi_{xm}/\phi_o)\cos(2\pi\phi_x^{(dc)}/\phi_o) \quad .$$

The internal flux response is in phase with the applied flux and the amplitude of ϕ_m is weakly modulated by $\phi_x^{(dc)}$; the magnitude of this modulation depending linearly upon $L_s i_c$ and periodically upon ϕ_{xm}. This solution was first derived by Hansma[10] in his original analysis of the low critical current rf SQUID.

To illustrate the nature of the SQUID ring response in this low critical current regime the peak amplitude of the internal flux response to $\phi_x = \phi_{xm}\sin\omega t$ as obtained from digital solutions of Eq. (1) is plotted in Fig. 5 versus ϕ_{xm} for the three cases $\phi_x^{(dc)} = 0, \frac{1}{4}\phi_o, \frac{1}{2}\phi_o$. This result was obtained assuming $Li_c = 0.02\phi_o$ and $\omega L_s/R_s = 0.1$. While the peak amplitude of the internal flux has both a slightly different periodicity than does ϕ_m, as given by Eq. (7c), and a somewhat larger amplitude at high ϕ_{xm} drive levels, Fig. 5 does indicate the basic nature of the ring response for very small critical current.

2. $L_s i_c \lesssim \phi_o/2\pi$

As $L_s i_c \rightarrow \phi_o/2\pi$ screening currents in the SQUID ring become important and Eqs. (7) no longer satisfactorily explain the SQUID ring response. Recently Soerensen,[11] Erne et al.[12] and Hansma et al.[13] have all investigated this regime through various analytical means and have been successful in obtaining

(a)

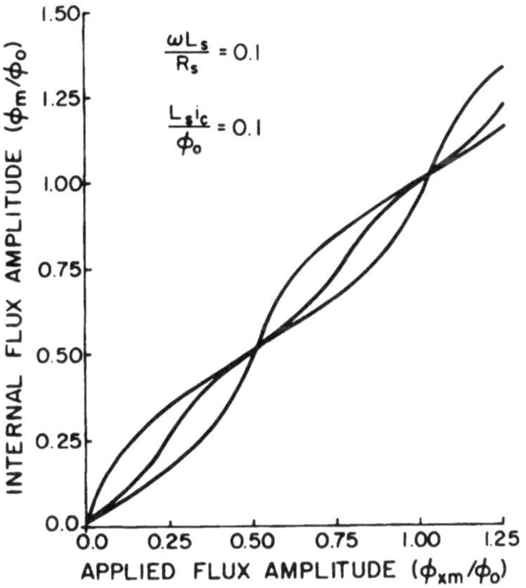

(b)

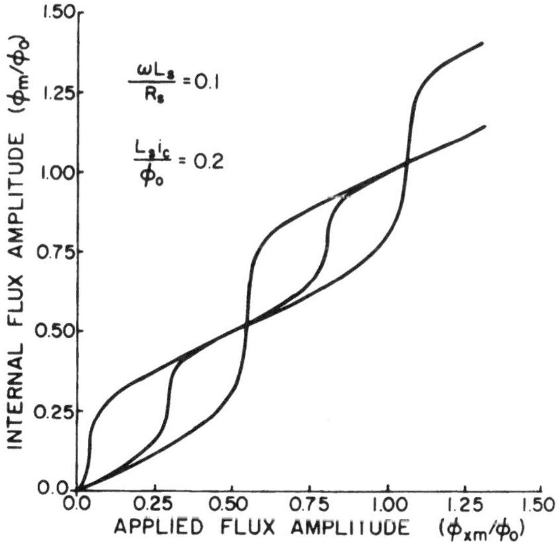

FIGURE 6 (a) and (b)

(c)

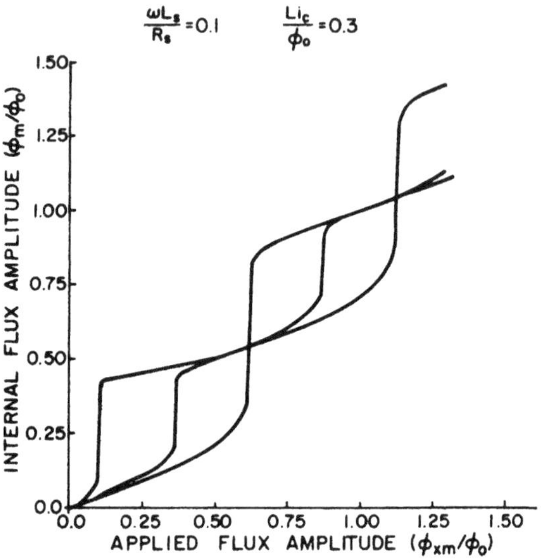

(d)

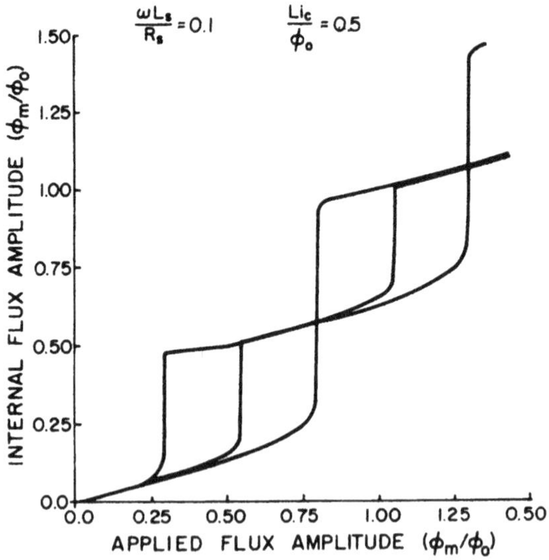

FIGURE 6 the peak amplitude of the rf response to the applied rf flux $\phi_x = \phi_{xm}\sin\omega t$ for (a) $L_s i_c = 0.1\phi_o$, (b) $L_s i_c = 0.2\phi_o$, (c) $L_s i_c = 0.3\phi_o$ and (d) $L_s i_c = 0.5\phi_o$. Due to the relatively high drive frequency $\omega L_s/R_s \approx 0.1$, the discontinuous regime actually begins at $L_s i_c \approx 0.2\phi_o$ rather than at $L_s i_c = \phi_o/2\pi$. The curves are plotted for $\phi_x^{(dc)} = 0, \frac{1}{4}\phi_o, \frac{1}{2}\phi_o$.

quite accurate descriptions of the SQUID ring behavior. For our purposes here, however, it is sufficient simply to discuss the general aspects of the ϕ vs ϕ_x response as deduced from Eq. (6) or from digital solutions to Eq. (1).

It is seen from Eq. (6b) that for sufficiently small $\omega L_s/R_s$ ϕ remains essentially in phase with ϕ_x for all $L_s i_c < \phi_o/2\pi$. The amplitude ϕ_m of the response at frequency ω can be obtained from numerical evaluations of Eqs. (6). Alternatively, the peak amplitude of this internal flux response to $\phi_x = \phi_{xm}\sin\omega t$ can be calculated from Eq. (1). Results of such calculations are shown for several values of $L_s i_c$ in Fig. 6. Again the distinction between the peak response and the component at frequency ω is slight: a somewhat different periodicity and a diminishing amplitude at higher drive levels.

3. $L_s i_c > \phi_o/2\pi$

Once $L_s i_c$ exceeds $\phi_o/2\pi$, the hysteretic nature of the quasi-static internal flux response to a periodic applied flux results in ϕ_m, as given by Eq. (6), becoming a discontinuous function of ϕ_{xm}. (This effect occurs at progressively larger values of $L_s i_c$ as $\omega L_s/R_s \to 1$. In Fig. 6 where $\omega L_s/R_s = 0.1$, the discontinuous response commences at $L_s i_c/\phi_o \simeq 0.2$.[9]) In this discontinuous regime it is most appropriate not to pursue further the analysis of the internal flux response from the point of view of a sinusoidal oscillation at frequency ω but rather to rely on digital solutions of the peak amplitude of the internal flux oscillation as a function of ϕ_{xm}. Examples of this peak response are given in Fig. 6.

II.3 Effect of dc Applied Flux

A major effect of the dc applied flux upon the ϕ-ϕ_x rf response curve is the periodic modulation of the values of ϕ_{xm} for which $d\phi_m/d\phi_{xm}$ is a maximum (see Fig. 6). This is the case regardless of the value of $L_s i_c$ and regardless of whether the peak amplitude of the response is detected on the component at frequency ω. In general, in an rf SQUID system the SQUID ring is coupled to a biasing circuit in such a way so that those regions of the ϕ_m-ϕ_{xm} response curve where $d\phi_m/d\phi_{xm}$ is a maximum present maximum opposition to changes in the flux applied by the biasing circuit, resulting in the "steps" in the SQUID circuit rf I-V characteristic. It is the periodic modulation by $\phi_x^{(dc)}$ of the location $\phi_{xm,max}$ of these regions of maximum opposition that yields the signal in the rf SQUID. This is discussed in somewhat more detail in section IV.

Thus the signal that can be obtained from the SQUID ring depends directly upon the amplitude of the modulation of the response curve about ϕ_{xm}, for constant ϕ_m. In the very low critical current regime, $L_s i_c << \phi_o/2\pi$, the maximum value of this modulation amplitude for a small signal flux $\delta\phi_x^{(dc)}$ can be obtained directly from Eq. (7). Assuming that the ring is biased near a point where $J_1(2\pi\phi_{xm}/\phi_o)$ is a maximum and that $\phi_x^{(dc)}$ has been set near the value, $\phi_x^{(dc)} = (n+\frac{1}{4})\phi_o$, for which maximum small signal sensitivity is obtained, then for a small signal $\delta\phi_x^{(dc)}$ the modulation is

$$8) \quad \delta\phi_{xm,max}(\delta\phi_x^{(dc)}) \simeq 2L_s i_c J_{1,m}(2\pi/\phi_o)\delta\phi_x^{(dc)}$$

where $J_{1,m}$ is the first maximum of the first order Bessel function J_1.

For larger critical currents the maximum modulation amplitude about ϕ_{xm} is more difficult to obtain exactly. However a value which is correct to within a factor of order unity can be obtained by employing the approximation which is shown graphically in Fig. 7. In this situation the ϕ-ϕ_x rf response curve is approximated by straight line segments whose slopes are equal to the local minima and maxima of $d\phi_m/d\phi_{xm}$. If it is the peak value of the rf flux, such as shown in Fig. 7,

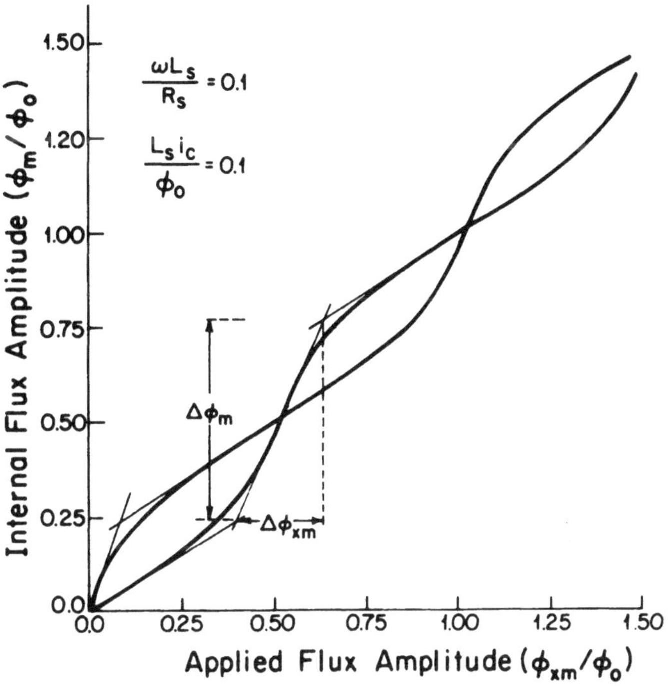

FIGURE 7 The straight line approximation to the ϕ-ϕ_x rf response curve. As $L_s i_c \to \phi_o/2\pi$ the accuracy of the approximation improves.

which is of interest, then the values of these slopes are $(1 + 2\pi L_s i_c/\phi_o)^{-1}$ and $(1 - 2\pi L_s i_c/\phi_o)^{-1}$ respectively. If the response component at frequency ω is being considered then the slopes of the first two straight line segments are $[1 + 2L_s i_c J_1'(0)]^{-1}$ and $[1 - 2L_s i_c J_1'(3.7)]^{-1}$. The differences between the two cases are slight and in the discussion that follows only the peak rf response function will be considered. The modulation of the response curve by $\phi_x^{(dc)}$ about ϕ_{xm} is also assumed to be triangular which is only strictly correct for $L_s i_c > \phi_o/2\pi$.

With these approximations the maximum amplitude of the modulation of the rf response curve is simply $\pi L_s i_c$ for $2\pi L_s i_c \lesssim \phi_o$ and $\phi_o/2$ for $2\pi L_s i_c > \phi_o$ (assuming $\omega L_s/R_s << 1$). Since the modulation is assumed to be linear for $\delta\phi_x^{(dc)} < \phi_o/2$ we have

$$\delta\phi_{xm,max}(\delta\phi_x^{(dc)}) \simeq \frac{2\pi L_s i_c}{\phi_o}\,\delta\phi_x^{(dc)} \quad , \quad \frac{2\pi L_s i_c}{\phi_o} < 1$$

9)

$$\simeq \delta\phi_x \quad , \quad \frac{2\pi L_s i_c}{\phi_o} > 1 \quad .$$

With this result the limiting sensitivity of the rf SQUID can be determined in a straightforward manner.

III INTRINSIC NOISE

In the context of the shunted junction model, the intrinsic noise of the SQUID originates with the current fluctuations in the weak link shunt resistor R_s. When the SQUID ring is biased at frequency ω it is simply the ratio of its response to these current fluctuations and its response to $\phi_x^{(dc)}$ that determines the maximum sensitivity of the rf SQUID. Formally, the effect of these fluctuations can be determined by adding a noise current term $L_s i_N$ to Eq. (1), where the spectral density of this noise current is $\langle i_N^2(\nu)\rangle^{\frac{1}{2}} = (4k_B T_s/R_s)^{\frac{1}{2}}$. The problem then becomes one of calculating the fluctuations of ϕ in the potential $U(\phi,\phi_x)$ for $\phi_x = \phi_{xm}\sin\omega t + \phi_x^{(dc)}$.

III.1 $L_s i_c << \phi_o/2\pi$

In this limit and for $\omega L_s/R_s << 1$ the problem reduces to the calculation of the spectral density of the fluctuating flux of a classical inductor L_s.

10) $$<\delta\phi^2(\nu)^{\frac{1}{2}}> = (4k_B T_s L_s^2/R_s)^{\frac{1}{2}} \quad .$$

As mentioned in Section II, in an rf SQUID system it is the modulation of the location of the rf response curve about the ϕ_{xm} axis by $\phi_x^{(dc)}$ that provides the SQUID signal. As a result it is necessary to cast the internal flux fluctuations into the form of equivalent applied flux fluctuations in order to obtain the intrinsic noise limited flux sensitivity. This can be achieved simply by employing the ϕ_m-ϕ_{xm} response function. In the very low critical current regime $\phi_m \approx \phi_{xm}$ for all ϕ_m, to zeroth order in $2\pi L_s i_c/\phi_o$. Thus the fluctuations of the SQUID ring are equivalent to an uncertainty in ϕ_{xm} of spectral density.

11) $$<\delta\phi_{xm}^2(\nu)>^{\frac{1}{2}} = (4k_B T_s/R_s)^{\frac{1}{2}} \quad .$$

Combining Eq. (10) with either Eq. (8) or Eq. (9), depending on whether the SQUID system is sensitive to the SQUID ring response at frequency ω or to the peak response, the maximum SQUID sensitivity or the minimum observable flux signal is

12) $$\delta\phi_{x,min}^{(dc)} \approx \frac{\phi_o}{2\pi L_s i_c 2J_{1,m}}\left(\frac{4k_B T_s L_s^2 \Delta f}{R_s}\right)^{\frac{1}{2}} \approx \frac{\phi_o}{2\pi L_s i_c}\left(\frac{4k_B T_s L_s^2 \Delta f}{R_s}\right)^{\frac{1}{2}} \quad ,$$

where Δf is the post-detection bandwidth of the SQUID system. This result was originally obtained by Danilov and Likharev[14] in their analysis of the limiting characteristics of the low critical current SQUID.

III.2. $L_s i_c \lesssim \phi_o/2\pi$

For larger values of critical current the fluctuation amplitude of ϕ is no longer independent of ϕ and ϕ_x. In the usual case of $L_s k_B T_s/\phi_o^2 << 1$ these fluctuations can still be treated however as small excursions about the minimum of the potential $U(\phi,\phi_x)$. Therefore the spectral density of the fluctuations in ϕ is given by

$$13) \quad <\delta^2\phi(\nu)>^{\frac{1}{2}} = L_s[1 + (2\pi L_s i_c/\phi_o)\cos 2\pi\phi/\phi_o]^{-1}(4k_B T_s/R_s)^{\frac{1}{2}} \quad ,$$

a quantity that depends directly on the internal flux ϕ and hence on the applied flux ϕ_x. As a result the uncertainty in the amplitude of the internal flux response ϕ_m depends on the position of the SQUID ring on the ϕ_m-ϕ_{xm} response curve. This can be clearly seen in Fig. 8 where ϕ_m-ϕ_{xm} response curves as calculated numerically from Eq. (1), with a noise current term included, are shown for various values of $L_s i_c$. In all cases the external flux amplitude ϕ_{xm} is increased sufficiently rapidly in the calculation so that fluctuations in ϕ_m are not averaged out on the scale of the drawing. The correlation between the fluctuation amplitude and the slope of the response curve is apparent. In general the spectral density of the fluctuations in ϕ_m is given, at least approximately by

$$14) \quad <\delta\phi_m^2(\nu)>^{\frac{1}{2}} \approx L_s(d\phi_m/d\phi_{xm})(4k_B T_s/R_s)^{\frac{1}{2}} \quad .$$

Since during rf SQUID operation the ring is biased on a region of maximum slope on the ϕ_m-ϕ_{xm} response curve, it is the maximum amplitude of this spectral density that must be considered

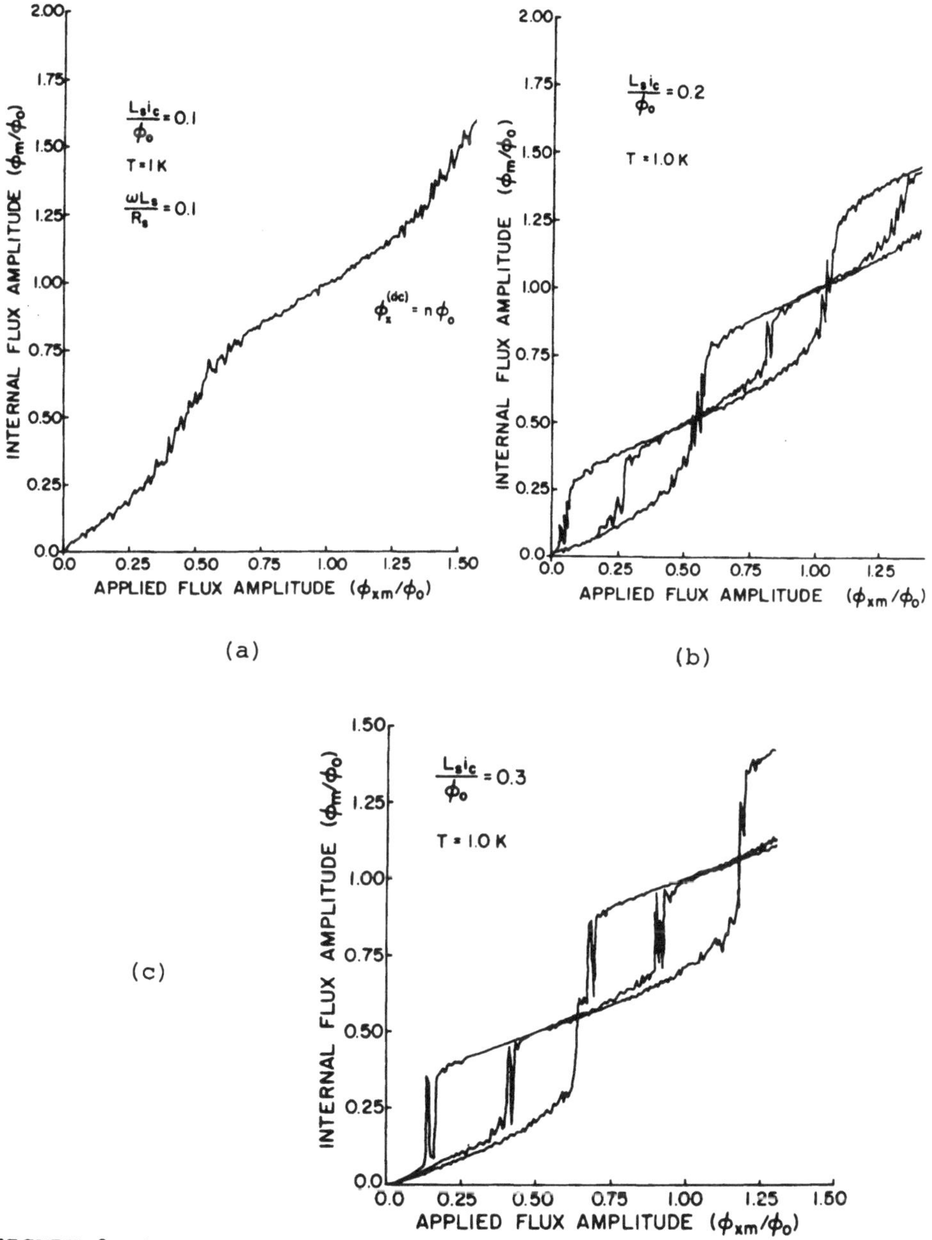

FIGURE 8 ϕ_m-ϕ_{xm} rf response curve calculated numerically from a noise term included vor various values of $L_s i_c$. Again the drive frequency $\omega \simeq 0.1 R_S/L_S$. For clarity only the curve for $\phi_x^{(dc)} = n\phi_o$ is shown in (a). The simulation assumes $T_S = 1K$ and $L_S = 10^{-9}H$.

15) $$<\delta\phi_m^2(\nu)>^{\frac{1}{2}} \approx L_s(1 - 2\pi L_s i_c/\phi_o)^{-1}(4k_BT_s/R_s)^{\frac{1}{2}} \quad .$$

To determine the maximum SQUID sensitivity in this larger critical current regime this fluctuation spectral density again must be converted into equivalent fluctuations in ϕ_{xm}. For $L_s i_c$ not too near to $\phi_o/2$ and for $L_s k_B T_s$ sufficiently small so that noise rounding of the time-averaged ϕ_m-ϕ_{xm} response is not too severe this can be accomplished with the use of the noise free response curve. The result is

16) $$\begin{aligned} <\delta\phi_{xm}^2(\nu)>^{\frac{1}{2}} &= <\delta^2\phi(\nu)>^{\frac{1}{2}} d\phi_m/d\phi_{xm} \\ &\approx L_s(4k_BT_s/R_s)^{\frac{1}{2}} \quad . \end{aligned}$$

Thus the equivalent fluctuation amplitude in ϕ_{xm} remains unchanged as $L_s i_c \to \phi_o/2$, showing that the simple approach of treating the noise current in R_s as the source of an applied noise flux $\delta\phi_{xm}$ is valid for quite large critical currents. Due to this constancy in $\delta\phi_{xm}(\nu)$ the minimum detectable flux signal remains as given by Eq. (12) and decreases as $L_s i_c \to \phi_o/2\pi$. While this result is no longer valid once $L_s i_c$ is sufficiently close to $\phi_o/2\pi$, for typical values of SQUID parameters, $L_s \sim 10^{-9}$H, $T_s \sim 4.2$K and $\omega << L_s/R_s$, it is at least approximately correct for $L_s i_c \lesssim \phi_o/4\pi$. Consequently the classical noise limit $\delta\phi_x^{(dc)} \approx (4k_BT_sL_s\Delta f/R_s)^{\frac{1}{2}}$ can be rather closely approached in the inductive or low critical current SQUID.

III.3 $L_s i_c \gtrsim \phi_o/2\pi$

As $L_s i_c$ approaches from below the value at which the ϕ_m-ϕ_{xm} response curve becomes discontinuous ($L_s i_c = \phi_o/2\pi$ for $\omega L_s/R_s << 1$) the amplitude of the internal flux fluctuations, when the ring is biased on a region of maximum slope of the response curve, eventually increases sufficiently that the intrinsic noise calculation can no longer be treated as a problem of classical fluctuations. Instead it must be treated as a problem of random transitions between quantum states. This effect can be seen in Fig. 8 and it commences somewhere below $L_s i_c = \phi_o/2\pi$ at a point dependent on $(L_s k_B T_s)^{\frac{1}{2}}$.

When the SQUID ring is biased at or near the region of maximum slope of the ϕ_m-ϕ_{xm} response curve, at the peak of the rf cycle the internal flux ϕ fluctuates back and forth between the two accessible states at a frequency which is much greater than the bias frequency ω, assuming $\omega << R_s/L_s$. This results in an uncertainty in the peak response ϕ_m which depends on the fluctuation frequency. As $L_s i_c$ exceeds $\phi_o/2\pi$ the probability of reverse transition diminishes and eventually at most only one forward transition in ϕ can be made at the peak of every half rf cycle. As a result of this pronounced decrease in the fluctuation frequency the minimum detectable flux of the rf SQUID increases until the maximum SQUID sensitivity reaches the value calculated by Kurkijärvi and Webb[15] for $L_s i_c >> \phi_o/2\pi$.

As an alternative to the exact calculation of Kurkijärvi and Webb the minimum detectable flux in the large critical current regime can be discussed along lines similar to those used here for the case $L_s i_c < \phi_o/2\pi$. Assuming $L_s i_c$ is sufficiently large and $\phi_x^{(dc)} = n\phi_o$, the peak amplitude ϕ_m of the rf response to ϕ_x is either ~0 or ϕ_o during any rf cycle, depending on whether or not a transition occurs during the peak of that cycle. If the probability that such a transition occurs every cycle is p, where p depends on the value of ϕ_{xm}, then the average value of the internal flux response is simply

17) $$\langle\phi_m\rangle = p\phi_o \quad ,$$

and for frequencies much less than ω the spectral density of the fluctuations in ϕ_m is

18) $$\langle\delta\phi_m^2(\nu)\rangle^{\frac{1}{2}} = [p(1-p)]^{\frac{1}{2}}\phi_o(2\pi/\omega)^{\frac{1}{2}} \quad .$$

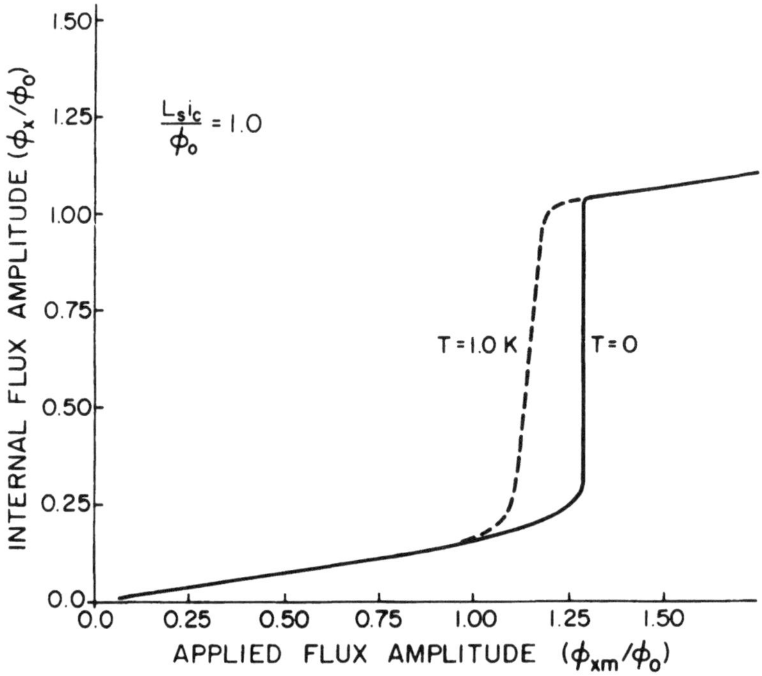

FIGURE 9 A numerically calculated time averaged rf response curve for the case where T = 0 and T = 1.0K. The slope of the dotted curve in the T = 1.0K case is equal to the fluxoid transition uncertainty as calculated by Kurkijärvi and Webb.

To convert these fluctuations in ϕ_m into equivalent fluctuations in ϕ_{xm} it is necessary to use the ϕ_m-ϕ_{xm} response curve which has been time-averaged over a period $\gg 1/\omega$. An example of such a time averaged response curve is shown in Fig. 9. The effect of the averaging is to remove the discontinuity from the response curve, replacing it with a curve of finite slope which is centered at a point on the ϕ_{xm} axis somewhat to the left of the discontinuity in the noise-free case. (In general the amount of shift depends upon i_s, T_s and ω.[8]) This curve, which is basically a straight line segment over most of its length represents the probability that a fluxoid transition will occur once every cycle for a particular value of ϕ_{xm}. The slope of the curve $d\phi_m/d\phi_{xm}$ is then simply the inverse of the width of the fluxoid transition uncertainty as calculated by Kurkijärvi.[8] Thus the equivalent spectral density of fluctuations in ϕ_{xm} where the SQUID ring is biased on the region of maximum slope of the time-averaged response curve is

$$19) \quad \langle\delta\phi^2_{xm}(\nu)\rangle^{\frac{1}{2}} \approx [p(1-p)]^{\frac{1}{2}}(2\pi/\omega)^{\frac{1}{2}}\sigma$$

where $\sigma \approx L_s i_c (k_B T_s 2\pi/\phi_o i_c)^{2/3}$.

This approximate result is of course only valid for biasing near the center of the region of maximum slope of the response curve where $p \approx 0.5$. In this case the limiting sensitivity is

$$20) \quad \delta\phi^{(dc)}_{x,min} \approx 0.5\left(\frac{2\pi\Delta f}{\omega}\right)^{\frac{1}{2}} L_s i_c \left(\frac{2\pi k_B T_s}{\phi_o i_c}\right)^{2/3} ,$$

which is essentially the result of K-W.

The limiting intrinsic noise for an rf SQUID is thus a minimum for i_c slightly less than the value required for the ϕ_m-ϕ_{xm} response curve to become discontinuous, which is $2\pi L_s i_c/\phi_o = 1$ for $\omega L_s/R_s \ll 1$. As i_c increases above this value the intrinsic noise increases rather rapidly toward the limit calculated by K-W for $L_s i_c \gg \phi_o/2\pi$. This is not the case however when the bias frequency is increased so that $\omega L_s/R_s \rightarrow 1$. For $\omega L_s/R_s \gtrsim 1$ it is straightforward to show that the intrinsic noise for the case of both a continuous and discontinuous response curve can very closely approach the limiting classical value. Therefore for low frequency operation $\omega L_s/R_s \ll 1$, if the intrinsic noise is of practical importance, which to date is seldom the case, then i_c should be adjusted so that $L_s i_c \lesssim \phi_o/2\pi$. For higher frequencies this constraint is largely removed.

IV CIRCUIT CONSIDERATIONS

In the usual case, the rf SQUID circuit consists of a current-driven high Q tank circuit of inductance L_t and capacitance C_t. The circuit is inductively coupled to the SQUID ring through mutual inductance M and the circuit is driven at or near its resonant frequency ω_o. When suitably biased and tuned, the tank circuit voltage is periodically modulated by $\delta\phi_x^{(dc)}$, yielding the SQUID signal. A schematic representation of such a SQUID circuit is shown in Fig. 10. The details of the tank circuit response in both the low critical current[4,10-14] and high critical current[4,6,9] regimes have been extensively discussed in the literature. Here only a few basic features will be sketched so as to establish a qualitative picture of the various noise sources and optimum coupling conditions for the rf SQUID

IV.1 rf I-V Characteristic

In general terms, the rf tank circuit is driven at a frequency ω such that changes of the rf flux amplitude within the SQUID ring present maximum opposition to change in the applied flux. As discussed by Hansma and others[10-14] in the very low critical current regime where the flux response at frequency ω

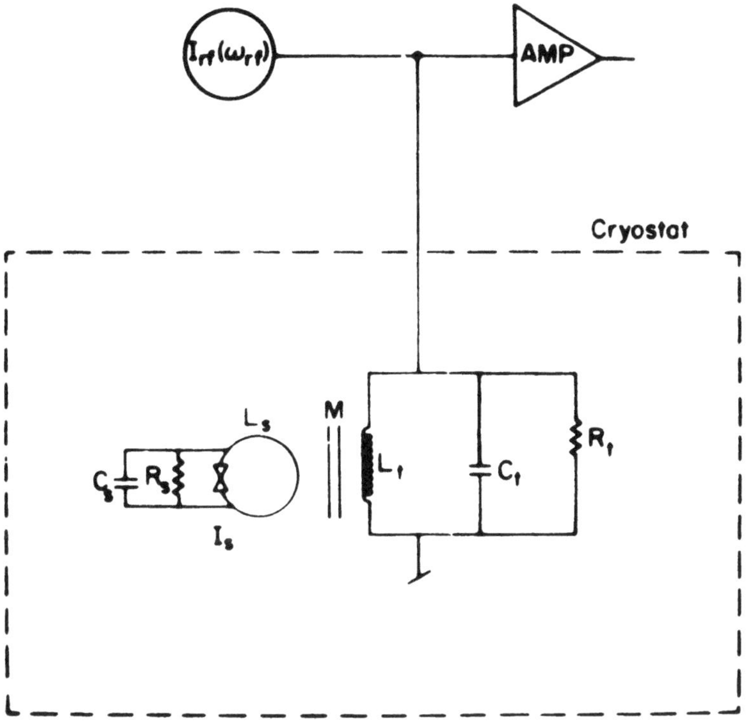

FIGURE 10 Schematic representation of a standard rf SQUID system.

is in phase with the drive flux this requires that the biasing frequencies to be either of the half power frequencies of the resonant circuit. For larger values of i_c when screening currents in the SQUID ring become important the circuit response is no longer anti-symmetric about ω_o and maximum signal is obtained by tuning to the larger half power frequency. As $L_s i_c \rightarrow \phi_o/2\pi$ the tuning becomes less critical and for $L_s i_c >> \phi_o/2\pi$ the tank circuit which is now sensitive to the peak ϕ_m-ϕ_{xm} response tunes broadly about its resonance frequency ω_o.

It is a relatively straightforward procedure to construct an approximate rf I-V characteristic of an optimally tuned SQUID system by using the simple linear approximation to the time-averaged ϕ_m-ϕ_{xm} response curve such as shown in Fig. 7. The basis for this construction is the following:

When the tank circuit current amplitude I_t is such that the rf flux applied to the SQUID ring, $\phi_{xm} = MI_t$, is brought to the point where maximum slope is found in the ϕ_m-ϕ_{xm} response curve, an additional impedance is brought into play through the induction by the SQUID ring of a current in the tank circuit. This current is directly counter to the current induced by the external drive current I_D. As a result a greater rf drive level is required to move ϕ_{xm} and hence I_t through the region of steepest slope in the ϕ_m-ϕ_{xm} response curve than through the region of minimum slope. Consequently a "step" is created in the rf I-V characteristic since the tank circuit voltage $V_t = \omega L_t = \omega L_t \phi_{xm}/M$. The length of this step in terms of the external drive current I_D is, to a fair approximation,

21) $$I_{D,step} \approx \frac{M}{L_t L_s} \Delta\phi_m + \frac{1}{MQ} \Delta\phi_{xm} \quad ,$$

where $\Delta\phi_m$ and $\Delta\phi_{xm}$ are the extent in ϕ_m and ϕ_{xm}, respectively, covered by the steeper straight line segments of the response curve. (See Fig. 7.) In the low critical current regime, as in Fig. 7, we have $\Delta\phi_m/\Delta\phi_{xm} = (1 - 2\pi L_s i_c/\phi_o)^{-1}$, although to be accurate the harmonic response function should be employed for which $\Delta\phi_m/\Delta\phi_{xm} = [(1 - 2L_s i_c J'_{1,m})(2\pi\phi_m/\phi_o)]^{-1}$ gives the slope of the first straight line segment of the approximate response curve. For $L_s i_c > \phi_o/2\pi$, $\Delta\phi_{xm}/\Delta\phi_{xm}$ is the reduced fluxoid transition uncertainty σ/ϕ_o. The rf voltage change along the step is, of course,

$$22)\quad V_{t,\ step} \approx \frac{\omega L_t}{M}\,\Delta\phi_{xm} \quad .$$

The various steps in the rf I-V characteristic are separated by "risers". The length of these risers in I_D can be estimated from the fact that from the end of one full step to the beginning of another I_D must increase by a minimum amount

$$23)\quad I_{D,riser} \approx \frac{1}{MQ}\,(\phi_o - \Delta\phi_{xm}) \quad , \quad .$$

while the tank circuit voltage changes by an amount

$$24)\quad V_{t,riser} \approx \frac{\omega L_t}{M}\,(\phi_o - \Delta\phi_{xm}) \quad .$$

By determining the changes in position of the various steps and risers on the rf I-V characteristic as $\phi_x^{(dc)}$ varies, it is quickly found that the voltage modulation by the applied flux signal $\delta\phi_x^{(dc)}$ of the positions of the rf steps at constant rf drive level is

$$\delta V_t(\delta\phi_x^{(dc)}) \approx \frac{\omega L_t}{M}\,(1 - \Delta\phi_{xm}/\Delta\phi_m)\,\delta\phi_x^{(dc)} \quad , \quad \delta\phi_x^{(dc)} < \phi_o/2$$

or

$$\delta V_t \approx \frac{\omega L_t}{M}\left(\frac{2\pi L_s i_c}{\phi_o}\right)\delta\phi_x^{(dc)} \quad , \quad 2\pi L_s i_c/\phi_o \lesssim 0.5$$

25)

$$\delta V_t \approx \frac{\omega L_t}{M}\,(1 - \sigma/\phi_o)\,\delta\phi_x^{(dc)} \quad , \quad 2\pi L_s i_c/\phi_o > 1 \quad .$$

While Eqs. (20)-(25) are clearly only approximate, particularly for the low critical current case, they can be used to construct rf I-V characteristics which agree reasonably well with I-V's which we have obtained from complete digital simulations of rf SQUIDs. Examples of such I-V's are shown in Figs. 11 and 12 for both the low critical current and high critical current cases. In all the I-V's shown the drive frequency is at or close to its optimum value. In general it is found from the simulations that the values of the maximum dc flux modulation amplitude and of the step and riser lengths are, at least for the first step, well within a factor of two of the values given above, once corrections have been made for the relatively large bias frequencies that were necessary in the simulations.

A rather important result, that can be extracted from either the arguments given above or directly from the simulations, is that the inductive SQUID, where $2\pi L_s i_c/\phi_o < 1$, can yield a signal that is within a factor of two of the maximum that is obtainable from the higher critical current device. Thus if intrinsic noise does become a practical problem, it might be appropriate for the critical current to be reduced from its usual value, $L_s i_c \sim \phi_o$, to $L_s i_c \simeq 0.5\phi_o/2\pi$, where the intrinsic noise is considerably less, assuming $\omega << R_s/L_s$.

IV.2 Circuit Noise and Optimum Coupling

In general there are three independent noise sources in an rf SQUID — the rf amplifier, the tank circuit, and the SQUID ring itself. Kurkijärvi[16] has analyzed the problem of tank circuit noise for the case of the large critical current device. Using the arguments given in the previous section, his results can be readily generalized to cover the lower critical current case.

The essential point of Kurkijärvi's analysis is that the amplitude of the tank circuit voltage fluctuations is determined by the bias position on the rf I-V characteristic; that is,

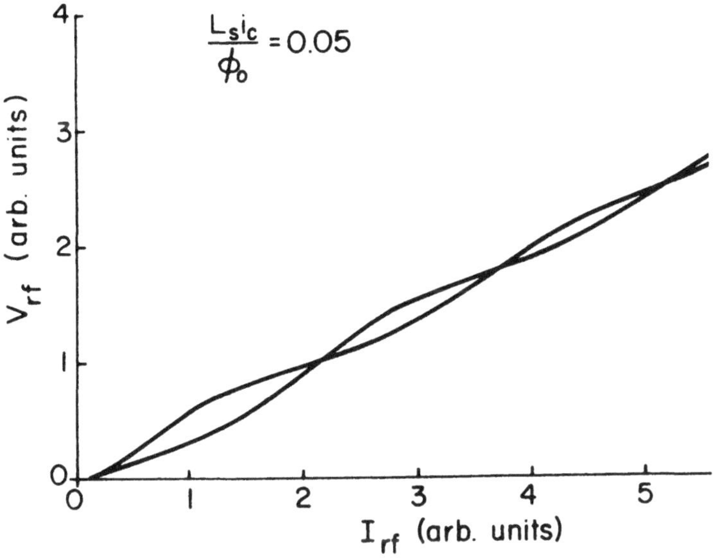

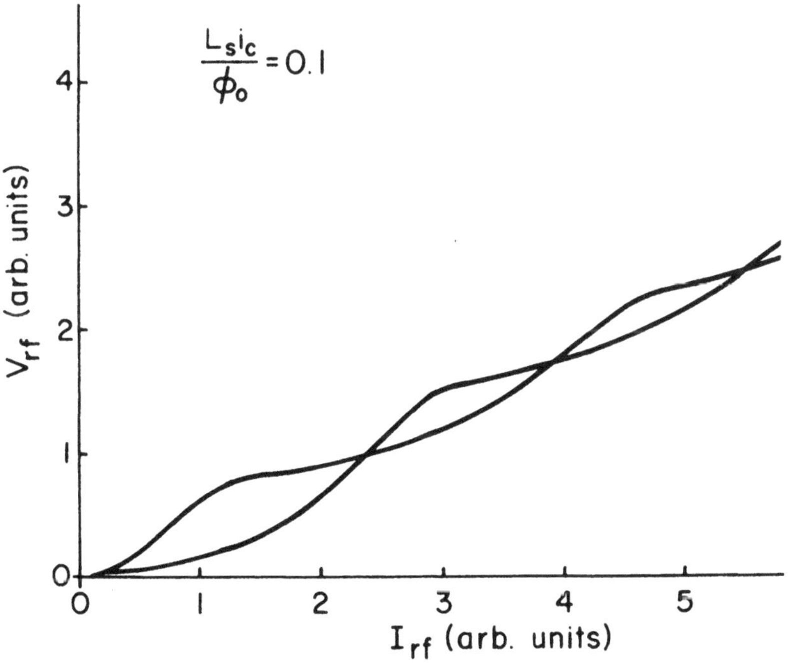

FIGURE 11 Simulated rf I-V characteristics for the low critical current SQUID. In the simulations $\omega L_S R_S = 0.1$, $K^2 Q = 4$. The tuning is close to optimum.

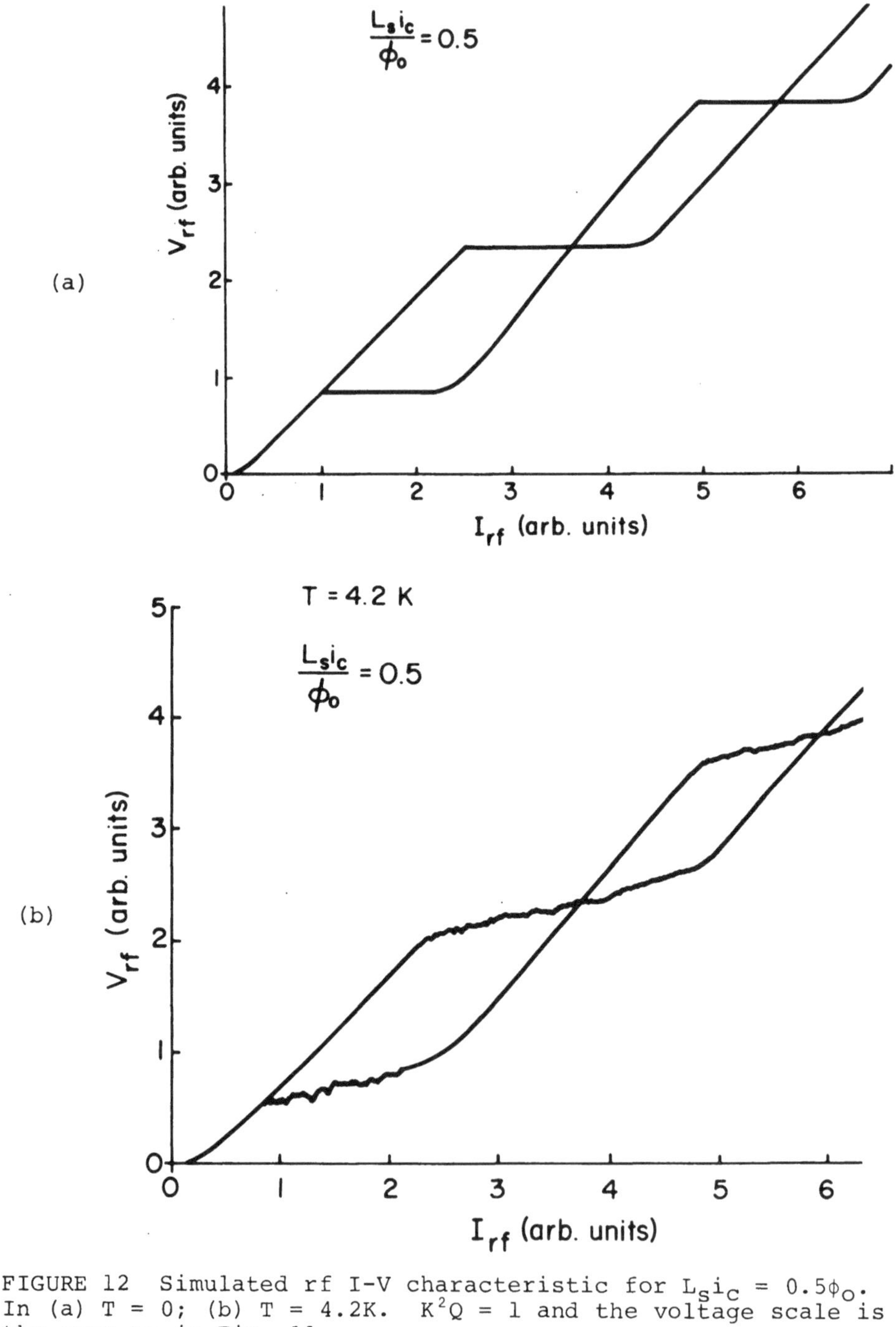

FIGURE 12 Simulated rf I-V characteristic for $L_s i_c = 0.5\phi_o$. In (a) T = 0; (b) T = 4.2K. $K^2Q = 1$ and the voltage scale is the same as in Fig. 11.

the origin of tank circuit noise is the current fluctuations in the equivalent tank circuit shunt resistor R_t. The response of the circuit to these fluctuations is determined by the local tank circuit impedance dV_t/dI_D. Since during SQUID operation the system is biased on an rf step, the tank circuit voltage fluctuation spectral density is

$$<\delta V_T^2(\nu)>^{\frac{1}{2}} = \frac{dV_{t,step}}{dI_{D,step}}\left(\frac{4k_B T_t}{Q\omega L_t}\right)^{\frac{1}{2}}$$

or

$$26) \quad <\delta V_T^2(\nu)>^{\frac{1}{2}} = \frac{\Delta\phi_{xm}}{\Delta\phi_{xm} + K^2 Q \Delta\phi_m} (4k_B T_t Q\omega L_t)^{\frac{1}{2}}$$

where $K^2 = M^2/L_s L_t$ and is the coupling constant between the SQUID ring and L_t. Now as $2\pi L_s i_c/\phi_o$ increases from zero to greater than one, $\Delta\phi_{xm}$ decreases from $\phi_o/2$ to its limiting value σ/ϕ_o while $\Delta\phi_m$ increases from $\phi_o/2$ to $\sim\phi_o$. Thus in the limit of large critical current, Eq. (26) reduces to the Kurkijärvi result.[7,16] Provided the critical current is not too small, $2\pi L_s i_c/\phi_o \gtrsim \frac{1}{4}$ and that $K^2 Q \gtrsim 1$, Eq. (26) can be approximated by

$$27) \quad <\delta V_T^2(\nu)>^{\frac{1}{2}} \approx \frac{(\Delta\phi_{xm}/\Delta\phi_m)}{K^2 Q} (4k_B T_t Q\omega L_t)^{\frac{1}{2}} \quad ,$$

where $\Delta\phi_{xm}/\Delta\phi_m \simeq 1-2\pi L_s i_c/\phi_o$ for small i_c and $\Delta\phi_{xm}/\Delta\phi_m \simeq \sigma/\phi_o$ for $L_s i_c >> \phi/2\pi$.

With this result the total equivalent noise power spectral density for the rf SQUID can be written as

$$28) \quad <\delta\phi_N^2(\nu)> = \frac{K^2 L_s}{\omega^2 L_t} \frac{<V_A^2(\nu)>}{(1 - \Delta\phi_{xm}/\Delta\phi_m)} + \frac{(\Delta\phi_{xm}/\Delta\phi_m)^2}{(1 - \Delta\phi_{xm}/\Delta\phi_m)^2} \times \left[\frac{k_B T_t L_s}{\omega} \frac{1}{K^2 Q}\right] + <\delta\phi_i^2(\nu)> \quad ,$$

where $<\delta V_A^2(\nu)>^{\frac{1}{2}}$ = spectral density of the voltage noise in the amplifier and where the intrinsic flux noise spectral density $<\delta\phi_i^2(\nu)>^{\frac{1}{2}}$ is

$$<\delta\phi_i^2(\nu)>^{\frac{1}{2}} \approx \frac{2\pi L_s i_c}{\phi_o}\left(\frac{4k_B T_s L_s}{R_s}\right)^{\frac{1}{2}} \quad , \quad L_s i_c < \phi_o/2\pi$$

29)

$$\approx \frac{1}{1 - \sigma/\phi_o}\left(\frac{0.5\pi}{\omega}\right)^{\frac{1}{2}}\sigma \quad , \quad L_s i_c >> \phi_o/2\pi \quad .$$

Since the effective tank circuit noise temperature T_t is usually greater than, and certainly never less than, T_s, Eqs. (28) and (29) show that for the inductive SQUID with coupling $K^2Q \sim 1$ the tank circuit noise term will invariably be larger than the intrinsic noise term provided $\omega << R_s/L_s$. Only for the impractical case of tight coupling does the tank circuit noise term approach the intrinsic noise level, which however is quite small. For the larger critical current case the intrinsic noise is greater and the ranking between the two terms depends on whether the ratio $L_s k_B T_t/\phi_o K\ Q \lessgtr \pi/2$.

In most practical SQUID systems the amplifier noise term dominates. To minimize this noise the coupling constant K is reduced to some minimum value. Usually this value is set by the requirement that, to obtain a continuous modulation of the step voltage by $\delta\phi_x^{(dc)}$, it is necessary that $I_{D,step} \gtrsim I_{D,riser}$. For $L_s i_c \gtrsim \phi_o/2\pi$ this requires that $K^2Q \gtrsim 1$, to within the accuracy of the approximations employed here. For $L_s i_c \lesssim 0.5\phi_o/2\pi$ the extensive rounding in the actual rf I-V characteristic which is not accounted for in the approximations used here require K^2Q to be somewhat greater than unity for maximum modulation of V_t by a flux change $\delta\phi_x^{(dc)} = \phi_o/2$.

The constraint $K^2Q \gtrsim 1$ is not rigorous and it is possible to increase the SQUID signal for small flux changes $\delta\phi_x^{(dc)} < \phi_o/2$ by decoupling beyond $K^2Q = 1$. This is most effective in the larger critical current regime. The range of $\delta\phi_x^{(dc)}$ for which the tank circuit voltage is responsive decreases as $\sim 1/K^2$. Still, some increase in the small signal response beyond the value

$$\delta V_t \approx \omega (QL_t/L_s)^{\frac{1}{2}} (1 - \Delta\phi_{xm}/\Delta\phi_m)\, \delta\phi_x^{(dc)}$$

is possible.

In some cases decoupling of the SQUID ring even to the point $K^2Q \approx 1$ may not be desirable as the tank circuit noise increases as $(K^2Q)^{-1}$. The ratio of the amplifier noise term to tank circuit noise term is

$$30)\qquad \frac{\langle \delta V_A^2(\nu) \rangle^{\frac{1}{2}}}{\langle \delta V_T^2(\nu) \rangle^{\frac{1}{2}}} \approx \frac{K^2Q \langle \delta V_A^2(\nu) \rangle^{\frac{1}{2}}}{(Q\omega L_t k_B T_t)^{\frac{1}{2}}} \left(\frac{\Delta\phi_m}{\Delta\phi_{xm}} \right) \quad .$$

Since the maximum value of $\Delta\phi_m/\Delta\phi_{xm}$ is $\lesssim 10$ for typical SQUID parameters, for sufficiently high tank circuit impedance the tank circuit noise may become a factor as improvements in rf amplifiers continue to be made. In this case tighter coupling than $K^2Q = 1$ may be required for maximum sensitivity.

V WEAK LINK EFFECTS AND ULTIMATE SENSITIVITY

Only the case of a sinusoidal current-phase relation for the weak link has been considered. In practice, weak links that are employed in SQUIDs often have distinctly non-sinusoidal current-phase relations. A typical non-sinusoidal current-phase relation is shown in Fig. 13. The result of such departures from ideal behavior is not drastic but it is detrimental to SQUID sensitivity.

The major effect of non-sinusoidal current-phase relations is to reduce the maximum signal amplitude that can be extracted from a SQUID system. As the critical phase angle θ_c of the current-phase relation increases beyond the optimum $\pi/2$ value

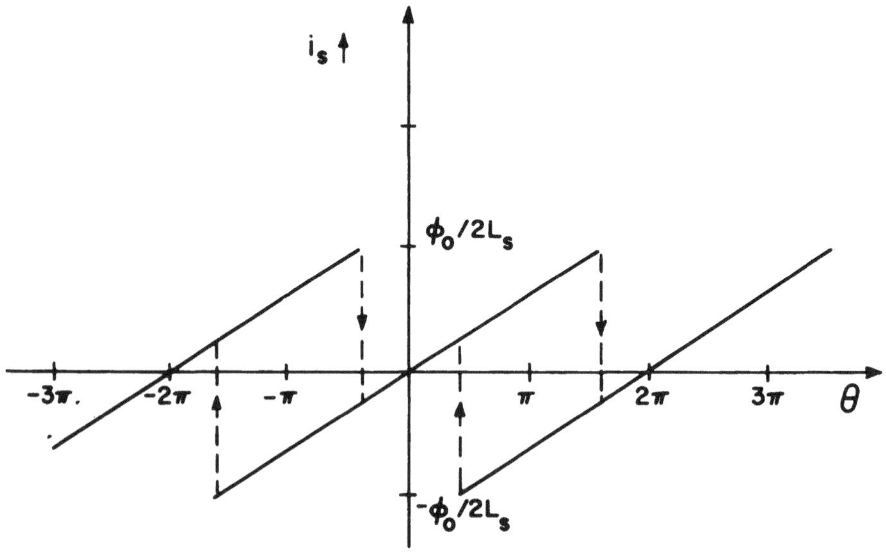

FIGURE 13 The measured non-sinusoidal current phase relation of a 1μ square tin microbridge.

the effect is to decrease the amplitude of $\Delta\phi_m$ on the steeper parts of the rf ϕ-ϕ_x response curve. In the low critical current regime the result is a corresponding increase in the minimum coupling constraint on K^2Q. For $\theta_c \gtrsim \pi$ the SQUID can only be operated in the discontinuous or hysteretic mode which has the higher intrinsic noise amplitude. Even then the minimum coupling constraint $I_{step} \gtrsim I_{riser}$ acts to decrease the available SQUID signal. In addition it has been found[7] that the fluxoid transition uncertainty also increases with critical phase angle θ_c. Besides increasing the intrinsic noise amplitude directly, this imposes a greater lower bound on the magnitude of $\Delta\phi_{xm}/\Delta\phi_m$. This results in turn in larger tank circuit noise and, if the transition uncertainty $\sigma \gtrsim 0.3\phi_o$, in a direct decrease in maximum signal strength. Experimentally[17] it has been found that even for extremely non-sinusoidal current-phase relations the transition uncertainty σ seldom increases beyond a factor of 5 of the value for a sinusoidal current-phase relation. It appears that local inhomogeneities in the larger weak links which have the non-sinusoidal relations restrict the fluctuations in the weak link. Consequently, practically any weak link which has sufficiently small critical current $i_c \approx \phi_o/L_s$ can be used in an rf SQUID in the discontinuous mode. While the performance is degraded if $\theta_c \gg \pi/2$ a usable signal can in fact usually be obtained, provided the link is underdamped and ω is not too large.

Finally it is interesting, although perhaps academic at present, to determine what the ultimate intrinsic SQUID sensitivity might be. It has been shown that in general when discussing SQUID sensitivity the appropriate sensitivity figure is not the total equivalent flux noise spectral density $\langle\delta\phi_N^2(\nu)\rangle^{\frac{1}{2}}$ but rather one which takes into account the efficiency by which flux signals can be coupled into the SQUID ring. In this case the SQUID sensitivity can be expressed in terms of an energy resolution δE where $\delta E(\nu) \approx \langle\delta\phi_N^2(\nu)\rangle/K'L_s$ where K' is the coupling constant between a signal coil and the

SQUID ring. Assuming that K' can be set to ≈1 for the most favorable case of $L_s i_c \lesssim \phi_o/2\pi$, the result is

$$31)\quad \delta E(\nu) \approx 4k_B T_s L_s/R_s \quad .$$

For an ideal Josephson junction $R_s \approx \pi\Delta(0)/2ei_c$ where $\Delta(0) = 1.76k_B T_c$ and is the T = 0 energy gap of the superconductor and where the temperature T_s of the weak link is assumed to be $\ll T_c$ of the superconducting material. In this case we have

$$32)\quad \delta E(\nu) \approx \frac{4k_B T_s L_s (2ei_c)}{1.76\pi k_B T_c}$$

but $L_s i_c \approx \phi_o/2\pi = \frac{h}{4\pi e}$, so

$$33)\quad \delta E(\nu) \approx \frac{4T_s}{1.76\pi T_c}\,\frac{h}{2\pi} \quad .$$

Thus for any rf SQUID system which has an ideal Josephson junction for the weak link, fabricated from any superconductor, and which is operated at $T_s \lesssim 0.5T_c$ the minimum intrinsic energy resolution is simply due to the zero point oscillations as set by the uncertainty principle:

$$34)\quad \frac{\langle\delta\phi_N^2(\nu)\rangle}{L_s} \gtrsim h \approx 6.6\times10^{-34}\ \text{Joules/Hz} \quad .$$

In terms of magnetic flux for a typical ring inductance $L_s \approx 10^{-9}$H this corresponds to an ultimate flux sensitivity

$$35)\quad \langle\delta\phi_N^2(\nu)\rangle^{1/2} \gtrsim 4\times10^{-7}\phi_o/\sqrt{\text{Hz}} \quad ,$$

which is somewhat more than a factor of ten lower than the best sensitivity that has ever been achieved in practice,[19] to the knowledge of this author. It would be rather impressive if this limiting noise figure is in fact ever closely approached.

This research supported by the Office of Naval Research. I am happy to acknowledge the very extensive and essential contributions of Dr. L. D. Jackel to the work presented here.

References

1. A. H. Silver and J. E. Zimmerman: Phys. Rev. 157, 317 (1967).
2. W. W. Webb: IEEE Trans. Magn. 8, 51 (1972).
3. J. M. Goodkind and D. L. Stolfa: Rev. Sci. Instrum. 41, 799 (1970).
4. J. Clarke: Proc. IEEE 61, 8 (1973).
5. R. P. Giffard, R. A. Webb and J. C. Wheatley: J. Low Temp. Phys. 6, 533 (1972).
6. J. E. Zimmerman: Cryogenics 12, 19 (1972).
7. L. D. Jackel and R. A. Buhrman: J. Low Temp. Phys. 19, 201 (1975).
8. J. Kurkijärvi: Phys. Rev. B 6, 832 (1972).
9. R. A. Buhrman and L. D. Jackel: IEEE Trans. Magn., in press.
10. P. K. Hansma: J. Appl. Phys. 44, 4191 (1973).
11. O. H. Soerensen,
12. S. N. Erne, H. D. Hahlbohm and H. Lübbig: J. Appl. Phys.
13. R. Rifkin, D. A. Vincent, B. S. Deaver and P. K. Hansma: J. Appl. Phys. 47, 2645 (1976).
14. V. V. Danilov and K. K. Likharev: Sov. Phys.-Tech. Phys. 20, 697 (1976).
15. J. Kurkijärvi and W. W. Webb: IEEE Pub. No. 72CHO682-5-TABSC (IEEE, New York, 1972), p. 581.
16. J. Kurkijärvi: J. Appl. Phys. 44, 3729 (1973).
17. L. D. Jackel and R. A. Buhrman: unpublished.
18. J. H. Claassen: J. Appl. Phys. 46, 2268 (1975).
19. M. R. Gaerttner operating at 440 MHz with a cooled preamplifier has obtained an equivalent flux noise spectral density $\langle\delta\phi_N^2(\nu)\rangle^{\frac{1}{2}} \simeq 7\times10^{-6}\phi_o/\mathrm{Hz}^{\frac{1}{2}}$ with a toroidal point contact device having $L_s \simeq 1\times10^{-9}$H.

OPTIMUM RESPONSE AND DAMPING OF JOSEPHSON JUNCTIONS

C.D. Andriesse
Kapteyn Astronomical Institute, University of Groningen
Groningen, The Netherlands

Abstract: An interpolation formula is discussed for current-voltage characteristics of capacitively loaded Josephson junctions with thermal noise. The wide-band response is found to peak for a particular value of β_c*, which is probably associated with critically damped oscillations of the phase.*

When polychromatic radiation falls on a Josephson junction, the effective noise temperature T^* is raised. The wide-band response of the junction can therefore be related to theoretical current-voltage characteristics for various noise levels. Some time ago Kurkijärvi and Ambegaokar have treated the thermal fluctuations in a resistively shunted junction with a capacitive load. They found that the capacitive load, which alleviates the damping, leads to a crossing of the characteristics computed for different T^*, and thus predicted a sign reversal in the wide-band response. Their results are of practical interest, because point-contact junctions, that optimally couple to external fields, are little damped. The characteristics can play a rôle in the optimization of the wide-band response, but, as they result from lengthy computations, the data are known for few parameter values only.

Let this be a justification of my attempt to approximate the results of Kurkijärvi and Ambegaokar by the following interpolation formula,

$$i/v = 1 + \beta_c^{3/2} \left[(1 + \beta_c^2/25)\ i\right]^{-\sqrt{\beta_c}/\Gamma} .$$

Here $i = \bar{I}/I_c$, $v = \bar{V}/(I_c R)$, $\beta_c = 2eI_cR^2C/\hbar$, $\Gamma = 2ekT^*/(\hbar I_c)$ and the other symbols have their common meaning. In Fig. 1 a comparison is made with the computed characteristics for $\beta_c = (1,4)$ at the noise level

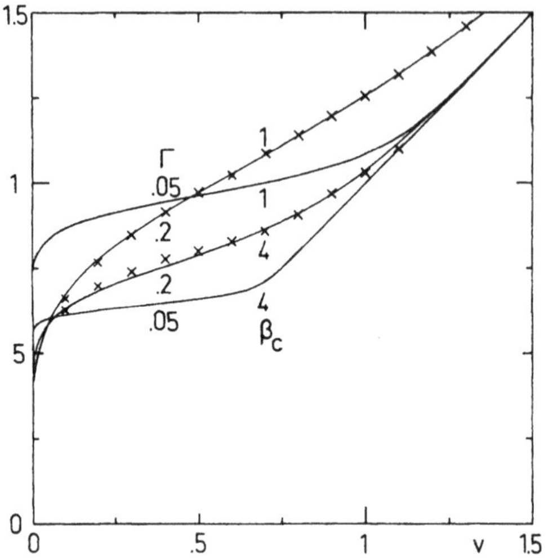

Fig. 1 Current-voltage characteristics according to the proposed formula for the given parameter values (curves) and according to Kurkijärvi and Ambegaokar (crosses).

$\Gamma = 0.2$. In particular for $\beta_c = 1$ the agreement is good. Note that the above mentioned crossing occurs at the correct position and that the apparent critical current is correctly reduced. When the noise is low (case $\Gamma = 0.05$), however, the connection with the pure resistive part of the characteristic is not represented very well. The approach to this asymptote is somewhat too fast, so that the negative response (at the high-voltage side of the crossing) is overestimated. On the other hand it can be expected that the formula gives a fair prediction of the positive response.

The corresponding quantity $(\partial v/\partial \Gamma)$ is shown in Fig. 2 as a function of v for various β_c at the noise level $\Gamma = 0.2$, which is a practical value at helium temperature and a critical current of about 1 μA. Maximum positive response occurs for $v < 0.22$ and it appears rather sensitive for the damping parameter. This is further shown in Fig. 3, where the maximum of $(\partial v/\partial \Gamma)$ is plotted versus β_c. The maximum response peaks for $\beta_c = 0.4$: any increase (decrease) of the damping below (above) $\beta_c = 0.4$ leads to a sizeable reduction of the response. Is this an artefact of the formula, or is it a real feature in the computations by Kurkijärvi and Ambegaokar?

To answer this question I first present as yet unpublished response measurements by Tolner, who experimented with series of niobium point contacts, in which the normal resistance R was varied such that $R \propto \beta_c$. The experimental responses also show a maximum. Data from three of these series are added to Fig. 3, where the R-values are scaled to fit the maximum at $\beta_c = 0.4$ and the response values to fit the wings of the curve. Although the agreement is not perfect for each of the series, the predicted dependence is generally confirmed. Then remains to discuss whether a maximum should occur at $\beta_c = 0.4$ or that this value is somehow due to the particular form of the formula.

Guessing that the value 0.4 contains at least a hint, we look for a possible special physical situation in the junction at this damping. Kurkijärvi and Ambegaokar already noted that the noise knocks phase oscillations into phase slips (and vice versa), which then increases

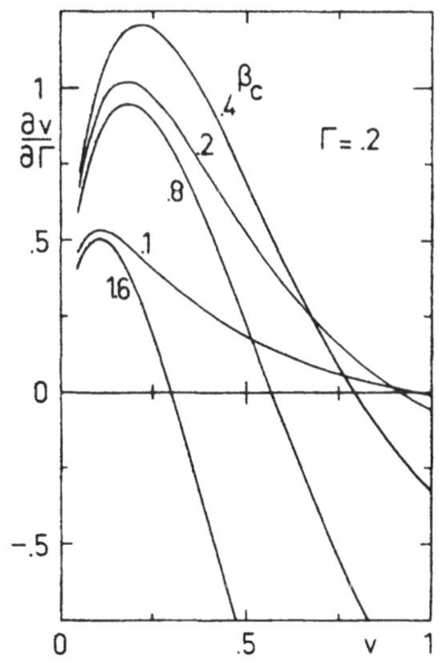

Fig. 2 Response as a function of the voltage for the given damping parameter β_c *at the noise level* $\Gamma = 0.2$.

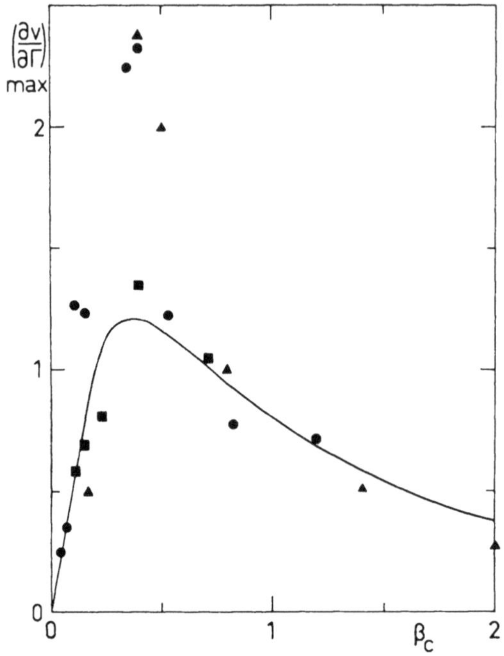

Fig. 3 Dependence of the maximum response on the damping parameter at the noise level Γ = 0.2 (curve). The three different symbols refer to experimental data by Tolner, fitted to the curve (see text); the discrepancy near $\beta_c = 0.4$ *can partly be ascribed to the enhanced coupling of the junctions (here R ∿ 100 Ω) to external fields.*

(decreases) the junction voltage. The ease of this process is evidently linked to the damping of the oscillations: heavy damping makes it difficult, so that the voltage change is small, and light damping so easy that the reservoir of the oscillations is readily drained in favour of that of the slips, so that again the voltage change is small. The process probably goes best for critical damping. Critical damping of the phase oscillations occurs at $\beta_c = 1/(4 \cos\langle\phi\rangle)$, where $\langle\phi\rangle$ is the thermal average of the phase difference. The phase average depends on the noise level by $\cos\langle\phi\rangle \simeq \Gamma^{1/3}$, which at $\Gamma = 0.2$ equals 0.58. This gives credit to the physical reality of the maximum at $\beta_c = 0.4$, for which $\cos\langle\phi\rangle$ would have the very acceptable value of 0.63.

Our game with the characteristics computed by Kurkijärvi and Ambegaokar leads thus to the following conclusion. The proposed formula gives a physically meaningful interpolation, despite its unphysical background, and suggests that maximum positive response to wide-band radiation is reached for critically damped oscillations of the phase. Requiring then that $\beta_c = 0.4$, a design criterion follows for the RC-time of point contacts. In the case of niobium, where in practice $I_cR = 1$ to 1.5 mV, it should be about 0.1 ps; having R of the order of 100 Ω for good matching, the desired capacitance C should be as low as about 1 fF.

I thank H. Tolner for the permission to present some of his measurements and J. de Vries for helping to prepare the figures.

Reference

J. Kurkijärvi and V. Ambegaokar, Physics Letters 31A, 314-315 (1970).

COMMUNICATION SYSTEMS

SQUID MAGNETOMETERS FOR SUBMARINE COMMUNICATIONS AT EXTREMELY LOW FREQUENCIES

John R. Davis and Martin Nisenoff
Naval Research Laboratory, Washington, D. C. 20375, USA

INTRODUCTION

Communications between shore stations and submerged submarines normally are effected at frequencies in the very low frequency (VLF) band, which extends from 3 to 30 kHz. Due to the relative opacity of seawater at these frequencies, however, receiving antennas must be placed within a few meters of the ocean surface for maximum signal-to-noise ratio (SNR), a circumstance that is achieved when noise of atmospheric origin is the controlling noise. At depths of more than a few meters, even rather large VLF receiving antennas are incapable of such sensitivity as will constrain receiver thermal noise to below average atmospheric noise.

The relative transparency of seawater for electromagnetic waves in the extremely low frequency (ELF) band, which extends from 30 to 3000 Hz, may permit submarines to operate much deeper if suitable receiving antennas can be provided. Based upon measured levels of ELF atomospheric noise [1],[2], the required receiver sensitivity is approximately 10^{-14} $T{\cdot}Hz^{-1/2}$ Present electric field (E-field) receiving antennas, which consist of electrode pairs towed at a separation of several hundred meters, can provide this required sensitivity. However, the dipole pattern exhibited by such an antenna imposes constraints on the maneuverability of the submarine, and either a different (omni-directional) antenna or an antenna with complementary pattern characteristics must be provided to eliminate these constraints. It is not practical to design a transverse E-field sensor with adequate sensitivity to complement the E-field trailing wire, and so a trailing wire magnetic field (H-field) antenna is necessary. However, the best demonstrated sensitivity for an H-field antenna trailed at speeds of operational interest is far below the required value. We are investigating the feasibility of using a superconducting quantum interference device (SQUID) in a magnetometer configuration to provide both the required sensitivity and the necessary omnidirectionality. As indicated in previous discussions of this program [3], [4], a number of engineering questions must be answered before the feasibility of a SQUID for this application can be determined.

One of the most serious problems relates to noise from rotation of the SQUID sensor in the earth's magnetic field of 5×10^{-5} T. Because the earth's field is uniform on the physical scale of the typical SQUID, a set of orthogonal sensors can be used to sum the vector components continuously and thus permit the earth's field to be removed as a DC component. This expedient makes the sensor omnidirectional, although using it thus as a scalar sensor also makes it sensitive only to the signal vector component parallel to the direction of the earth's DC magnetic field.

In order to use a triaxial SQUID magnetometer for this application, two principal problems must be solved: (a) providing enough dynamic range in the parts of the system ahead of the vector summing stage to prevent the SQUID electronics from being saturated, and (b) achieving orthogonality among the sensors. Assuming completely free rotation of the SQUID in the earth's field at, for example, a bandwidth of 1 Hz, the total dynamic range requirement is about 196 dB. This figure is far in excess of the available dynamic range of electronic components and, in fact, of SQUIDs themselves, and thus a certain degree of motional stability must also be provided. Assuming that electronics can be provided with linear dynamic range approximately equal to that which is available in SQUIDs, a figure of 140 to 150 dB, rotational noise of the sensor must be constrained to less than 10^{-3} rad·$Hz^{-1/2}$ to permit the necessary performance.

These circumstances suggest that there are six major requirements for demonstrating the feasibility of SQUIDs for underwater reception of ELF communications:

- Providing adequate electronic and sensor sensitivity to permit detection of the ELF signal in noise of roughly 10^{-14} T·$Hz^{-1/2}$.
- Providing 140 to 150 dB linear dynamic range.
- Constraining sensor motion noise to 10^{-3} rad·$Hz^{-1/2}$.
- Providing adequate orthogonality among vector sensors to permit motional noise to be removed by vectorial summation.
- Developing a signal processing technique to reduce the motional noise to an acceptable level in real time.
- Demonstrating that a suitable cryogenic environment can be provided for a SQUID, without unacceptable logistical complications, physical danger to the submarine, and unacceptable risk to submarine security.

It is our purpose in this paper to describe the SQUID magnetometer that was designed to satisfy these requirements and to describe the tests that were performed on the magnetometer and other components of the ELF receiving system to demonstrate the feasibility of this approach to submarine communications reception.

SQUID AND CRYOGENIC REQUIREMENTS

When the procurement of the SQUID system was started about two years ago an evaluation was made of the performance available from commercial systems and these characteristics were compared to what was required for ELF reception. In addition to obvious specifications on sensitivity and frequency response, it was decided to impose requirements on stability and drift of the system under the influence of environmental parameters, such as barometric pressure, temperature and humidity. The goal was to obtain a well-designed and engineered system that could be located at relatively remote, magnetically quiet sites and operated unattended for periods of days. Such stringent operational requirements had not been demanded of any previous SQUID system.

SQUID System Specifications

A summary of the required specifications for the triaxial SQUID magnetometer system needed for demonstrating the feasibility of a SQUID ELF H-field antenna is given in Table I. In the following sections, a brief rationale will be presented for each of these specifications.

TABLE I - Specifications for Triaxial SQUID Magnetometer

Sensitivity [dc to 200 Hz]	10^{-14} $T \cdot Hz^{-1/2}$
Analog Dynamic Range	140 dB (dc to 65 Hz) 130 dB (at 130 Hz)
Drift [24 hour period]	10^{-12} T (Max.)
Stability 0 to 40 C 0 to 80 % relative humidity	10^{-12} T (Max.)
Linearity	130 dB or greater
Mechanical Orthogonality of Coils in each channel	10^{-4} rad or less

Sensitivity. An instrument noise level of 10^{-14} $T \cdot Hz^{-1/2}$ is the required sensitivity for an H-field antenna to satisfy the overall ELF system needs as outlined earlier in this paper.

Frequency Response. The frequency response needed for ELF communications is stated to be from 30 to 130 Hz. Because of the motion-induced signals resulting from the rotation of the vector SQUID sensor in the earth's magnetic field, the frequency response for the individual SQUID magnetometers was specified to be DC to 200 Hz. The vibrations and oscillations of most mobile platforms, such as those suitable for supporting the SQUID antenna, occur at frequencies near 1 Hz, and thus the wider frequency response would allow the monitoring of the motion-induced signals and permit suitable data processing to be performed to cancel these undesired signal components.

Drift and Stability Requirements. In order to be able to operate the system unattended at relatively remote sites, the requirement was established that during any 24 hour period, the maximum permitted drift of the output of an individual channel of the SQUID system, with the magnetic field sensing coils surrounded by a superconducting shield should not exceed 10^{-12} T for the case of a freely vented helium dewar and for ambient temperature variations between 0 and 40 C and for relative humidity variations between 0 and 80 percent.

System Configuration. In order for personnel to make electrical adjustments to the system, record output signals, etc. without interfering either magnetically or seismically with the operation of the system, it was required that controls for adjusting the SQUID electronics and the output terminals for each channel had to be located at the ends of cables that were at least 100 m long.

Intermodulation Distortion. As mentioned above, the SQUID is a vector device, and there will be components in the output of a SQUID magnetometer system associated with the motion of the sensor in the earth's magnetic field. These motion induced signals can be orders of magnitude larger than the signal of interest. Therefore, a specification was imposed that the intermodulation products produced by the mixing of two intense signals within the response band of the SQUID electronics shall not have an amplitude greater than minus 130 dB relative to the sum of the amplitudes of the two intense signals.

Dynamic Range and Slewing Rate. For a stationary SQUID magnetometer used for receiving signals in the band from 30 to 130 Hz, a large dynamic range--the ratio of incident signal amplitude that would result in SQUID system resetting or "flux-jumping" to the rms instrument noise--is not needed. However, for a SQUID magnetometer moving in the earth's magnetic field the dynamic range must be large enough to handle motion-induced signals without resetting. Since the specified instrument sensitivity level is 196 dB below the magnitude of the earth's magnetic field, a freely rotating SQUID system would require a dynamic range with a magnitude comparable to this value. Such linear dynamic ranges are beyond current state of the art electronics. Thus a compromise had to be reached between the largest linear dynamic range that could be obtained and imposing some restriction on the orientational stability of the platform on which the SQUID antenna might be mounted. Platform stabilities of the order of 10^{-3} rad have been achieved by a (passive) hydrodynamically designed towed body, as will be shown below. Accordingly the requirement on linear dynamic range for the electronics was established to be 140 dB. SQUID systems that had been built earlier were reported to have had dynamic ranges of the order of 110 to 120 dB and thus requiring 140 dB was not an excessive request. In principle, this required that the dynamic range should extend out to 130 Hz, the upper end of the specified frequency range for ELF communications. However, the difficulty of building a SQUID feedback system with 140 dB linear dynamic range from DC to 130 Hz was too severe, and furthermore there is little likelihood that the motion-induced noise at high frequencies will exceed the system noise level by 140 dB. Thus the requirement was eased to a dynamic range of 140 dB from DC to 65 Hz and then rolling off with higher frequencies but exceeding 130 dB at 130 Hz. This relaxation of the dynamic range was not considered a serious degradation in system performance.

Orthogonality of Pick-up Loops for the Three SQUID Channels. As will be described in detail elsewhere in this paper, the technique selected for cancelling motion induced signals from the output of the SQUID antenna was to form the sum of the squares of the outputs of three orthogonal SQUID magnetometers. This procedure would give a rotationally invariant quantity proportional to the square of the magnitude of the earth`s field. If the three sensors are not "precisely" orthogonal, terms in addition to the squares of the outputs of the individual channels must be included. These additional terms are "cross-products" between the outputs of two of the channels. It can be shown that these "cross-terms" can be made negligible compared to the square of the output of each channel if a restriction is placed on the maximum departure from orthogonality between various SQUID sensors. If the departure from orthogonality is represented by angle Ω where $\Omega \ll 90^{\circ}$ then the following inequality must be satisfied

$$\Omega < \frac{1}{\theta_{max}{}^{2}} \quad \frac{H_{IN}}{H_{e}}$$

where θ_{max} is the maximum rotational instability of the platform on which the SQUID system is mounted (expressed in radians), H_{IN} is the rms instrument noise of the SQUID sensor and H_e is the earth`s magnetic field (approximately 5×10^{-5} T.)

For a platform orientational instability of the order of 10^{-3} rad, and for SQUID instrument rms noise of $10^{-14} T \cdot Hz^{-1/2}$ the above relation imposes an upper limit of 10^{-4} rad on the non-orthogonality among the three channels. Since verification of magnetic orthogonality requires very elaborate and expensive facilities not readily available to most commercial companies, the only requirement placed on orthogonality was a mechanical one, that is that the support structure on which the pick-up loops for each of the three channels were mounted must be constructed such that the physical orientations of the loops were orthogonal to each other to within 10^{-4} rad. For a support block approximately 8 cm in dimension this requirement meant that the pickup loops had to be located to within plus/minus 4 micrometers of the specified position.

Cryogenics. There were two distinct requirements for dewars to be used with this SQUID antenna. For field tests and feasibility studies with the sensor stationary, a vertical dewar is satisfactory. To allow for operation in remote sites on the surface or for extended test periods below the surface of the ocean, a seven day hold time dewar is considered adequate.

In addition to an experimental dewar, a prototype dewar that might be suitable for sea tests was also acquired. For sea tests, the dewar would have to be mounted in a buoy with its long dimension horizontal. In addition, since replenishment of the helium supply would be difficult if the buoy were deployed behind the towing vehicle, a hold time between refilling of 90 days was arbitrarily chosen in order to demonstrate that a static helium dewar could be suitable for these assumed operational conditions.

Description of Systems

The original intent was to purchase a single three axis SQUID magnetometer system for evaluation of the SQUID ELF H-field antenna concept. However, in the proposals received from the several commercial SQUID manufacturers, two rather distinct design concepts evolved and thus two systems were purchased in order to evaluate and compare the performance that could be achieved with each of these designs.

Point Contact SQUID System. One system, built by the SHE Corporation, San Diego, CA, USA used a bulk toroidal niobium SQUID sensor with a "permanently" adjusted point contact [5]. This type of sensor had just become commercially available, and there was very little experience on reliability of operation of such a sensor under field conditions. However, preliminary tests in the laboratory indicated that the performance of this type of sensor was at least comparable to the thin film SQUID sensor that had been available commerically for several years.

In the SHE system, the rf bias circuitry was located at the top of the dewar and coaxial transmission lines connected them to the SQUID sensor located in the liquid helium reservoir. (See Fig. 1) A 100 m cable connected the

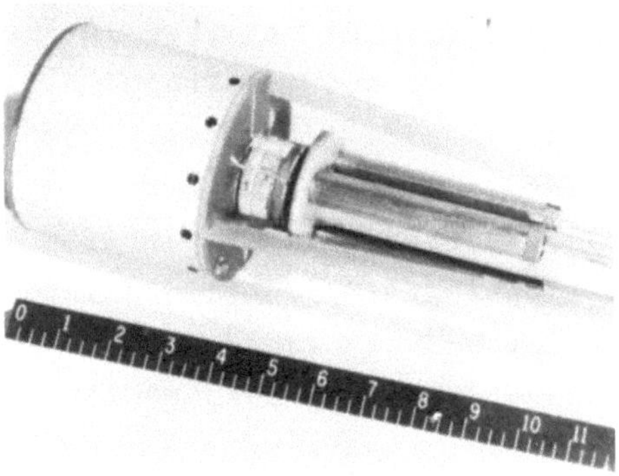

Figure 1 Lower end of probe from SHE system showing transmission lines, superconducting shield surrounding SQUID sensors and fiberglass shield around flux transformer assembly. Scale is in inches.

electronics at the dewar top to the control chassis that contained the feedback circuitry, the output terminals for each of the channels and the remote controls for adjusting and tuning the rf bias circuits. The interconnect cable carried the detected rf signal from the dewar to the control box, the processed feedback signal back to the dewar as well as the control voltages used to tune and adjust the rf circuits located at the top of the dewar. The various leads in the interconnecting cables had to be very carefully shielded in order to reduce possible interferences between the various sets of leads in the cable and from external noise sources.

The magnetic sensing coils for the three SQUID channels were wound on a quartz cube (See Fig. 2). The cube, approxi-

Figure 2 Flux transformer assembly for SHE system showing precision ground quartz cube and primaries of flux transformers for each channel (dark lines). Scale is in inches.

mately 8 cm (3 inches) on edge was ground and polished with adjacent faces orthogonal to one another to better than 10^{-4} rad. Grooves to accept the pickup coils were ground into the quartz block midway between each set of parallel faces. When the niobium pickup coil was laid into a given groove, the plane of the loop was parallel to the corresponding set of faces of the cube to better than 1 part in 10^{-4}. In order to insure maximum magnetic orthogonality between these coils all superconducting lead wires, superconducting shields, SQUIDs etc. had to be positioned about a symmetry axis of the cube to insure minimum mutual inductance between the coils.

The last major distinguishing feature of the SHE system was that the feedback signal was injected in series into the flux transformer by means of a superconducting DC transformer. (See schematic circuit diagram in Fig. 3). Accordingly when

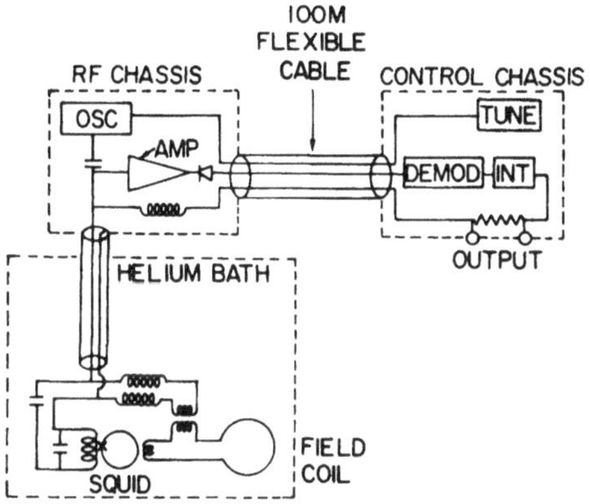

Figure 3 Schematic circuit diagram of individual channel of SHE system.

the feedback circuit was closed, the feedback signal maintained the current flowing in the flux transformer at zero, both in the primary sensing coil and the secondary coil that coupled to the SQUID sensor.

Thin Film SQUID System. The second SQUID system, built by Develco, Inc., Mountain View, CA, used thin film SQUID sensors. This type of device is fabricated from a cylindrical thin film of an indium-tin alloy deposited onto a quartz tube. At one point along the circumference of the film a very narrow constriction had been sculptured. This type of sensor had been used in a large number of SQUID systems built during about three years prior to this procurement, and its characteristics were fairly well documented [6].

In the Develco system, both the rf bias and feedback circuitry were located at the top of the dewar. (See Fig. 4). A 100 m cable connected the electronics at the top of the dewar to a

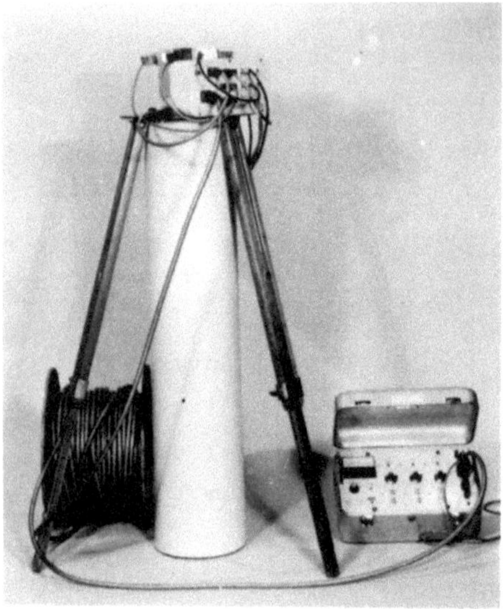

Figure 4 Photograph of triaxial SQUID magnetometer system built by Develco, Inc.

terminal box where the output signals from the three channels and voltage control leads for adjusting the rf bias circuitry were located. By placing the feedback circuitry at the dewar, possible pickup of external noise by the leads between the feedback circuitry and the dewar was minimized compared to the configuration where the feedback signal had to be transmitted down a 100 m interconnecting cable as in the case of the SHE system. (External pickup by the feedback leads of the SHE system was never a problem as these leads had been adequately shielded against external pickup.)

Develco proposed to use Helmholtz coils for the primaries of each of the three flux transformers. A quartz cube was ground and polished to a precision of better than one part in 10^4. Quartz discs were positioned at the center of each face of the cube and cemented in place. Niobium wires were wound around each of these discs and the loops on opposite faces were connected in series to form the primary of a given flux transformer. The construction tolerances required for achieving a given degree of orthogonality between three sets of Helmholtz coils mounted on the faces of a cube is much less stringent than required for the same degree of orthogonality for three simple coils mounted on adjacent faces of the cube. As in the other system, great care had to be taken in positioning superconducting lead wires, superconducting shields and SQUID assemblies so as to maintain symmetry along one axis of the cube and thus minimize mutual coupling among the set of pickup coils.

The last distinguishing feature of the Develco system was that the feedback signal was applied in parallel to the secondary, and thus also to the primary, of the flux transformer. The sense of the feedback signal was such that, when the feedback loop was closed, the current flowing in the secondary was zero while the current in the primary of the flux transformer was twice the current that flowed into the primary when the feedback circuit was closed.

Evaluation of the SQUID Systems

The performance characteristics of each channel of the two systems were checked in a magnetically shielded environment--first in a superconducting shield mounted in the experimental dewar and then in the Low Magnetic Field Facility at the National Magnet Laboratory in Cambridge, MA, USA--and verified, whenever feasible, at a remote magnetically quiet site. The performance of both systems was in compliance with the specifications listed in Table I.

The only serious problem encountered in using these systems for measurements in the surface environment was the problem of "flux-jumping" caused by electromagnetic pulses associated with thunderstorm activity in the vicinity. Although these pulses had frequency components that were well above the upper frequency cut-off of the system, the leading edges of these pulses were so abrupt and the amplitudes so intense that the signal leaked into the SQUID system and exceeded the "slewing" rate of the feedback system, resulting in random resettings. To reduce the occurrence of these resets to an acceptable level, an eddy current shield (an aluminum can with 0.2 cm thick walls), which provided a 3 dB roll-off at 100 Hz, was placed around the dewar. (For operation below the surface of seawater, the aluminum eddy current shield was not needed as the seawater effectively attenuated these electromagnetic pulses.)

Both systems have been used for surface reception tests at a magnetically quiet remote site near Washington, D. C.. In addition, the SHE system has been operated at a depth of about 100 m below the ocean surface off the coast of California for underwater reception tests. These tests demonstrated that this type of equipment can be operated for extended periods of time unattended except for weekly replenishing of the liquid helium and periodic replacing of the data tapes.

Magnetic Orthogonality of Coils. The mechanical orthogonality of the magnetic field sensing coils for each system was better than 1 part in 10^4, the specified requirement. The magnetic orthogonality of the coils in the SHE system was measured at the NASA Goddard Space Flight Center, Greenbelt, Md., USA. This facility consists of three nominally orthogonal sets of 13 m diameter Helmholtz coils. A feedback system connected to the coil system can cancel the earth's magnetic field at the center of the coil system to better than 10^{-9} T over a 2 m diameter spherical volume. In addition, a magnetic field of amplitude 5×10^{-5} T known to an accuracy of at least 10^{-9} T can be applied with any desired orientation. The SQUID system was placed at the center of the Helmholtz coil system and the ambient field cancelled to less than 10^{-9}T, and then a field of $5\ 10^{-5}$T was applied and its orientation varied in order to determine the magnetic axis of the primary (magnetic field sensing) coil of each of the three flux transformers. The result of these measurements was that the magnetic orthogonality of the three channels of the SHE system was of the order of or less than $2x10^{-4}$ radians, the uncertainty in the orthogonality of the coils in the Greenbelt facility. With this degree of orthogonality between the channels and stability of a towed body comparable to the one to be described later, it should be possible to treat this triaxial system as if it were orthogonal and not have to take into account explicitly "cross-terms" involving the product of the outputs of two SQUID channels.

Cryogenics

Experimental Dewar. For evaluation of the SQUID systems and for demonstrating the feasibility of the antenna concept, a superinsulated fiberglass dewar was supplied with each SQUID system. The dewars were 1.3 m tall and 0.3 m in outer diameter and held helium for approximately 7 days with the SQUID probe in position. An eddy current shield was built into the vacuum space of the dewar to attenuate high frequency signals, and provided a frequency response flat to within 0.5 dB from DC to 200 Hz.

Long Hold Time Dewar. A study was made of the various methods that could be used to cool a SQUID antenna in an application in which it would be buoy-mounted and towed behind a submarine. [7] The conclusion was that the simplest and most reliable cooling technique would be to mount the SQUID antenna in a superinsulated fiberglass dewar with a helium hold time of 90 days, which would be long enough to exceed the duration of a typical submarine mission.

To verify that such a dewar could be built with the dimensions and configuration that would enable it to be mounted in a typical towed buoy, a prototype dewar was built by Superconducting Technology, Inc. of Mountain View, CA. USA. A photograph of the dewar is shown in Fig. 5, and a diagram of it is shown in Fig. 6. The outer vacuumn jacket,

Figure 5 Photograph of 90 day hold-time dewar suitable for sea tests.

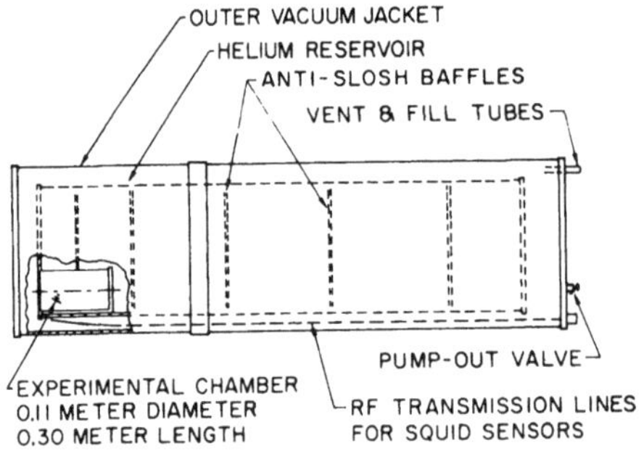

Figure 6 Schematic diagram of long-hold time dewar.

helium reservoir and experimental chamber are made from G-10 fiberglass using epoxied joints. The space between the helium reservoir and the outer vacuum jacket is filled with layers of a special cryogenic insulation. The insulation consists of alternating layers of aluminized mylar, which serve as thermal radiation shields, and a soft spacer material of spun fiberglass known as Dexiglass. Interspersed between layers of superinsulation are three copper strip radiation shields that are thermally anchored to the helium gas vent tube. In order to support the weight of the helium reservoir, fiberglass "pads" are judiciously spaced between the reservoir and the outer walls. The primary heat leak into the helium reservoir is along these support members.

In addition to the fiberglass support pads, the inner chamber is connected to the outer wall by a fill tube and a vent tube. The vent manifold is above the liquid-gas interface for all combinations of roll, pitch and yaw, each up to at least 15 degrees relative to the nominal upright orientation. The antislosh baffles prevent the build-up of larger oscillations in the liquid helium during movement of the dewar.

The experimental chamber, shown in the cutaway view in Fig. 6, is large enough to accept the SQUID sensor package. Coaxial transmission leads running from connectors on the end plate to a point near the experimental chamber were installed during construction for eventual use with the SQUID sensor package.

Hold times were established by a carbon resistor thermometer mounted in the experimental chamber. Temperatures deduced from the resistance of this thermometer were corrected for daily variations in barometric pressure. An end to the useful hold-time of the dewar was considered to have occurred when the temperature rose by 10 millikelvin. The dewar achieved a hold-time of 102 days, exceeding the required 90-day hold-time by 12 days. A summary of the characteristics of this dewar are given in Table II.

These results are indeed encouraging that a 90 day hold time dewar can be built for use with the SQUID ELF antenna. However, the dewar described here was only designed to withstand "normal handling". The next dewar must be designed to comply with the shock and vibration specifications required for operational towed buoy systems.

TABLE II Summary of Dewar Characteristics

Physical	
Size	0.56 m diameter
	1.83 m length
Liquid Helium Volume	180 liters
Weight	190 kg (empty)
	209 kg (full)
Performance	
Measured Hold-Time	102 days
Average Boil-Off Rate	1.8 liters/day
Frequency Response ($\pm$0.5 dB)	DC to 300 Hz

AUXILIARY SYSTEMS

Motion Noise Control

The ELF noise environment of submarines has not yet been measured thoroughly enough to determine whether a SQUID could be operated in a location on the hull or superstructure. However, submarine-generated electromagnetic noise may make it necessary to tow the SQUID in a buoy at a distance of as much as one hundred meters behind the submarine. The degree of hydrodynamic stability that can be engineered into the buoy then will determine whether the sensor motion noise can be constrained to less than 10^{-3} rad·Hz$^{-1/2}$ as specified above. To verify the achieveability of this seemingly rotational noise requirement, a series of rigorous motion tests were performed on a buoy that has been designed by the U.S. Naval Ship Research and Development Center to be hydrodynamically stable. This buoy, shown in Figure 7, has high-lift hydrofoils that position it well above the submarine wake. It is 3.7 m long and 0.5 m in diameter, and thus encloses a large enough volume to house the large dewar described above as adequate for 102 days continuous operation.

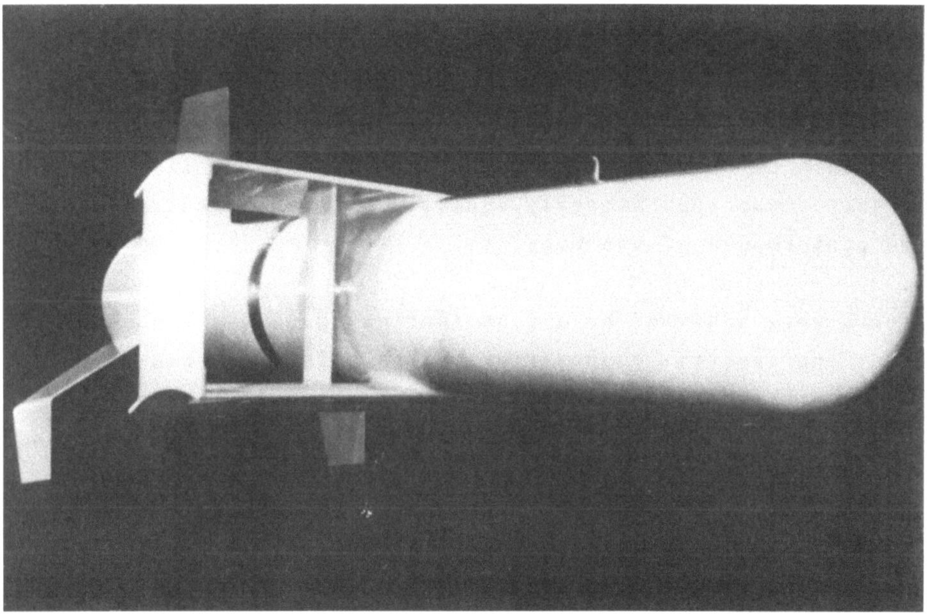

Figure 7 Photograph of towed communications buoy used for motion spectrum measurements.

The noise bandwidth of concern for ELF signal reception extends from 30 to 130 Hz. Within this band, motion noise must be constrained by a combination of sensor motion stabilization and further signal processing (in this case, the vector-summation procedure that removes the earth's magnetic field motion noise component) to below the nominal receiver noise of 10^{-14} T·Hz $^{-1/2}$. In addition, saturation of the receiver must be prevented at all frequencies within its passband. For a SQUID magnetometer, which has essentially a flat frequency response from near-DC to an upper limit determined by electronic filtering in its feedback control system, this requirement can be achieved only if motion noise is constrained to less than 10^{-3} rad·Hz$^{-1/2}$ from DC to 130 Hz. Thus the low-frequency motions of a buoy designed to carry such a SQUID are quite important.

The buoy was instrumented with a triaxial rate gyroscope and a triaxial linear accelerometer. The former device was selected for its high sensitivity at low frequencies (DC to 20 Hz) and the latter was selected for the higher frequencies (10-100 Hz). In the case of the rate gyroscopes the sensitivity limit at low frequencies was approximately equivalent to the required motion-noise performance of the buoy.

The data were analyzed by a Fast Fourier Transform technique to display the spectral composition of the noise, and examples of noise spectra from the pitch sensor are shown in Figure 8 for towing speeds from 2 to 7 knots. These data, which correspond in each case to spectral analysis of about 25 seconds of data, show low frequency motions of the buoy that increase in bandwidth as speed is increased. There is no consistent evidence of increase in the amplitude of these motions with increasing tow speed. The maximum amplitude in the data is about 0.7 millirad/sec. and occurs at the lower speed. Figure 9 is an integrated version of the data in Figure 8 and therefore represents motional excursions of the buoy in the 0 to 10 Hz band. These integrated data show the maximum excursions of pitch to be between 0.1 and 1 millirad in the 0-10 Hz band, depending on speed. Note that the data appear very near instrument noise, and thus the true motion noise may be considerably less.

Typical data from the accelerometers, which provide information at higher frequencies, are shown in Figure 10a. These are also pitch data and are shown as angular acceleration (the accelerometer output parameter) in millirad/sec^2 in Figure 10a; a doubly integrated version, indicating angular exclusion in millirad, appears in Figure 10b. The spectral peaks at 42 and 56 Hz are due to vortex shedding from blunt trailing edges of the buoy hydrofoils and indicate that buoy motion appraoched 10^{-2} millirad at these turbulence-generated structural resonant peaks. Fairings were attached to the horizontal control surfaces to evaluate the

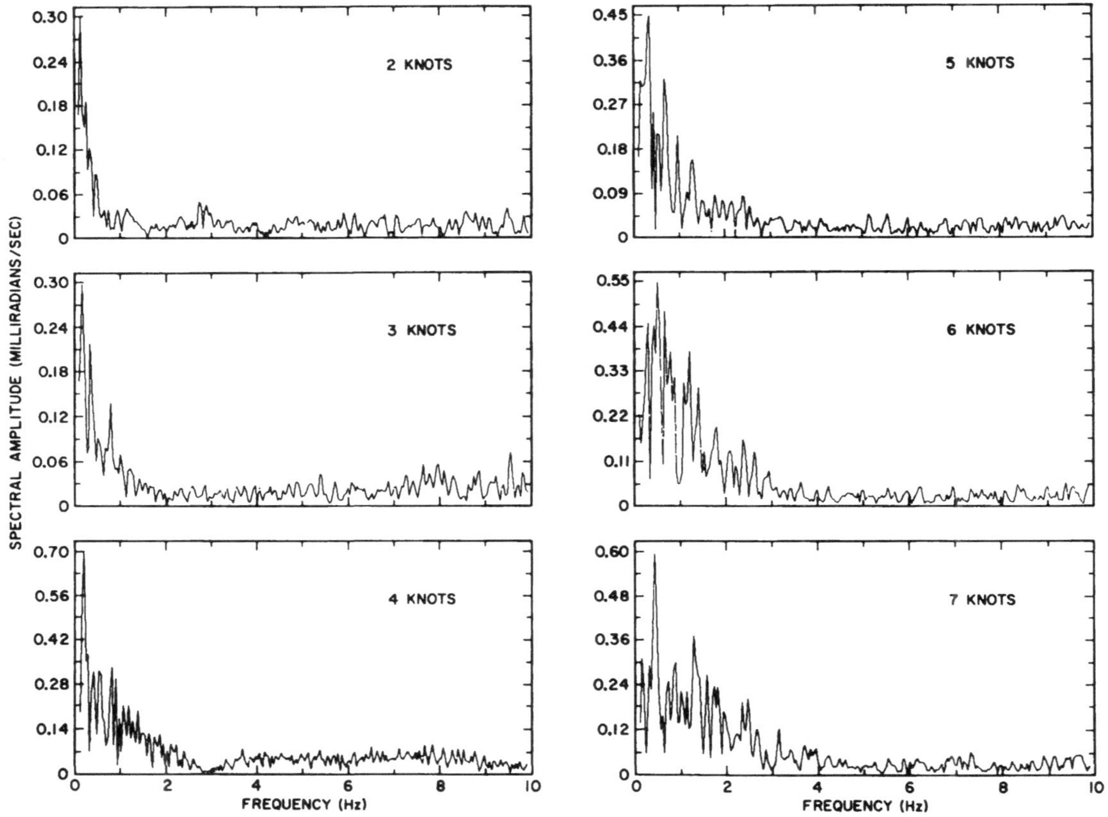

Figure 8 Spectra of the pitch rate gyroscope output for towing speed of 2 to 7 knots showing low-frequency buoy rotation rates.

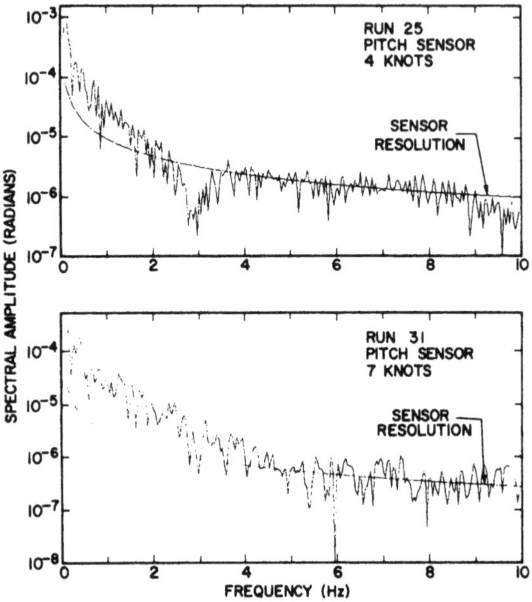

Figure 9 Integrals of two of the spectra shown in Figure 2, showing pitch angular excursions. In each case the sensor resolution curve indicates the sensor noise level.

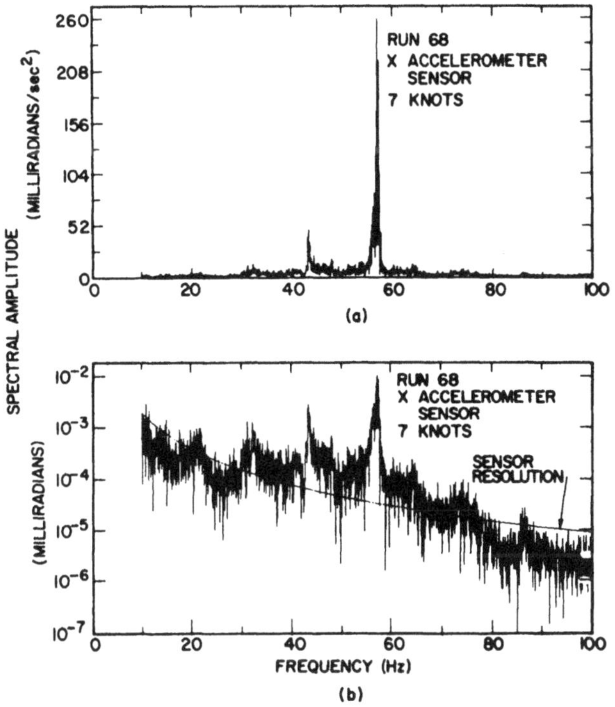

Figure 10a Spectrum of the pitch-sensitive accelerometer output for 7 knots towing speed, showing high frequency buoy accelerations.

Figure 10b Spectrum of doubly integrated pitch-sensitive accelerometer output for 7 knots towing speed, showing pitch angular excursions.

feasibility of reducing these vibrational resonances, and Figure 11 shows the resulting accelerometer output. Compared with Figure 10a, it may be seen in Figure 11 that the main resonant peak at 56 Hz has been reduced to below noise level and the peak at 42 Hz has been reduced slightly. The addition of fairings to other control surfaces and supporting struts would reduce these components still further, but it is readily apparent from Figures 10 and 11 that motional excursions within the 30-130 Hz ELF bandwidth can be constrained to be less than 10^{-6} rad. Data from the roll and yaw

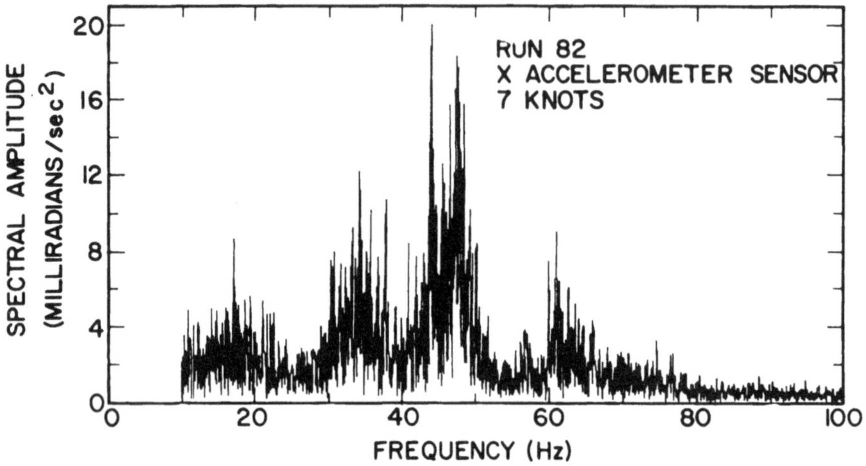

Figure 11 Spectrum of the pitch-sensitive accelerometer output for 7 knots towing speed after insertion of fairings. The spectrum noise floor can be observed between 80 and 100 Hz.

sensors at these and other towing speeds gave similar results to those shown above. The static angle of attack of the buoy, between limits of $\pm 3.5^{o}$, was shown to have no discernible effect on the motion spectra.

The motion stability measurements may be summarized as follows: The low frequency excursions, which are 1 millirad at most, are the largest in amplitude and hence establish the necessary dynamic range. Considering the 25 second integration time of these data, the maximum corresponding noise level would be 0.5 to 5 millirad$\cdot Hz^{-1/2}$ and would decrease with increasing speed. These findings are interpreted as confirming the adequacy of the dynamic range goal of 140-150 dB established above. The in-band motion of 10^{-6} rad generates a maximum signal of $2x10^{-10}$ T, which is 86 dB above the receiver sensitivity in the noise bandwith of 0.06 Hz used for our analysis. The signal processing technique that is discussed below is capable of reducing the in-band motion noise by this amount.

These low levels of motion noise were achieved in the artificial environment of a towing channel and cannot be claimed as proof that similar performance can be expected in the open sea with the buoy at the end of a long cable. Indeed, large amplitude yawing instabilities can occur in buoys towed with cables of certain resonant lengths. Previous sea trials of this buoy, with somewhat less sensitive motion sensors than the apparatus we used, showed no evidence of such yawing motion. Furthermore, the high-lift hydrofoils are designed to keep the buoy clear of the submarine wake, and its intended towing depth on the order of one hundred meters is well below the oceanic regime that is affected by turbulence from surface wave motions.

It has been our purpose in this section to illustrate the motion noise measurements by example and not to present an exhaustive proof that the motion stability requirements can be met under all conditions of buoy speed, depth and environmental turbulence. For a more extensive description of the motion-noise measurements and further discussion of problems anticipated in ocean operations the reader is encouraged to consult Reference [8].

Motion Noise Signal Processing

The towing tests described above indicate that buoy motion in the 30-130 Hz band can be constrained to be less than 4×10^{-6} rad. However, additional motion compensation must be employed to reduce the level of motion-induced noise still further. The signal induced in a magnetic vector sensor rotated through angle Ω in an earth's magnetic field H_e from an initial orientation θ_o of the sensor axis relative to H_e is

$$(1) \quad H_M = H_e \left[\cos \theta_o \left(\cos \Omega - 1\right) + \sin \theta_o \sin \Omega\right]$$

For $H_e = 5x10^{-5}$ T and $\Omega = 10^{-6}$ rad, in the 0.06 Hz bandwidth that typifies the motion spectrum data described above the resulting H_M is 86 dB above the sensitivity requirement of 10^{-14} $T \cdot Hz^{-1/2}$. We discuss in this section an adaptive signal processing scheme that can provide this remaining degree of motion compensation.

The usual technique for removing motion noise in three orthogonal vector magnetic sensors consists of summing the squares of the sensor outputs to produce a quantity proportional to the field magnitude. This simple technique will not work for SQUID sensors because their output is independent of the absolute value of the magnetic field along the sensor axis. That is, the SQUID sensor follows external magnetic field <u>changes</u> very accurately, with a low noise level and true DC response, but the actual magnetic field value along the axis of the sensor is never known. However, this magnetic field value must be known at least approximately in order to remove motion noise. The essence of the technique described here is that by sampling the changes in magnetic field that occur as the SQUID sensors undergo motion in the earth's field, and knowing the functional form that these changes must assume, then the absolute values of the magnetic field components can be determined by parameter estimation methods. The technique must be continuously adaptive because the absolute values of the magnetic field components are not constant, but vary as the sensors move through the earth's field. The adaptation method that has been developed to provide this behavior is described in Ref. [9] and will be summarized here, together with results of a computer simulation of the adaptive process based on the motion-noise data described above. Some results of actual motion-processing tests of the SQUID ELF receiver will be described in a subsequent section of the paper.

For sensors that are perfectly orthogonal, a combination of the three sensor outputs may be formed that is independent of sensor orientation. This combination is simply a vector sum.

Defining $\vec{H}_M$ the sensor output due to motion in the earth's DC field $\vec{H}_e$, the quantity $|\vec{H}_e + \vec{H}_M{}^2|$ is invariant and equal to $\vec{H}_e{}^2$ because in the absence of signal and noise there is no other source of magnetic field strength than $\vec{H}_e$. Thus,

(2) $$|\vec{H}_e + \vec{H}_M|^2 = 2\vec{H}_e \cdot \vec{H}_M + \vec{H}_M^2 + \vec{H}_e{}^2 = \vec{H}_e^2$$

(3) $$2\vec{H}_e \cdot \vec{H}_M + \vec{H}_M^2 = 0$$

In addition to an output generated by motion in the uniform field of the earth, the SQUIDs also respond to the ELF signal vector, $\vec{H}_S$, so that the total sensor output is

(4) $$\vec{H}_o = \vec{H}_M + \vec{H}_S$$

Computing $2\vec{H}_e \cdot \vec{H}_o + \vec{H}_o^2$ and making use of Eq. (3) gives

(5) $$2H_e \cdot \vec{H}_o + \vec{H}_o{}^2 = 2\vec{H}_e \cdot \vec{H}_S + 2\vec{H}_M \cdot \vec{H}_S + \vec{H}_S^2$$

The ELF signal amplitude $\vec{H}_S$ is 10^{-14} T, as a lower limit, and the maximum in-band value of $\vec{H}_M$ is 5×10^{-11} T. Substitution of these values into the above equation indicates that the second and third terms on the right hand side are nearly six orders of magnitude lower than the first term. Hence, there remains an error quantity S defined as

(6) $$S \triangleq 2\vec{H}_e \cdot \vec{H}_o + \vec{H}_o^2 \sim 2\vec{H}_e \cdot \vec{H}_S$$

The quantity S is proportional to the signal vector and, when properly evaluated, is independent of sensor orientation. In Eq. (6), the signal vector is seen to be projected in the direction of the earth's magnetic field, and hence it should be borne in mind that the SQUID ELF sensor, while omnidirectional (i.e. it is equally sensitive to signals arriving from a given direction regardless of sensor orientation) has a response that will vary with location of the SQUID receiving system on the earth's surface.

The inclusion of the effects of sensor nonorthogonality, sensitivity differences, and offset errors does not affect the form of the equation for S. The values of the components of $\vec{H}_e$ are modified by these deviations from the ideal case; these modified components of $\vec{H}_e$ are relabelled and become the components A_1, A_2, A_3 of a vector $\vec{A}$ that is only approximately equal to $\vec{H}_e$. The equation for S takes the form

$$(7) \quad S = 2\vec{A} \cdot \vec{H}_o + \vec{H}_o^{\,2}$$

Note from Eq. (6) that when $\vec{H}_S$ is removed by filtering, the error quantity S is zero when the proper value of $\vec{A}$ is used.

The adaptive processing then proceeds as follows: the SQUID outputs are low pass filtered to remove $\vec{H}_S$ and the value of $\vec{A}$ found that minimizes S. The value of $\vec{A}$ found in this manner is then combined with the unfiltered SQUID outputs to give an output proportional to $\vec{H}_S$. This final output is free of motion-related noise.

An error analysis shows that for motion noise of $4\mathrm{x}10^{-6}$ rad$\cdot$Hz$^{-1/2}$ (the peak value of motion noise within the receiver bandwidth observed in the motion stability measurements) the components of the adaptive vector $\vec{A}$ must be determined with an accuracy of one part in 10^4. In order to determine $\vec{A}$ to this accuracy the fractional sensitivity errors of the SQUID outputs must be less than 10^{-2}; the deviations of the sensor axes from orthogonality have only to be less than 10^{-2} rad. As indicated above, orthogonality errors in the test device are no greater than $2\mathrm{x}10^{-4}$ rad. The fractional sensitivity errors in the SQUID electronics can easily be maintained to within one part in 10^5 over long periods of time.

The motion noise removal technique is summarized in Fig. 12. For

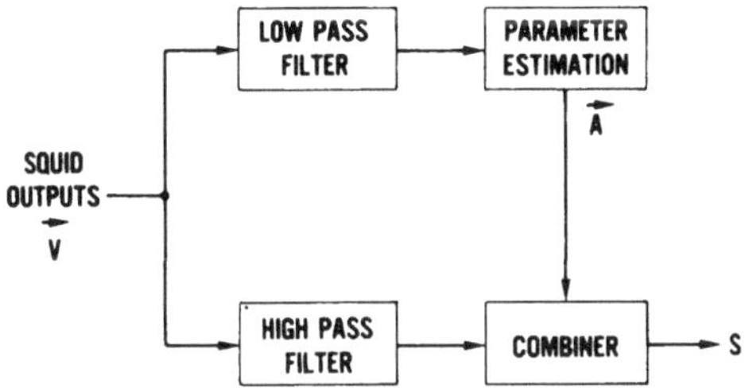

Figure 12 Block diagram of the motion processing method.

the determination of the parameter vector $\vec{A}$ in Eq. (7), the signal vector $\vec{H}_S$ is removed by low pass filtering of the SQUID outputs, leaving a residual SQUID output V. The signal $\vec{H}_S$ has a frequency in the region of 40 to 80 Hz, but in practice the filter cutoff frequency can be set considerably lower at about 10 Hz. It is the large amplitude excursions of the host buoy occurring below 10 Hz that allow $\vec{A}$ to be evaluated rapidly.

The vector $\vec{A}$ is determined so that Eq. (7) is satisfied. This value of $\vec{A}$, which is continuously evaluated and updated, is then combined with the high pass filtered outputs of the SQUIDs to form again the quantity S. This value of S is proportional to $2\,\vec{H}_e \cdot \vec{H}_S$, as indicated by Eq. (6).

The determination of $\vec{A}$ in Eq. (7) can be achieved by using techniques that have been developed for adaptive control. Two distinct techniques have been investigated to evaluate $\vec{A}$: a Least Squares Technique and a Steepest Descent Technique. The least squares methods of determining process parameters that were developed in [10] and [11] have been applied to the present problem. Steepest descent techniques are also discussed in [10], and in more detail in [12] and [13]. The least squares technique determines $\vec{A}$ by minimizing the integrated value of S^2 over a time interval T_o, while a continuous steepest descent method updates the current value of $\vec{A}$ according to the integrated value of S.

These two adaptive techniques have been investigated using a computer simulation of the SQUID sensors and the adaptation. A block diagram of the simulation is shown in Fig. 13. The

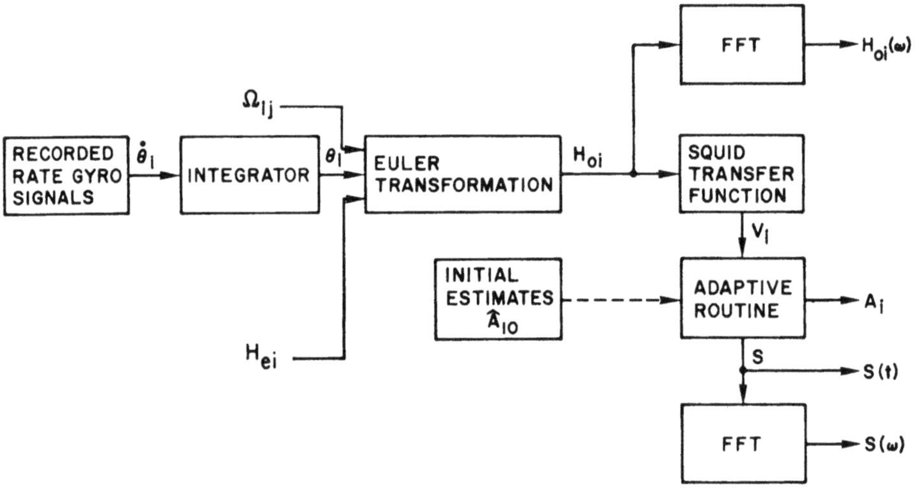

Figure 13 Block diagram of the computer simulation of the least squares and the steepest descent processing techniques. The A_{io} are required only for the steepest descent method.

input signals, labelled θ_i in Fig. 13 are the tape-recorded outputs of the three rate gyroscopes that were mounted on the buoy during the motion stability measurements described above. The outputs of the rate gyros are the angular time rates of change of the buoy orientation in three orthogonal directions. Integration of the rate gyro signals gives the angular excursions of the buoy. By combining the buoy excursions with assumed values of the components of the earth's magnetic field in an Euler transformation matrix, magnetic field signals are generated that simulate the motion-induced magnetic field variations that would have been observed if SQUID sensors had been mounted in the buoy. The effects of nonorthogonal sensors are included by a transformation from an orthogonal to a nonorthogonal coordinate system whose deviations from orthogonality are specified by the angles Ω_{ij}. The SQUID transfer function is then applied to give the simulated SQUID outputs V_1, V_2, V_3. The outputs V_1, V_2, and V_3 are the inputs to a computer simulation of the least squares fit and steepest descent processing schemes. The output of the adaptive processing, S, was Fourier analyzed by a standard Fast Fourier Transform program. The simulated SQUID outputs V_1, V_2, and V_3 were also Fourier analyzed for comparison with the Fourier transform of S.

The main emphasis of the computer simulation was to establish that the least squares and steepest descent processing techniques could indeed provide sufficiently accurate estimates of the A_i. The frequency response of the rate gyros was limited to a maximum of about 20 Hz, so that a detailed investigation of the noise in the output S in the 30 to 130 Hz ELF band was not possible. For the purposes of the simulation, the information below 20 Hz is sufficient. Indeed, it should be noted that in actual operation, as in the simulation, the large amplitude motion excursions that occur below 10 Hz provide the information by which the adaptive technique can determine the values of A_i and thereby remove motion noise above 30 Hz.

Although a number of runs at various speeds were employed in the computer simulation, the results of only one run are presented here. This run was typical of all runs analyzed. The speed was 7 knots and the run duration approximately 90 seconds, although only about 30 seconds of the run could be processed at one time because of computer limitations.

Least Squares Adaptation. Two different aspects of the least squares technique were investigated:

(1) the dependence of the derived estimates of $\vec{A}_i$ on the data length T_o, and

(2) the effect of nonorthogonal sensors.

The top trace on Fig. 14 (labelled A) displays SQUID output

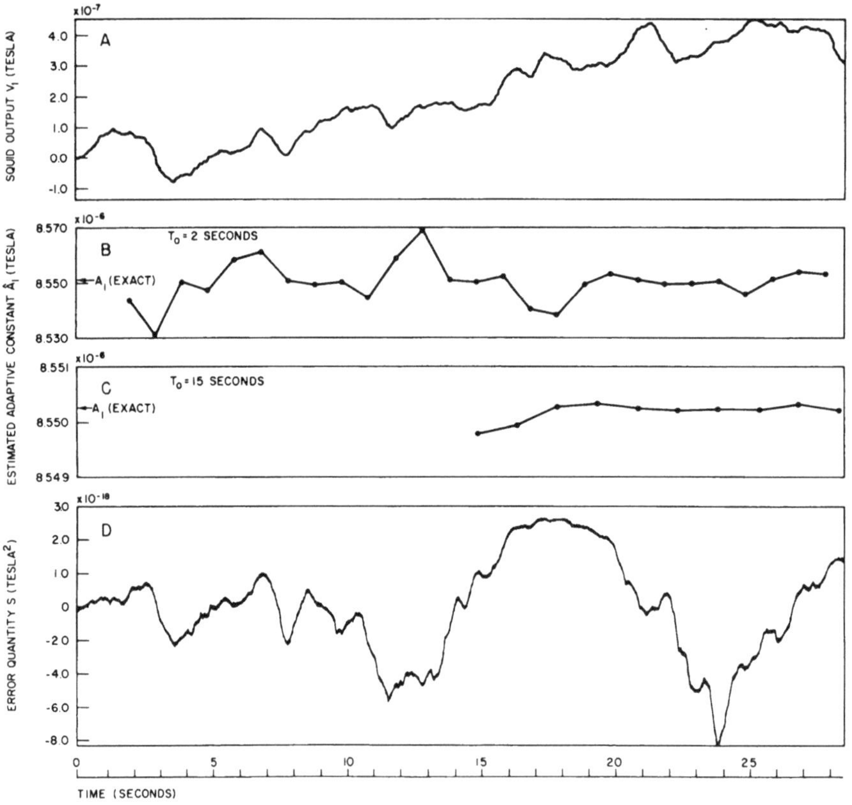

Figure 14 Computer simulation results for the least squares technique . Trace A is the time variation of the SQUID output V_1. Traces B and C display the time variation of the estimated values of V_1 for values of T_o of 2 and 15 seconds, respectively. The exact value of A_1 is also shown on the left hand scale in both traces. Trace D is the time variation of the error quantity S.

V_1 as an example of the inputs to the least squares adaptive routine. The values of the earth's field components H_{ei} were held constant during the duration of Fig.14 with φ_e = 60 degrees and θ_e = 20 degrees. The nonorthogonality angles Ω_{ij} were set equal to zero. The plots labelled B and C in Fig. 14 show the variation of the estimates of A_1 as given by the least square routine. In plot B each estimate of A_1 uses the previous 2 seconds of data. Note that since the points are spaced by only one second, the data blocks used to compute A_1 overlap by 50 percent. In plot C of Fig. 14 each estimate of A_1 uses the previous 15 seconds of data, with an overlap of 90 percent from one data block to the next. Clearly, the longer span of data provides a better estimate of A_1. Similar results are obtained for A_2 and A_3. The optimum value of T_o is approximately 10 seconds for the input data used in this simulation; a longer length of time gives only a small improvement. This value of T_o is, of course, optimum only for the simulation. In an actual SQUID receiver, longer or shorter values of T_o may be optimum. For T_o = 15 seconds, results of simulation indicate that the least squares technique is capable of determining the adaptive constant A_1 to within 1 part in 10^5. It was indicated above that the values of the A_1 must be determined to within 1 part in 10^4 (we emphasize that this value relies on an in-band buoy motion amplitude no greater than the $4x10^{-6}$ rad$\cdot$ $Hz^{-1/2}$ described above). Hence, the least squares technique is capable of estimating the A_1 with the necessary accuracy. Trace D in Fig. 14 is a plot of S using the values of A_1 estimated at t = 15 seconds with T_o = 15 seconds. Figure 15 compares the spectrum of V_1 with the spectrum of S. The spectrum amplitude above 10 Hz is seen to be about a factor of 25 below the specified value of $2\,\vec{H}_e\cdot\vec{H}_S$, which is $10^{-18}\ T^2$ for the minimum signal level of 10^{-14} T.

<u>Steepest Descent Simulation</u>. The parameters that are of importance

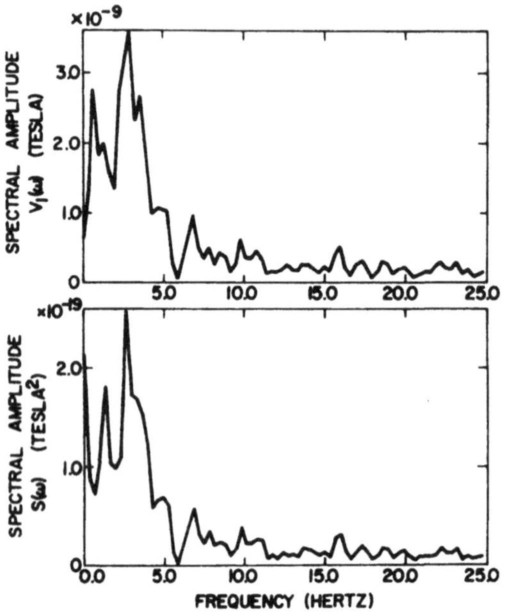

Figure 15 (TOP) Frequency spectrum of the SQUID output V_1 (trace A in Fig. 4) (BOTTOM) Frequency spectrum of the error quantity S (trace D in Fig. 4).

in steepest descent adaptation are a rate control parameter K, which controls the rate at which the approach to the minimum error is approached, and an averaging time T, over which the error function is averaged.

The purposes of the steepest descent simulation were to determine optimum values of K and T, to determine how accurately the initial estimates A_{io} must be known, and to compare the performance of two alternative error functions:

$$\varepsilon_1 = S^2 \quad , \text{ and}$$

$$\varepsilon_2 = \int_t^{t+T} S^2 \, dt$$

To determine the optimum value of K, the same input block of data V_i to the adaptive routine was used repeatedly, K was varied, and the fluctuations in S were then observed and compared. The optimum values of K were found to lie in a range from 10^{15} sec^{-1} T^{-2} to $5x10^{15}$ sec^{-1} T^{-2}. For values of K smaller than 10^{15} sec^{-1} T^{-2} the noise fluctuations in the simulated SQUID outputs V_1 were not followed, and the value of S drifted away from zero. Values of K which were larger than $5x10^{15}$ sec^{-1} T^{-2} would often send S into oscillation. The same value of K was found to be optimum for both the ε_1 and ε_2 error functions.

In general, once the optimum value of K was found, both error functions were capable of producing an error quantity S that was lower in spectral amplitude at 10 Hz (and above) than 10^{-18} T^2. However, in a number of runs, it was observed that S would unexpectedly and precipitously go into oscillation when the ε_1 error function was used. This phenomenon is shown in Fig. 16, which displays S as determined by ε_1 and ε_2. The input data for the steepest descent routine are the same as were used for the least squares routine. The dashed trace in Fig. 16 shows the sudden onset of oscillations at t = 4 sec. The point of onset is not observed to have any obvious correlation with any disturbance in the V_i that are the inputs to the routine. Reducing the value of K prevents the oscillations from occurring, but the overall performance is degraded because the value of S tends to drift from zero. Most of the steepest descent simulation concentrated on the ε_2 error function, which seems to be immune from such oscillations as long as K is not too large.

The integration time T for ε_2 could vary over a large range without affecting the noise level of S. The minimum value of T was found to be approximately 15 milliseconds.

An investigation of the convergence of the $\vec{A}_i$ to their exact values and hence S to zero revealed that unless the initial values A_{io} were quite close to the exact values, convergence could not always be insured. It is possible that the time

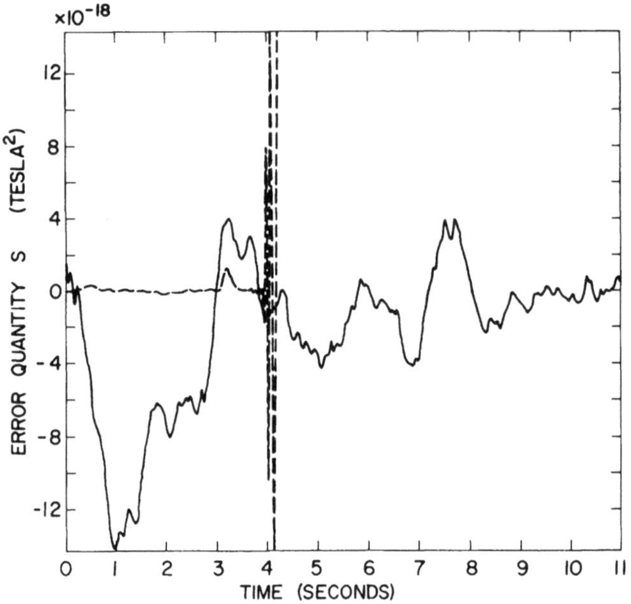

Figure 16 Time variations of the error quantity S, as determined by the steepest descent technique. The solid curve is the result of a simulation using the error function ε_2 with $K = 5\ 10^{15} sec^{-1} T^{-2}$. The dashed curve is a result of a simulation using the error function ε_1 with $K = 5x10^{-15} sec^{-1} T^{-2}$. The scale sensitivity for the dashed curve has been reduced by a factor of 10^3. Oscillations that occur in the dashed curve beyond 4.25 seconds have not been shown, for clarity.

limit of 30 seconds imposed by the memory size of the computer is not sufficiently long to permit convergence to be obtained and that eventually the A_i would approach their minimum if enough time were allowed. This possibility seems even more likely when the significance of the value of K is considered. A value of 5.10^{15} sec^{-1} T^2 implies that a 10^{-7} T change in V_i and a 10^{-17} T^2 change in S in one second (typical values from Fig. 14) produce only a 10^{-8} T change in the value of A_i.

A larger value of K would be desirable, but, as indicated earlier, a larger value can cause oscillations to occur in S. With $K = 5x10^{15}\ sec^{-1}\ T^{-2}$ the value of A_{io} had to be within approximately 0.1 percent of the exact value of A_i in order for S to converge towards zero.

It is quite likely that the convergence rate can be considerably improved by the use of a more sophisticated gradient minimization technique. Such techniques include the generalized Newton-Raphson method, the Davidon-Fletcher-Powell method, and the projected gradient method [14]. These methods require more computational machinery than the steepest descent technique and lack its appealing simplicity.

The spectrum of S, as determined by the steepest descent technique, was very similar in amplitude and other features to the spectrum of S computed from the least squares technique. The behavior of the spectral amplitude as a function of the non-orthogonality angles was also similar.

The simulation demonstrates that the least squares technique is able to provide accurate estimates of the adaptive constants without the need for any initial estimates. However, the fact that the method is not a closed-loop technique and requires up to 10 seconds of data in order to evaluate the adaptive constants is a decided handicap. The requirement of a digital microprocessor might be considered a disadvantage in some applications, but in the submarine communication system under consideration here, a microprocessor can probably be provided. The essentially continuous determination of the adaptive constants offered by the steepest descent technique and its closed loop nature make it a very appealing technique. In applications where a microprocessor or digital computer is not available, the steepest descent technique is clearly superior, because it can be implemented by purely analog computing machinery. On the other hand, the requirement of an initial value of each of the adaptive constants, which must apparently be rather close to the true value, is a disadvantage.

An optimum system might consist of a hybrid combination of the two techniques. Processing could be initiated by zeroing the SQUID outputs, entering ten seconds of SQUID output into a computer memory, and solving for the adaptive constants using the least squares technique. This value would then be used to begin the integration of the steepest descent equations, which could occur by analog or digital techniques. Continuous monitoring of the output S by a threshold detector would allow verification of the tracking of changes in the earth's field. If the noise in S became excessive, the SQUID outputs could be rezeroed and the determination begun anew.

FIELD TESTS AND DEMONSTRATIONS

Preliminary Field Test

As verification of its performance in a field environment the SQUID receiver, in the 7-day dewar that was provided by the vendor, was placed in a rural location near Washington, D. C., operated for several days and compared with a conventional ELF receiving loop antenna. During this period, the SQUID system was adjusted to have sensitivity comparable to the conventional device by short-circuiting part of its superconducting pickup loops. The ELF noise level of the surface atmosphere is as much as 40 dB higher than the design environment of the SQUID receiver.

A crucial part of this test was to determine whether the highly impulsive atmospheric noise that characterizes the ELF spectrum would cause a serious degradation in SNR due to flux-counting discontinuities. There was some doubt that the slewing rate of the SQUID system would be adequate to accommodate this noise under a diversity of conditions.

An ELF test transmitter located in the midwestern U.S. was used to transmit low level continuous wave signals for this test. Figure 17 shows this signal as received both on the

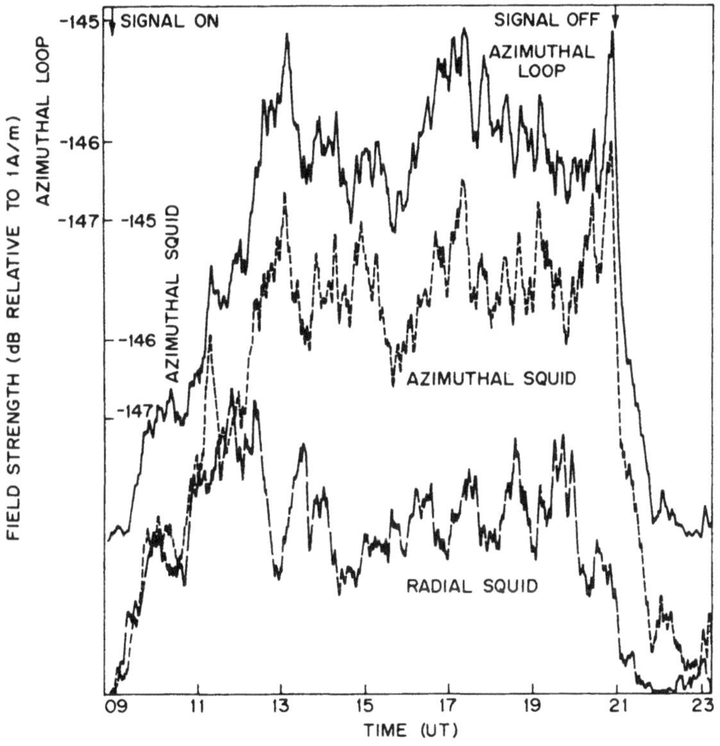

Figure 17 Comparison of transmitted ELF signals received on the SQUID antenna and on a conventional air-core loop antenna. The azimuthal loop and azimuthal SQUID traces are offset vertically for display purposes.

conventional antenna, designated "Azimuthal Loop", and on two of the SQUID axes, designated "Azimuthal SQUID" and "Radial SQUID". The two SQUID axes are shown independently because in this stationary test no motion-noise processing was necessary. The azimuthal signal is the main component of the transmitted energy, but at receiving locations near the transmitter a radial component also exists. This radial component is shown for illustrative purposes on one SQUID axis in Fig. 17 but is unscaled. It is apparent that the signal-to-noise ratios of the loop and SQUID azimuthal channels are similar (the SQUID and loop channels are offset on the figure to permit convenient comparison -- their actual signal levels are the same), and that the temporal behavior of the signals received on them is qualitatively similar. Slight differences in frequency response and signal processing between the two systems are responsible for the small differences in the appearance of the two signals.

Sea Tests

Following the preliminary tests, a pressure vessel and support system were designed so that the SQUID receiver could be tested in its design environment below the ocean surface. A concrete, pancake-shaped stabilizing base was designed to accommodate the SQUID in the 7-day dewar, plus inclinometers, a compass, a closed circuit television camera, and fittings for the electronic cables and helium vent hose (the boil-off helium was vented to the surface). This assembly is shown in Fig. 18 just prior to its insertion into the water. It was lowered approximtely 100 m below the surface of the Pacific Ocean at San Clemente Island off the coast of California, where it rested on the bottom for several weeks of measurements with periodic retrieval for replenishment of the liquid helium. The cables and vent hose were supported by a large buoy moored near the SQUID system, and data were re-

Figure 18 Photograph of SQUID system in pressure vessel being lowered to 100 m below ocean surface for submerged performance tests.

corded on an instrumentation boat moored nearby. During these measurements, both ELF noise and signals from the U. S. Navy test transmitter were recorded and verified that the SQUID system functioned satisfactorily. This series of measurements demonstrated that SQUID-based electronic systems can be operated satisfactorily in the hostile environment of the ocean depths without heroic support and service measures.

Motion-Noise Test

When the performance of the SQUID, its auxiliary systems, and the methods for dealing with motion noise had been verified, the effectiveness of the motion-noise processing technique was evaluated with actual SQUID motion data (Note that the previous motion-noise processing tests were conducted with simulated data from the buoy towing tests). It was not possible to mount the SQUID system in a buoy for these tests due to the expense of engineering suitable mounting and towing apparatus for the horizontal dewar in Fig. 5. A pneumatic vibration table was constructed to support the SQUID system in its vertical dewar. With this vibration table, the SQUID system was subjected to controlled low frequency vibrations of the approximate magnitude shown in the towing tests, and the resulting SQUID motion noise recorded on digital magnetic tape with a linear dynamic range of about 66 DB. While this figure is less than the 86 DB required for a thorough test ofthe motion-noise processing schemes, it was expected that these tests would permit an evaluation of the relative merits of the adaptive methods and give an estimate of the processing system storage and speed requirements.

The adaptive algorithms were implemented on a Digital Electronic Corp. PDP-11 minicomputer, and initial attempts to perform the motion-noise processing have been encouraging. We believe that refinements in the implementation and adaption of improved versions of the adaptive algorithms will permit real-time adaptation to be successful for this application.

CONCLUSION

The cryogenic environment necessary for SQUID use in a submerged receiving application can be provided simply and reliably by suitable packaging of a liquid helium dewar. A horizontal dewar, of an appropriate size to fit in a typical communications buoy, has been procured and tested and has demonstrated a 102-day hold time.

Motion stability measurements have been made on a hydrodynamically stabilized buoy in a towing channel and have shown that motion in the 0-30 Hz band can be constrained to 10^{-3} rad. Motion within the 30-130 Hz ELF reception band was held to less than 10^{-6} radian.

A triaxial test SQUID was procured with 140 dB linear dynamic range, a value that will permit the SQUID to be used aboard a towed buoy without electronic saturation due to motion of the SQUID in the earth's magnetic field. The SQUID also was configured with a triaxial pickup assembly in which the three pickup loops were orthogonal to within 0.7×10^{-4} radian. This high degree of orthogonality simplifies the adaptive signal processing scheme that is necessary to employ a SQUID in this application.

Because a SQUID cannot sense the absolute value of the earth's magnetic field, the earth's field strength must be determined adaptively to a degree of precision that will permit it to be removed from the sum of the squares of the individual sensor outputs to below noise level. Two adaptive techniques were studied and evaluated with the aid of simulated motion data from the motion stability measurements. A hybrid of open-loop (least squares) and closed loop (steepest descent) adaptation methods was found to be the most attractive approach to solving the signal-processing problem.

These investigations have demonstrated that a SQUID has the potential capability to serve as an extremely sensitive ELF radio-frequency receiving element on a mobile platform. Provision of the cryogenic environment and an adequate combination of motional stability and adaptive signal processing have been found to be readily achievable to permit a SQUID to be atmospheric noise limited for a submarine receiving application. A test of the complete, buoy-mounted system has not been made. However, the ability of the SQUID to operate at 100 m depth below the ocean surface has been proven, and the ability of the motion-processing method to deal with anticipated buoy motions has been demonstrated.

ACKNOWLEDGEMENTS

The authors would like to acknowledge the contributions made by their colleagues at NRL, R. J. Dinger, J. Goldstein and S. A. Wolf, and would like to express their appreciation to Dr. David Cohen of the National Magnet Laboratory and to Mr. B. Brown of the NASA Goddard Space Flight Center for the use of their respective Low Magnetic Field Facilities.

REFERENCES

1. Evans, J. E., and Griffiths, A. S.; Design of a Sanquine Noise Processor Based Upon World-Wide Extremely Low Frequency (ELF) Recordings, IEEE Trans. Comm. COM-22, 528-539, (1974).

2. Ginsberg, L. H., Extremely Low Frequency (ELF) Atmospheric Noise Level Statistics for Project Sanguine, IEEE Trans. Comm., COM-22, 555-561, (1974).

3. Wolf, S. A., Davis, J. R., and Nisenoff, M., Superconducting Extremely Low Frequency (ELF) Magnetic Field Sensors for Submarine Communications, IEEE Trans. Comm., COM-22, 549-554, (1974).

4. Wolf, S. A., Nisenoff, M., and Davis, J. R., Superconducting ELF Magnetic Field Sensors for Submarine Communications, NRL Report 7720, May 1974.

5. Park, J. G., The Physics of Resistive SQUIDs Modulated at Low Frequency, J. Phys. F: Metal Phys. 4, 2239-2263 (1974).

6. Goodman, W. L., Hesterman, V. W., Rorden, L. H., and Goree, W. S.: Superconducting Instrument Systems, Proc. IEEE, 61, 20-27, (1973).

7. Cox, J. E.: Potential Cooling Methods for an ELF SQUID, NRL Memorandum Report 2899; October 1974.

8. Dinger, R. J. and Goldstein, J.: Motion Stability Measurements of a Submarine-Towed ELF Receiving Platform, IEEE Jour. Oceanic Eng. (in the press), 1976.

9. Dinger, R. J., and Davis, J. R. : Adaptive Methods for Motion-Noise Compensation in Extremely Low Frequency Submarine Receiving Antennas, Proc. IEEE 64, No. 10, 1976.

10. Young, P. C.: The Determination of the Parameters of a Dynamic Process, Radio and Electronic Engineer, (J. IERE), 29, 345-361, (1965).

11. Young, P. C.: Process Parameter Estimation and Self Adaptive Control, Proc. 1965 IFAC Conf. In Theory of Self Adaptive Control Systems, (P. H. Hammond, Ed. Plenus Press: New York, 1966) pp. 108-139.

12. Eveleigh, V. W.: Adaptive Control and Optimization Techniques, (McGraw-Hill Book Company: New York, 1967), pp. 104-130.

13. Bekey, G. A. and Karplus, W. J., Hybrid Computation, (John Wiley and Sons: New York, 1968) Ch. 9, pp. 244-300.

14. Tabak, D. and Juo, B. C., Optimal Control by Mathematical Programming, NJ: (Prentice-Hall, 1971), pp. 41-56.

SIGNAL PROCESSING

COMPLETE LINEAR EQUIVALENT CIRCUIT FOR THE SQUID

G.J. Ehnholm
Low Temperature Laboratory, Helsinki University of Technology,
SF-02150 Espoo 15, Finland

The squid is extensively used for detecting signals at very low frequencies from magnetometers and susceptibility measurement coils. This type of signal source is almost purely inductive and has a very low impedance at small frequencies so that dissipative terms in the squid input impedance do not show up. At the Low Temperature Laboratory in Otaniemi I and my co-workers are presently applying the squid for noise thermometry. Our signal source is essentially a resistor which is connected in series with a capacitor that tunes out the input inductance of the squid at the resonance frequency. This can be made as small as 1 kHz by using a matching transformer between the capacitor and the squid. The maximum Q value with this arrangement is approximately 200. A similar type of tuned arrangement could be used for monitoring signals from string viscometers, gravitational wave detectors, nmr measurement systems, and so on.

Using the squid with a tuned input offers a great improvement in sensitivity. If the signal frequency is for instance 1 kHz and the desired bandwidth is 10 Hz then a tuned circuit with a Q value of 100 can be used at the input of the squid. This will improve the current sensitivity by a factor of 100. Such a tuned circuit will, however, be very sensitive to the input parameters of the squid. A small variation in the squid input inductance might shift the resonance frequency away from the desired value, or a change in the input resistance will change the Q value causing a large change in the amplification. The input resistance is, in fact, negative so that a badly designed input circuit will start to oscillate!

It is clear, that if the squid is to be used in the described way a thorough understanding of its properties is needed, including the internal interactions that govern the dependence of the electronic in- and output parameters on the parameters of the signal source and output load. What is needed, in fact, is a complete linear equivalent circuit of the squid which can be used in a reasonably straightforward way to compute the properties of circuits containing squids. An equivalent circuit containing nonlinear elements is not too useful: It merely substitutes a new component, an automatic switch or something else, for the Josephson junction without really making the system simpler. In the following a linear equivalent circuit will be derived for the first time, and as an important example of its use, the input impedance will be computed. It consists of a negative resistance in parallel to an inductance, both of which are functions of the tank circuit parameters. The line of argument of the theory will be drawn without going too much into the details. The detailed computations are given in the Reference.

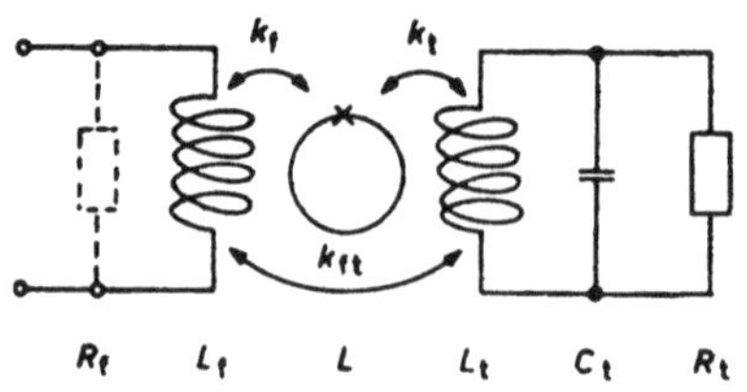

Fig. 1

In Fig. 1 is shown an rf squid system consisting of the signal coil L_f, squid ring with inductance L, and tank circuit with inductor L_t, capacitor C_t and resistor R_t. The losses of the tank circuit are contained in the resistor. The k:s denote the coupling coefficients between the different inductors. The resistor R_f shortcircuits the signal coil at rf frequencies and thus prevents the rf flux from penetrating through the signal coil. In practice it can consist of a resistor connected across the input to shield it from spurious rf interference or it could be part of the signal source. It will cause a decrease in the inductances L and L_t at high frequencies. The new values will be called L_p and L_{tp}. The coupling coefficient k_t will also change at the rf frequency; the new value will be called k_p.

We now introduce ideal transformers to designate the impedance transformations between on one hand the signal and the tank circuit coils and, on the other hand, the squid ring. The real transformers remain with a 1:1 transforming ratio: for these we substitute their equivalent circuits consisting of three inductors in a "T" configuration. The result can be seen in Fig. 2, where the "boxes" represent the ideal transformers.

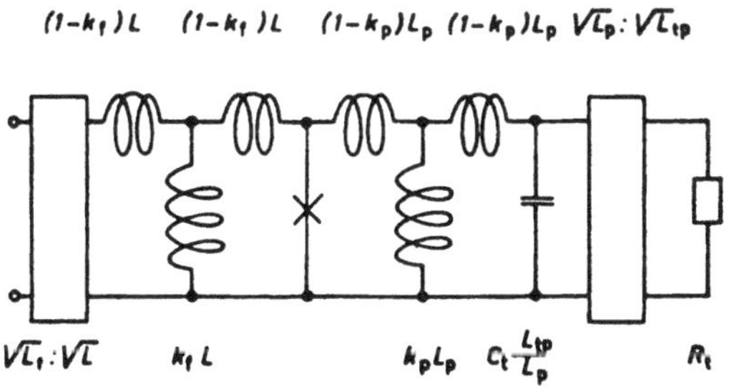

Fig. 2

The part of the circuit at the output side containing three inductors and one capacitor can be further simplified. At the tank circuit frequency

$$(1) \qquad \omega_p = (L_{tp}C_t)^{-\frac{1}{2}}$$

it can be exchanged for an ideal transformer with the transforming ratio k_p:1 and an inductor $(1 - k_p^2)L_p \approx L_p$. The result is seen in Fig. 3. For rf frequencies differing from ω_p by the

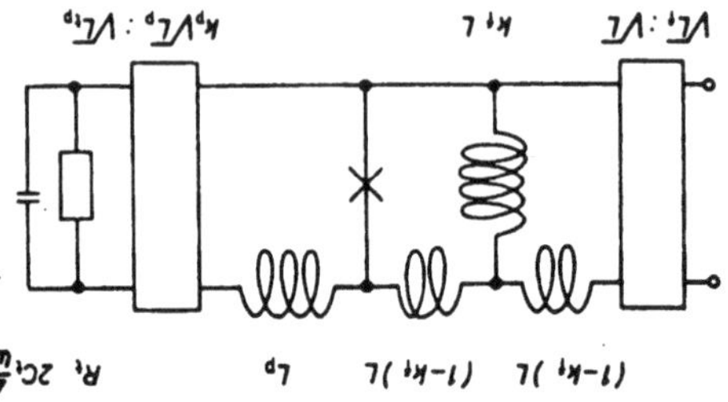

Fig. 3

amount $\Delta\omega$ some of the tank circuit reactance will remain; this can be represented by the capacitance $2C_t\Delta\omega/\omega_p$. The two ideal transformers at the output side have been combined to a single one in the Figure.

The essential parts of the squid, the heart of the squid so to speak, is between the input and output transformers. These are in a way extrinsic implements for impedance matching. The tank circuit impedance, on the other hand, can be looked upon as an external load.

In Fig. 4 we see the essential parts of the squid together with the symbolds used for the currents and fluxes at the in-

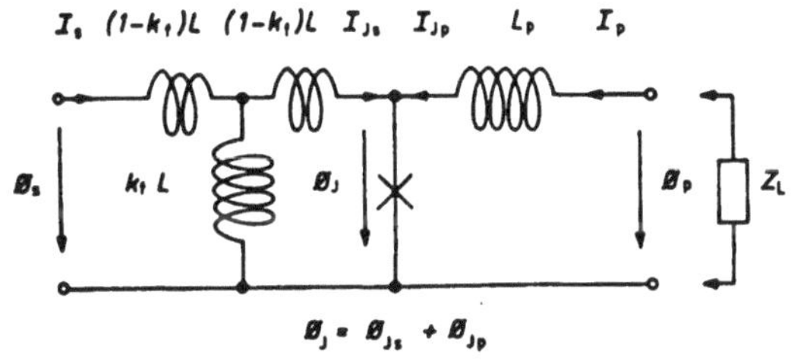

Fig. 4

put, output, and weak link, respectively. We now want to characterize the squid by writing two of the variables I_s, Φ_s, I_p, or Φ_p as functions of the remaining two. It turns out that the best choise of independent variables at this stage is I_s and Φ_p, the input signal current and the output flux, which is proportional to the rf voltage of the tank circuit. We combine these two variables to a single one, which we call the external flux Φ_x:

$$(2) \qquad \Phi_x = I_s k_f L + \frac{\Phi_p}{1 - k_f^2 \sin^2\theta} ,$$

where

$$(3) \qquad \theta = \mathrm{arctg}\sqrt{\Phi_o/4i_c L}$$

approaches zero for big values of i_c. The denominator under Φ_p in Eq. (2) is caused by the resistor R_f in Fig. 1. Using Φ_x as the independent variable we can write down the equation for the "flux" Φ_j of the weak link:

$$(4) \qquad \Phi_J = n\Phi_0 \cos^2\theta + \Phi_x \sin^2\theta \quad .$$

The equation is written in a simplified linearized form. The variable n takes integral values only and makes a transition to a new value when the critical current of the weak link is exceeded. The equation can be graphically presented in a well known manner, c.f. Fig. 5. In the Figure only two branches,

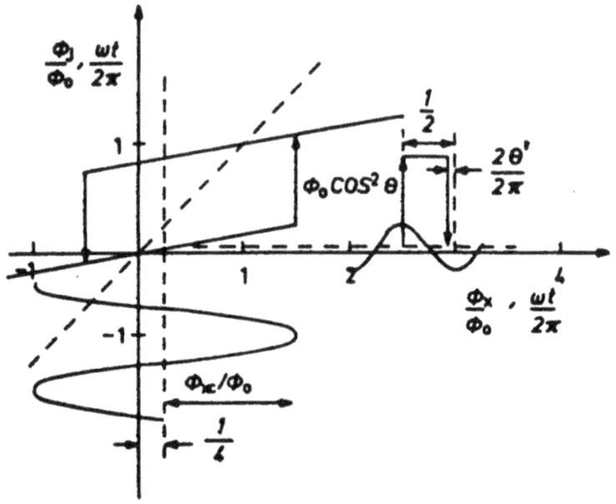

Fig. 5

for n = 0 and n = 1, are drawn; those are the two used in normal operation of the squid. We see that the overlap between the two branches produces a hysteresis curve in the weak link flux.

The external flux used during normal operation is shown at the bottom of the Figure as a function of time in units $2\pi/\omega$. The signal flux, i.e. the part of the external flux proportional to the signal current, is allowed to change in the vicinity of the value $\Phi_0/4$, which is plotted as a vertical dashed line. The rf part of the external flux varies sinusoidally around this dashed line. It usually has an amplitude such that the righthand maximum just touches the end of the squid equation branch through the origin, i.e. it induces a current into the weak link which just reaches the critical value. This combination of signal current and rf amplitude we define as the normal working point of the squid.

The flux Φ_J of the weak link as a function of time is plotted on the right. It can be resolved into a sum of three components. The dashed line is the linear reaction on the signal flux, and the sinusoidal curve around it the linear reaction on the rf flux. The signals are formed by reflection of the external flux from the squid equation branch through the origin. In addition to the linear reactions we have a nonlinear one which represents the interaction between the signal and the rf flux in the weak link. This has the form of flux pulses formed when the system travels around the hysteresis loop. During the excursion on the upper equation branch the linear reaction signals are elevated by the amount $\Phi_o \cos^2\theta$; the net output is thus formed by the sum of the three indicated signals.

We next investigate the probability of obtaining a flux pulse during one cycle of the rf wave, i.e. the probability p of a flux transition. If the righthand maximum of the external flux is called Φ_{xa} we realize that $p = 1$ if $\Phi_{xa} > \Phi_{xc}$ and $p = 0$ if $\Phi_{xa} < \Phi_{xc}$, because the first case corresponds to exceeding the critical current and the second one not to reaching it. If, however, the system is noisy, as is usually the case, the point of transition is blurred. We can then find a new value of the external flux Φ'_{xc}, such that $p > 1/2$ if $\Phi_{xa} > \Phi'_{xc}$ and $p < 1/2$ if $\Phi_{xa} < \Phi'_{xc}$. Small variations dp around the point where $p = 1/2$ can be written:

$$(5) \qquad dp = \frac{2}{\alpha}(\Phi_{xa} - \Phi'_{xc})/\Phi_o = 2\psi_x/\alpha\Phi_o \quad ,$$

where α is a constant measuring the transition probability around $p = 1/2$.

Next the influence of a flux pulse on the in- and output of the squid will be discussed. The weak link flux Φ_J contains two important components: Φ_{Js}, which is the time average of Φ_J and effects the input, and Φ_{Jp}, which is the Fourier component of Φ_J at the rf frequency; this part affects the ouput. The input flux and the output current can easily be computed once Φ_{Js} and Φ_{Jp} are known. The result of the necessary computations, using suitable approximations, turns out to be:

$$(6) \qquad \Phi_s = I_s L(1 - k_s^2) + p\Phi_o k_s^2/2k_f \quad ,$$

where $k_s = k_f \cos\theta\sqrt{1 - 2\theta'/\pi}$, and

$$(7) \qquad \bar{I}_p = \frac{2p\Phi_o}{\pi L_p}\cos^2\theta\cos^2\theta' e^{j\theta'} \quad ,$$

where $j = \sqrt{-1}$. The meaning of the symbol θ' is seen in Fig. 5. Around the working point θ' is approximated with a constant value.

We see, that the input flux Φ_s contains one part which depends only on the input current I_s; this part can be looked upon as caused by a parasitic inductance. The second part of Φ_s is a function of p, the flux transition probability. The output current $\bar{I}_p$, on the other hand, is written with only a p dependent part. Actually there is a parasitic inductance at the output as well, but this can be compensated by a slight returning of the tank circuit and has therefore been omitted as unimportant. $\bar{I}_p$ is not in phase with the output voltage U_p, defined as the time derivative of the output flux Φ_p. It has therefore been written as a complex number where the real part is defined as being in phase with U_p.

The phase shift of $\bar{I}_p$ with respect to U_p can be compensated for by operating the tank circuit slightly above its resonance frequency. Both I_p and U_p will then have the same phase and we don't have to work with complex numbers any more. Assuming this to be the case we have plotted I_p vs. U_p in Fig. 6. The

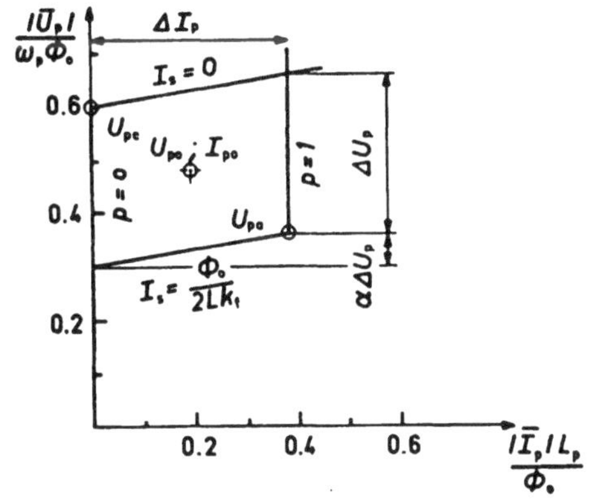

Fig. 6

plotted curves can be called the characteristics of the squid. To each value of the input current corresponds a specific curve at the output. It consists of a straight line segment starting from the y-axis, where p = 0, and ending at the value ΔI_p that corresponds to p = 1. The slope of the line segments is determined by the noise parameter α, which is taken as constant in this plot for the sake of simplicity. Actually α is constant only in the middle of the area covered by the line segments so that the corners of the Figure are rounded. The two external line segments, corresponding to zero input and a signal of half a flux quantum have been plotted. The Figure resembles the "staircase" pattern for squids, but the difference is that the latter contains the influence of the tank circuit so that it is more tilted.

The tank circuit can be taken into account in the present plot with the aid of a load line. This line is the graphical counterpart of the equation for the current I_L through the load:

(8) $$I_L = U_p/Z_L = I_b - I_p \quad .$$

The equation tells that the load current equals the output voltage U_p divided by the load impedance Z_L, and at the same time it equals a constant bias current I_b, which is being fed to the system, minus the current I_p that goes into the output of the squid.

The load line that extracts the largest change in output power from the squid is the diagonal with a negative slope of approximately 45° through the Figure. Because of the choice of unnits on the co-ordinate axes the slope of this line corresponds to

(9) $$Z_L \simeq \omega_p L_p \quad .$$

Taking the ideal transformer at the output in Fig. 3 into account this becomes

(10) $$Z_L \simeq \omega_p L_{tp} k_p^{-2} \quad .$$

If we are close to the tank circuit resonance frequency Z_L can be written

(11) $$Z_L \simeq Q\omega_p L_{tp} \quad ,$$

so that Eq. (10) becomes

(12) $$k_p^2 Q \simeq 1 \quad ,$$

a familiar expression for the optimal tank circuit coupling.

The input also has its characteristics, formed by the set of curves that give Φ_s as a function of I_s for different values of the output voltage U_p, i.e. it is the graphical counterpart of Eq. (6). It has been plotted on Fig. 7. To every value of

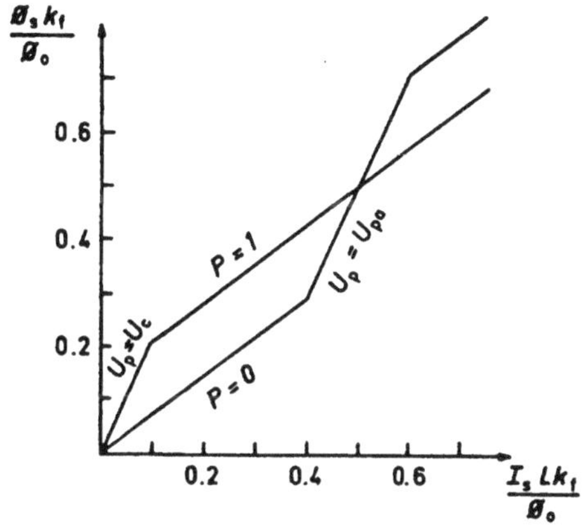

Fig. 7

U_p corresponds a sloping line segment starting from the p = 0 line and ending on the p = 1 line. The slope of a line in this diagram corresponds to a specific value of the input inductance. A 45° line, in particular, corresponds to L, the inductance of the squid ring.

The input inductance is generally a function of the load impedance produced by the tank circuit. This can be seen from Fig. 8, which shows both the input and the output characteristics. Clearly a specific point on one characteristic has a counterpart on the other one. In fact the two characteristics are linear transformations of each other. Therefore a straight line, e.g. the load line, plotted into one characteristic will be mapped as a straight line into the other one. The corner points A, B, C, and D are labelled in both characteristics and with the aid of these it is easy to find out the input inductance that corresponds to some specific load line. The line AC, for instance, which is the one that corresponds to maximum change in output power, i.e. to $k_p^2 Q \approx 1$, corresponds to the inductance L. If we take the ideal transformer at the input into account we find that this inductance becomes L_f, the inductance of the signal coil. If the coupling between the tank circuit and the squid ring is made tighter we get the dashed load line, which is reflected to the input as an inductance less than L_f. Note, however, that in this case there is a local increase in the inductance around points A and C in the input characteristic, because here the flux has to follow the curve corresponding to constant U_p for some distance. The result is that the input inductance averaged over large signals of several fluxquanta always is equal to L_f.

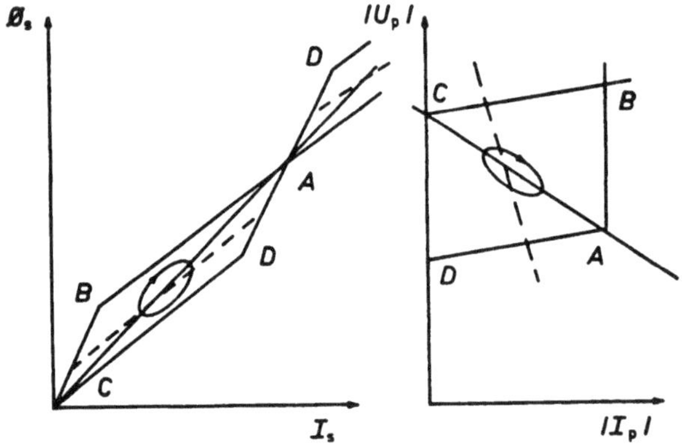

Fig. 8

From Fig. 8 we can also conclude something about the input resistance. If a rapidly changing signal is fed to the input Φ_s tends to follow the U_p = constant lines because it takes time to change the tank circuit voltage. A sinusoidal input signal therefore makes a loop around the line corresponding to the static input inductance. Because the U_p = constant lines are steeper than the input inductance line the loop is traversed clockwise. The input flux is thus bigger than in the static case for an increasing input current. Connecting a positive resistance across the input would in this case decrease the input flux because part of the supplied current would flow through the resistor. The equivalent input parallel resistance for the squid is therefore negative.

The value of this resistance can be computed with the aid of the small signal equivalent circuit of the squid. The small signals are defined as local variations around a central working point in the characteristics. The small signal equations for the squid can be derived with the help of Eqs. (2), (5), (6), and (7). The result is

$$\phi_s = i_s L(1 - k_s^2) + \frac{k_s^2 L}{\alpha}(i_s + u_p/r) \quad , \tag{13}$$

and

$$\bar{i}_p = \frac{1}{\alpha \bar{Z}_p}(u_p + i_s r) \quad , \tag{14}$$

where we have introduced

$$\bar{Z}_p = \frac{\pi \omega_p P_p (1 - k_f^2 \cos^2\theta)}{4\cos^2\theta \cos^2\theta'} e^{-j\theta'} \tag{15}$$

which also is equal to $\Delta U_p/\Delta \bar{I}_p$, cf. Fig. 6, and

$$r = \omega_p L k_f (1 - k_f^2 \cos^2\theta) \quad . \tag{16}$$

The equivalent circuit corresponding to Eqs. (13) and (14) can be drawn on inspection, cf. Fig. 9. The circuit contains a current source with the strength u_p/r of the input which produces feedback within the squid and makes the input impedance a function of the load across the output. The rf voltage generator at the output gives a voltage of amplitude $i_s r$ in phase with that of the tank circuit. When the signal current is changed the voltage generator affects the tank circuit through the impedance $\alpha \bar{Z}_p$. If the tank circuit is operated a bit above its resonance so that its impedance gets the same phase angle as $\bar{Z}_p$ the change in tank circuit voltage will be in amplitude only, not in phase.

The equivalent circuit is a complete model of the squid and can thus be used to compute the in- and output impedances for different signal source and tank circuit impedances. In particular, the input resistance can be determined. As this is rather a lengthy calculation only the result will be given. For the solid load line of Fig. 8, corresponding to $k_p^2 Q \simeq 1$ and $L_{in} = L_f$, we have

$$(17) \qquad R_{in} = \omega_p L_f / 2Qk_s^2 \cos^2\theta'(1 - \alpha) \quad .$$

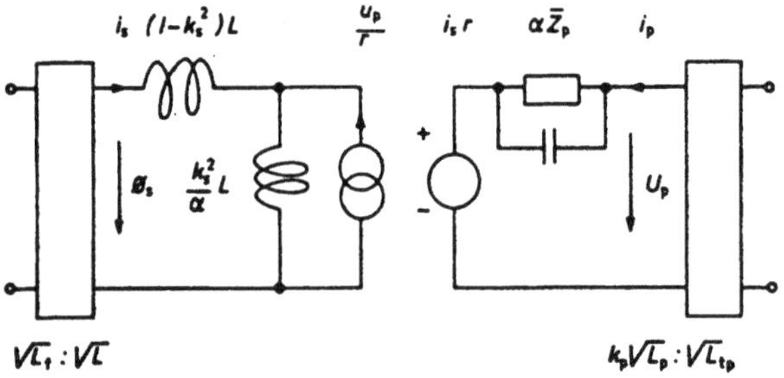

Fig. 9

If one puts typical values for the squid parameters Eq. (17) becomes

$$(18) \qquad R_{in} \simeq -5\omega_p L_f / Q \quad .$$

As a summary of this paper: Using the squid in demanding applications, especially with tuned circuits at the input, requires a precise knowledge of its electronic properties. These are presented for the first time as a complete analog circuit containing only linear elements. It is shown, that the input inductance of the squid depends on the load, and that there is a negative input resistance. An expression for this is given. A complete derivation of the presented results is given in the

Reference

G.J. Ehnholm, Derivation of a Complete Linear Equivalent Circuit of the Squid. Report no. TKK-F-B48 (1976)

Two Junction SQUID using a Sampling Technique.

Peter Gutmann

Physikalisch-Technische Bundesanstalt
Braunschweig und Berlin
Bundesallee 100
3300 Braunschweig
W.-Germany

Abstract:

An electronic system is described, which is based on a weakly damped double tunnel junction device and utilizes the transition from the zero voltage into the voltage carrying state. The measuring system has a magnetic flux resolution of $2.6 \cdot 10^{-5}\ \Phi_0/\sqrt{Hz}$.

Most of the work concerning the application of DC-SQUID devices apply to junctions with a non-hysteretic dc-current-voltage (IU) characteristic /1,2,3/. The standard technique of the electronic measuring system was introduced by Forgacs and Warnick /4/. They used SQUIDs which are biased at a constant dc current larger than the maximum critical supercurrent I_o. The voltage across the junctions is periodic with the applied magnetic field. The read out system functions as a linear amplifier, because it is driven in a feedback mode on the basis of a static IU-characteristic.

By way of contrast, a double tunnel junction device (DTJD) behaves as a memory device, the switching point of which is very sensitive to a change in the applied magnetic field. Here an electronic measuring system is proposed and demonstrated which makes use of a sampling technique and is based on a weakly damped DTJD. Fig. 1 shows the principle of obtaining the information from the DTJD and the subsequent signal processing by a room temperature circuitry. The device with a tunnel junction IU-characteristic is supplied by a triangular wave current i (t) - denoted test current - of amplitude $I_B > I_o$ and frequency $f_o = 1/T_o$. The time-dependent voltage response u (t) of the DTJD is also shown in Fig. 1. During the time interval T_1, the test current is smaller than I_o and the DTJD remains in the zero voltage state. Whenever I_o is reached, a voltage appears across the device with an extremly short transition time. On account of the hysteresis and the nonlinearity a voltage of about $u \approx U_g$ (gap voltage) remains across the junction for decreasing i. As for most devices a rest current has to be considered, so that u already vanishes for times $t < T_o$. The information (thick solide line in Fig. 1) regarding I_o is contained in the time interval $T_1 \sim I_o$, where T_1 is defined by the test current values i = 0 and i = I_o within a period. In Fig. 1

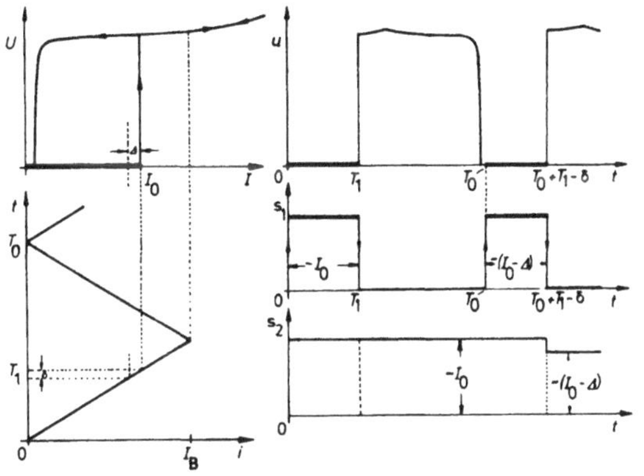

Fig. 1 The detection of information from the DTJD and the signal processing by a room temperature circuitry

below the DTJD response is drawn the time histories of the realized read out system. To begin with a square wave pulse $s_1(t)$ is produced with I_o proportional to the pulse duty factor T_1/T_o. The pulse period T_o is derived from the test current, whereas the pulse width T_1 depends only on the value of I_o. By further processing the pulse width is transformed into an analogue amplitude measured at times $t_n = n\ T_o + T_1$, where n is an integer number. Fig. 1 shows an example of how an assumed change Δ of I_o modifies the signals u (t), $s_1(t)$ and $s_2(t)$, where δ means a corresponding change of the time interval T_1.

Fig. 2 shows a simplified block diagram of the measuring system which works on the basis of the principles

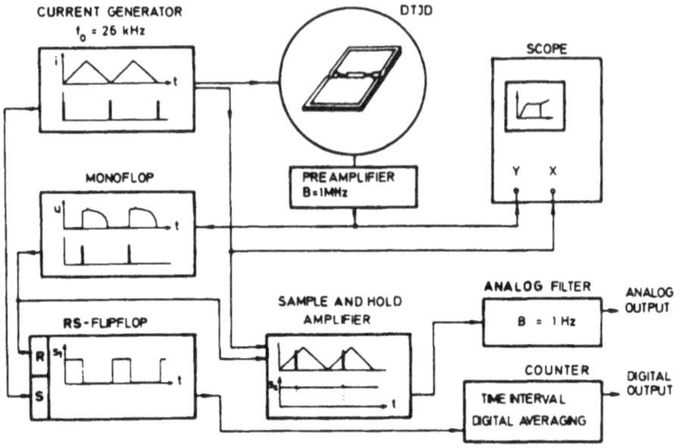

Fig. 2 Simplified block diagram of the measuring system

mentioned /5/. The test current fed into the DTJD produces a voltage pulse with the qualities already discussed. Since I_o is exceeded within each period, discrete samples are obtained at a repetition rate f_o. At the beginning of each cycle a "reference pulse" is derived from the zeros of the test current. After the time interval T_1 a voltage is produced across the DTJD. The voltage rise is converted into a short pulse by a monoflop. This pulse which eliminates the hysteresis is denoted "transition pulse". The information contained in the time delay T_1 of two pulse trains is transferred to the signal $s_1(t)$ by an RS-flipflop, which is set periodically by the reference pulse, and is reset by the transition pulse. By use of a suitable counter, further digital processing can be carried out for measuring time intervals, time interval jitter and averaged time intervals. The analogue signal $s_2(t)$ is produced by means of a sample and hold amplifier. Its analogue input is supplied by a signal proportional to the test current. The transition pulse train is connected to the digital input. Within the short duration of this trigger pulse at the analogue input a signal amplitude proportional to I_o is sampled, the amplitude of which is held at the output until the consequent trigger pulse arrives. The transfer function of the sample and hold amplifier is roughly proportional to $|(\sin \pi T_o f)/(\pi T_o f)|$, so that an input signal $T_1(t)$ is filtered for frequencies f larger than about $f \approx f_o/2$. A further decrease in the bandwidth is achieved by using an analogue filter.

The DTJD were produced by means of an oblique evaporation technique developed by Niemeyer /6/. A glass substrate with two parallel scratches crossed by a glass fibre is subjected to oblique evaporation in a vacuum chamber from two evaporation sources, one of which is moved to the right and then to the left in the direction of the scratches. After the first evaporation the layer is oxidized by a thermal oxidation process. After the deposition of the second metal film, the junctions are formed at the overlap of the scratches and the fibre. A planar superconducting ring of rectangular shape is obtained, interrupted by two small tunnel junctions of the type $Pb-Pb_xO_y-Pb$. The external field penetrates the area A enclosed by the superconducting ring. If desired, this may be enlarged by removing some parts of the lead film.

For the optimal operation of the DTJD the following discussion may be of some interest. Fig. 3 shows the IU-characteristics of two different DTJDs with corresponding $I_o(\Phi)$. $I_o(\Phi)$ is the periodic characteristic of I_o versus a linear increasing external flux Φ penetrating the area A, measured at the sample and

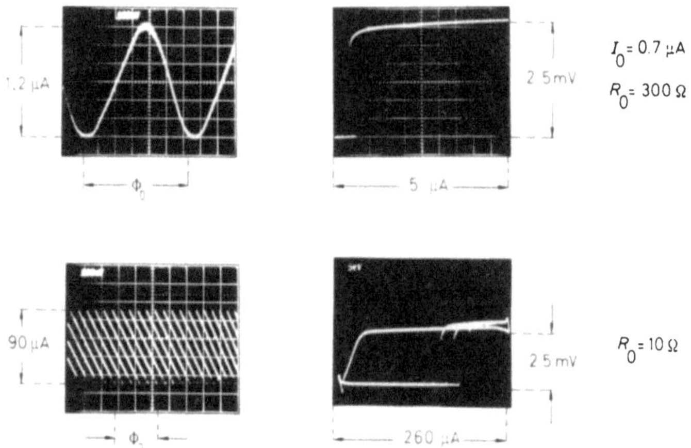

Fig. 3 IU-characteristics and corresponding I_o (Φ) characteristics of two different DTJD

hold amplifier output. I_o is modulated between 0.7 μA and 1.9 μA by Φ for one device, and between 130 μA and 220 μA for the other. R_o is the junction resistance which is measured at a dc bias voltage of about 25 mV. In both cases A was $3 \cdot 10^{-3}$ mm^2. The inductance L of the DTJD can be estimated by using the limiting condition /7/ $L = \Phi_o / \Delta I$ valid for $L \cdot I_o \simeq \Phi_o$ and which yields L = 20 pH when a modulation depth ΔI = 90 μA is used. The slide shows that a limit to larger I_o is set by the appearance of multiple critical supercurrents, a consequence of multiple quantum states /8/. There is also a limit to smaller currents. It is well known that according to the BCS theory the product of I_oR_o is constant; this results in an experimentally obtained value 1.5 mV for lead at a temperature of 4.2 K. It was found that this relation is valid for resistances $R_o \leqslant 75\,\Omega$. For increasing resistances $75\Omega < R_o \leqslant 1\ k\Omega$, I_oR_o decreases because of the suppression of I_o by noise currents; in particular I_o becomes zero for $R > 1\ k\Omega$.

The minimum resolvable magnetic flux ϕ_r, referred to the flux quantum ϕ_o, is given by the equations:

$$(1) \qquad \Delta I_o/\phi_o = |dI_o/d\phi|_{max}$$

$$(2) \qquad \phi_r/\phi_o = I_r/\Delta I_o \text{ for the bandwidth } B \approx f_o/2,$$

where the DTJD sensitivity $\Delta I_o/\phi_o$ depends on $I_o(\phi)$ /8/. The maximum change ΔI_o per ϕ_o is given by equation (1) where the absolute value of the right hand term gives the maximum slope of I_o versus ϕ. This value depends on the DTJD properties, such as the critical supercurrent of each junction and the loop inductance. The sensitivity of small critical supercurrent DTJD is also decreased by additional noise currents, the amplitudes of which are comparable with I_o and the frequencies $f \gg f_o/2$. Equation (2) relates the flux resolution per flux quantum ϕ_r/ϕ_o to the sensitivity $\Delta I_o/\phi_o$ and the RMS value of the critical supercurrent fluctuations I_r for a $B \approx f_o/2$. The peak-to-peak jitter of the critical supercurrent fluctuations in the IU-characteristic, or the peak-to-peak variations at the output of the sample and hold amplifier was measured and divided by a factor of five to derive I_r. I_r depends essentially on the intrinsic noise of the DTJD and the noise of the read out system, which means that noise currents are present at the output of the current generator and at the input of the preamplifier. One also has to consider the phase noise of the transition pulse caused by trigger uncertainties, which depend on the rise time of the preamplifier and on the noise level at the input comparator of the monoflop. It should also be taken into account that the noise content of I_r above $f_o/2$ is aliased into the signal band $0 \leq B \leq f_o/2$, so enhancing the total output noise.

In order to obtain the flux resolution per unit bandwidth a filter with 1 Hz bandwidth was used, achieving a further noise reduction by a factor of $\sqrt{f_o/2 \text{ Hz}}$ when a constant noise power density up to $f_o/2$ is assumed. Contrary to the read out system described before, digital signal averaging improves the signal-to-noise ratio by a factor of $\sqrt{f_o/1 \text{ Hz}}$, averaging over a time interval of T = 1s. The electronic system described was used to test various DTJDs with $A \approx 3 \cdot 10^{-3}$ mm^2. The best results, as can be seen from Fig. 3, were obtained with small critical supercurrents. Specific data for the DTJD with $R_o \approx 300\,\Omega$ are:

$$f_o = 830 \text{ Hz}, \; \Delta I_o/\Phi_o = 4.3\,\mu A/\Phi_o, \; I_r = 3.8 \text{ nA for } B = 415 \text{ Hz}$$

$$f_o = 25.9 \text{ kHz}, \; \Delta I_o/\Phi_o = 3.4\,\mu A/\Phi_o, \; I_r = 10 \text{ nA for } B = 13 \text{kHz}$$

giving an optimum performance $\Phi_r/\Phi_o = 2.6 \cdot 10^{-5}$ for a 1 Hz bandwidth. Measurements have been carried out using a toroidal field coil with an inductance of 13 μH, in order to couple a magnetic field to the DTJD. The mutual inductance between field coil and DTJD was typically in the range $2\,\Phi_o/\mu A$ to $0{,}5\,\Phi_o/\mu A$, an experimental result which could be considerably improved using a more sophisticated arrangement.

References

/1/ Decker, S.K., Mercereau, I.E.: Appl.Phys. Letters 23, 347 (1973)

/2/ Gallop, J.C.: J. Phys. E.: Sci. Instrum. 7, 855 (1974)

/3/ Clarke, J., Goubau, W.M., Ketchen, M.B.: Appl. Phys. Lett. 27, 155 (1975)

/4/ Forgacs, R.L., Warnick, A.: IEEE Trans. IM-15, 113 (1966)

/5/ It should be noted that all pulse modulation techniques can be used to transfer the information from the DTJD to a room temperature electronic circuitry. Since a frequency measurement can usually be carried out more easily than an interval measurement, one would rather prefer to control the amplitude of the test current by I_o. In this case I_o becomes inversly proportional to the frequency of the test current.

/6/ Niemeyer, J.: PTB-Mitt. 84, 251 (1974)

/7/ Zimmermann, J., Silver, A.H.: Phys. Rev. 141, 367 (1966)
Clarke, J.: Phil. Mag. 13, 115 (1966)

/8/ Fulton, T.A., Dunkleberger, L.N., Dynes, R.C.: Phys. Rev. B 6, 855 (1972)

RF-SQUID IN THE NONHYSTERETIC MODE: THE PHASE MODULATION OF THE TANK VOLTAGE

S.N. Erné
Physikalisch-Technische Bundesanstalt, Institut Berlin,
Abbestr. 2-12, D-1000 Berlin 10, Germany

The usual mode of operation of an RF-SQUID is to couple the SQUID loop inductively to a cavity or a tank circuit either at or near their resonant frequency. The purpose of this paper is to study the influence of an applied dc flux on the phase of the ac voltage at the tank circuit in the non-hysteretic mode. Nonhysteretic mode means that the value of $\beta = \frac{LI_c}{\Phi_0}$, which is a characteristic parameter of the SQUID loop, does not exceed the limit $\frac{1}{2\pi}$. Thereby L denotes the SQUID loop inductance, I_c the critical current of the weak link and Φ_0 the flux quantum $\Phi_0 = h/2e$.

We assume the tank circuit to be a high Q resonant circuit with capacitance C_T, inductance L_T and resonance frequency $\omega_T/2\pi$ and coupled to the SQUID loop by the mutual inductance M. The following determination of the dependence of the flux Φ on a flux Φ_{dc} starts from the solution of the adiabatic equation

$$(1) \quad \frac{\Phi}{\Phi_0} + \beta \sin\left(2\pi \frac{\Phi}{\Phi_0}\right) = \frac{\Phi_a}{\Phi_0}$$

for the internal flux Φ of the SQUID loop due to an applied flux Φ_a:

$$(2) \quad \Phi = \Phi_a + \Phi_0 \sum_{n=1}^{\infty} \left(-\frac{1}{n\pi}\right)^n J_n(2\pi n\beta) \sin\left(2\pi n \frac{\Phi_a}{\Phi_0}\right)$$

This solution is introduced into the system of differential equations describing the two coupled circuits.

The complete evaluation carried out elsewhere [1] results in the following conditions for the amplitude $\hat{I}_T$ and the phase angle ϑ of the current $I_T = \hat{I}_T \sin(\omega t + \vartheta)$ flowing in the inductance L_T:

$$(3) \quad \frac{\hat{I}_{RF}}{\hat{I}_T} \sin\vartheta = -\frac{\zeta_T}{Q} - k^2 \zeta_T^2 \frac{\omega\tau_q}{a} \left\{ 1 - \left[1 - \left(1 - \zeta_s^2\right) \frac{\alpha}{2\pi\beta} \right] F \right\}$$

$$(4) \quad \frac{\hat{I}_{RF}}{\hat{I}_T} \cos\vartheta = \left(1 - \zeta_T^2\right) - \frac{k^2 \zeta_T^2}{a} \left\{ \zeta_s^2 \left(1 - \zeta_s^2\right) - \omega^2\tau_q^2 - \left[1 - \zeta_s^2 + \omega^2\tau_q^2 \frac{\alpha}{2\pi\beta} \right] F \right\}$$

These are valid for the low frequency range ($\omega \leq 10^8 s^{-1}$). $\zeta_T = \omega\sqrt{L_T C_T}$ is the working frequency normalized to the tank circuit resonance frequency; $\zeta_s = \omega\sqrt{LC_j}$ is the ratio of the working frequency and the SQUID loop resonance frequency (C_j = junction capacitance). $k = M/\sqrt{L_T L}$ denotes the coupling coefficient between SQUID loop and tank circuit, $1/\tau_q = R_q/L$ is the conventional time constant of the SQUID loop and α the coefficient of the quasiparticle interference ($\cos\varphi$)term, a is defined to be $a = (1 - \zeta_s^2)^2 + \omega^2\tau_q^2$, $\hat{I}_{RF}$ the amplitude of the bias current. The nonlinear action of the SQUID is described by the function F

$$(5) \quad F = \frac{2}{\pi} \left[\sum_{m=1}^{\infty} \frac{(-1)^n}{m} J_m(2\pi n\beta)\, J_1\!\left(2\pi n \frac{M\hat{I}_T}{\Phi_0}\right) \cos\left(2\pi n \frac{\Phi_{dc}}{\Phi_0}\right) \right] \frac{\Phi_0}{M\hat{I}_T}$$

where J_n denotes Bessel functions of the first kind and integer order.

From the equations (3) and (4) which determine amplitude $\hat{I}_T$ and phase ϑ of the current can be seen that both are influenced by an applied dc flux. Just so does the voltage at the tank circuit $U_T = \hat{U}_T \sin(\omega t + \xi)$ which is related to $\hat{I}_T$ and ϑ in the following way:

$$(6)\quad \hat{u}_T = L_T \omega_T \zeta_T \left[1 - 2\frac{\hat{I}_{RF}}{\hat{I}_T}\cos\vartheta + \left(\frac{\hat{I}_{RF}}{\hat{I}_T}\right)^2\right]^{1/2}\hat{I}_T$$

$$(7)\quad \zeta = \mathrm{arctg}\left[-\mathrm{cotg}\,\vartheta + \left(\frac{\hat{I}_{RF}}{\hat{I}_T}\right)\frac{1}{\sin\vartheta}\right]$$

Consequently there exists a non vanishing amplitude-modulation-depth U_M defined by

$$(8)\quad U_M = \hat{u}_T\left(\Phi_{dc} = 0\right) - \hat{u}_T\left(\Phi_{dc} = \frac{\Phi_0}{2}\right)$$

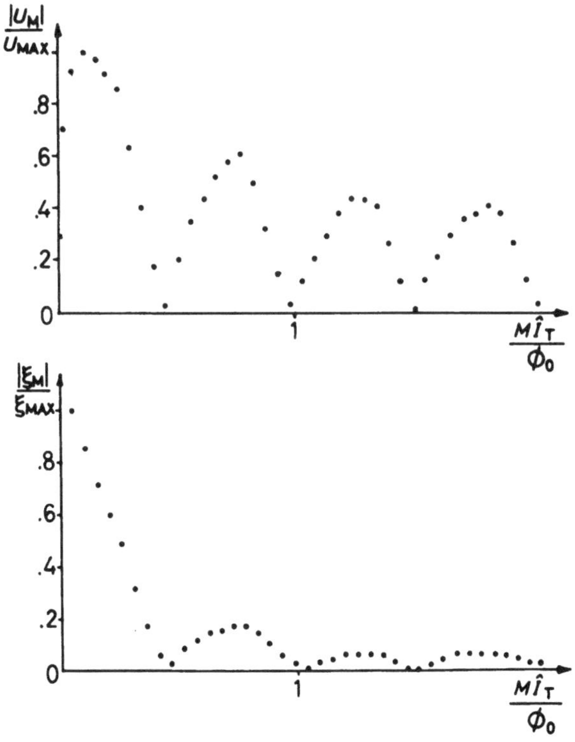

Fig. 1

Typical dependence of the normalized absolut value of the modulation depth on the normalized amplitude of the SQUID excitation: in the upper part the amplitude modulation $|U_M|$ in the lower part the phase modulation $|\xi_M|$

as well as a non vanishing phase-modulation-depth defined by

(9) $$\xi_M = \xi\left(\phi_{dc}=0\right) - \xi\left(\phi_{dc}=\frac{\phi_0}{2}\right)$$

In fig. 1 the calculated dependences of amplitude- and phase-modulation-depth on the excitation amplitude for a typical set of circuit parameters are shown. It confirms the Bessel function like dependence of the amplitude modulation depth proportional to $F \cdot \hat{I}_T$ and the phase modulation depth directly proportional to F what can be evaluated by a convenient approximation of equations (3) and (4).

An essential characteristic of the phase modulation depth is its dependence on the off tune parameter $\delta = \left(\frac{\omega}{\omega_T} - \frac{\omega_T}{\omega}\right)Q$

In fig. 2 the amplitude-modulation-depth U_M is compared with the combination of amplitude and phase modulation, eq. (10). We find a maximum of ξ_M near the resonance case whereas for the same argument the amplitude modulation depth is very small. By this is confirmed what first has been pointed out by Silver and Zimmerman[2]: In the nonhysteretic mode the maximum variation of the amplitude of the tank voltage occurs approximately at the frequency corresponding to the - 3 db point of the

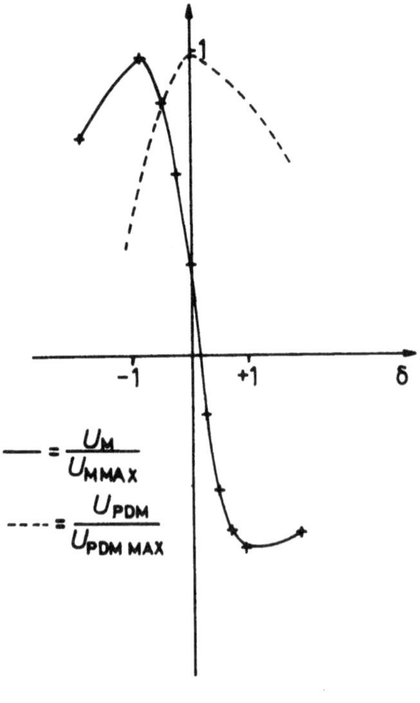

Fig. 2 Dependence of the normalized modulation depth on the off tune parameter: (——) the amplitude modulation U_M, (--------) the combination of amplitude and phase modulation U_{PDM}

tank circuit and moreover U_M is nearly an antisymmetrical function of the off tune parameter for this mode.

Furthermore, the results of the numerical computation of the equations (4), (5), (7) and (8) show that one can accomplish phase demodulation too in operating with an RF SQUID in the non hysteretic mode: in other words, in this way it is possible to combine the information of amplitude and phase demodulation . The modulation, which is obtained in this way, is defined as

$$(10) \quad U_{PDM} = \hat{u}_T(\phi_{dc}=0)\cos\xi(\phi_{dc}=0) - \hat{u}_T\left(\phi_{dc}=\frac{\phi_0}{2}\right)\cos\xi\left(\phi_{dc}=\frac{\phi_0}{2}\right)$$

A series of measurements on a toroidal SQUID working at 20 MHz was performed. The results confirm the theoretical results.

To estimate the value of the SQUID parameter ß, in fig. 3 a comparison between theoretical and experimental results of the dependence of the amplitude modulation depth on the off tune parameter is shown. The measured data exhibit the typical non hysteretic behaviour of the SQUID since under resonant conditions only a small modulation depth is observed. Under this assumption for the given set of SQUID parameters, the product demodulation, in the mean-

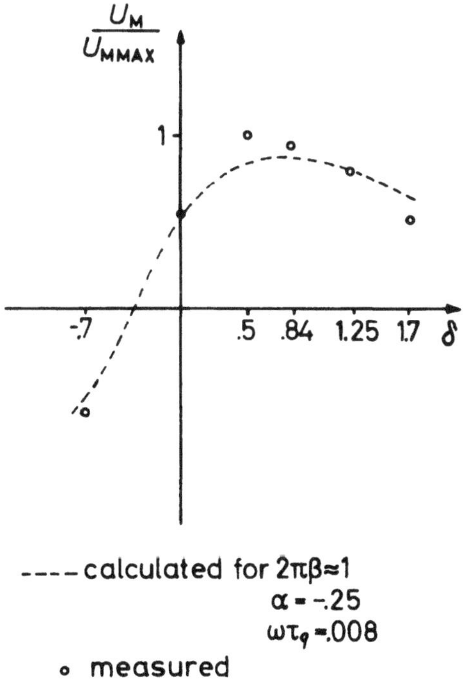

Fig. 3 Comparison between calculated and measured dependence of the amplitude modulation depth on the off tune parameter δ

ing of eq. (10), has been calculated and compared with the corresponding experimental data. Both, theoretical and experimental data are in agreement, showing that U_{PDM} has a well defined maximum, which is higher than the maximum value of the simple amplitude modulation depth.

In fig. 4 a comparison between calculated and measured depen-

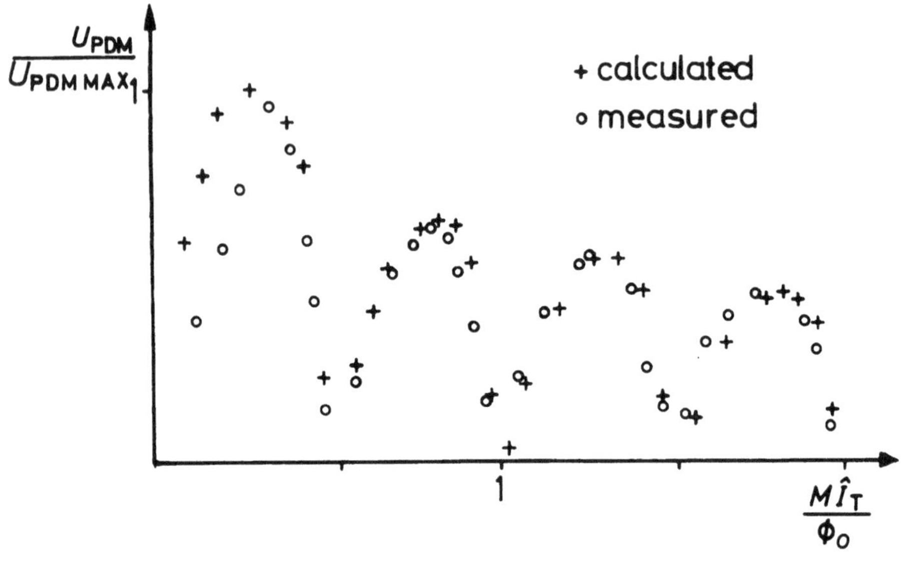

Fig. 4 Comparison between calculated and measured dependence of U_{PDM} on the normalized amplitude of the SQUID excitation; $\delta = 0$ the other SQUID parameters correspond to fig. 3

dences of U_{PDM} on the excitation current is shown. Both are normalized with respect to the first maximum. The deviations for the first three points result from a non-linearity of the attenuator, which controls the bias amplitude.

A difficulty of this comparison arises from the different absolut scaling factors. Normalized to the maximum of the calculated amplitude modulation depth, the calculated values of the maximum of U_{PDM} are increased by a factor 1.5, whereas the experimental results for U_{PDM} compared to the measured maxima of U_M are increased by a factor 2.5. This means, we find an effect larger than expected and, up to now, it was not possible to find any explanation for this disagreement.

I thank Mr. Scheer for helpfull assistance in performing the experiments.

REFERENCES

[1] S.N. Erne, H.-D. Hahlbohm and H. Lübbig: J. Appl. Phys. 47, (1976), 5440

[2] A.H. Silver and J.E. Zimmerman: Phys. Rev. 157, (1967), 317

COMPUTER ELEMENTS

SQUIDS as Computer Elements

P. Wolf
IBM Zurich Research Laboratory, 8803 Rüschlikon, Switzerland

I INTRODUCTION

In recent years, applications of Josephson junctions in digital circuits have received considerable attention. For a review see references (1,2,3). The interest is based on the combination of high switching speed and very low power consumption of these devices. Junctions with theoretical switching speeds from the superconducting to the resistive state of some ps(= 10^{-12}s) have been fabricated (4). This turns out to be faster than measureable with available equipment, which has a time resolution of 30 - 40 ps. Due to the operating voltage in the millivolt range, power consumption is in the microwatt region. The power-delay product, a figure of merit for switching devices is several orders of magnitude smaller for Josephson devices as compared to transistor circuits. This combination of properties is essential for applications in high-speed computers where short device-switching time is required as well as small transmission delay on interconnection lines between circuits. The latter demands dense packing of devices which makes heat extraction difficult if power consumption per device is not very small. Here Josephson devices are at an advantage compared with transistors. In fact, today, heat extraction is a primary design problem in transistorized high-speed computers (5).

In digital circuits, the Josephson junction is used basically as a magnetically-actuated switch from the superconducting to the resistive state or vice versa. As an example, a logic circuit is shown in Fig. 1a (1,6,7).

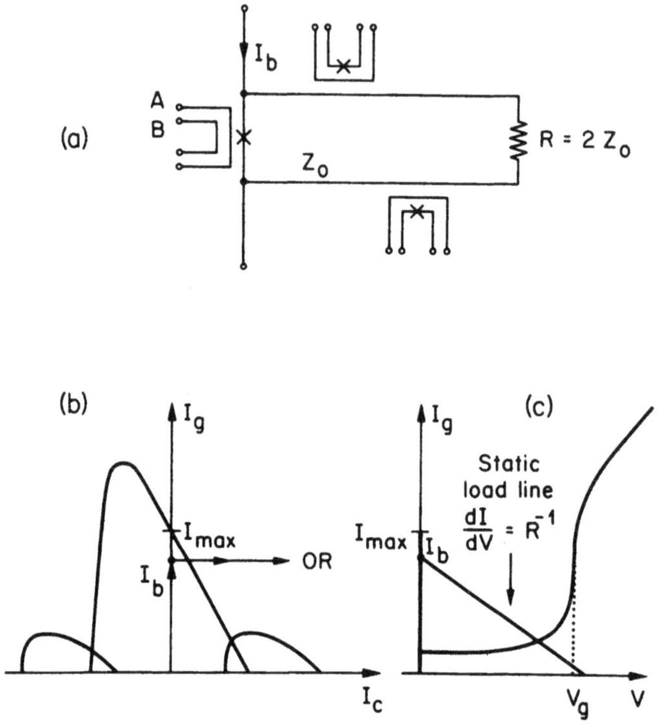

Fig. 1 Logic circuit with Josephson junctions. a)Schematic diagram, the crosses symbolize junctions; the input has two control lines A and B. b) Threshold characteristics with operating currents for performing a logic OR function. c) I-V characteristics with static load line of output circuit.

The Josephson junction is biased with a dc current I_b in its superconducting region. The junction has two (or more) control lines, and currents through them generate a magnetic field which influences the Josephson current. If the maximum Josephson current is decreased below the bias current I_b, then the junction switches to the voltage state. How an OR function can be performed by the superposition of control currents is depicted in Fig. 1b.

Parallel to the junction, the output circuit is connected, thus allowing other logic circuits to be controlled. It consists of a transmission line with characteristic impedance Z_0, and an output resistor R. If $R = 2Z_0$, then the transmission line is matched, which gives minimum delay time because reflections are avoided. In its superconducting state no voltage occurs over the junction, and consequently, there is zero current in the output circuit. A current only develops in the resistive state of the junction. With an output impedance as in Fig. 1c, most of the bias current I_b is transferred into the output circuit. More properties of this circuit will be discussed later, here it served mainly to show how a Josephson junction is used as a switching element.

An important design parameter for Josephson junctions is the ratio junction length L_j over Josephson penetration depth $\lambda_j = \sqrt{\phi_0/(2\pi\ \mu_0\ d\ j_{max})}$ with ϕ_0 the flux quantum, $d = t_{ox} + \lambda_1 + \lambda_2$ the sum of tunnel oxide thickness and London penetration depths in the electrodes, and j_{max} the Josephson current density. For instance, in logic circuits, the ratio of output current to input current, the "gain", has to be unity or slightly larger. This requirement can best be met by $L_j/\lambda_j \gtrsim 3$.

II INTERFEROMETERS AS SWITCHING ELEMENTS

1) Advantages of SQUIDS

As is well known, Josephson interferometers or SQUIDS have found extensive use in various measurement applications, especially in magnetometry (8). Recently, they have also been applied in digital circuits for basically two reasons:

a) The static threshold characteristics (diffraction patterns) are different from those of single junctions. This gives the circuit designer an additional base to choose from and makes it easier for him to find a suitable device for a particular circuit requirement. Moreover, interferometer static characteristics can be tailored over a wide range by selecting the number of junctions, (two, three or more), their maximum Josephson currents, and the values of the interferometer inductances.

b) The control mechanism in interferometers is different. In a single junction its area is not only used for generating the tunnel current, but also, by applying a magnetic field, for controlling the Josephson current. In interferometers these two tasks are separated. The junctions are sources of the Josephson currents only, and control is achieved by applying the magnetic field to the interferometer inductances. How this increases design flexibility is shown by a simple example in Fig. 2. Usually, one wants to operate with currents as small as possible. As mentioned earlier in a single junction the length L_j is determined by the L_j/λ_j ratio required for a particular circuit application. The minimum width w is fixed by the required number of control lines, and their minimum line width allowed by photolithography. This leaves one free parameter only, the

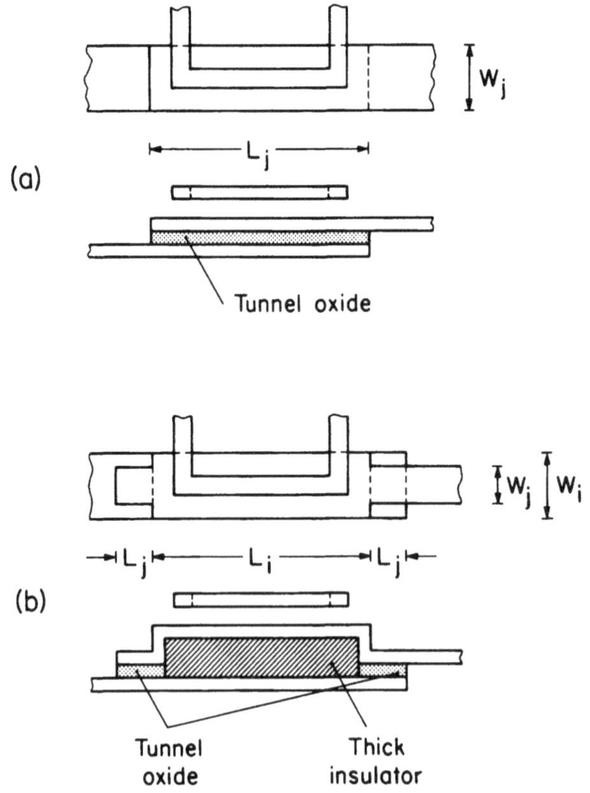

Fig. 2 Comparison of a single junction with a two-junction SQUID.
a) Single junction; b) two-junction SQUID (simplified).

Josephson current density j_{max}, the range of which is often narrowed by requirements such as the necessary switching speed of the device.

In an interferometer, the width w_i of the inductance as shown in Fig. 2b is again determined by the number of control lines and their minimum linewidth. However, the junction area $w_j \cdot L_j$ can be chosen rather freely to give a Josephson current I_0 suitable for the intended application. As will be discussed later, optimal operation of the interferometer in a circuit requires a certain product $L \cdot I_0$ of inductance L and Josephson current I_0. In most cases $L \cdot I_0 \approx (0.1 \ldots. 1) \cdot \phi_0$. In many respects, the $L \cdot I_0$ product is analogous to the L_j/λ_j ratio of single junctions. The inductance L also depends, in addition to the width w_i, on the length L_i and the effective oxide thickness $d_i = t_{ox,i} + \lambda_1 + \lambda_2$; two parameters which can still be chosen in a rather wide range.

As is obvious from the foregoing discussion, an interferometer has more free geometrical parameters than a single junction, and in most cases these can be chosen to minimize operating currents. In logic circuits, improvements by nearly a factor of ten have been realized. Smaller currents not only mean lower power consumption but, as will be seen later, improve packing density of interconnection lines between circuits as well.

As a disadvantage, especially with more than two junctions, interferometers have more complex structures than single junctions. The Josephson current, in general, will be less well defined than in single junctions because there are more geometrical parameters which can be influenced by fabrication tolerances.

2) Structure and Fabrication of SQUIDS for Digital Circuits

SQUIDS for digital circuits differ from those for magnetometers in that the former generally have higher currents and lower inductances than the latter. Fabrication methods are not very different from those for tunnel junctions.

In the circuits to be described later the fabrication methods as reported in (9) have been used. The electrodes consist of evaporated lead with additions of gold and indium (10). The tunnel barrier is obtained by rf-plasma oxidation (11) and the thick insulators are mostly evaporated SiO. Structures are patterned with photolithography. In order to supply a good ground for strip transmission lines, a ground plane of Nb is provided.

From the various structures realized, two are shown schematically as examples in Fig. 3. The two-junction SQUID in Fig. 3a is basically a crossline junction with a hole in the bottom electrode and most of the SQUID inductance is situated in the area within or close to the hole. A photograph of such a SQUID is given in Fig. 4. The control line above the top junction electrode can also be seen. For simplicity, it has been omitted in Fig.3a. Four- and six-junction SQUIDS with the same technique have been realized too (12). A somewhat more complex but more compact structure is sketched in Fig. 3b. Here, the inductance is obtained by placing a thick spacer insulator ($t_{ox} \approx 2000$ Å) between part of the junction electrodes. Prior to evaporation of the spacer insulator, the base electrode was heavily rf-plasma oxidized in order to prevent shorts by pinholes in the spacer layer (13). A realized structure is shown in Fig. 7 ; dimensions of the junctions with 4×7 µm and the inductance with 6×7 µm are quite small.

Several other examples of fabricated structures can be found in references on circuits to be quoted later.

3) Equivalent Circuit and Electrical Properties

Circuit design requires a good understanding of the devices used. This is the case for Josephson tunnel junctions as well as for interferometers. The electrical properties can be deduced if the equivalent circuit is known. A simplified equivalent circuit for a two-junction SQUID is shown in Fig. 5. The extension to SQUIDS with three or more junctions is straightforward.

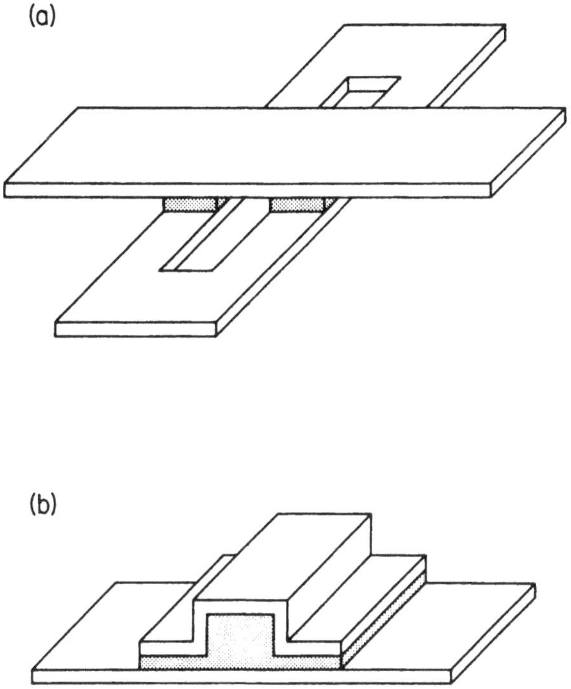

Fig. 3 Examples of structures of two-junction SQUIDS. a) Inductance realized by slot in base electrode (for simplicity, control lines omitted). b) Inductance realized by thick spacer insulator between electrodes (for simplicity, current feeds and control lines omitted).

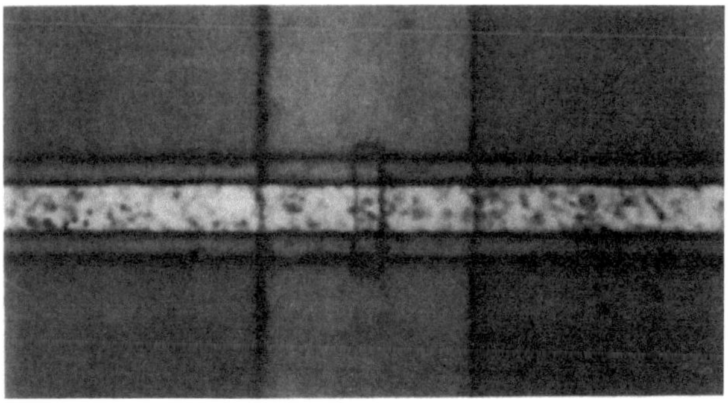

Fig. 4 Photograph of two-junction SQUID with inductance realized by slot in base electrode. Size of junctions is about 8×9 μm and slot size is about 12×2.5 μm. The control line on top of the electrode running from left to right is also visible.

For the tunnel junctions, the well-known point junction model of Stewart-McCumber can be used with a sine-Josephson term, with a nonlinear resistor given by the single particle current and with the junction capacitance (14,15). The interferometer inductance L is composed of L_1 and L_2, the ratio of which depends on how the current is fed into the interferometer. In some applications, it is necessary to take into account the skin losses in the electrodes, which can be done by the equivalent circuit for the inductance as shown in Fig. 5. Coupling of the control lines to the inductances can be modelled in the simplest case by an ideal transformer.

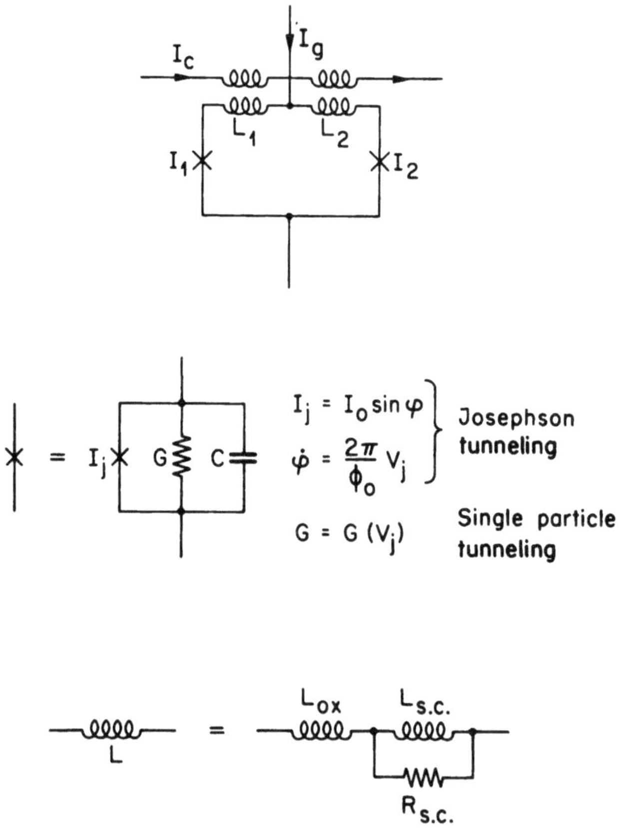

Fig. 5 Equivalent circuit of two-junction SQUID. The junction model as well as the model for a lossy inductance is shown (L_{OX}: inductive contribution of fields in insulators, $L_{S.C.}$ and $R_{S.C.}$: inductive and loss contributions of electrodes).

Real SQUIDS are often more complicated either because the control line couples in only part of the interferometer inductance or because coupling is not ideal; cases which will not be dealt with here.

4) Static Threshold Characteristics (Diffraction Patterns)

The theory of diffraction patterns has been treated by various authors, for some more recent references see (12,16,17). For the two-junction interferometer, an analytic solution is known in parametric form with the Josephson phase φ of one of the junctions as the independent parameter (12, 16,18). From Kirchhoff's rules, it is easy to find relations for the gate current I_g through the interferometer, and for the control current I_c which sets up the magnetic field in the inductances:

$$I_g = \sqrt{I_1 I_2}\,(i \sin \varphi_1 + \frac{1}{i} \sin \varphi_2)$$

$$I_c = \sqrt{I_1 I_2}\left[\frac{\ell}{i} \sin \varphi_2 - i\,(1-\ell) \sin \varphi_1 + \frac{1}{\lambda}(\varphi_2 - \varphi_1 + 2\pi n)\right].$$

Here, the following abbreviations have been used with parameters as in Fig. 5:

$$\lambda = \frac{2\pi}{\phi_0} \sqrt{I_1\, I_2}\,(L_1 + L_2) \qquad i = \sqrt{\frac{I_1}{I_2}}$$

$$\ell = \frac{L_1}{L_1 + L_2} \qquad n = \text{mode number.}$$

The phase angles φ_1 and φ_2 at the vortex mode boundaries are given by a quite simple expression, which is obtained from stability considerations:

$$\cos \varphi_2 = - \frac{i \cos \varphi_1}{\lambda \cos \varphi_1 + 1/i} \quad .$$

To construct the mode boundaries, one phase angle is used as an independent parameter, then the stability equation is solved for the other phase and both are inserted into the equation for currents I_g and I_c. To obtain a full mode boundary, a convenient choice for the angles φ_1 and φ_2 is the following:

a) φ_1 is independent and varies between -90° and $+90^o$. The stability equation is solved for two solutions of φ_2 with values restricted to the range -180° and $+180^o$. This yields one half of the vortex boundary.

b) The second half is obtained as before but φ_1 replaced by φ_2 as independent variable.

The mode number n determines the number of flux quanta in the mode.

Unfortunately, analytic solutions for three- and more-junction interferometers are usually not available. For three-junction interferometers, methods for the calculation of mode boundaries are given in (12,16,17).

5) Dynamics of Interferometers

The switching behavior of interferometers is similar to that of single junctions, and the switching speed in both cases is mainly determined by the ratio C/I_0 of capacitance to Josephson current of the junctions.

Self-induced dc steps caused by device resonances exist in interferometers too, and have to be taken into account in design of circuits. Single junctions have a multitude of cavity resonances, the number of which is proportional to junction length. Interferometers are lumped devices with a good approximation and have a small number of resonances. The lumped two-junction SQUID has one resonance, the lumped three-junction SQUID has two, and so on. The resonance frequency ω of the two-junction SQUID is $\omega = 1/\sqrt{(L_1 + L_2)\ C_1\ C_2/(C_1 + C_2)}$, with the parameters as defined in Fig. 5. The measured self-induced steps in two-junction SQUIDS can be quite high, up to 0.65 of $I_{max} = I_1 + I_2$.

The switching characteristics of SQUIDS and their dynamic behavior in circuits is, in general, difficult to treat, with analytic solutions rarely available. In most cases one has to resort to numerical methods. For analysis of the circuits to be described later, computer programs like CSMP (19) for the solution of coupled, nonlinear differential equations or ASTAP (20) for the treatment of nonlinear electrical networks, have been used successfully.

III DIGITAL CIRCUITS WITH SQUIDS

Memory cells and logic circuits have been made with SQUIDS. In principle, after suitable redesign, it is possible to replace single junctions in circuits by interferometers. However, some interferometer circuits are hard to realize with single junctions, e.g., the memory cell with its particular readout to be described next.

1) Memory Cells

An appealing method of storing information in a superconducting environment is to code it as persistent ring currents in superconducting loops. Cells based on this principle have been realized (21,22,23). Accompanied by a persistent ring current is a magnetic flux which is quantized and which has to be an integer multiple of the flux quantum $\phi_0 = h/2e = 2.07 \cdot 10^{-15}$ Vs. The aforementioned cells contain so many flux quanta that flux appears essentially as a continuous quantity. It is, of course, tempting to consider information storage with single flux quanta because it promises memory cells with minimum size and energy. It turned out that such storage can be achieved quite simply, in principle, in all Josephson devices which have overlapping vortex modes (24,25). To this large class of devices belong the long Josephson junction and interferometers with one or more junctions and a sufficiently high $L \cdot I_0$ product. Vortex modes are superconducting states of these devices and the modes are distinguished by the number of flux quanta stored in them. As an example, in Fig. 6, the well-known vortex modes of the two-junction SQUID are plotted in the $I_g - I_c$ plane. For $L \cdot I_0$ products larger than about $0.2\ \phi_0$ the vortex modes overlap one another considerably. For the situation as shown in Fig. 6, this means that in the overlap region either of two superconducting states are possible, differing by just one flux quantum. It is this bistable region which is well suited to store one binary bit of information. Storage

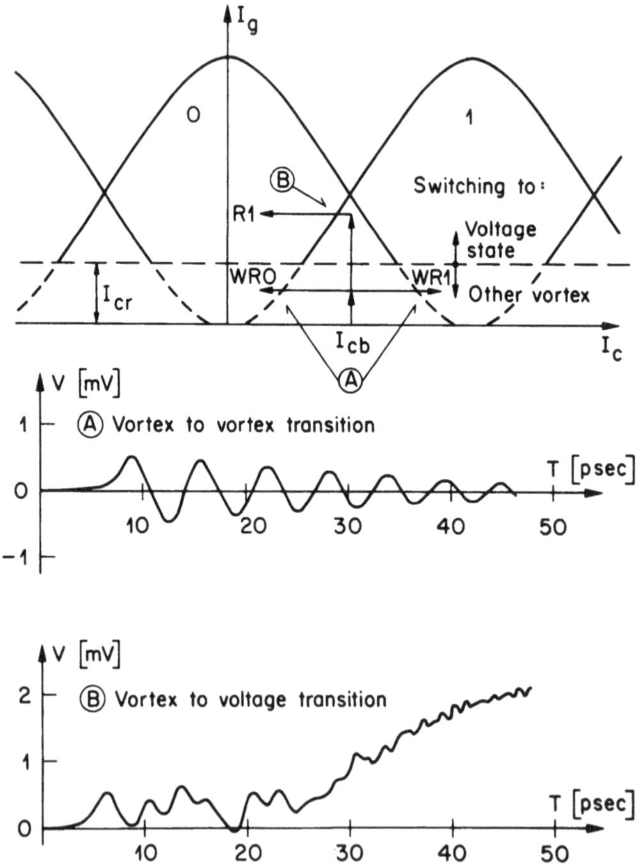

Fig. 6 Vortex modes and vortex transitions in a two-junction SQUID, used for single flux quantum storage. Operating currents for writing (WRO and WR1) and reading (R1) are given. Computer simulations of voltage transients for vortex-to-vortex transitions and vortex-to-voltage transitions are shown (30).

into overlapping vortex modes was first theoretically predicted and then experimentally demonstrated for two-junction interferometers (24) and long Josephson junctions (25). Because they operate with single flux quanta, these storage devices are also called single flux-quantum (SFQ) memory cells.

The two-junction SQUID offers some advantages over other devices for single flux-quantum storage: because all vortex modes have equal shape in this SQUID, the two storage states are equivalent, which has advantages with respect to operating margins; the length-to-width ratio is more square than for long Josephson junctions; perhaps most important, the two-junction SQUID allows a read mode, to be described later, which is probably difficult to achieve with other Josephson devices. In the following, the write and read operations are described. In the quiescent state, in order that stored information is not lost, the device has to be biased into an overlap region by an external magnetic field or a dc-control current I_{cb}.

For writing, use is made of vortex-to-vortex transitions. It is assumed that vortex states $n = 0$ and 1 are used for storing a "0" or "1" of the binary system, and further, that the SQUID is originally in the state $n = 0$. In order to write a one (WR1), one applies currents as shown in Fig. 6, which leave the overlap region and bias the device into the $n = 1$ vortex. When the boundary of the state $n = 0$ is reached, this mode is no longer stable and a transition occurs to the $n = 1$ vortex state. After the write currents are removed, one returns into the overlap region, but stays in the state $n = 1$. In this way a "1" is stored in the device. If one starts originally from the "1" state, nothing happens because the WR1 currents stay within the $n = 1$ vortex. In order to write a "0" (WR0) one simply has to use a write current I_c with opposite polarity. This is how writing can be accomplished.

In vortex-to-vortex transitions, the flux in the device is changed by one flux quantum. Therefore, a voltage spike with $\int V \cdot dt \approx \pm \phi_0$ appears over the SQUID. A transient obtained from a computer simulation is shown in Fig. 6. It is oscillatory with a period given by the SQUID resonance mentioned earlier. Decay time, determined by internal losses is quite short. For Pb-alloy devices with $j_{max} = 2\ kA/cm^2$, as Fig. 6 shows, a value of about 50 ps is predicted. In principle, one can utilize the voltage spike for reading by using a small junction as a detector. When biased close to its threshold current I_{max} it will be switched by the spike to the voltage state. This reading method has been demonstrated experimentally (24,25).

However, another simple reading method is possible which is also shown in Fig. 6. Vortex-to-vortex transitions usually occur only if the gate current I_g in the device is zero or quite small. If I_g is large, then when a vortex boundary is traversed, the device no longer switches into another vortex state but instead goes permanently to the voltage state. This effect was first observed and explained in (26). It only occurs if the SQUID is little damped by internal or external losses. The use of this vortex-to-voltage transition for reading was proposed and experimentally shown in (18). Reading is simplified with this transition because now a dc voltage has to be detected over the device and not a very short voltage spike. How reading (R1) is done is indicated in Fig. 6. From the bias point in the overlap region, pulse currents I_c and I_g with high enough I_g are applied so that the vortex boundary is crossed in a region where vortex-to-voltage transition occurs. When the device is originally in the "1" state, then switching to the voltage state occurs; when it is in the "0" state, then nothing happens. In this way, a stored "0" or "1" can be distinguished. After reading, the cell has to be reset to the superconducting state by switching off I_g for a brief period. In the read process the "1" state is destroyed and has to be rewritten. The critical value I_{cr} of the gate current I_g, above which safe switching to the voltage state occurs according to simulations and measurements, is at about 0.25 of I_{max} and only depends weakly on the $L \cdot I_0$ product and on external damping, provided the latter is not too strong. A computer simulation of a vortex-to-voltage transition is shown in Fig. 6, too. It is very fast and takes place in about 50 psec. More details about the vortex transitions can be found in reference (13).

In Fig. 7, the smallest of several realized devices is shown. Its structure and fabrication has already been discussed. The device has a size of only about 150 $(\mu m)^2$; the inductance L is about 0.6 pH. With current densities between 6 and 12 kA/cm^2 the cell was operated with $L \cdot I_0$ products between 0.5 .... $1\phi_0$.

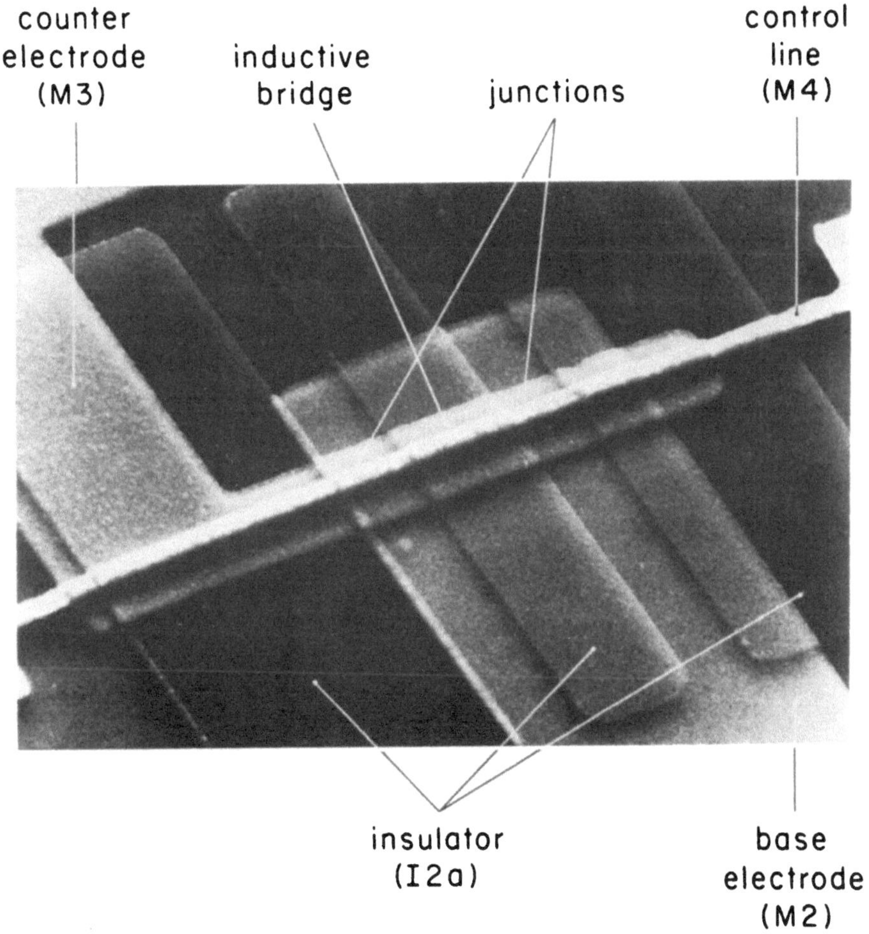

Fig. 7 SEM photograph of two-junction SQUID with inductance realized by thick insulator between junction electrodes. Junction and inductance dimensions are 4 × 7 μm and 6 × 7 μm, respectively. The control line is about 2 μm wide. This device was operated as SFQ cell.

In Fig. 8, the vortex transitions measured in the overlap region of the device in Fig. 7 are shown. The modes are tilted because as shown in Fig.7, the current I_g is fed asymmetrically into the SQUID. The measurements were carried out with pulse sequences. The device was always biased in the overlap region by a suitable dc current I_{cb}. To measure the vortex-to-voltage transitions, a sequence of two pulses was used. With the first part consisting of simultaneous pulse currents of I_g and I_c, the device was set into a definite vortex state. Then the second part of the sequence was applied with amplitudes of I_g and I_c, adjusted so that the mode boundary was just crossed in a region where vortex-to-voltage transitions occurred. To trace out the vortex-to-vortex transitions, a sequence of three pulses had to be used because the voltage spike of the transition was too small to be detected with the equipment available. Once more, with the first pulse the SQUID was set in a definite vortex state (for instance n = 0), then a disturb pulse with variable amplitude for I_g and I_c was applied

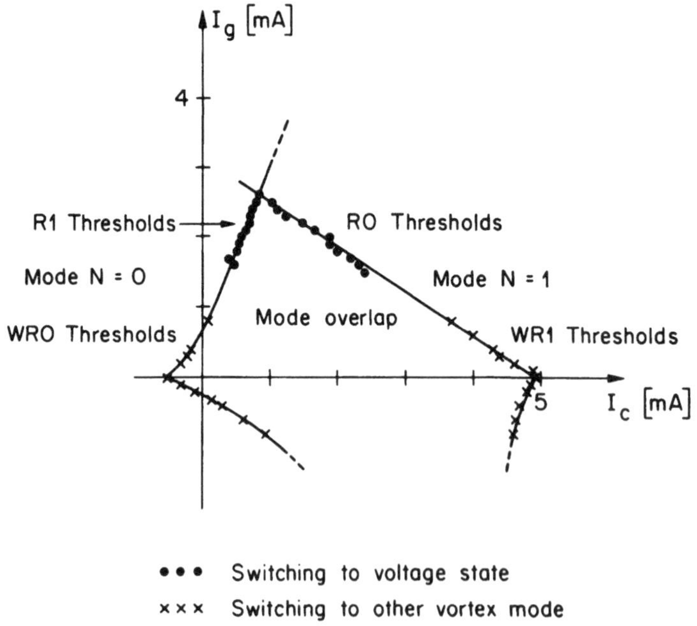

Fig. 8 Measured mode boundaries of the device in Fig. 7. The boundaries in the overlap region have been determined via vortex-to-vortex and vortex-to-voltage transitions.

which when sufficiently high led to vortex switching from the previous state (for instance n = 0) to another state (for instance n = 1). The third pulse used I_g and I_c values which safely led to vortex-to-voltage switching provided the device had already switched into the second state (for instance n = 1). By varying the amplitudes of I_g and I_c of the second pulse, it was possible to trace out the otherwise invisible vortex boundaries. As soon as the second pulse reached the vortex boundary of the original state, vortex-to-vortex switching to another state occurred, the presence of which was then detected by the read pulse. In Fig. 8, the measured vortex-to-vortex transitions are marked by crosses. In general, the measured vortex boundaries agree well with theory, although refinements of the simple equivalent circuits were required to bring measurements and theory together (13).

The vortex-to-voltage switching is sometimes quite complicated, because first a transition to the resonance state of the SQUID can occur, which becomes apparent by the dc voltage over the device being much smaller than the gap voltage. However, if one moves away from the vortex boundary, for instance, by increasing I_c then the device leaves the resonance state and switches to the gap. Experimentally it was found that SQUIDS with higher j_{max} very often showed this behavior, whereas SQUIDS with lower j_{max} tended to switch directly to the gap. In memory applications, switching to a resonance state is undesirable, because of the small dc voltage for reading involved and because of the high-amplitude rf oscillation in resonance which could lead to disturb effects in an array environment.

The simulations shown in Fig. 6 indicate that SFQ cells should be quite fast. The measured rise time of a vortex-to-voltage transition for the SQUID in Fig. 7, was about 100 psec (13). It is slower than the simulation result, mainly due to parasitics of the chip on which the SQUID was situated. It is obvious that such cells have good potential for a high-speed memory. Another attractive feature is that the device consumes no power in the storage states because these are superconducting. A more detailed discussion of properties of SFQ cells can be found in (13).

2) Logic Circuits

Logic circuits with SQUIDS have been discussed extensively in (27). Therefore only a brief description is given here. In principle, in the logic circuit shown in Fig. 1, one can replace the single junction with an interferometer. A two-junction SQUID is not a very good choice for such a purpose, because between the vortex modes there is little space left for the voltage state. A better choice is a three-junction SQUID, with, in addition, a middle junction having twice the Josephson current as the outer junctions (28). Calculated characteristics are shown in Fig. 9 (29). As mentioned earlier, owing to the greater number of geometrical parameters, SQUIDS can be designed easier for small operating currents than single junctions. This, of course, means lower operating power of the circuit. Even more important, higher impedances of the output circuits are possible. The value of the required output impedance Z_0 is $V_j/2I_{out}$ where V_j is the voltage over the junction in the switched state and $I_{out} \lesssim I_{max}$ the current in the output loop. As is obvious from the expression, smaller I_{out} means higher Z_0. The impedance Z_0 of strip lines is inversely proportional to their width w. Accordingly, a small width w can be chosen for the interconnection lines between the circuits which, of course, leads to a very desirable increase in packing density (28).

The characteristics of the logic circuit in Fig. 1 depend very much on the choice of the output resistor which influences the Stewart-McCumber factor β_c (14,15):

$$\beta_c = 2\pi \frac{I_{max} R^2 C}{\phi_0} .$$

Here, C is the junction capacitance, I_{max} the maximum Josephson current of the device and R is given by the external load resistor and the single-particle characteristics below the gap. In general, tunnel junctions show a high β_c, which leads to the large hysteresis of these devices. The hysteresis behavior of SQUIDS is to first order the same as that of single junctions with very little influence from the SQUID inductances.

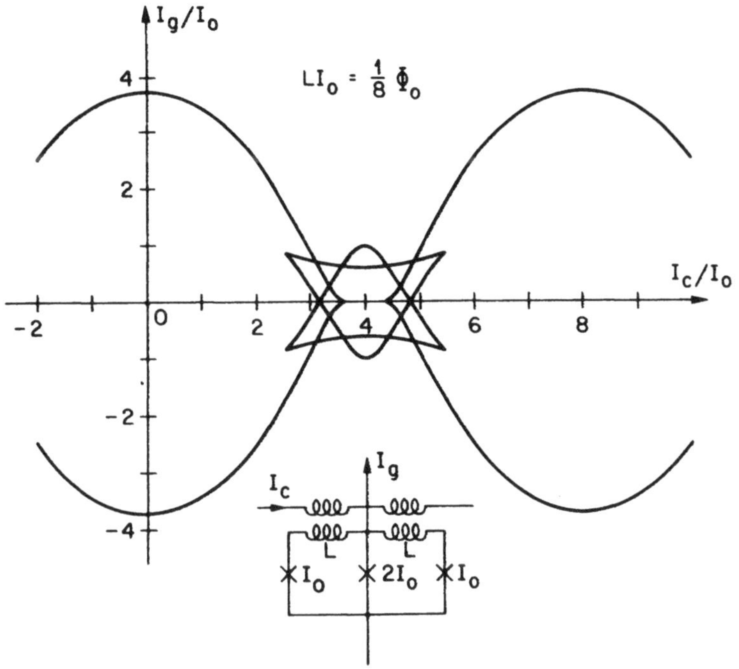

Fig. 9 Computed mode boundaries of a three-junction SQUID (29).

For sufficiently high values of the output impedance, for instance those shown in Fig. 1 , the value of β_c stays high. This means that the logic circuit, after it has switched, remains in the voltage state even if the control currents at the input have been removed. In order to reset the circuit, the supply current has to be switched off. This is known as a latching logic circuit.

However, by making the output resistor low, β_c can be decreased to values below about two, where junction hysteresis is quite small. Then the device is non-latching and resets to the superconducting state when the control currents are removed. This is a mode of operation similar to most semiconductor logic circuits. In logic circuits with single junctions, the output impedance would be unpractically low; interferometers with their low operating currents are much better suited for non-latching operation.

A circuit realized with I_{max} = 0.7mA required output resistors of 0.2 ohm which is still quite low but closer to a practical range than possible with single junctions (27). Owing to the steep load line the interferometer had an operating voltage in the switched state of 70 microvolt only. This, of course, together with the small bias current led to an extremely small power consumption in the device of less than 40 nanowatt. Low output impedance combined with low operating voltage has the disadvantage that the output circuit becomes sensitive to the unavoidable inductive disturbances given by the control windings over interferometers arranged along these output lines. The main penalty is that propagation delays increase. Despite this, the aforementioned circuit had a delay between logic states of only about 235 psec.

IV SUMMARY

The previous discussions have shown that SQUIDS are not only important devices for established applications like magnetometry but that they have a place in digital applications, too. With SQUIDS one has realized memory cells which store information as single flux quanta. Logic circuits built with SQUIDS have the advantage of very low operating currents. The application of interferometers in digital circuits is quite new, therefore one can expect considerable advances in the future.

LITERATURE REFERENCES

(1) Anacker, W.: Josephson tunneling devices - a new technology with potential for high-performance computers. AFIPS Conference Proceedings 41, 1269 (1972).

(2) Matisoo, J.: Josephson computers. IEEE Trans. Magn. (to be published in spring 1977).

(3) Anacker, W.: Josephson junctions as computer elements. in 1976 ESSDERC Proceedings, Conference Series of the Institute of Physics, London (to

be published in March 1977).

(4) Jutzi, W., Mohr, Th.O., Gasser, M., and Gschwind, H.P.: Josephson junctions with 1 μm dimensions and with picosecond switching times. Electronic Letters 8, 589 (1972).

(5) As an example , an air-cooled package for high-speed transistor circuits can be found in: Electronics, October 30, 1975, p.86.

(6) Henkels, W.H.,: An elementary logic circuit employing superconducting Josephson tunneling. IEEE Trans. Magn. MAG-10, 860 (1974).

(7) Herrell, D.J.: Femto Joule Josephson tunneling logic gates. IEEE J. Solid-State Circuits SC-9, 277 (1974).

(8) Clarke, J.: Low-frequency applications of low-frequency superconducting quantum interference devices (this issue).

(9) Greiner, J.H., Basavaiah, S., and Ames, I.: Fabrication of experimental Josephson tunneling circuits. J. Vac. Sci. Technol., 11, 81 (1974).

(10) Lahiri, S.K.: Metallurgical considerations with respect to electrodes and interconnection lines for Josephson tunneling circuits. J. Vac. Sci. Technol., 13, 148 (1976).

(11) Greiner, J.H.: Josephson tunneling barriers by rf sputter etching in an oxygen plasma. J. Appl. Phys. 42, 5151 (1971).

(12) Schulz-DuBois, E.O., and Wolf, P.: Static characteristics of Josephson interferometers (to be published).

(13) Guéret, P., Mohr, Th.O., and Wolf, P.: Single flux-quantum memory cells with Josephson junctions. IEEE Trans. Magn. (to be published in spring 1977).

(14) Stewart, W.C.: Current-voltage characteristics of Josephson junctions. Appl. Phys. Letters 12, 277 (1968).

(15) McCumber, D.E.: Effect of ac impedance on dc voltage-current characteristics of superconductor weak-link junctions. J. Appl. Phys. 39, 3113 (1968).

(16) Tsang, W.T., and Van Duzer, T.: Dc analysis of parallel arrays of two and three Josephson junctions. J. Appl. Phys. 46, 4573 (1975).

(17) Landman, B.S.: Calculation of threshold curves for Josephson quantum interference devices. IEEE Trans. Magn. (to be published in spring 1977).

(18) Zappe, H.H.: A single flux-quantum Josephson memory cell. Appl. Phys. Letters 25, 424 (1974).

(19) Continuous system modelling program III (CSM: III), Program reference manual, IBM publication SH19-7001-2 (1972).

(20) Advanced statistical analysis program (ASTAP), IBM publication SH20-11180 (1973).

(21) Anacker, W.: Potential of superconductive Josephson tunneling technology for ultrahigh performance memories and processors. IEEE Trans. Magn. MAG-5, 968 (1969).

(22) Zappe, H.H.: A subnanosecond Josephson tunneling memory cell with nondestructive readout. IEEE J. Solid-State Circuits SC-10, 12 (1975).

(23) Broom, R.F., Jutzi, W., and Mohr, Th.O.: A 1.4 mil^2 memory cell with Josephson junctions. IEEE Trans. Magn. MAG-11, 755 (1975).

(24) Guéret, P.: Experimental observation of switching transients resulting from single flux-quantum transitions in superconducting Josephson devices: Appl. Phys. Letters 25, 426 (1974).

(25) Guéret, P.: Storage and detection of a single flux quantum in Josephson-junction devices. IEEE Trans. Magn. MAG-11, 751 (1975).

(26) Fulton, T.A., Dunkleberger, L.N. and Dynes, R.C.: Quantum interference properties of double Josephson junctions. Phys. Rev. B., 6, 855 (1972).

(27) Zappe, H.H.: Josephson quantum interference computer devices. IEEE Trans. Magn. (to be published in spring 1977).

(28) Zappe, H.H.: Quantum interference Josephson logic devices. Appl. Phys. Letters 27, 432 (1975).

(29) Schulz-DuBois, E.O., (private communication).

(30) Guéret, P. (private communication).

SWITCHING BETWEEN TWO DIFFERENT VORTEX MODES IN JOSEPHSON JUNCTIONS

S. Hasuo, T. Imamura and K. Dazai
Fujitsu Laboratories Ltd.
1015, Kamikodanaka, Nakahara-ku, Kawasaki 211, Japan

ABSTRACT

This paper presents computer simulated switching behaviors between two different vortex modes in Josephson junctions. It has been shown that the switching is successfully achieved by changing the external magnetic field or the total current through the junction. Transient voltage appears across the junction at the moment of the switching, then the junction settles to another supercurrent state again if the damping of the AC Josephson oscillation is suitably controlled.

I INTRODUCTION

A Josephson tunnel junction is attractive as a new switching device, because of its high switching speed and low power consumption. Logic circuits hitherto reported are mostly based on the switching between supercurrent state and voltage state of the junction [1,2,3]. Recently a new mode of the switching has been reported [4]. In this mode, the junction transfers between different vortex modes in the supercurrent state, so that no stand-by power is necessary although a small amount of power is consumed in the switching process. Therefore, this type of the switching is useful for logic circuits especially for memory circuits.

In this paper the discussions are focused on the switching between two different vortex modes, which we call "the inter-mode switching" for simplicity. Since the switching is based on the nucleation and the anihilation of one flux quantum, we describe these processes in detail, concentrating on the flux flow in the junction.

In section II, basic equations and boundary conditions are described. Supercurrent states for a given set of total current through the junction I and external magnetic field H_e are calculated in III, and their stabilities are also discussed. In section IV, switching processes between two different vortex modes are demonstrated in two types of the switching procedure. Section V describes how to operate the switching successfully, since the junction sometimes transfers to the voltage state instead of another supercurrent state.

II Basic equations

The model of a Josephson tunnel junction used in the analyses is assumed to be an in-line gate type [5] as shown in Fig.1. A superconductive ground plane is provided under the junction. In order to discuss the current density and magnetic field distributions in the junction, we should apply a distributed circuit model as an equivalent circuit of the junction.

Using the distributed circuit model of the Josephson junction in a one dimensional model, we can get basic equations to describe dynamic behaviors of the junction.

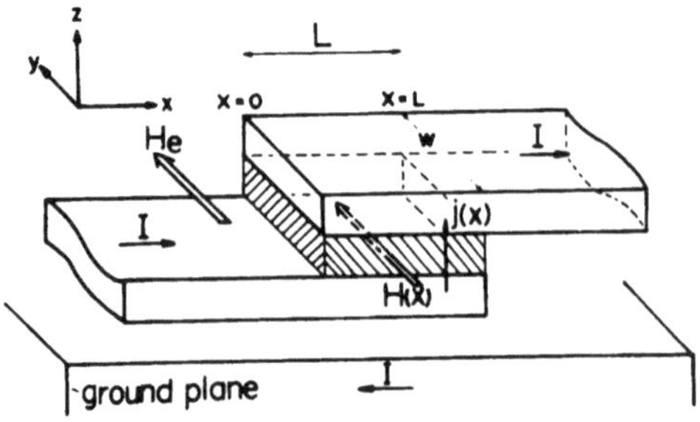

Fig.1 A model of a Josephson tunnel junction used in the analyses

$$\begin{cases} j = \text{rot}\ H = j_J \sin\phi + \frac{\varepsilon}{d_o}\cdot\frac{\partial V}{\partial t} + g\cdot V , & (1) \\ H = \frac{\hbar}{2e\mu_o d}\cdot\frac{\partial\phi}{\partial x} , & (2) \\ V = \frac{\hbar}{2e}\cdot\frac{\partial\phi}{\partial t} . & (3) \end{cases}$$

where notations are as follows; $\phi(x,t)$: the phase difference between two superconductors, $j(x,t)$: the current density in the junction at position x and time t, $H(x,t)$: the magnetic field, $V(x,t)$: the voltage, j_J: the maximum Josephson current density, ε: the permittivity of the tunneling oxide barrier, μ_o: the permeability of free space, g: the conductance per unit area of the junction, d_o: the thickness of the tunneling barrier, d: the sum of d_o and twice the London penetration depth λ_L, $\hbar$: Planck's constant devided by 2π, -e: the electronic charge.

The applied current I (per unit width of the junction) and the total magnetic flux within the junction Φ are given by

$$I = \int_0^L j(x,t)\ dx , \tag{4}$$

$$\Phi = \mu_o d \int_0^L H(x,t)\ dx , \tag{5}$$

where L is the junction length. Using eqs.(1)-(3), we obtain

$$\lambda_J^2\cdot\frac{\partial^2\phi}{\partial x^2} - t_J^2\cdot\frac{\partial^2\phi}{\partial t^2} - \Gamma t_J\cdot\frac{\partial\phi}{\partial t} = \sin\phi , \tag{6}$$

where λ_J is the Josephson penetration depth and t_J is the plasma oscillation period devided by 2π, and Γ denotes the normalized loss parameter given by

$$\Gamma = \sqrt{\frac{\hbar d_o}{2ej_J\varepsilon}}\cdot g . \tag{7}$$

The magnetic field at the edges of the junction is assumed to be the sum of the external magnetic field and the self-induced field caused by the applied current and its image current. The image current flows in the ground plane in the opposite direction as shown in Fig.1. Therefore the boundary conditions are written as

$$H(0,t) = \frac{\hbar}{2e\mu_o d}\cdot\frac{\partial\phi}{\partial x}\bigg|_{x=0} = H_e \quad , \tag{8}$$

$$H(L,t) = \frac{\hbar}{2e\mu_o d}\cdot\frac{\partial\phi}{\partial x}\bigg|_{x=L} = H_e + \frac{I}{W} \quad . \tag{9}$$

Here it is assumed that the junction is operated by a constant current source and no power is radiated from the junction. The effect of the power radiation will be discussed later.

III STABILITY OF STATIC DISTRIBUTIONS

Prior to discussions on dynamic behaviors of the junction, it is necessary to obtain static distributions of current density and magnetic field in the junction. Although the static distributions have been reported by several authors [6,7], the stability of the solution has not been discussed. We describe in this section the static distributions for a given set of H_e and I and also discuss the stability of them.

The static distributions are calculated with a following equation under the boundary conditions given by eqs.(8),(9).

$$\lambda_J^2\cdot\frac{d^2\phi}{dx^2} = \sin\phi \quad . \tag{10}$$

Since the boundary condition at x=0 (eq.(8)) is determined by H_e and it is independent of I, the solution for a given H_e can be calculated if we assume the value of the phase difference at x=0, $\phi(0)$, as a parameter. Using the spatial distribution of ϕ obtained from eq.(10) in this manner, we can calculate the total magnetic flux Φ and the total current I with eqs.(4),(5). It must be noticed that another boundary condition at x=L (eq.(9)) is selfconsistently satisfied by

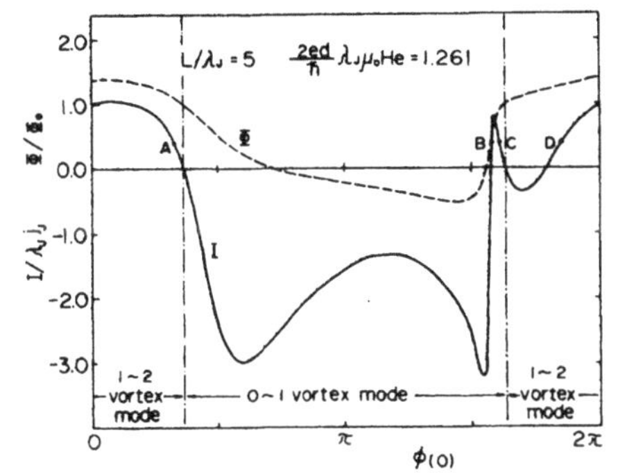

Fig.2 The phase difference at x=0, $\phi(0)$, dependence of the total magnetic flux within the junction Φ and the total current through the junction I. A, B, C and D are the states corresponding to the same current $I/\lambda_J j_J$ = 0.4.

the given H_e and the calculated I.

Figure 2 shows the normalized I and Φ as functions of $\phi(0)$ for a junction with $L=5\lambda_J$, where the normalized H_e is 1.261. This figure indicates that there are multiple solutions satisfying the boundary conditions for a given set of H_e and I. For example, there are four solutions A, B, C and D for $I/\lambda_J j_J=0.4$.

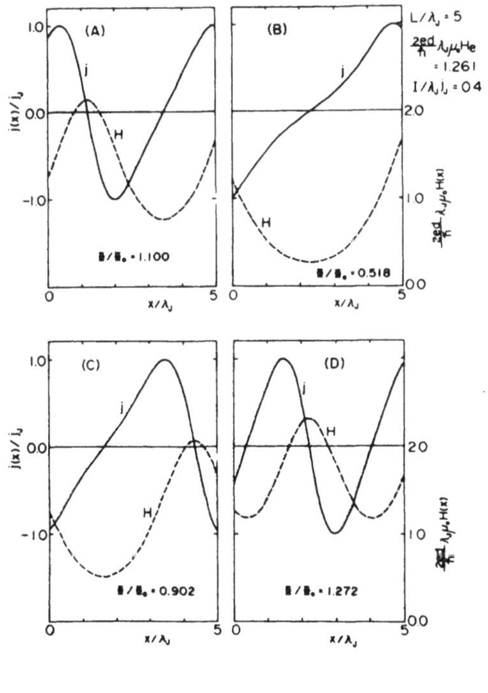

Fig.3 Four supercurrent states with different distributions of current density j and magnetic field H for a given set of H_e and I. These states correspond to A, B, C and D in Fig.2.

The spatial distributions of current density and magnetic field are plotted in Fig.3 for these solutions. Among these states, A and D are in the 1∿2 vortex mode ($1 \leq |\Phi/\Phi_0| \leq 2$), and B and C are in the 0∿1 vortex mode ($0 \leq |\Phi/\Phi_0| \leq 1$).

In order to check the stability of these distributions, we have numerically simulated their time variations using eq.(6). The distributions A, B, C and D have been used as initial conditions of the simulation. Figure 4 shows the time variations of Φ under the condition that H_e and I are kept constant at their initial values. The distributions A and C are unstable and both converge to the distribution B, which is stable in the 0∿1 vortex mode ($\Phi/\Phi_0 = 0.518$). The distribution D stays in the 1∿2 vortex mode ($\Phi/\Phi_0 = 1.272$). We have confirmed that the stable distributions of current density and magnetic field are, in general, uniquely determined in each vortex mode for a given H_e and I. The fact that there exist unstable solutions such as A and C can be understood from the following analogous situation. In the lumped circuit model there are two solutions $\phi = \phi_0$ and $\phi = \pi-\phi_0$ satisfying the equation $I = I_c \sin\phi$, where I_c is the critical current and ϕ_0 is a principal value of $\sin^{-1}(I/I_c)$. But the latter solution $\phi = \pi-\phi_0$ is unstable because of the potential energy of the junction E_p being maximum, when the junction is operated by a constant current source. E_p is given by the following equation.

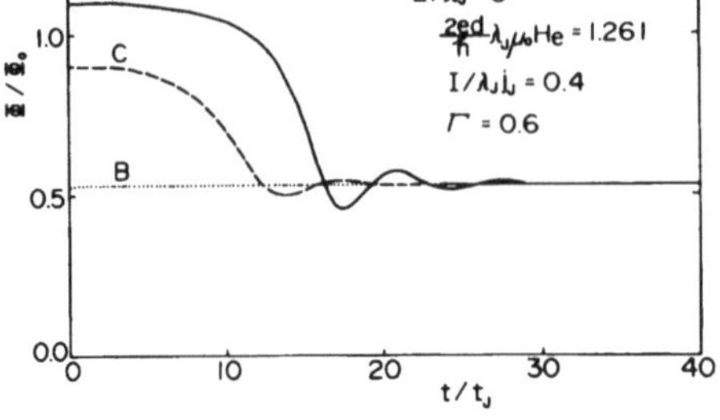

Fig.4 Time variations of Φ. Four distributions A, B, C and D are used as the initial states of the solutions A, B, C and D, respectively. H_e and I are kept constant with time.

$$E_p = \frac{\hbar}{2e}[I_c(1-\cos\phi) - I\phi] \qquad (11)$$

The external magnetic field dependence of the critical supercurrent (I_c-H_e characteristic) can be calculated as a maximum current in the I-ϕ(0) relation for a given H_e, and the I_c-H_e characteristic is shown in Fig.5 for a junction with $L=5\lambda_J$. This chart is deviated from the familiar Fraunhofer diffraction pattern and it exhibits overlapped regions of different vortex modes. From the discussions above, there are two stable states for a given set of H_e and I in the overlapped region. The stable distributions B and D shown in Fig.3 are given for the set of H_e and I marked by a cross in Fig.5.

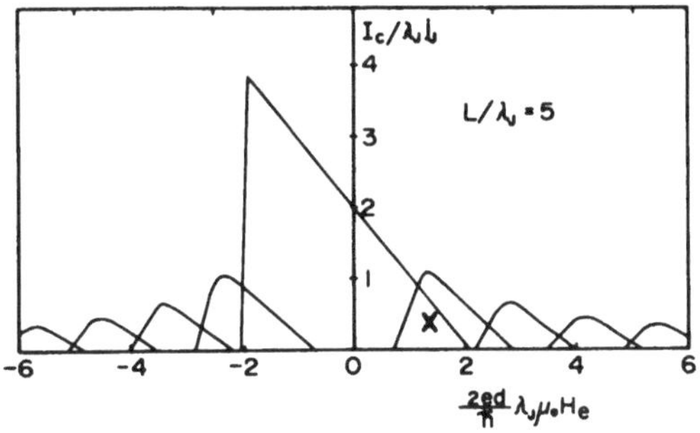

Fig.5 The I_c-H_e characteristic of a device with $L=5\lambda_J$. The cross indicates a given set of H_e and I exhibiting distributions A, B, C and D in Fig.3.

IV. SWITCHING BETWEEN DIFFERENT VORTEX MODES

Since two different vortex modes are possible to exist for the same set of H_e and I in the overlapped region of the I_c-H_e characteristic, it is possible to change the mode by moving H_e and/or I momentarily outside the overlapped region. Figure 6 shows the simulated inter-mode switching between two supercurrent states, whose distributions are given by B and D in Fig.3. The initial state (a) is set to be distribution B in the 0∿1 vortex mode. In this simulation, I is kept constant and H_e is varied with time as shown in the figure, and their

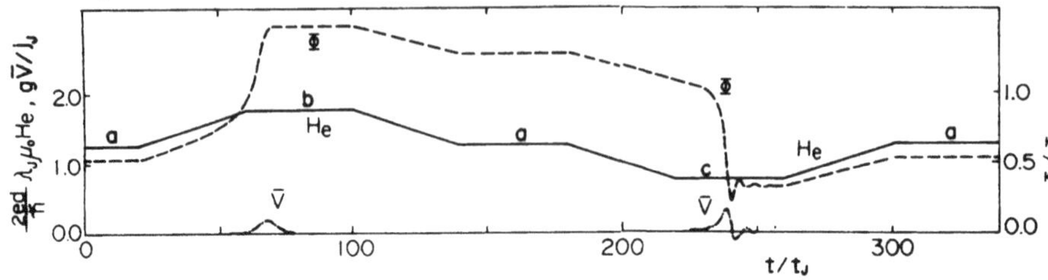

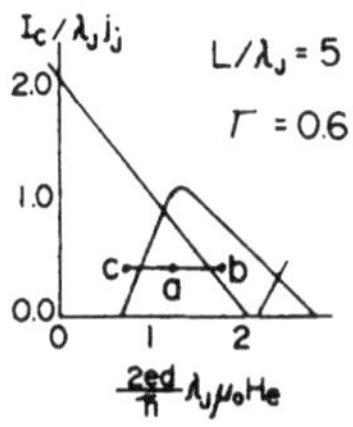

Fig.6 (above) The total magnetic flux Φ and the spatially averaged junction voltage $\bar{V}$ corresponding to a given external magnetic field variation shown by the solid line. (left) The variation of H_e is shown on the I_c-H_e chart.

variations are also depicted in the I_c-H_e chart. In the process of moving from state (a) to (b), the total magnetic flux within the junction Φ increases gradually with increasing H_e. When H_e intersects the threshold line of the 0∿1 vortex mode, a transient voltage appears across the junction, and simultaneously the vortex mode changes into the 1∿2 mode accompanying an abrupt increase of Φ. After the junction settles in the 1∿2 vortex mode, H_e decreases to the initial value. But the junction still remains in the 1∿2 vortex mode, where distributions of the current density and the magnetic field coincide with D in Fig.3. Therefore the switching from distribution B to D is achieved by increasing H_e momentarily to state (b) in the 1∿2 vortex mode. The reverse switching from D to B is also possible by decreasing H_e once to state (c) in the 0∿1 mode.

Time variations of the magnetic field distribution in the switching process from the 0∿1 to the 1∿2 mode and vice versa are shown in Fig.7. The left figure indicates that roughly one flux quantum is nucleated at the right edge of the junction and it propagates toward the center of the junction until it is trapped there. This is the switching process to the higher vortex mode. The right figure shows that the trapped vortex is driven out from the left edge of the junction and the junction settles in the lower vortex mode.

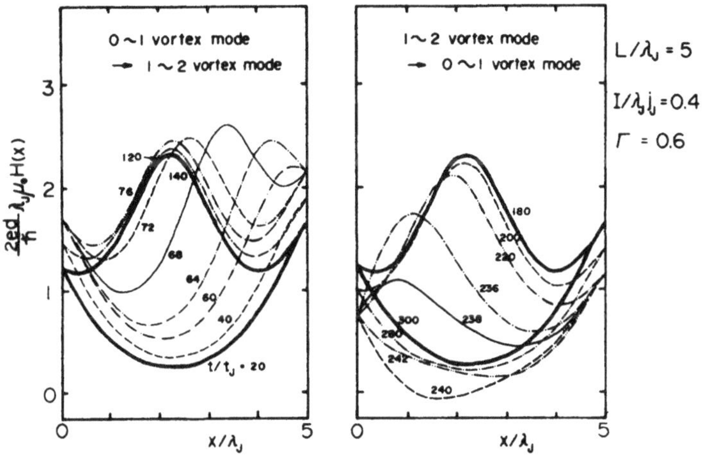

Fig.7 The magnetic field distributions in the process of the inter-mode switching. The left figure indicates the switching from the 0∿1 to the 1∿2 mode, and the right one from the 1∿2 to 0∿1 mode.

The inter-mode switching demonstrated above is achieved by changing H_e under the condition that I is kept constant, but a similar switching is also possible by changing I under constant H_e. Figure 8(a) shows the time variations of Φ

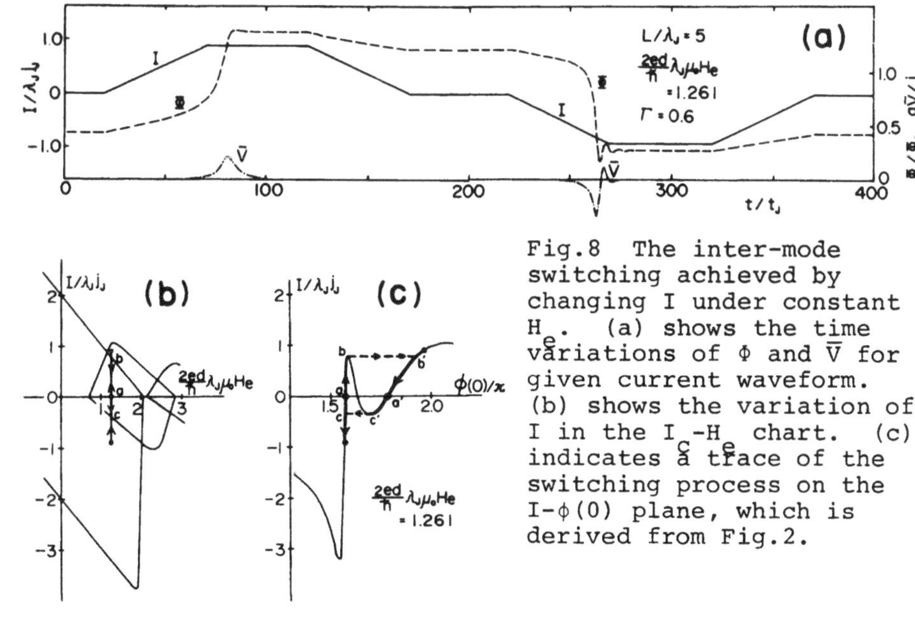

Fig.8 The inter-mode switching achieved by changing I under constant H_e. (a) shows the time variations of Φ and $\bar{V}$ for a given current waveform. (b) shows the variation of I in the I_c-H_e chart. (c) indicates a trace of the switching process on the I-φ(0) plane, which is derived from Fig.2.

and junction voltage $\bar{V}$ when I is changed as shown in the I_c-H_e chart of Fig.8(b). Figure 8(c) shows a magnified I-$\phi(0)$ chart of Fig.2, and the trace of this switching process is plotted on it. When I is increased from the point a, the trace moves on the I-$\phi(0)$ curve until it reaches b, where I is equal to I_c of the $0\sim1$ vortex mode. If I exceeds the point b, $\phi(0)$ jumps to b' in the $1\sim2$ mode with an abrupt increase of Φ. After that, by decreasing I, the trace moves from b' to c' and then jumps to c in the $0\sim1$ mode. As shown in this figure, the switching process exhibits a hysteresis loop in the I-$\phi(0)$ chart.

Time variations of the magnetic field distribution in this switching process is plotted in Fig.9. The left figure shows the process from the $0\sim1$ to $1\sim2$ mode and the right one shows vice versa. In this case, the trapped vortex is driven out from the right edge of the junction instead of left edge. This is simply because the direction of motion of the vortex depends on the direction of the applied current.

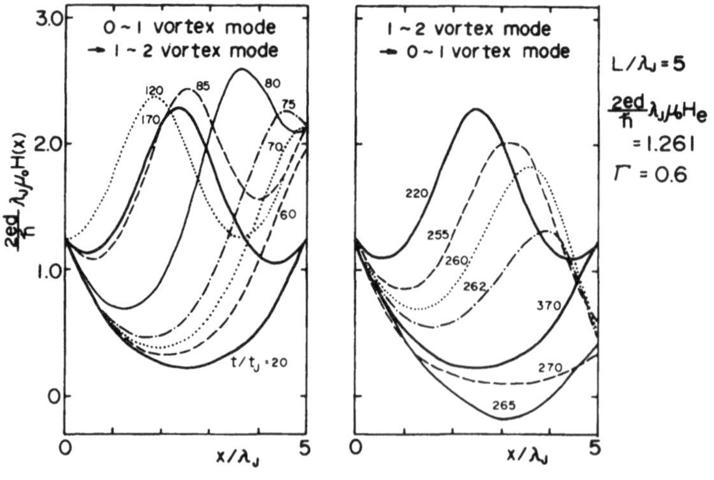

Fig.9 Time variations of the magnetic field distributions in the switching processes from the $0\sim1$ to $1\sim2$ mode and vice versa.

V THE EFFECT OF THE LOSS PARAMETER ON THE INTER-MODE SWITCHING

The loss parameter Γ, which is directly proportional to the tunnel conductance due to the normal electron and the quasi-particle, plays an important role when we design the logic circuits utilizing the inter-mode switching. Since Γ works as a damping factor of the switching transient, the junction transfers to the voltage state instead of another supercurrent state when the mode is changed, if Γ is small enough to sustain the AC Josephson oscillation. A typical example of erroneous operations of the inter-mode switching is shown in Fig.10, which indicates time variations of the total magnetic flux Φ and the junction voltage $\overline{V}$. H_e is varied in the same way as shown in Fig.6. In this case the value of Γ is equal to 0.1, while it has been equal to 0.6 in Fig.6. After the mode is expelled from the 0∿1 mode, the junction does not settle in the 1∿2 mode but the oscillatory voltage appears across the junction because of the AC Josephson effect. When H_e is decreased to the initial value the AC Josephson oscillation changes into the plasma oscillation and finally the junction settles in the 0∿1 mode. In this situation, the junction fails to be changed to the higher mode by increasing H_e momentarily.

In order to prevent the erroneous operation in the inter-mode switching caused by the transfer to the voltage state, a load

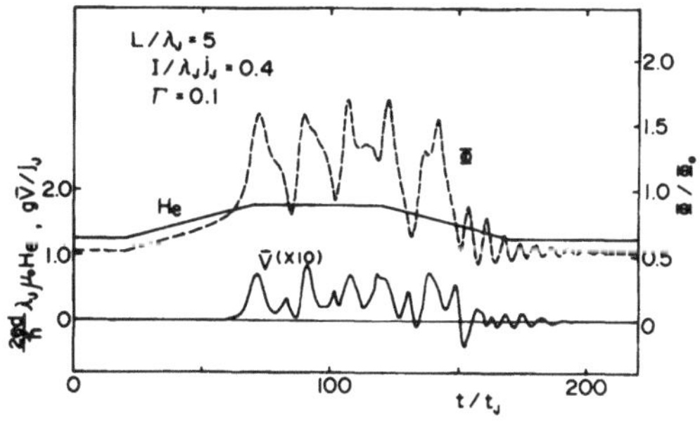

Fig.10 An example of an erroneous operation of the inter-mode switching. The switching procedure is the same as that in Fig.6. But the value of the loss parameter Γ is set to be 0.1 instead of 0.6.

for the AC Josephson oscillation should be increased. One of the ways is to connect a shunt conductor parallel to the junction, since the conductor absorbs the oscillation power from the junction and it increases the effective tunnel conductance. Another way is to connect a transmission line in order to absorb the radiated power from the junction. In the discussions above, we have assumed that no power is radiated from the junction. But if we take into account the effect of power radiation, the boundary conditions are modified as follows,

$$H(0) = H_e + \frac{\hbar}{2ed_o} \cdot \frac{\partial \phi}{\partial t}\bigg|_{x=0} \cdot \frac{1}{Z_{ol}} , \tag{12}$$

$$H(L) = H_e + \frac{I}{W} - \frac{\hbar}{2ed_o} \cdot \frac{\partial \phi}{\partial t}\bigg|_{x=L} \cdot \frac{1}{Z_{or}} . \tag{13}$$

Where Z_{ol} and Z_{or} are characteristic impedances outside of the junction at the left and right edges, respectively. Generally, Z_{ol} is different from Z_{or} because of the device structure shown in Fig.1, but we have set them equal to $Z_i \sqrt{d/d_o}$ in the inter-mode switching shown in Fig.11. Here Z_i is given by $\sqrt{\mu_o/\varepsilon}$. In this case the switching is successfully achieved in spite of the small loss parameter.

As demonstrated here, the energy loss for the oscillation plays an important role, so that we should suitably control it for the purpose of utilizing the inter-mode switching.

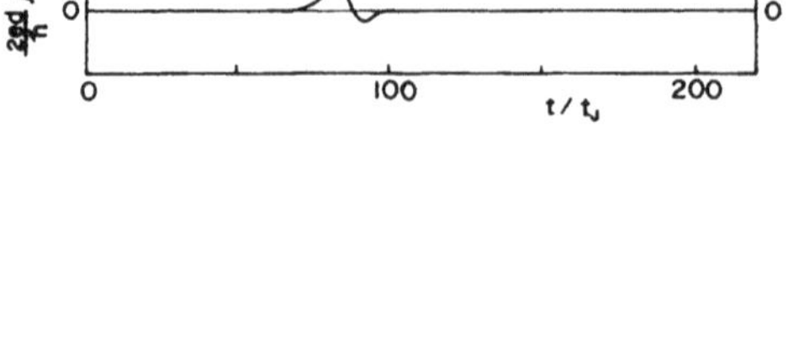

Fig.11 The inter-mode switching when the power radiation from the junction is taken into account. The characteristic impedance outside of the junction is set to be $Z_i \sqrt{d/d_o}$, where Z_i is the field impedance of the junction given by $\sqrt{\mu_o/\varepsilon}$. The switching procedure is the same as that in Fig.6, and the loss parameter Γ is set to be 0.1.

VI CONCLUSION

In the present paper, we have numerically simulated switching processes between two different vortex modes in Josephson tunnel junctions, which we have called the inter-mode switching. Prior to the simulation of the switching, we have confirmed that the state of the junction is uniquely determined in each vortex mode for a given set of external magnetic field H_e and total current through the junction I. This fact means that the number of the state realizable in the junction coincides with a degeneracy of the vortex mode for the same set of H_e and I.

The inter-mode switching has been demonstrated in this paper, and the vortex mortion in the junction has been visualized. It has been shown that the switching is successfully achieved by expelling the set of H_e and I from the overlapped region of the two different vortex modes. However, the junction transfers to the voltage state instead of settling down in another vortex mode when the energy loss caused by the normal electron tunneling is small enough to sustain the AC Josephson oscillation. This erroneous operation of the inter-mode switching can be prevented by introducing an additional energy loss mechanism to the junction. One way is to connect a shunt conductor parallel to the junction, since the normal current flows in the conductor and it effectively increase the energy loss for the AC Josephson oscillation. Another way is to absorb radiated power from the junction by connecting a transmission line. This is because the power radiation from the junction causes the damping of the oscillation.

ACKNOWLEDGEMENT

The authors are grateful to Drs. K. Kurokawa and T. Isobe for their encouragements.

REFERENCES

[1] Herrell, D. J.: A Josephson tunneling logic adder. Digest INTERMAG, 864-867 (1974).

[2] Herrell, D. J.: An experimental multiplier circuit based on superconducting Josephson devices. IEEE J. Solid State Circ., SC-10, 360-368 (1975).

[3] Baechtold, W.: A flip-flop and logic gate with Josephson junctions. Digest Int'l. Solid State Circuits Conf. 164-165 (1975).

[4] Gueret, P.: Storage and detection of a single flux quantum in Josephson junction devices. IEEE Trans. Mag., MAG-11, 751-754 (1975).

[5] Anacker, W.: Potential of superconductive Josephson tunneling technology for ultrahigh performance memories processers. IEEE Trans. Mag., MAG-5, 968-975 (1969).

[6] Owen, C. S., Scalapino, D.J.,: Vortex structure and critical currents in Josephson junctions. Phys. Rev., 164, 538-544 (1967).

[7] Basavaiah, B., Broom, R.F.: Characteristics of in-line Josephson tunneling gates. IEEE Trans. Mag., MAG-11, 759-762 (1975).

SWITCHING DYNAMICS OF JOSEPHSON JUNCTION LOGIC CIRCUITS

H. W. Chan, B. T. Ulrich,[+] and T. Van Duzer
Electrical Engineering and Computer Sciences
University of California, Berkeley, California 94720, U.S.A.

High switching speed and low power dissipation make Josephson junction devices very attractive for computer applications.[1,2] The most basic logic circuit consists of a Josephson junction connected by output lines to a load. Input to the junction is via control lines which initiate switching by suppressing the junction critical current to a value below the supply current bias to the junction. Similarly, the output line from the junction can be used as a control line for junctions in the next stage. The basic circuit can operate in different modes, "latching", "nonlatching", and "self-resetting", depending on circuit parameters. We describe here, briefly, the switching dynamics of the basic logic circuit in terms of a lumped circuit model.

BASIC LUMPED CIRCUIT MODEL

The elements of a basic Josephson junction logic circuit are shown in Fig. 1. We begin by illustrating the principle of operation for a convenient, simple, choice of circuit parameters for the elements. Suppose that the output lines are terminated with a matching resistor $2Z_o$, so there are no reflections. Then the impedance of the output line may be lumped in parallel with the junction shunt resistance, and represented by a single resistor R. Suppose also that the junction shunt capacitance can be neglected, so the entire junction logic circuit, including the output line, reduces to the resistively shunted junction model shown in Fig. 2. Initially, the junction is in the zero-voltage state carrying a constant supply current I_S. With no current in the control lines, the junction critical current is $I_C(0)$ and the current-voltage curve is connected directly from $I_C(0)$ to the finite voltage part, without hysteresis, as shown by the upper curve of Fig. 2. When a current is applied to the

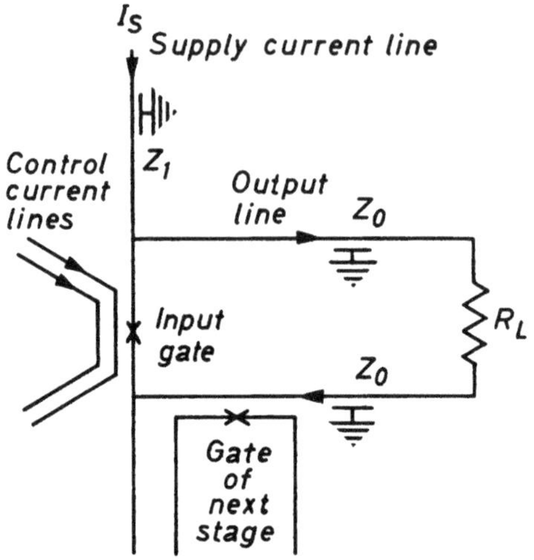

Fig. 1 Basic Josephson junction logic circuit.

control lines, the magnetic field H of the current depresses the junction critical current to $I_C(H)$, shifting the entire current-voltage curve downward. Now, with the supply current remaining constant, a finite non-zero voltage appears across the junction. This voltage is the output of the logic circuit. Depending on the magnetic field produced by each control line, relative to that needed to produce a reduction in $I_C(H)$ sufficient for an output voltage, the circuit can generate logic functions, including OR and AND. When the control line currents are removed, the current-

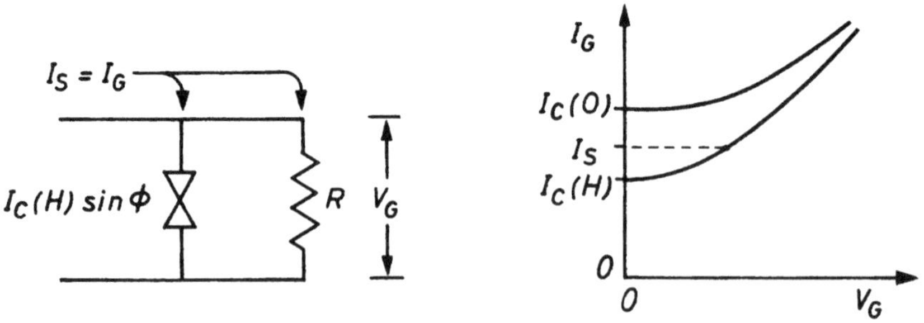

Fig. 2 Principle of operation of a Josephson junction logic circuit, as illustrated by the ideal resistively shunted junction model. Such a circuit would operate in the nonlatching logic mode. The effect of shunt capacitance, important in tunnel junctions, is not included.

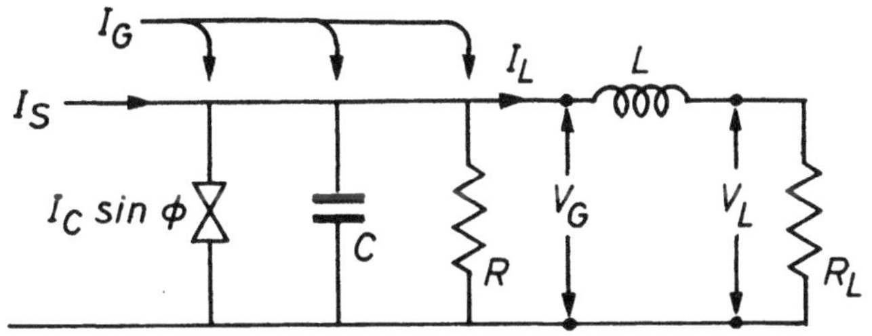

Fig. 3 Basic lumped circuit model used to illustrate the three modes of logic operation: latching, nonlatching, and self-resetting. The gate junction is modeled within the resistively shunted junction model including a shunt capacitance.

voltage curve returns to its former position, and the output voltage returns to zero. The logic circuit is "nonlatching": the output voltage remains only during the input from the control lines.

A more realistic lumped circuit model for the basic logic circuit is shown in Fig. 3. The gate junction is again represented by the resistively shunted junction model, but now with the inclusion of capacitance, which is significant in tunnel junctions. The inductance L represents a short transmission line leading to a load resistance $R_L < Z_o$. The circuit model for the gate junction does not include effects of finite junction size, nor the voltage-dependence of the quasi-particle conductance of a tunnel junction. This circuit is adequate, however, to illustrate the different modes of logic operation possible with the basic logic circuit: latching, nonlatching, and self-resetting.

To analyze the circuit, it is useful to consider its current-voltage characteristic in the absence of the output portion of the circuit, L and R_L. This I_G-V_G characteristic is the I-V curve of a capacitively shunted junction, analyzed by Mc Cumber[3] and Stewart.[4] The current I_G is the gate current, the portion of the supply current that does not flow out toward the load. The capacitively shunted junction also represents a gate connected to a terminated output line ($R_L = 2Z_o$), if the junction resistance R is replaced by its parallel combination with the output lines,

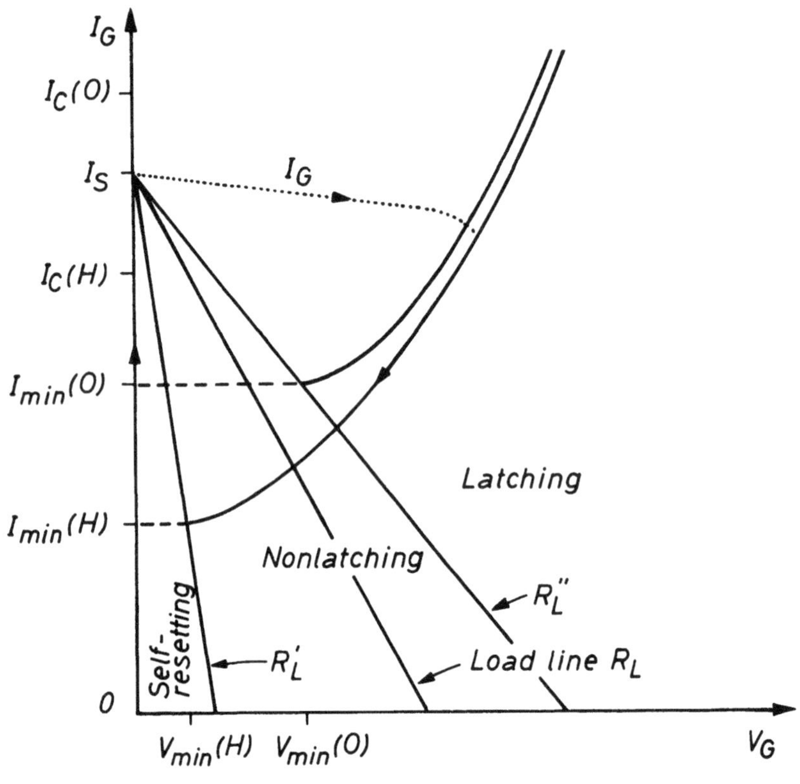

Fig. 4 Quasi-static current-voltage diagram for the gate junction of the Fig. 3, illustrating the regions for the different logic modes for different load resistances R_L, and fixed values of the other circuit parameters. The resistance R_L'' forms the boundary between latching and nonlatching operation, and R_L' forms the boundary between nonlatching and self-resetting operation. In this diagram, I_G and V_G are averaged over times comparable to the period of the fundamental Josephson oscillation. Two I_G-V_G curves are shown, corresponding to control line magnetic fields of 0 and also $H \neq 0$. The current-voltage curve in the region of subharmonic oscillations for voltages below V_{min} is not shown.

$R_{EFF} = 1/(1/R + 1/R_L)$. The current-voltage curve now contains hysteresis. For supply current I_S between I_C and $I_{min} = \alpha I_C$, with $0 < \alpha < 1$, where α depends on the junction capacitance, the junction can be biased either at zero or at finite voltage. The minimum current $I_{min}(H)$ also depends on the magnetic field of the control lines. If we choose the supply current so $I_{min}(0) < I_S < I_C(0)$, the gate junction will remain in the finite output voltage state after the control line magnetic field returns to zero.

This is the "latching" mode of logic operation. Before the gate junction can function again for a logic operation, the output voltage must be reset to zero, such as by removing the supply current I_S momentarily. Alternatively, if we choose $I_S < I_{min}(0)$, the gate junction voltage resets to zero, and the gate still operates in the nonlatching mode. In both cases, for the gate to switch, the control lines must suppress I_C to $I_C(H) < I_S$.

When an inductive output load circuit is included, as in Fig. 3, a third, "self-resetting" mode of logic operation becomes possible.[5] In Fig. 4 are shown the two I_G-V_G curves for zero and finite control field, H, for the complete circuit of Fig. 3. These I-V curves are similar to those for the junction with shunt capacitance, but with an important difference. The finite voltage curves no longer extend smoothly down to the zero-voltage axis, but only to a finite voltage V_{min} which can be determined numerically. The dashed lines indicate the average gate current I_{min} corresponding to V_{min}. At voltages below V_{min}, the time averaged I_G-V_G curves extend abruptly upward toward I_C in a region where the junction oscillates at a subharmonic of the fundamental Josephson frequency, in a relaxation oscillation mode.[6] The operating point of the gate junction is determined by the intersection of the subharmonic portion of the I-V curve (not shown in Fig. 4) with the load line of the load resistor R_L. For the supply current I_S indicated in Fig. 4, when the load resistance $R_L < R_L'$, the operating point lies in the subharmonic region when the control field is applied: the gate junction emits a train of relaxation oscillation voltage spikes as output. When the control field is removed, the entire subharmonic portion of the curve joining $I_C(0)$ with the point ($V_{min}(0)$, $I_{min}(0)$ lies above the load line, so the only operating point is $V_G = 0$ at current I_S. The input to the control lines should be timed so that only one output spike is emitted.

Thus, the lumped circuit model of Fig. 4 exhibits three modes of logic circuit operation, latching, nonlatching, and self-resetting, depending on how the gate junction returns to the zero-voltage state. Which of these modes occurs depends on the values of the circuit elements in the logic circuit, <u>including</u> the parameters of the output circuit. We

TABLE I

I_S	C	R	$I_C(H)$	L	R'_L	R''_L
1 mA	10 pF	0.5 Ω	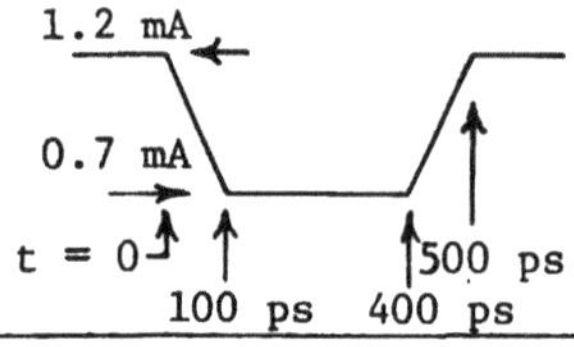	25 pH	0.18 Ω	0.29 Ω
1 mA	10 pF	0.5 Ω		100 pH	0.14 Ω	0.39 Ω

studied the dynamics of the basic logic circuit through digital computer simulation for different values of R_L at fixed values of the other lumped circuit elements in Fig. 3. Two boundary values of load resistances, R'_L and R''_L, separate the regions of latching, nonlatching, and self-resetting operation, as shown in Fig. 4. For $R''_L < R_L$ the gate operates in the latching mode. For $R_L < R'_L$ the gate operates in the self-resetting mode. The final mode of operation, with $R'_L < R_L < R''_L$ is nonlatching. It most closely resembles the mode of logic operation familiar in semiconductor logic. The output voltage is present ($V_G \neq 0$) only while the control current lines maintain I_C at $I_C(H)$. During this output state, the junction oscillates at the fundamental Josephson frequency, rather than at a subharmonic. We determined the boundary values of load resistance, R'_L and R''_L, for typical values of load inductances as shown in Table 1. We find that an increase in L increases R'_L and decreases R''_L, thus widening the region for load resistances for nonlatching operation.

We illustrate the typical switching dynamics for the nonlatching mode of the circuit in Fig. 5. The upper diagram shows the gate critical current, which is decreased linearly with time by the magnetic field of the control lines. The lower diagram shows the gate voltage, and the current transfered into the output line as a function of time. These curves can be interpreted in terms of the gate I_G-V_G curves of Fig. 4. As the gate critical current falls below the supply current of 0.8 mA, the gate current and voltage point switches rapidly along the dotted line labeled I_G to the current-voltage curve of a capacitively shunted junction with that critical current. Then, as current is transfered outward through the inductor into the load resistance, the operating point slides down the current-voltage

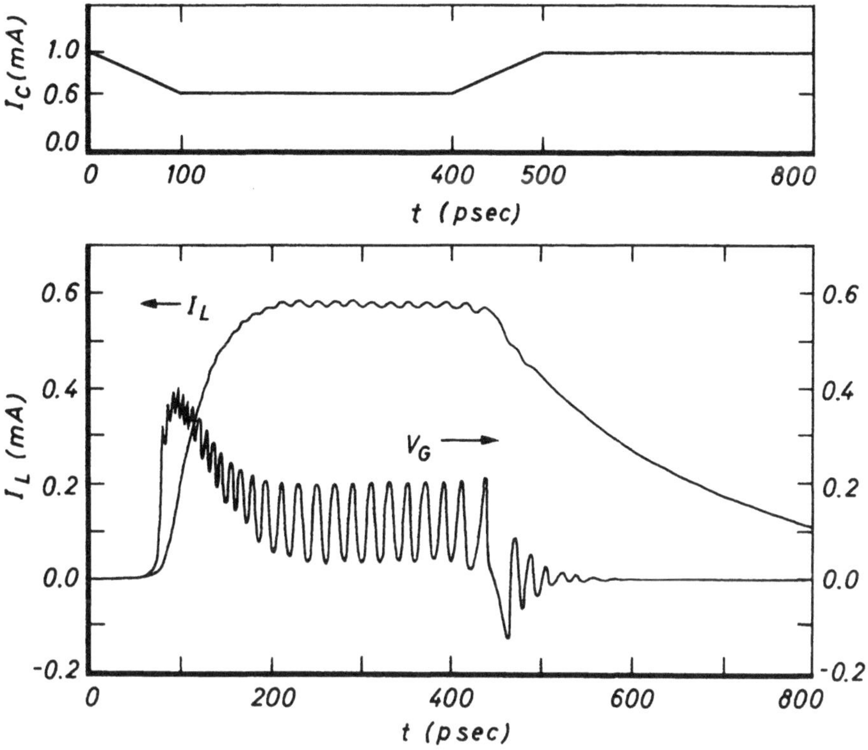

Fig. 5 Switching characteristics illustrating nonlatching operation for the logic circuit model of Fig. 3 with $I_S = 0.8$ mA, $R = 0.6\ \Omega$, $C = 20$ pF, $L = 40$ pH, and $R_L = 0.18\ \Omega$. The boundary load resistances for nonlatching operation are $R_L' = 0.14\ \Omega$ and $R_L'' = 0.21\ \Omega$. The upper diagram shows the time evolution of the junction critical current $I_C(H)$ as controlled by the magnetic field of the input control lines, and the lower diagram shows the time evolution of the gate voltage V_G and the current switched into the load I_L.

curve of the gate alone, until it reaches the intersection of this curve with the load line R_L. This picture illustrates the evolution of the point (V_G, I_G) when averaged over times comparable to the period of the Josephson oscillation. The instantaneous trajectory of this point is shown in Fig. 6 for the same parameters as in Fig. 5. The qualitative features are described well in terms of Fig. 4. The amplitude of the voltage oscillation at the Josephson frequency during the switching process is emphasized in Fig. 6. The switching dynamics of the different modes of logic operation have been discussed recently.[7]

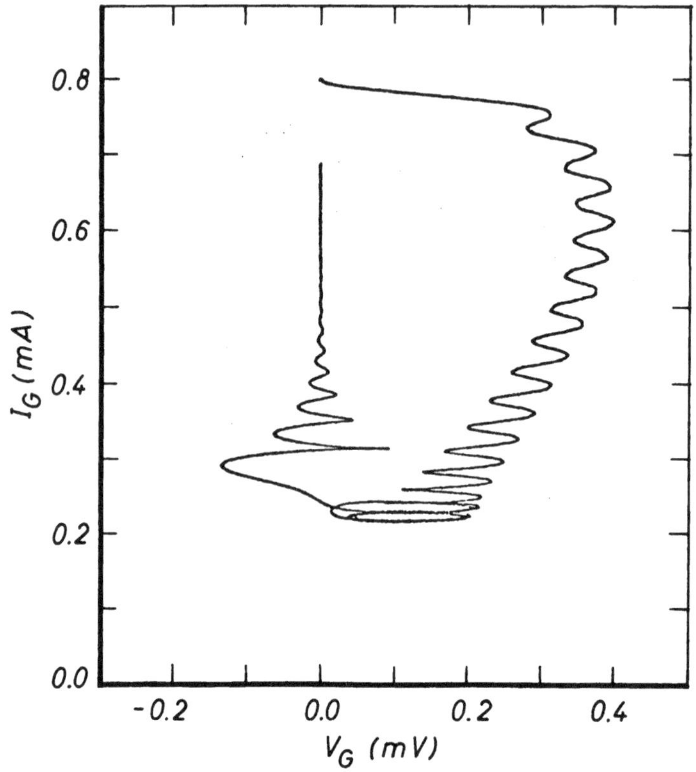

Fig. 6 Trajectory in the (V_G, I_G) plane during the switching operation illustrated in Fig. 5.

LONG OUTPUT LINE: THE IMPORTANCE OF A MATCHED LOAD

The lumped circuit model of Fig. 3 adequately represents the basic switching circuit of Fig. 1 only when the characteristic impedance Z_o is much larger than the load resistance R_L, and the line is short compared to a wavelength at $f = 0.35/t_r$, where t_r is the rise time of the input gate. To represent a longer output line, we used a 10 section lumped element representation of a transmission line as shown in Fig. 7, and varied the impedance of the line. To compare the properties of the two circuits, the inductance of Fig. 3 was divided into ten equal parts for each of the sections of Fig. 7, and the capacitance of each section was chosen according to $2Z_o = (L/C)^{1/2}$ to give the desired impedance. For large line

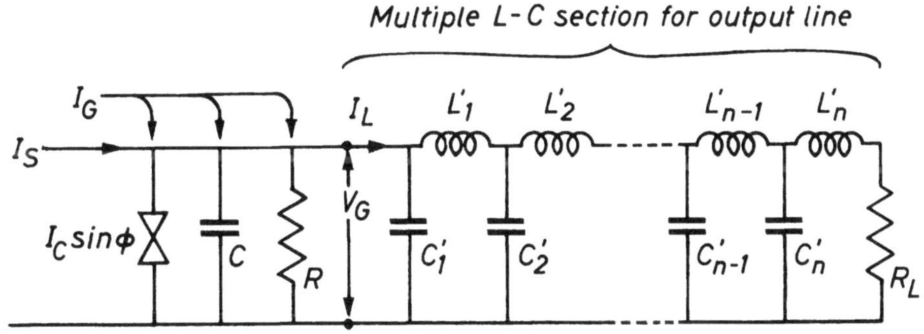

Fig. 7 Lumped circuit model used to study the properties of logic gates with long output lines. The text describes how the L'_i and C'_i were chosen for each section.

impedance, ($2Z_o \gg R_L$), the switching behavior closely resembles that for the circuit of Fig. 3. For example, with $2Z_o = 10\ \Omega$, and the other circuit parameters the same as in Fig. 4, the voltage across the gate, and the current I_L flowing away from the gate as shown in Fig. 8 is nearly the same as in Fig. 5. Note that the current I_L shown in Fig. 8 is the current flowing away from the gate junction, as defined in Fig. 7, rather than the current flowing in the load resistor.

For smaller values of output line impedance, the switching behavior with an unterminated long output line begins to deviate more and more from that obtained using the circuit of Fig. 3. Successive current and voltage reflections appear in $V_G(t)$, and in $I_L(t)$. If the current reflections are large enough, the circuit may even self-reset instead of operating in the nonlatching mode. Thus, for a circuit with an unmatched output line, even the mode of logic operation depends on the details of the output circuit. As with any unterminated transmission line, the time required for the reflections to damp out depends on the length of the line, and the energy loss per reflection. Because the logic mode, and switching time depend on output line parameters, circuit design in such a case becomes difficult.

Switching is faster, and design simpler, if the output line is terminated ($2Z_o = R_L$). Normally, the output line will include control lines of junctions at later logic stages. The output control lines are inductive

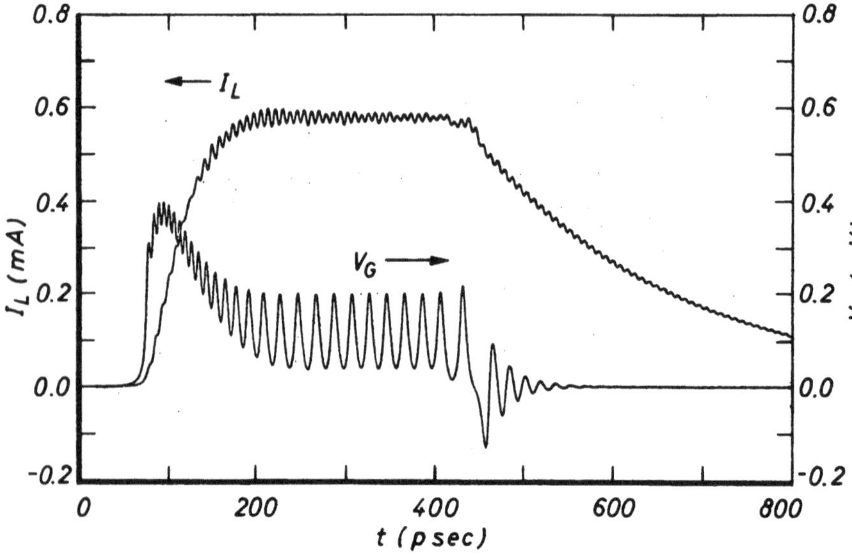

Fig. 8 Switching characteristics with the same circuit parameters as in Fig. 4, but with a high impedance ($2Z_o$ = 10 Ω) ten L-C section model for the output line as in Fig. 7. For each section, L_i' = 4 pH, and C_i' = 0.04 pF. The current I_L is the current switched away from the gate junction.

because additional energy is stored in the magnetic field which links the next gate junction. Reflections at the output control lines can be reduced with shunt capacitors.[8]

When the output line is well matched by these procedures, the operation of the basic gate resembles that of the simpler circuit mentioned earlier: the parallel combination of a Josephson supercurrent $I_C(H)\sin\phi$, a capacitance C, and an effective resistance $R_{EFF} = R \| R_L$. In such a circuit, no relaxation oscillations occur, and only the latching and nonlatching modes of operation are possible.

Research sponsored by National Science Foundation Frant ENG75-12834.

+Present address: Physics Department, University of Nijmegen, Toernooiveld, Nijmegen, NETHERLANDS.

REFERENCES

1. Matisoo, J.: The tunneling cryotron -- a superconducting logic element based on electron tunneling. Proc. IEEE 55, 172-180 (1967).
2. Henkels, W.H.: An elementary logic circuit employing superconducting Josephson tunneling gates. IEEE Trans. on Mag. MAG-10, 860-863 (1974).
3. McCumber, D.D.: Effect of ac impedance on dc voltage current characteristics of superconducting weak-link junctions. J. App. Phys. 39, 3113-3118 (1968).
4. Stewart, W.C.: Current-voltage characteristics of Josephson junctions. Appl. Phys. Lett. 12, 277 (1968).
5. Chan, H.W., Lum, W.Y., Van Duzer, T.: High-speed switching and logic curcuits using Josephson devices. IEEE Trans. Mag. MAG-11, 770-773 (1975).
6. Dempsey, D.G., Levinsen, M.T., Ulrich, B.T.: Energy Storage and subharmonic oscillations in Josephson junctions. IEEE Trans. Mag. MAG-11, 811-814 (1975).
7. Lum, W.Y., Chan, H.W., Van Duzer, T.: Memory and logic circuits using semiconductor-barrier Josephson junctions. Proc. 1976 Applied Superconductivity Conf., IEEE Trans. Mag. (1977, to be published).
8. Zappe, H.H.: Josephson quantum interference computer devices. Proc. 1976 Applied Superconductivity Cong., IEEE Trans. Mag. (1977, to be published).

SUPERCONDUCTING NEUROMIME: A JOSEPHSON JUNCTION MODEL NEURON

B. T. Ulrich[+]
Electrical Engineering and Computer Sciences
University of California, Berkeley, California 94720, U.S.A.

A neuromime is a device which mimics some of the properties of a biological neuron. In this paper we are concerned with the signal processing properties of the neuron. The superconducting neuromime based on the Josephson effect, or Josephson junction model neuron processes signals in the same way as does the biological neuron, in several important respects. This paper explores the similarities and differences between the neuromime and its biological counterpart. The most important similarity is in signal processing properties. The most important difference is in speed. The time scale of the superconducting neuromime is 10^{-11} sec, in contrast to the 10^{-3} sec time scale of biological neurons. The superconducting neuromime is not meant to model the internal workings of a biological neuron. It is meant to be a system which processes signals in a way analogous to that of biological neurons, but on a time scale 100 million times faster.

This paper is interdisciplinary in that it creates a new link between two widely separated fields. As with all interdisciplinary discussions, before the subject becomes a discipline in itself, as has occurred with biochemistry, the potential audience most likely is to be found in one discipline or in the other. Here, in physics, we need to introduce a biological vocabulary. Accordingly, this paper is divided into three parts: first, we introduce the basic properties of the biological neuron which we model; second, we present the neuromime, and describe how it corresponds to the biological neuron; and third, we comment briefly on the potential applications of the neuromime.

BIOLOGICAL NEURON

We begin with a simplified description of the biological system we wish to

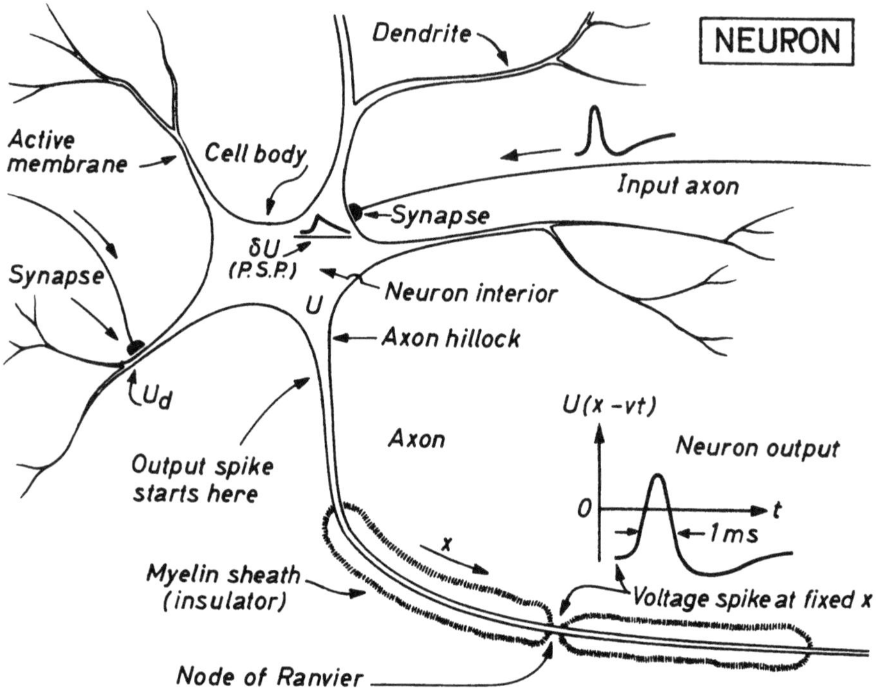

Fig. 1 Schematic representation, not to scale, of a motoneuron, which has output to a muscle. The myelin sheath increases the spike propagation velocity in the axon. The axons of other neurons are the input axons to this neuron. Input axons may branch locally (not shown), and form several synapses on a given neuron. The size of the synaptic buttons is greatly exaggerated in this drawing. Redrawn after Eccles, and Katz, Ref. 1.

model. The most important signal processing part of a nervous system consists of cells called neurons.[1] Throughout this paper, "neuron" will refer to the biological neuron, and "neuromime" to the superconducting neuromime, or Josephson junction model neuron. A neuron is considered to be a basic functional unit of the nervous system. A neuron has a cell body of diameter ranging from 2 μm to 50 μm, which branches into a treelike structure, the branches of which are called dendrites, as illustrated in Fig. 1. The cell body and dendrites form the input part of the neuron. A neuron has a single output line called an axon, which begins at a place on the cell body called the axon hillock, and which is physiologically part of the same cell body. The axon is a cylindrical fiber a few micrometers in diameter, of length ranging from 20 μm, to meters, for those axons extending from the brain toward distant muscles.

A membrane separates the inside of the neuron, (ie. the inside of the dendrites, cell body, and axon), from the outside. The electrical conductivity of the fluid filling the membrane is high enough, and the conductance across the membrane is low enough, that the interior of the dendrites and cell body behaves like a leaky, overdamped, passive transmission line. The spatial decay length of the potential in the dendrites ranges from a fraction of the distance from a location on a dendrite to the cell body, to greater than this distance. A potential U_d in one of the dendrites additively influences the potential U inside the cell body near the axon hillock. This potential near the axon hillock is of crucial importance because its value determines the output state of the neuron.

So far, we have been discussing a system that is entirely passive as far as the input potentials U_d in the dendrites influence the potential U at the axon hillock. However, the membrane is an active chemical system. Without any inputs to the dendrites or cell body, the internal potential U remains at a resting potential, U_r = -70 mV, produced by a difference in ion concentrations across the semipermeable membrane. This difference in concentrations is maintained by an active dissipative chemical process. Most of the studies of this active membrane have been done on the membrane of an axon, such as the giant axon of the squid, which has an unusually large diameter of 300 μm. The membrane is also "excitable". The axon can propagate a fixed shape voltage pulse, or spike, U(x - vt), along its length, where x is the distance along the length of the axon, and the velocity v ranges from 5 m/sec to 100 m/sec depending on the diameter of the axon, and the way in which it is electrically insulated. This spike, propagating away from the cell body of a neuron is the output signal of the neuron.

The axons of vertebrate animals generally end on the dendrites or on the cell bodies of other neurons, or on muscles, as synapses. The output spikes along axons from one neuron become the inputs to the next neuron or

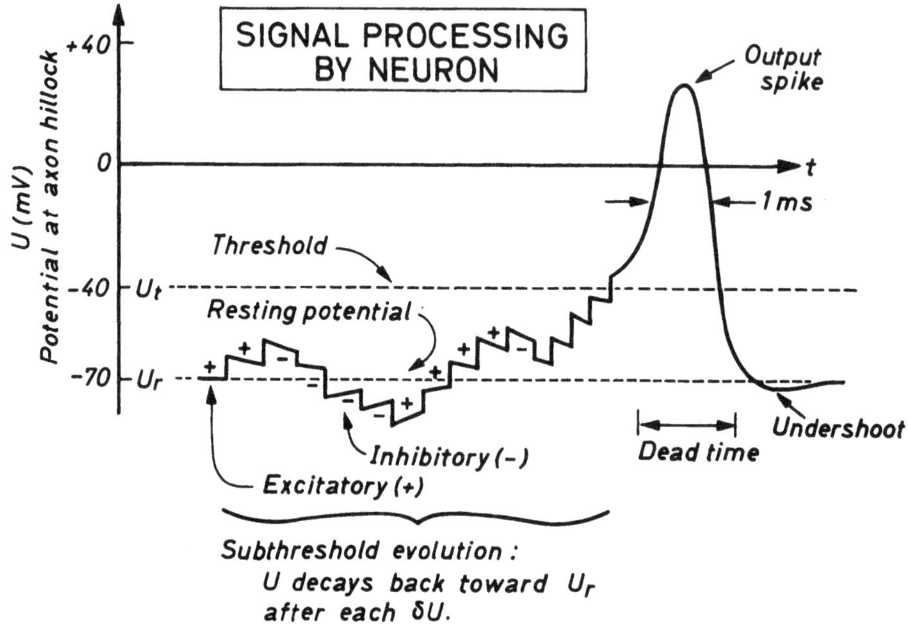

Fig. 2 Neuronal signal processing combines the linear process summing the small potential changes δU produced by input spikes, with the nonlinear threshold process of output spike emission whenever U exceeds U_t. The time evolution of U is shown schematically. In a nervous system, the arrival times of δU's often appear to be random, rather than periodic. For neurons with a large number of inputs, individual δU's are not distinguishable, and U(t) appears as a noisy, fluctuating voltage.

muscle at the synapses. Generally, the spike which arrives at a synapse does not simply propagate into the next neuron, but produces a small step change δU in the internal potential of the next neuron just beyond the synapse. This is called a Post Synaptic Potential (P.S.P.). Depending on the synapse, δU may be positive (excitatory synapse), increasing the local potential U_d above U_r, or negative (inhibitory synapse), decreasing U_d. When U_d deviates from U_r, it decays back toward U_r with a time constant of about 4 ms to 10 ms, as shown in Fig. 2. Together, all the U_d values at the dendrites and cell body which receive synaptic terminations of input axons, influence the potential U at the axon hillock. We simplify our discussion to neglect the effects of distance between a synapse on a dendrite and the axon hillock, but these effects could be included in an obvious way in the neuromime. We focus our attention on U, and use a single time constant to express its decay toward the resting potential U_r.

Each time the cumulative effect of the δU's from inputs causes U to exceed a threshold $U_t = -40$ mV, a single output pulse of voltage, or spike, $U(x - vt)$, of about 1 ms duration, begins at the axon hillock, and travels with constant velocity and amplitude away from the cell body, along the axon. This summation of positive and negative contributions from multiple input lines, and the emission of a single spike whenever the internal parameter U exceeds a threshold, is the basic signal processing property of a neuron, as illustrated in Fig. 2. During the output spike, it is not possible to initiate another spike. This interval is called a dead time. There is also a time after the spike during which it is a little more difficult to trigger another spike.

The geometry of neuronal interconnections is complicated. Each output axon may branch several, or 10^2, or even 10^4 times to arrive at that number of terminal synapses on other neurons. Several of the branches may arrive at the same target neuron, or may arrive at several different target neurons. On the average, a neuron, in turn, receives from 10^2 to 10^4 input synapses.

In the simplified biological picture we present here, the dendrites transmit passive, additive voltage signals from the post synaptic potentials produced by each input spike. The axons transmit a train of output spikes, without modification, from one neuron to the next. Thus, the axons function simply as transmission lines. The pulse rate of spikes on axons ranges from 10 to 100 spikes per second in peripheral parts of the nervous system and from 1 to 10 spikes per second in more central parts. In the present view, signal processing takes place in the neuron, and the state of the neuron is characterized by the instantaneous value of the potential U at the axon hillock. It is this potential which determines when a neuron emits an output spike. An excellent review of the properties of active membrane is given by Scott.[2] He treats both the local membrane properties described by the Hodgkin-Huxley equations as well as signal transfer in dendrites and axons.

SUPERCONDUCTING NEUROMIME

The neuromime which exhibits properties analogous to those of a neuron is based on the Josephson junction circuit shown in Fig. 3. In the following we describe the analogy between the circuit and the biological system. Inside the neuron, the internal state parameter is the voltage U near the axon hillock. In the neuromime, the internal parameter is the loop current I_p circulating through the Josephson junction, the input resistance R_p, and the inductance L. As long as $I_p < I_o$, where I_o is the Josephson junction critical current, almost no voltage develops across the junction. The relaxation of the internal parameter in the neuron occurs with a time constant of 4 to 20 ms, and in the neuromime, with a time constant L/R_p, which with reasonable values for a neuromime fabricated by microcircuit techniques, with $L = 3.5 \times 10^{-11}$ H, and $R_p = 1$ Ohm is 3.5×10^{-11} sec. Input to the neuromime is in the form of current pulses, $I_i(t)$, which can be positive or negative to correspond to the excitatory or inhibitory synapses of the neuron. The resting value of U in the neuron is -70 mV, and the resting value of I_p in the neuromime is 0.

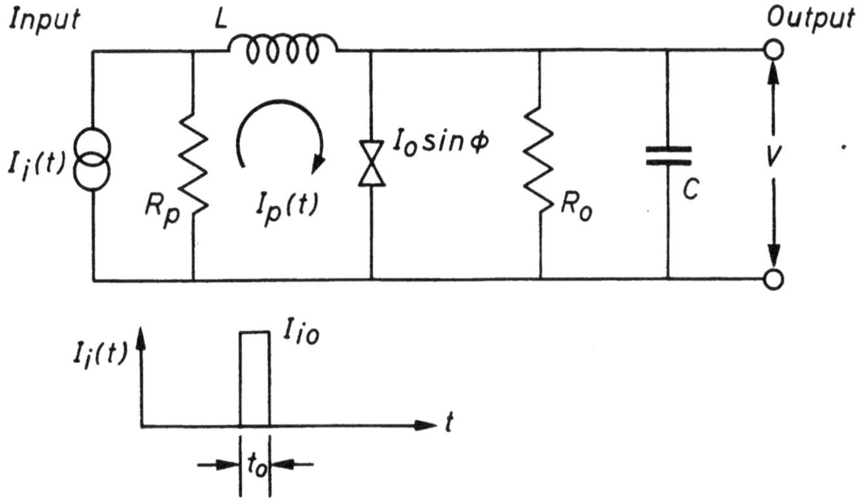

Fig. 3 Circuit for the superconducting neuromime, or Josephson junction model neuron.

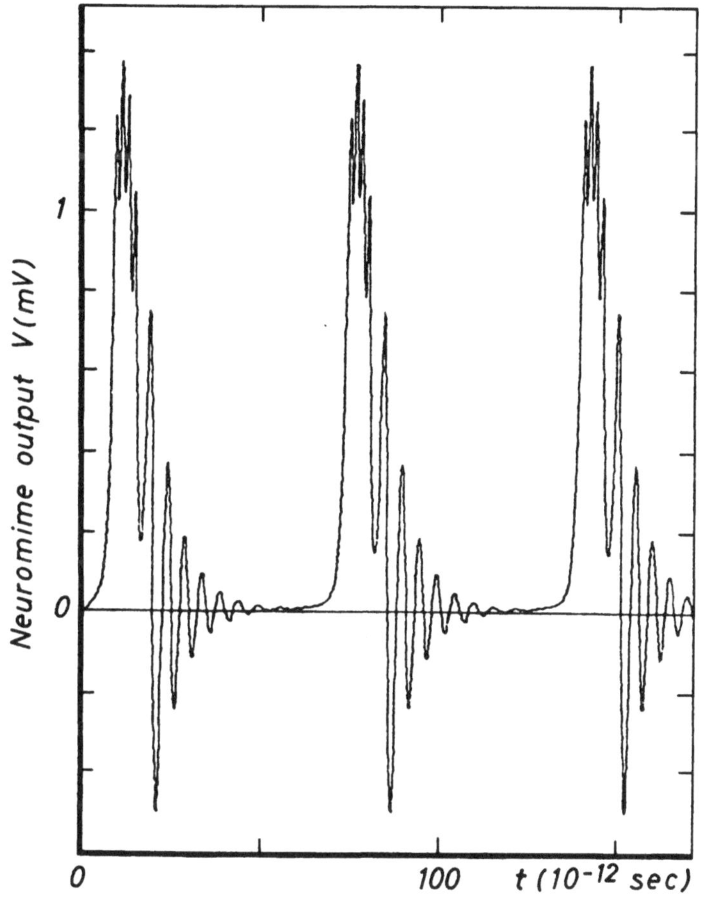

Fig. 4 Neuromime output spikes, calculated numerically for representative values of circuit parameters: R_p = 1 Ohm, L = 3.5 x 10^{-11} H, I_o = 2.0 x 10^{-4} A, R_o = 10 Ohms, and C = 3.5 x 10^{-13} F. Output spike duration is ~10^{-11} sec. A steady input current, $I_i(t)$ = 2.8 x 10^{-4} A begins at t = 0, with initial conditions, ϕ = 0, V = 0, and dV/dt = 0, at t = 0.

Both the neuron and the neuromime emit an output pulse when the internal parameter exceeds a threshold. The threshold of the neuron is -40 mV, and the threshold of the neuromime is I_o. We now describe the process of spike emission. When $I_p(t)$ exceeds I_o, the current I_p no longer can all flow as a supercurrent $I_o \sin\phi$, although $\sin\phi = 1$. Some of the current must pass through C and R_o, requiring the voltage across the junction to become non zero. Without loss of generality, we take V > 0. The voltage

TABLE I

ANALOGY BETWEEN NEURON AND NEUROMIME

		Neuron	Neuromime
1.	Internal state parameter	U Voltage near axon hillock	I_p Circulating current
2.	Relaxation time of internal parameter	$\sim$4 to 20 ms	$\sim L/R_p$
3.	Threshold for output spike	$U_t \sim -40$ mV	I_o
4.	Excitatory input	$\delta U > 0$	$\delta I_p > 0$
5.	Inhibitory input	$\delta U < 0$	$\delta I_p < 0$
6.	Duration of spike	$\sim$1 ms	$\sim L/(R_p + R_o)$
7.	Value of internal parameter after spike	$U_r \sim -70$ mV	$I_p = I_{reset} \gtreqless 0$
8.	Dead time	$\sim$During spike	$\sim$During spike

across the junction causes the phase ϕ to increase via $d\phi/dt = 2eV/\hbar$, which in turn reduces the supercurrent $I_o \sin\phi$ through the junction. Now the voltage must increase faster to force more current through C and R_o. This process initiates the output voltage spike. In fact, the supercurrent $I_o \sin\phi$ through the junction begins to oscillate so rapidly that its time average is nearly zero, forcing all of I_p through C and R_o. Energy stored in the inductor is dissipated now in both R_p and R_o in a time scale $L/(R_p + R_o)$, whereas during the subthreshold evolution, energy was dissipated only in R_p. The parameter $I_p(t)$ decays now more rapidly until the voltage spike terminates with $I_p < I_o$. The neuromime is again ready to receive input spikes. One of the conditions[3] for a voltage spike in which the junction supercurrent oscillates several times is $R_p < R_o$. Thus the spike duration in the neuromime is less than the relaxation time of the internal parameter, just as in the neuron. The value to which I_p resets after a spike depends on the circuit parameters. There is a dead period in the neuromime, as in the neuron. A train of output spikes for a neuromime with constant input is shown in Fig. 4. Individual neuromimes can be interconnected by passive, or by active (neuristor) transmission lines[4], which transmit spikes nearly without attenuation. The analogy between the neuromime and the neuron is summarized in Table 1.

There is an important difference between the neuromime and the neuron. The neuromime emits a pulse whenever $I_p > I_o$, as well as whenever $I_p < -I_o$, whereas the neuron has only one threshold. The neuromime can emit either positive spikes, or negative spikes, whereas the neuron can emit only positive going spikes. A large number of negative inputs to a neuromime produces a negative spike. Many inhibitory inputs to a neuron do not produce a spike during the inhibition. What are the implications of this feature of the neuromime? Although neurons give only a positive output, with a train of positive spikes, the biological system introduces negative numbers in several ways. For example, if the output of a neuron ends at an inhibitory synapse, this output can be considered as negative. However, with this sign convention, the output still cannot change sign: it remains negative or zero. In another example, the average firing rate of a neuron could be considered as "zero", and positive and negative numbers introduced through positive, and negative deviations around the average rate. In yet another example, opposing muscles are used to give motions in positive and negative directions, with "positive" input to each of the muscles. The author does not know whether the single signed nature of the output of a neuron is fundamental to the proper operation of a nervous system, or whether it is an artifact arising because chemically powered axons transmit pulses of only one sign. Certainly, the mathematics of the biological nervous system is more difficult than if the output firing rate of a neuron could change sign. This difficulty arises because in most mathematical techniques it is difficult to constrain a function in a way that prevents sign change. Negative numbers arise naturally in one of the simplest of mathematical operations: subtraction, and it is just this operation that expresses the local effect of an inhibitory synapse. Yet, in the neuron, when there is an excess of inhibitory (negative) inputs, there is zero output during the inhibition, rather than a negative output.

POTENTIAL APPLICATIONS

We conclude by commenting briefly on the potential applications of the superconducting neuromime. There are two, and both are highly speculative.

The first is a biological one, in which arrays of interconnected neuromimes might be used to deduce properties of a corresponding array of neurons. The objective here would be to observe the dynamical properties of the neuromime system to further the understanding of the biological system.

The second application is in the area of artificial intelligence and computing. It may prove possible to construct systems of neuromimes which process signals and information in the same manner as a biological system. Such an application would require further understanding of how neuronal systems process information, either via the technique just mentioned, or from continued progress directly in the neurosciences. A system of superconducting neuromimes could process information up to 100 million times faster than its biological counterpart.

The author gratefully acknowledges helpful discussions with many persons including G. D. Bittner, D. G. Dempsey, P. I. M. Johannesma, M. T. Levinsen, E. R. Lewis, and R. S. Zucker, and thanks P. I. M. Johannesma for a critical reading of the manuscript.

+Present address: Physics Department, University of Nijmegen, Toernooiveld, Nijmegen, NETHERLANDS.

REFERENCES

1. Eccles, J.C.: The physiology of the synapses. Academic Press, New York, 1964; Katz, B.: Nerve, muscle, and synapse. McGraw-Hill, New York, 1969; Griffith, J.S.: Mathematical neurobiology. Academic Press, New York, 1971.
2. Scott, A.C.: The electrophysics of a nerve fiber. Rev. Mod. Phys. 47, 487-533 (1975); see also MacGregor, R.J. and Lewis, E.R.:

Neural modeling, quantitative descriptions of electrical signal processing in nervous systems. (1977, to be published).

3. Dempsey, D.G., Levinsen, M.T., Ulrich, B.T.: Energy storage and subharmonic oscillations in Josephson junctions. IEEE Trans. Mag. MAG-11, 811-814 (1975).

4. Crane, H.D.: Neuristor -- a novel device and system concept. Proc. IRE 50, 2048 (1962); Reible, S.A., Scott, A.C.: Pulse propagation on a superconductive neuristor. J. Appl. Phys. 46, 4935-4945 (1975); Scott, A.C., Chu, Y.F., Reible, S.A.: Magnetic-flux propagation on a Josephson transmission line. J. Appl. Phys. 47, 3272-3286 (1976), where further references on active transmission lines may be found.

DYNAMICS OF FLUX TRANSITIONS IN THE SQUID FOR THE HYSTERETIC CASE

S.N. Erné and H. Lübbig
Physikalisch-Technische Bundesanstalt, Institut Berlin,
Abbestr. 2-12, D-1000 Berlin 10, Germany

Several authors have pointed out that for operating with an RF SQUID the best condition for the product of the ring inductivity L and the maximum critical current I_c of the junction is a value near to the flux quantum $\emptyset_o$.

Under this conditions the well known SQUID equation

$$\frac{\Phi}{\Phi_0} + \beta \sin\left(2\pi \frac{\Phi}{\Phi_0}\right) = \frac{\Phi_a}{\Phi_0} \qquad \beta = \frac{L I_c}{\Phi_0} \tag{1}$$

is no longer a suitable basis for the description of the operation of the SQUID since it describes only stationary, equilibrium states whereas the RF SQUID is in principle a dynamic device. For such a choice of the parameters L and I_c one says the SQUID is working in the hysteretic mode, because the enclosed flux $\emptyset$ is a multivalued function of $\emptyset_a$, the applied flux.

To understand the dynamical behaviour of the SQUID one must generalize the adiabatic equation: here we wish to consider the resistively shunted model of the weak link taking account of the conventional circuit elements, i.e. the junction capacitance C_j and the normal resistance R_q.

In this way one obtains the following differential equation

$$\left(\frac{1}{\omega_s}\frac{d^2}{dt^2} + \tau_q \frac{d}{dt} + 1\right)\frac{\Phi}{\Phi_0} + \beta \sin\left(2\pi \frac{\Phi}{\Phi_0}\right) = \frac{\Phi_a}{\Phi_0} , \tag{2}$$

$$\omega_s = \sqrt{1/LC_j} , \quad \tau_q = L/R_q .$$

Based on this equation, it is the intention of this paper to study the transition of the flux in the SQUID between two quantum states: especially the effect of the current phase-relationship on the flux passage time and on the energy which is dissipated during the transition.

We start our approachment to the problem of flux transition considering a typical time dependence of the flux in the SQUID as shown in fig. 1. The curves are the result of the numerical integration of the system equation (2): as initial condition we have set $\emptyset(0) = \emptyset_0/4$ and the applied flux is set according to the equilibrium condition ($\dot{\emptyset} = 0$, $\ddot{\emptyset} = 0$) plus a small quantity $\delta\Phi_a$ to start an irreversible transition.

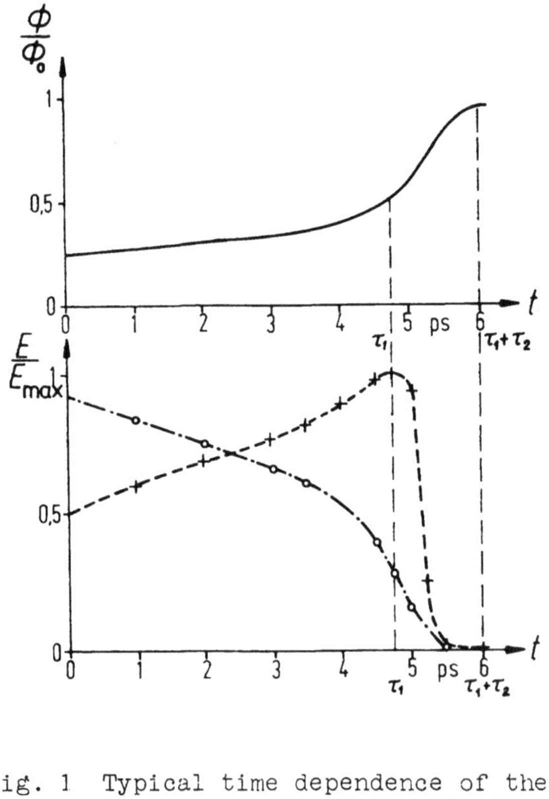

Fig. 1 Typical time dependence of the flux in the SQUID loop during a transition; the lower part contains the corresponding dependence of the barrier energy (-+-) and magnetic energy (-o-)

In the lower part of the figure we can observe the time dependence of two energy terms: the barrier energy

$$E_b(t) = \frac{I_c \Phi_0}{2\pi}\left[1 - \cos\left(2\pi \frac{\Phi}{\Phi_0}\right)\right] \qquad (3)$$

and the magnetic energy $E_L(t) = L\, I^2(t)/2$. The energies are referred to the current-free state $I = 0$, $\phi = 0$, $\phi_a = 0$. By comparison of the curves we can distinguish two characteristic times:

- τ_1 is the time during which a part of the total energy is transferred from the inductance to the junction; during this time only negligible dissipation occurs;
- τ_2 is the time during which the barrier energy drops like the magnetic energy to the next lower level while the main dissipation occurs.

The total time for the flux passage is the sum of both times. The times τ_1 and τ_2 are related to different physical processes and so they depend on the system parameters in different ways. In fig. 2 the dependence of τ_1/τ_q on the SQUID parameter ß is shown for different values of the critical current I_c as a result of the numerical integration. The de-

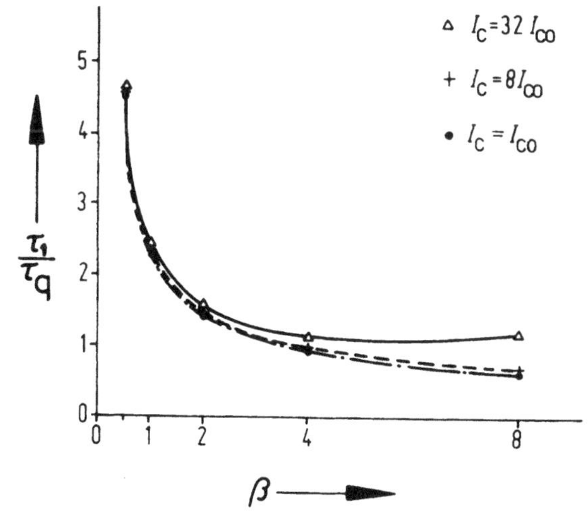

Fig. 2

Dependence of τ_1/τ_q on the SQUID parameter ß for different values of the junction critical current I_c, $\delta\phi_a = 2 \cdot 10^{-2}\phi_0$

pendence on I_c for very large values of ß is a consequence of the nonvanishing junction capacitance (here the junction capacitance is set in the order of magnitude of 10^{-14} F). To understand the ß-dependence we simplify equ. (2) by setting $C_j = 0$, ϕ_a = const. and obtain

$$\ddot{\Phi}/\dot{\Phi} = -\frac{1}{\tau_q}\left[1 + 2\pi\beta \cos\left(2\pi \frac{\Phi}{\Phi_0}\right)\right] \tag{4}$$

In agreement with the calculated time dependence we assume that the flux in the SQUID increases according to the expression

$$\Phi(t)/\Phi_0 = \frac{1}{2} - \frac{1}{4}\cos\left(\frac{\pi}{2}\frac{t}{\tau_1}\right) \qquad 0 \leq t \leq \tau_1 \tag{5}$$

and we obtain as self-consistent solution of equ. (4) the following expression for $\dot{\phi}(t)$

$$\ln\left(\dot{\Phi}/\dot{\Phi}(0)\right) = \frac{\tau_1}{\tau_q}\left[\frac{t}{\tau_1}\left(2\pi\beta\, J_0\left(\frac{\pi}{2}\right) - 1\right) + 4\beta\sum_{k=1}^{\infty}\frac{(-1)^k}{k} J_{2k}\left(\frac{\pi}{2}\right)\sin\left(\frac{\pi k t}{\tau_1}\right)\right] \tag{6}$$

Here J_n denotes the Bessel functions of the first kind of integer order. Now one determines τ_1 from its definition

$$\int_0^{\tau_1} \dot{\Phi}(t)\, dt = \Phi_0/4 \tag{7}$$

by using equ. (6) and the initial condition $\dot{\phi}(0) = \delta\Phi_a/\tau_q$.

The approximated solution, $\tau_1(\beta)$, of the equation in the range between $1 \leq \beta \leq 10$ shows the following dependence on the SQUID parameter:

$$\tau_1/\tau_q = \frac{2}{2\pi\beta\, J_0\left(\frac{\pi}{2}\right) - 1} \ln\left\{\frac{\left[2\pi\beta\, J_0\left(\frac{\pi}{2}\right) - 1\right]\Phi_0}{4\,\delta\Phi_a} + 1\right\} \tag{8}$$

This is the wanted expression for the ratio between τ_1 and τ_q as function of the SQUID parameter ß, and it is in good agreement with the numerical results.

To understand the dependence of the second characteristic time, τ_2, on the system parameters, in fig. 3 the ratio τ_2/τ_j is shown as function of the SQUID parameter ß. Here $\tau_j = \emptyset_o/R_q I_c$ denotes the intrinsic gap response time.
It can be seen that with increasing ß, τ_2 becomes independent

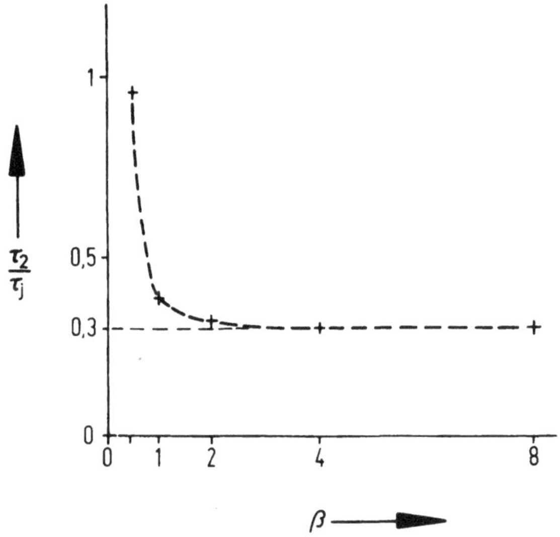

Fig. 3

Dependence of τ_2/τ_j on the SQUID parameter ß

from ß and depends only on the intrinsic response time τ_j of the junction.

This property follows from the definition of τ_2: during the time interval τ_2 the main part of the energy dissipation occurs and the junction voltage increases pulselike. Therefore τ_2 is determined by two conditions: the first one involving the dissipated energy

$$\int_{\tau_1}^{\tau_1+\tau_2} \frac{\dot{\Phi}^2}{R_q}\, dt = E_d \qquad (9)$$

and the second one involving its definition

$$\int_{\tau_1}^{\tau_1+\tau_2} \dot{\Phi}\, dt = \frac{\Phi_0}{2} \qquad (10)$$

To solve this system of equations one needs an Ansatz for the time dependence of the junction voltage during τ_2. In agreement with the results of the numerical integration of the system equation (2) we can assume

$$\dot{\Phi}/\dot{\Phi}_{MAX} = \sin\left[\frac{\pi}{\tau_2}(t-\tau_1)\right] \quad ; \quad \tau_1 \leq t \leq \tau_2 \qquad (11)$$

and one obtains for equ. (10) the following result

$$\dot{\Phi}_{MAX}\,\tau_2 = \Phi_0/1.28 \qquad (12)$$

Now, to eliminate the maximum junction voltage $\dot{\emptyset}_{max}$ one needs an expression for the dissipated energy E_d. At the time t = 0 the energy involved in the inductance L is

$$E_L = \frac{L\, I_c^2}{2} \qquad (13)$$

At $t = \tau_1$ this is reduced by the amount by which the barrier energy is increased and by the amount which has been dissipated during τ_1:

$$E_L = \frac{1}{2} L I_c^2 - \frac{I_c \Phi_0}{2\pi} - \frac{1}{32\beta}\left[2\pi\beta\, J_0\!\left(\frac{\pi}{2}\right) - 1\right] I_c \Phi_0 \qquad (14)$$

For large values of β the SQUID current drops during τ_2 by an amount $\emptyset_0/L$ and so one obtains for the first contribution to the dissipated energy the following expression

$$\Delta E_L = \left\{2\pi - 1 - \frac{\pi}{16\beta}\left[2\pi\beta\, J_0\!\left(\frac{\pi}{2}\right) - 1\right]\right\} \frac{I_c \Phi_0}{2\pi} - \frac{\Phi_0^2}{L} \qquad (15)$$

During the same time the barrier energy is changed and its

contribution to the dissipated energy is

$$\Delta E_b = 1.95 \frac{\Phi_o I_c}{2}$$

Using these expressions for the dissipated energy one can finally determine τ_2 for large values of ß.

As asymptotic value one obtains

$$\tau_2 = 0.29\, \tau_j = 0.29\, \Phi_o / R_q I_c$$

the dissipation time is approximately 1/3 of the intrinsic junction response time τ_j. Furthermore, as can be seen from the numerical results, this relation holds even for finite ß.

The conclusions for this analysis are: the flux transitions in the SQUID involve two processes, first a partial rearrangement of the energy from the inductance to the junction barrier, second the energy dissipation. Though the total transition time may be long, since it is related to the classical time τ_q of the system, the second process always needs time intervals in the order of magnitude of the intrinsic junction response time τ_j.

We conclude that an electrodynamical investigation of the flux dynamics in the hysteretic case can no longer be founded on quasistatic current-phase-relationships like the resistively shunted model. For further investigations one needs a dynamical current-phase-relationship which describes the dissipation process in the weak link .

SUMMARY AND CONCLUSIONS

SUMMARY AND CONCLUSIONS

John Clarke
Department of Physics, University of California, Berkeley, and Materials and Molecular Research Division, Lawrence Berkeley Laboratory, Berkeley, California 94720.

In his rather elegant talk on high frequency detectors, Paul Richards likened his survey to visiting cages in a zoo: one had to make brief inspections of many different animals.' In retrospect, I think his problem was much simpler than mine because his zoo was merely 1-dimensional. As I looked over the program for this conference, and wondered how I could possibly summarize such a diversity of topics, I realized that my own zoo was 8-dimensional! Rather than visit in turn each cage of this enormous zoo, I decided that I would try to summarize the conference in a general way. Therefore, I shall refer to very few specific papers, and, instead, try to give an overview of the present status of Josephson junctions and devices, and to speculate on future prospects. I have divided the summary into two parts: first, I will briefly discuss the junctions that are the building blocks of devices, and second, I will talk about the devices themselves.

The three basic configurations currently used as Josephson devices are the tunnel junction (oxide barrier or semiconductor barrier), the microbridge (constant thickness, variable thickness, or proximity effect), and the point contact. The properties of a point contact can be close to those of a tunnel junction, close to those of a microbridge, or lie somewhere between, depending on how the contact is made. I shall confine myself to a brief discussion of the tunnel junction and the microbridge, since it seems to me that, ultimately, Josephson devices will use thin film technology. How well do we understand the behavior of tunnel junctions and microbridges? - that is, to what extent do present theories have any correspondence with the properties of the real junctions? In the case of tunnel-junctions, the theory is well understood. The critical current can be accurately predicted from the junction resistance. If the junction is shunted, the resistively shunted junction model makes an excellent prediction of the shape of the current-voltage characteristic and whether or not it is hysteretic. In the case of microbridges, the situation is very much less clear. My own understanding of the theory of microbridges is in a rather confused state. We have heard a number of theoretical discussions on microbridges based on either GL or TDGL theories. To some extent these models explain some features of some types of microbridges. However, there are many additional factors to be considered, some of which may need to be included in a more successful description, for example: heating, nonlinear quasiparticle conductance, quasiparticle relaxation processes (there are several to choose from!), gap relaxation processes, vortices, and non-uniform current distribution. It remains to be shown which of these factors are essential in a proper description. As evidenced by the many papers at this conference, this area of research is very active at present, and hopefully we shall see a resolution of these problems in the next two or three years. However, one important point seems to be generally agreed upon: to obtain a sinusoidal current-phase relationship, one must make the length of the bridge shorter than the coherence length.

Let me turn to junction fabrication. The most reproducible tunnel junctions are undoubtedly the lead (alloy)-oxide-lead (alloy) junctions produced by IBM. Niobium-oxide-lead and particularly niobium-oxide-niobium tunnel junctions are also very reliable, and are attractive for devices. The barrier is grown by thermal oxidation or by a plasma discharge in oxygen. Junctions with semiconductor barriers are also a very promising development. With the aid of mechanically machined masks, one can obtain junction dimensions down to perhaps 25 μm. Photolithographic techniques enable one to achieve dimensions of about 1 μm. In the case of microbridges, the favored materials are niobium, tantalum, tin, and indium. We have heard about very sophisticated electron beam-lithographic and ion-milling techniques that can achieve dimensions of 0.2 μm or even less. The technology of microbridge fabrication seems to be in good shape, although it is certainly very expensive.

The reproducibility and reliability of tunnel junctions are excellent. Microbridges are reproducible if the appropriate technology is available, and for the hard superconductors, at least, the reliability is excellent. On the other hand, microbridges made of soft superconductors tend to burn out rather easily.

What are desirable properties for junctions? For SQUIDs and high frequency detectors, I_cR should be as large as possible, and RC (i.e. C) should be as low as possible (I_c, R, and C are the critical current, resistance, and capacitance). The current-voltage characteristic should be non-hysteretic ($2\pi I_cR^2C/\Phi_o \lesssim 1$). It is desirable for the junctions to operate well below the transition temperature without developing large critical currents, and for dI_c/dT to be small. In the case of high frequency detectors, one wants the junction to couple as efficiently as possible to rf radiation. The computer memories and logic circuits developed so far have used tunnel junctions with hysteretic current-voltage characteristics ($2\pi I_cR^2C/\Phi_o >> 1$). The main requirement is to achieve the highest possible value of I_cR.

I would like to make some suggestions on junctions. If you are in the business of making tunnel junctions or microbridges (or point contacts for that matter), it is obviously important to state the value of I_cR that is achieved. However, in addition, one would like to know the range over which I_c and R can be varied. For example, in a microbridge it might be relatively straightforward to obtain $I_c \sim 10$ mA and $R \sim 0.1\ \Omega$ so that $I_cR \sim 1$ mV. But can one also achieve $I_cR \sim 1$ mV with $I_c \sim 1\ \mu$A and $R \sim 1\ k\Omega$, the values one would like to use in a SQUID? Another important parameter to report is the maximum frequency at which one can obtain a response from the junction: a useful criterion is the highest voltage at which a microwave-induced step can be observed. A final suggestion is for those who have developed these beautiful techniques for producing very small microbridges. I would like to see someone make an earnest attempt to fabricate tunnel junctions with dimensions $1 \times 1\ \mu$m or (preferably!) less, and to test these out in SQUIDs and high frequency detectors. (IBM has made junctions of this size, but, as far as I know, has used them only in computer elements.) A tunnel junction with dimensions $1 \times 1\ \mu$m would have a capacitance of about 0.03 pF. One could then make $I_c \sim 1\ \mu$A (the critical current density, 10^2 A cm^{-2}, is easily achievable), and use a shunt resistance of 100 Ω without having hysteresis. SQUIDs made with these junctions should be exceedingly sensitive. For high frequency detectors, one would probably want to use a somewhat higher critical current, and a correspondingly lower resistance. In my opinion, small area tunnel junctions are likely to have a considerable impact on the device technology.

For the second part of the summary, I will review the three main areas of application covered by the conference: high frequency detection, computer elements, and SQUIDs. Paul Richards reviewed high frequency detectors, and I will just repeat the list of cages in his zoo. He began with the video or square law Josephson detector. The sensitivity of this device compares somewhat unfavorably with the superconducting transition edge bolometer, the best semiconductor bolometers, and the super-Schottky diode, and is unlikely to be used very much (except, possibly, when a very rapid response time is essential). The Josephson mixer can be used with either an internal local oscillator (that is, the Josephson self-oscillator) or an external local oscillator. Of these, the mixer with an external local oscillator seems to be the more promising, but is subject to strong competition from the super-Schottky diode mixer. Lastly there are the Josephson parametric amplifiers, either internally or externally pumped; the externally pumped parametric amplifier in a doubly degenerate mode with zero bias current has the best performance.

All of these devices have been made with point contacts. In addition, the externally pumped parametric amplifier has been made recently with an array of microbridges. The arrays have considerable promise as high frequency detectors, although the junctions in an array appear to be coherent only at low frequencies. Coherence is not necessary for externally pumped devices but is essential for internally pumped devices. This is an application in which small area tunnel junctions could have a considerable impact, since arrays could be made with a high degree of reproducibility and stability.

It is by no means obvious which of the Josephson high frequency detectors will finally have the best performance. It is also not clear that any of the Josephson detectors is vastly superior to the competition. It seems that a better microscopic understanding of the properties of junctions is required in order to achieve successful operation at the higher frequencies (above, say, 50 GHz) where improved performance is most urgently needed. One may hope for steady progress over the next two or three years.

At present, computer elements are the most rapidly developing application of Josephson devices. As far as I know, nobody has yet made a superconducting computer, but the potential of Josephson junctions for high speed computers seems enormous. This work has been largely dominated by IBM. Memory cells with writing times of less than 100 ps and several types of latching and non-latching logic circuits have been successfully operated. One of the most important aspects of the Josephson devices is their extremely low power dissipation, as a result of which one can obtain a very high packing density. It is essential to have closely spaced elements and consequently short transit times if one is to take advantage of the very fast switching times of these devices. I find it impossible to make any predictions of developments in this field over the next couple of years. I would just make a plea to IBM that they keep us informed of progress!

The main topic of this conference was SQUIDs. This is one area in which one can safely say that the Josephson devices are clearly superior to the competition. Numerous applications were described at this meeting: measurements of susceptibility, flux creep, thermopower, rf power, and ion-current, a precision resistance comparator, current stabilization, calorimetry, and even submarine communications. (Two major areas that for some reason were not mentioned here are geophysics and medical physics.) The steady growth in the applications is due at least in part to the commercial availability of reliable SQUIDs. Thinking back over the conference, I realize that there has been relatively little development in SQUID hardware over the last couple of years. The reason for this static situation is probably that the devices have had a performance that was at least adequate (and in many situations much more than adequate) for the applications in hand. However, there are now at least two applications in which more sensitivity is required, namely gradiometers and gravity-wave detectors. I suspect that the next two years will see a substantial improvement in the sensitivity of SQUIDs.

The theory of noise in both the rf and dc SQUID is now well understood, and is in good agreement with the measured noise limits. The optimum energy resolution per unit bandwidth for the dc SQUID is on the order of $k_BT/(R/L)$, where T is the device temperature, and L is the SQUID inductance. This result assumes that the noise temperature of the preamplifier is much less than T. The same result applies to the rf SQUID provided that the pump frequency, ω, is R/L; however, in this case, it is very difficult to obtain a preamplifier noise temperature that is less than the device temperature. How might one improve the performance of SQUIDs? One could obviously lower T. However, for practical reasons, it is usually convenient to run the SQUID at 4.2 K. In a few specialized applications one might operate the SQUID at a few mK, but in order to take advantage of this low temperature, one would need a cryogenic preamplifier with a very low noise temperature, probably another SQUID. In the present generation of devices, L is typically 1 nH, or somewhat less. Although a reduction in L (accompanied by an appropriate increase in ω in the case of the rf SQUID) would evidently improve the performance of the SQUID, it would most likely decrease the coupling efficiency to the signal coil, so that the energy resolution referred to the signal coil would probably not be substantially improved. Thus, the most likely way to improve performance is to increase the junction resistance, R: one would really like to have a value of 100 Ω, or more. Point contact junctions with high resistances have certainly been made, but whether they are sufficiently stable for long-term use is not altogether clear to me. It seems very difficult to produce microbridges with resistances of 100 Ω or more, unless one makes them extremely long. Unfortunately, long bridges have a very non-sinusoidal current-phase relationship, and are likely to contribute excess noise to the SQUID. Thus, one is left with small area shunted tunnel junctions as the most likely means to achieve higher performance. A properly optimized dc SQUID with a junction resistance of 100 Ω and an inductance of 1 nH should have an energy resolution per Hz approaching 10^{-32} JHz^{-1} (but still greater than $\hbar$, the ultimate limit!). An rf SQUID with the same values of R and L, and pumped at a frequency $R/2\pi L \sim 10$ GHz should have the same performance, provided that the preamplifier has a low enough noise temperature.

One final plea for SQUID makers. In the literature, one still occasionally finds the flux resolution quoted as a figure of merit. However, for most applications, this is not a useful quantity unless one knows the inductance, L_s, of the signal coil coupled to the SQUID, and the mutual inductance, $M = \alpha(LL_s)^{1/2}$ between the SQUID and the signal coil. At least in the zero frequency limit, a more useful figure of merit is $S_\Phi(f)/2\alpha^2L = S_\Phi(f)/2(M^2/L_s)$, where $S_\Phi(f)$ is the spectral density of the flux noise. (In some specialized applications, for example, a parametric amplifier coupled to a gravity wave detector, this figure of merit is also inadequate, as one needs some measure of the dissipation that occurs at a non-zero signal frequency.)

I should like to briefly refer to another paper in this conference, dealing with a topic that is of growing importance, namely refrigeration. If one is to see a widespread use of Josephson devices in applications outside the laboratory, in my opinion it will be necessary to develop a reliable type of closed-cycle refrigerator. Jim Zimmerman and his colleagues have designed a very ingenious Stirling refrigerator that presently achieves 14 K. This temperature is not yet low enough to enable one to operate the presently available devices, and there remain problems of magnetic noise generated by the refrigerator. Nevertheless, this refrigerator represents a considerable advance in low noise closed cycle refrigerators, and I believe that this is an important area for further development. Without trying to wish too much competition on Jim Zimmerman, I hope that a few other people become involved in this work.

I gather that there may well be another IC-SQUID in two or three years. I feel that this is an appropriate time scale, and I hope very much that a second conference does take place. I have a minor suggestion for the repeat performance. All of you have undoubtedly noticed the rather ingenious symbol for the conference:

The symbol obviously stands for IC, for flux, Φ, and for an rf SQUID threaded by flux. However, if you look at the program cover, you will notice that the symbol has apparently been printed upside-down. In order to avoid any confusion at a future conference, it occurs to me that one can readily introduce some symmetry into the symbol so that one does not have to worry about which way up to print it. Thus, for the next conference, I suggest that the symbol be slightly modified to look like this:

To conclude: I should like to express my thanks, and I am sure also the thanks of everyone here, to the organizers and sponsors of this conference. IC SQUID was sponsored by the European Physical Society and the Physikalisch-Technische Bundesanstalt, to whom we are grateful for support. We should thank the program committee for their hard work in the selection of the papers and the organization of the sessions. We are also grateful to the many people behind the scenes who helped make the conference a success: those ladies and gentlemen who made sure that the projectors and loudspeaker systems worked, who supplied coffee when we needed it, and who took care of travel arrangements, visits to the PTB, and the other miscellaneous details of a scientific conference. Many of us have enjoyed the hospitality of the local scientists who have acted as our guides. Finally, our warmest thanks are due to the conference chairman, Professor Hahlbohm, who has labored for a year to make this conference a success, and who has taken overall responsibility for the organization of the meeting. Thank you very much.

SUBJECT INDEX

AUTHOR INDEX

Bei Fragen zur Produktsicherheit wenden Sie sich bitte an:
If you have any questions regarding product safety,
please contact:

Walter de Gruyter GmbH
Genthiner Straße 13
10785 Berlin
productsafety@degruyterbrill.com